Physical Nonequilibrium in Soils

Modeling and Application

Edited by

H. Magdi Selim
Liwang Ma

Ann Arbor Press
Chelsea, Michigan

Library of Congress Cataloging-in-Publication Data

Physical nonequilibrium in soils : modeling and application / edited by H. Magdi Selim, Liwang Ma.
p. cm.
Includes bibliographical references and index.
ISBN 1-57504-049-2
1. Soil physics. 2. Soil chemistry. I. Selim, Hussein Magdi Eldin, 1944– . II. Ma, Liwang.
S592.3.P476 1998
631.4--dc21 97-48516
CIP

ISBN 1-57504-049-2

ANN ARBOR PRESS, INC.
121 South Main Street, Chelsea, Michigan 48118
Ann Arbor Press is an imprint of Sleeping Bear Press

PRINTED IN THE UNITED STATES OF AMERICA
10 9 8 7 6 5 4 3 2 1

Dedication

R. Jeff Wagenet
(1950–1997)

It is our privilege to dedicate this book on *Physical Nonequilibrium in Soils: Modeling and Application* to Dr. R. Jeff Wagenet, an internationally known soil physicist, who has inspired and cooperated in the publication of this book. Dr. Wagenet died on July 31, 1997.

Dr. Wagenet was born on August 10, 1950 in Pittsburg, California and obtained his Ph.D. from the University of California, Davis in 1975, following his B.S. (1971) and M.S. (1972) from University of California, Davis and University of Oklahoma, respectively. Prior to his appointment at Cornell University in 1982, Dr. Wagenet was a professor of soil science at Utah State University, in the Department of Soil Science and Biometeorology (1976–82), where he was twice recognized by the Professor of the Year teaching award. While at Cornell, he also chaired the Soil, Crop, and Atmospheric Sciences Department for several years.

Dr. Wagenet made significant contributions to the understanding of water and solute transport through the unsaturated zone to groundwater. His LEACHM model has been used worldwide to simulate the behavior of nitrogen fertilizers, inorganic salts, pesticides, and manure under field conditions. The model has been requested by more than 300 research institutes, and a large number of short-term visitors have been hosted at Cornell University. Dr. Wagenet's research interests extended to field measurement of crop response to transient water and solute conditions, modeling of soil spatial variability, and GIS application.

Interest in international relationships led Dr. Wagenet to collaborate with scientists in Belgium, Brazil, China, England, France, India, Israel, The Netherlands, and South Africa. Dr. Wagenet published numerous articles in scientific journals, book chapters, and SSSA and ASA special publications. He served as Editor of the *Journal of Environmental Quality* from 1989-95 and Associate Editor of *Advances in Soil Science*. He was elected as a Fellow of the SSSA and ASA in 1989.

Jeff embodied the highest standards of integrity and character. He was open and uncomplicated in demeanor, heart, and mind. He was friendly, unassuming, unpretentious and generous in dealing with all people. He left a loving family. His wife Linda, a scientist and homemaker, fully supported his professional goals and activities. The scientific community worldwide has lost an inspiring colleague and a friend.

The Editors

THE EDITORS

H. Magdi Selim is Professor of Soil Physics at Louisiana State University, Baton Rouge. He received his M.S. and Ph.D. in Soil Physics from Iowa State University, Ames, Iowa, in 1969 and 1971, respectively, and his B.S. in Soil Science from Alexandria University, Alexandria, Egypt, in 1964. Professor Selim has published numerous papers and book chapters, and is a coauthor of one book and several monographs. His research interests concern the modeling of the mobility of contaminants and their retention behavior in soils and groundwaters. His research interests also include saturated and unsaturated water flow in multilayered soils. Professor Selim served as associate editor of *Water Resources Research* and the *Soil Science Society of America Journal*. He has received several professional awards. He was the recipient of the 1980 Phi Kappa Phi Young Faculty Award, the 1982 First Mississippi Research Award, and the 1991 Gamma Sigma Delta Outstanding Research Award. He is a Fellow of the American Society of Agronomy and the Soil Science Society of America.

Liwang Ma is a research associate at Louisiana State University, Baton Rouge, Louisiana. Dr. Ma received his B.S. and M.S. in Biophysics from Beijing Agricultural University, Beijing, China in 1984 and 1987, respectively, and his Ph.D. in Soil Physics from Louisiana State University in 1993. He has authored and coauthored several papers and book chapters. His research interests include water and solute transport modeling, agricultural systems modeling, plant growth modeling, and nutrient cycling in agricultural systems. Dr. Ma is the recipient of the 1994 Prentiss E. Schilling Outstanding Dissertation Award in the College of Agriculture, Louisiana State University.

PREFACE

The shell of the Earth, which we call soil, is not uniform in structure and composition. Such a nonuniformity extends spatially and temporally, and significantly affects water flow and solute transport in soils. Phenomena resulted from soil nonuniformity or heterogeneity include instability of water flow, preferential water flow, and nonequilibria of solute transport. Experimentally, soil heterogeneity was investigated by mapping soil pores, characterizing soil structure, and studying the characteristics of water flow and solute transport in soils under various conditions. Mathematically, conceptual models were proposed to account for soil heterogeneity, including mobile-immobile model, two-flow domain model, multiple flow domain model, transfer function model, and cellular automaton fluid approach. New experiments are also developed to quantify parameters associated with these models. Most model applications are at the laboratory scale, with only few designed and tested for the field scale.

Physical nonequilibrium concepts of water and solute transport are one of the fascinating research areas in soil physics and hydrology disciplines. Although no single model has been universally accepted and/or validated, physical nonequilibrium concepts advanced our understanding of the behavior of water and solute in the soil environment. These concepts also revealed the limitation of traditional approaches, proposed valuable guidelines to future research, and significantly impacted on many scientific disciplines, such as soil science, agricultural engineering, civil engineering, as well as others. It is also of significance as it affects our understanding of several issues concerning groundwater contamination, land treatment, and remediation. In spite of their shortcomings and difficulties in application, the models outlined in this book describe, in part, physical nonequilibrium phenomena such as preferential flow, multiple peaks, excessive tailing, and irregular solute distribution in soils. Future advances in experimental methodologies, mathematical solutions, and conceptual innovations are likely to make physical nonequilibrium approaches useful tools in the interpretation of observed water and solute behavior under field conditions.

This book was compiled for the purpose of providing cutting-edge knowledge on physical nonequilibrium phenomena in soils, and to provide insight to the complexity of our physical world. The book is organized in 18 chapters; the first five chapters describe various approaches yielding coupled physical and chemical nonequilibrium models. The next three chapters provide laboratory and field evaluation of multiregion models and methods of parameter estimation. The remaining chapters deal with stochastic approaches, and nonaqueous phase liquid (NAPL) dissolution figuring and preferential flow.

Chapter 1 provides an evaluation of coupled physical nonequilibrium (mobile-immobile model) with two chemical nonequilibrium models (two adsorption site model and redistributed sorption rate model). Chapter 2 focuses on basic concepts of soil pore classes; macro-, meso-, and micropores, and illustrates such concepts through field experimental measurements and numerical modeling. Chapter 3 discusses multiprocess nonequilibrium modeling and its applications to several laboratory and field studies. In Chapter 4, emphasis is on chemically based nonequilibrium models with an extensive evaluation of coupled chemical models to the mobile-immobile physical nonequilibrium approaches. Chapter 5 summarizes various analytical solutions to physical nonequilibrium models with linear solute retention mechanism.

The primary focus of Chapters 6 through 11 is on physical nonequilibrium behavior under field conditions. Efforts of modeling physical nonequilibrium phenomenon based on soil structure information are described in Chapter 6. In Chapter 7, applications of a multiregion model (MACRO) to simulate preferential flow and solute transport as affected by soil aggregate size are presented. Evaluation of two physical nonequilibrium models (PEDAL and CRACK) to laboratory and field experimental data sets is given in Chapter 8. In Chapter 9, experimental techniques for confirming and quantifying physical nonequilibrium processes and associated model parameters are presented, whereas Chapter 10 deals with difficulties in parameterizing the mobile-immobile model in soils characterized by aggregates or macropores. The limitations of multiple tracer approach in estimating parameters associated with the mobile-immobile model are discussed in Chapter 11. An extensive review on transfer function approaches is presented in Chapter 12, which is followed by a review of factors affecting solute transport in the field scale and limitations associated with current experimental methods (Chapter 13).

In Chapter 14, an overview of the effects of fluid density and viscosity on flow instability and contaminant transport is presented. In Chapter 15, a numerical model describing NAPL dissolution figuring in multiphase flow systems and a comparison to a linear stability analysis approach is discussed. Experimental demonstrations of preferential flow of NAPLs in different soil formations are illustrated in Chapter 16. In Chapter 17, analysis of preferential flow using the continuum macroscopic and the cellular automation fluids approaches are discussed. In Chapter 18, simulated hydraulic properties at the farm scale based on a geostatistical approach are described.

We wish to thank all the authors for their contributions. Special thanks are due to Skip DeWall and his staff at Ann Arbor Press for their efforts in the publication of this book.

H. Magdi Selim
Liwang Ma

CONTRIBUTORS

T.M. Addiscott
Rothamsted Experimental Station
Harpenden, Herts AL5 2JQ
United Kingdom

A.C. Armstrong
ADAS Land Research Centre
Gleadthorpe, Meden Vale
Mansfield, Notts NG20 9PF
United Kingdom

H.W.G. Booltink
Wageningen Agricultural University
Department of Soil Science and Geology
6700 AA Wageningen
The Netherlands

J. Bouma
Wageningen Agricultural University
Department of Soil Science and Geology
6700 AA Wageningen
The Netherlands

Mark L. Brusseau
Department of Soil, Water and Environmental Science and
Department of Hydrology and Water Resources
University of Arizona
Tucson, AZ 85721

W. Chen
Novartis Environmental Safety Department
P.O. Box 18300
Greensboro, NC 27419-8300

Brent E. Clothier
Environment Hort Research
PB 11-030
Palmerston North
New Zealand

J.H. Dane
Department of Agronomy and Soils
Auburn University
Auburn, AL 36849-5412

Liliana Di Pietro
Unité de Science du Sol
INRA
Domaine St. Paul
Site Agroparc
84914 Avignon, Cedex 9
France

P. Droogers
Wageningen Agricultural University
Department of Soil Science and Geology
6700 AA Wageningen
The Netherlands

Robert B. Edis
Department of Agriculture and Resource Management
The University of Melbourne
Parkville, Victoria 3052
Australia

Markus Flury
Department of Crop and Soil Sciences
Washington State University
Pullman, WA 99164

Simon N. Gleyzer
Department of Environmental Sciences and Engineering
University of North Carolina at Chapel Hill
Chapel Hill, NC 27599

Steve R. Green
Environment Group
HortResearch, PB 11-030
Palmerston North
New Zealand

Jin-Ping Gwo
Oak Ridge National Laboratory
P.O. Box 2008
Oak Ridge, TN 37831

Lee K. Heng
Department of Agriculture and Resource Management
The University of Melbourne
Parkville, Victoria 3052
Australia

Robert Horton
Department of Agronomy
Iowa State University
Ames, IA 50011

Tissa H. Illangasekare
Department of Civil Engineering
University of Colorado
Boulder, CO 80309-0428

Paul T. Imhoff
Department of Civil and Environmental Engineering
University of Delaware
Newark, DE 19716

Philip M. Jardine
Earth and Atmospheric Sciences Section
Oak Ridge National Laboratory
P.O. Box 2008
Oak Ridge, TN 37831-6038

Nicholas Jarvis
Department of Soil Science
Swedish University of Agricultural Science
Box 7014
S-750 07 Uppsala
Sweden

Dan B. Jaynes
USDA-ARS
National Soil Tilth Laboratory
2150 Pammel Drive
Ames, IA 50011

William A. Jury
Department of Soil and Environmental Sciences
University of California
Riverside, CA 92521

Eileen J. Kladivko
Agronomy Department
Purdue University
West Lafayette, IN 47907

P.B. Leeds-Harrison
Cranfield University School of Agriculture Food and Environment
Silsoe, Beds MK45 4DT
United Kingdom

Feike J. Leij
USDA-ARS
U.S. Salinity Laboratory
450 West Big Springs Road
Riverside, CA 92507-4617

H.H. Liu
Environmental Systems Engineering Department
Clemson University
Clemson, SC 29634

R.J. Luxmoore
Oak Ridge National Laboratory
P.O. Box 2008
Oak Ridge, TN 37831-6038

Liwang Ma
Agronomy Department
Louisiana State University
Baton Rouge, LA 70803

Robert S. Mansell
P.O. Box 110290
Department of Soil and Water Science
University of Florida
Gainesville, FL 32611-0290

Cass T. Miller
Department of Environmental Sciences and Engineering
University of North Carolina
Chapel Hill, NC 27599

Rachel O'Brien
Department of Geology
Washington State University
Pullman, WA 99164-2812

Andrew S. Rogowski
Department of Agronomy
445 A.S.I. Building
The Pennsylvania State University
University Park, PA 16802-3504

Dave R. Scotter
Soil Science Department
Massey University
Palmerston North
New Zealand

H.M. Selim
Agronomy Department
Sturgis Hall
Louisiana State University
Baton Rouge, LA 70803

Dilip Shinde
Soil and Water Science Department
University of Florida
Gainesville, FL 32611

Nobuo Toride
Department of Agricultural Sciences
Saga University
Saga 840
Japan

Iris Vogeler
Environment Group
HortResearch, PB 11-030
Palmerston North
New Zealand

R.J. Wagenet
Department of Agronomy
Cornell University
Ithaca, NY 14853

Robert E. White
Department of Agriculture and Resource Management
The University of Melbourne
Parkville, Victoria 3052
Australia

Glenn V. Wilson
Desert Research Institute
755 E. Flamingo Rd.
Las Vegas, NV 89119-0040

CONTENTS

CHAPTER ONE

Coupling Sorption Rate Heterogeneity and Physical Nonequilibrium in Soils

R.J. Wagenet and W. Chen

INTRODUCTION

Solute transport under nonequilibrium conditions is often complicated by the heterogeneity in both physical and chemical properties of natural porous media. Mathematical modeling of nonequilibrium transport has been impeded largely by the current incomplete understanding of the influence of heterogeneity on transport, sorption/desorption, and intra-aggregate diffusion processes. The influence of heterogeneity in hydraulic properties on transport has been a central issue in the debate as to whether the classical convection-dispersion theory is valid or whether alternatives are more appropriate for describing these types of flow regimes. While the influence of heterogeneity on nonequilibrium processes is primarily the result of the structural variations of natural porous media and the multiplicity of sorbing components, it is not clear to what extent this heterogeneity is responsible for the current confusion of the widely divergent time scales reported in the literature for nonequilibrium processes to reach equilibrium (from a few hours to many years).

One of the consequences of nonequilibrium transport in structurally heterogeneous soils are the extreme events of contaminant transport from soil to groundwater. Rapid pulses of freshly added chemicals or those in a relatively labile status may result from a small fraction of fast flow pathways. By contrast, extended leaching of a very small amount of mass can occur when transport is controlled by the relatively slow component of the nonequilibrium process. Whether the solute flux is large over short times or small over long times, the impact on groundwater quality is important, given that toxicological tolerances for many organic contaminants are often below the level of one part per billion. It is particularly important therefore, for legal as well as toxicological reasons, that the extremes of solute transport be accurately quantified and predicted.

This work focuses on further understanding of the mechanics of nonequilibrium processes influenced by soil heterogeneity in both transport-related and sorption-related properties. The new knowledge obtained should benefit the future development of simulation models that are not only able to describe flow in relatively homogeneous (uniform matrix) soils, but also to accurately describe the extremes of transport, which is an essential concern for groundwater protection.

LITERATURE REVIEW

The basic principles of water flow and solute transport as embodied in the Richards equation and the convection-dispersion-diffusion equation (CDE), have been found successful in describing experimental results obtained from well-controlled laboratory studies of uniformly packed soil columns. When coupled with terms describing sorption (usually as an equilibrium isotherm), degradation, and volatilization, the CDE has well-described the transport of weakly sorbed inorganic and organic compounds in soil (Nielsen et al., 1986; Wagenet and Rao, 1990). Perplexing results, however, arise from attempts to describe the transport of strongly sorbed organic compounds or the field-scale transport phenomenon. In these cases, extremes of transport such as the protracted tailing of very low concentration levels due to nonequilibrium sorption, or the rapid displacement of solute mass due to macropore flow, are beyond the processes currently represented in most existing transport models.

Soil Sorption-Site Heterogeneity and Nonequilibrium Processes

Classical concepts of sorption of organic compounds consider the soil as a homogeneous material and sorption as a unique and reversible equilibrium process. Although this classical assumption has successfully described experimental results obtained from batch studies (Lavy, 1968; Dao and Lavy, 1978; Huang et al., 1984; Kookana et al., 1992), there is increasing evidence that this assumption is not always valid (Pignatello, 1989). Nonequilibrium usually occurs when these compounds move through soil with a fast flow velocity relative to the rate of the sorption/desorption processes. Many early soil column studies of pesticide transport showed substantial deviation between calculated and measured effluent curves using the classical assumption (Kay and Elrick, 1967; Davidson et al., 1968; Davidson and McDougal, 1973). In these studies, a retarded peak and pronounced tail were observed in the same breakthrough curve, indicating that instantaneous and time-dependent sorption/desorption processes occurred simultaneously.

A pragmatic assumption for the observed heterogeneous sorption behavior is the two-site (TS) approach which assumes that the soil solid phase is composed of two sorption fractions: type-1 on which sorption is assumed to be instantaneous, and type-2 on which the process is assumed to be time-dependent (Selim et al., 1976; Cameron and Klute, 1977; van Genuchten and Wagenet, 1989). These two types of sorption sites have been further specified by Lafleur (1979), who related the instantaneous sites to mineral fractions, and the time-dependent sites to the soil organic matter. Under relatively high water flow conditions (i.e., usually >1.0 cm/hour), a limited number of experimental studies (Rao et al., 1979; Lee et al., 1988; Gamerdinger et al., 1990) have shown that the TS model better fits much of the asymmetry (tailing) in the breakthrough curves (BTCs) obtained for pesticide displacement through packed soil columns. The better fit, however, was achieved at the cost of parameter values which differ according to experimental conditions. For example, an increase in flow velocity tended to increase the sorption rate constant for the single time-dependent site (Gamerdinger et al., 1991; Brusseau, 1992). However, real soils contain many different sorptive components that may react with solute at different rates and intensities (Garbarini and Lion, 1986; Rutherford et al., 1992; Weber et al., 1992). Presumably, high velocity flow results in "bypass" of the slower sorption site(s), while emphasizing the interaction with the faster site(s). Batch-type desorption experiments by Karickoff and Morris (1985) showed that the TS model failed to describe the final release profile of sorbed mass at a longer time-scale. A recent study of Connaughton et al. (1993) further showed that the entire naphthalene release profile from a contaminated soil was better described when the first-order desorption rate coefficient was assumed continuously distributed, indicating the involvement of an array of sorption-sites in the desorption process.

An alternative interpretation of sorption nonequilibrium has been the diffusion process, often known as intra-particle diffusion (Wu and Gschwend, 1986; Ball and Roberts, 1991; Pignatello et al., 1993). The diffusion assumption emphasizes the length-scale effect for sorption, and it is equivalent to a large number of identical first-order sites in series. The overall sorption process is expressed by a diffusion equation. To solve the diffusion equation, the sorption process is usually visualized as a mass transfer process of sorbate molecules moving through the interstices of sorbent particles, which are further hypothesized as spheres of either soil mineral grains or natural organic matter. Although the physical basis of this approach is straightforward, there still remain substantial questions regarding its practical application, due to natural heterogeneities in particle geometry, sizes, and composition. The definition of sorbent geometry is also ambiguous given that the basic particle size (mineral grains or organic amorphism) has never been clearly distinguished from the relatively large aggregates in which physical nonequilibrium of a nonsorptive chemical such as chloride occurs. Little is also known about the degree to which new surfaces created by reduction in size of mineral grains or organic amorphism during sieving affect the original sorption/desorption characteristics.

Sorption nonequilibrium in aggregated soils may be complicated by the simultaneous occurrence of physical nonequilibrium processes resulting from soil aggregate structure (multiprocess nonequilibrium). Physical nonequilibrium processes affect both sorptive and nonsorptive chemicals. For a sorptive chemical such as pesticide to substantially interact with the solid phase in such soils, the sorbate molecules must first transfer physically from the relatively fast flowing mobile region (inter-aggregate) to the immobile region (intra-aggregate), and the overall process is also referred to as intra-aggregate diffusion. There have been many studies in which the intra-aggregate diffusion was independently investigated using nonsorptive chemicals such as chloride in packed column experiments of either sieved soil or synthesized porous aggregates. A unique first-order process (van Genuchten and Dalton, 1986; van Genuchten and Wagenet, 1989) or a diffusion process in uniform spheres/cubes (Passioura, 1971; Rao et al., 1980a; 1980b) has been assumed to characterize this process. Heterogeneity in both shapes and sizes of aggregates (or the rate coefficient in the first-order kinetic model) in these studies has been simplified by using an equivalent shape such as spheres, or a weighted mean aggregate size (or a unique value for the first-order rate coefficient). Several studies on molecular sieves, however, have shown that distributions of shapes and sizes of aggregated particles can significantly affect the physical process (Ruthven and Loughlin 1971) and chemical transport in packed beds (Rasmuson, 1985).

Relatively few studies have been accomplished for transport of multiprocess nonequilibrium. Brusseau et al. (1989) combined the two-region (i.e., the first-order intra-aggregate diffusion) and the TS approaches into one model, and were able to show that the multiprocess model well predicted independent experimental data obtained from different velocity cases. There still remains a paradox, however, that the TS assumption worked well when it was coupled with the intra-aggregate diffusion process, while it became velocity-dependent when no intra-aggregate process was involved. It is very likely that when the TS model is coupled in series with an intra-aggregate diffusion process, the overall nonequilibrium may become controlled by the latter. Given the wide existence of aggregate structure in field soils, more attention needs to be focused on the coupling effect of the multiple nonequilibrium processes.

Soil Spatial Variability and Modeling of Flow

Soil spatial heterogeneity of macroscopic properties can be defined in terms of the size of their representative elementary volumes (REV) (Bear, 1972; Wagenet, 1985). In large part, variability derives from using a sampling instrument that cannot sample a sufficiently large volume (<< REV) to obtain a representation of the soil properties at a macroscopic scale that embodies all possible variations at the microscopic level. This means that almost all current sampling programs will provide field data of a "point" or two-dimensional (soil profile) measurement under the unique conditions of the situation. To represent the statistical nature of the measurement, a stochastic approach is usually employed. In this approach, it is assumed that the basic flow principles (e.g., the Darcy's law and the CDE) hold at a local scale, but parameters such as the hydraulic conductivity are variable at a larger scale. The random processes or spatial random fields resulting from such spatial variability of parameters is further assumed as stationary and egodic to facilitate the limitation of measurements that are often a single realization of the system.

There have been an increasing number of studies over the last two decades dedicated to the influence of local fluctuations of hydraulic properties on solute transport at large scales (Dagan, 1986; Gelhar, 1986). Research has focused on demonstrating the asymptotic validity of the CDE for large scale (time/space) transport, and to establish relationships between the local heterogeneity in hydraulic conductivity and the field-scale dispersion coefficient, i.e., macro-dispersivity, which are often reported as several orders of magnitude larger than those obtained from laboratory column studies (Gelhar et al., 1979; Dagan, 1984). While the validity of the CDE may be true in situations such as aquifers where the space dimension is large, it becomes questionable for field soils since the time and space scales are often limited (usually a few days, and several meters deep in a soil profile). Substantial questions also remain regarding the flexibility of the stochastic approach to further incorporate mechanistic descriptions of solute nonequilibrium processes that are often found for many organic contaminants in soil.

Recognizing the above, transport in spatially heterogeneous soils is described deterministically using the continuum theory (Bachmat and Bear, 1986). Whether explicitly or implicitly, this approach always presumes the existence of an REV in the heterogenous porous medium. Within an REV, a limited number of continua (or regions) are assumed as overlapping and interacting with each other. Variables and parameters of the various continua are averaged over an REV, providing continuous functions of the spatial coordinates. Such continuity enables description of flow and other phenomena by means of partial differential equations such as the Richards equation and CDE. Studies via this approach have been well documented in a large body of literature (Nielsen et al., 1986). Among them, the simplest and most popular approach has been the two-region model, which assumes that the porous medium is composed of two interactive regions coexisting in the same volume (REV), with one as a mobile region and the other completely stagnant (Coats and Smith, 1964; van Genuchten and Wierenga, 1976; van Genuchten and Wagenet, 1989; Brusseau et al., 1989). Alternatives to the two-region approach have been developed based upon different mechanistic processes prevailing in each region. It has been shown that when the mobile region is viewed as mainly composed of structural cracks, macropores, fissures, and other preferential pathways, water flow through this region is no longer characterized by the conventional Darcy-Richards equation (Wagenet and Germann, 1989; Germann, 1990). In such a case, boundary layer flow theory assuming viscous flow may apply, and a kinematic wave equation becomes appropriate (Germann and

Beven, 1985). There are also other studies that replace the completely stagnant region with a less permeable pore system, resulting in a dual-porosity model, in which both regions are mobile, and can be described by the same Richards equation but with different characteristic coefficients (Gerke and van Genuchten, 1993), or with one region described by other flow mechanisms (Chen and Wagenet, 1992). Recent extension of this approach has led to the development of multiregion models that are intended to more closely represent the high heterogeneity of field soils (Hutson and Wagenet, 1995).

While the continuum approach provides the flexibility to deterministically incorporate more processes, the number of parameters in the model also increases steadily as more flow regions are needed to represent the heterogeneity. Moreover, the assumption of the existence of an REV that contains all possible flow regions (or continua) cannot be verified for a field soil. A compromise between the stochastic and continuum approaches assumes the macroscopic transport as an ensemble of continuously distributed flow pathways or streams (Jury and Roth, 1990; Simmons, 1982). On each of the pathways, nonequilibrium processes may be further allowed (Cvetkovic and Shapiro, 1990). This approach has been long adopted in chemical reaction engineering and chromatography (Aris, 1982; Villermaux, 1981), and has been recently reviewed by Sardin et al. (1991). In the present work, the concept of the residence time distribution (RTD) (Nauman and Buffham, 1983) was used to specify the velocity ensemble in a heterogeneous porous medium.

A PROBABILITY DISTRIBUTION APPROACH

In the current work, we use probability distributions to simultaneously consider heterogeneity in both hydraulic properties and the local solute physical/chemical reactions. A stochastic-convective representation was first employed to describe hydrodynamic transport, with the solute residence time distribution (RTD) to specifically represent the velocity heterogeneity resulting from the spatially variable hydraulic properties. The RTD approach, compared to the CDE, is general in the sense that the heterogeneity of hydraulic properties may also be incorporated using various RTDs of specific interest. In addition to the RTD, a probability distribution function is also proposed to describe sorption rate heterogeneity. Intra-aggregate diffusion is considered separately in the case of multiprocess nonequilibrium transport in aggregated porous media. A more detailed description of the model development is available through the recent works by Chen (1994), and Chen and Wagenet (1995, 1997).

The Residence Time Distribution of Hydrodynamic Processes

The Residence Time of a Single Solute Particle

Consider a reactive solute particle tracing through an aggregated porous medium in the mean flow direction *x*. The porous system is divided into two regions: an inter-aggregate (mobile) space and an intra-aggregate (immobile) space. Each space is further composed of a fluid phase (θ_m and θ_{im}, L^3/L^3) and a sorbed (solid) phase (S_m and S_{im}, M/M), where the subscripts *m* and *im* denote the mobile and immobile regions, respectively. In terms of water content, the relationship between these two regions is $\theta = \theta_m + \theta_{im}$, where θ is the total porosity, (L^3/L^3). A solute molecule can be either sorbed onto the sorption sites that are directly exposed to the mobile fluid (S_m), or can undergo a series of intra-aggregate diffusion (θ_{im}) and sorption (S_{im}) processes in the immobile region during transport along a pathway as in Figure 1.1. Neglecting local

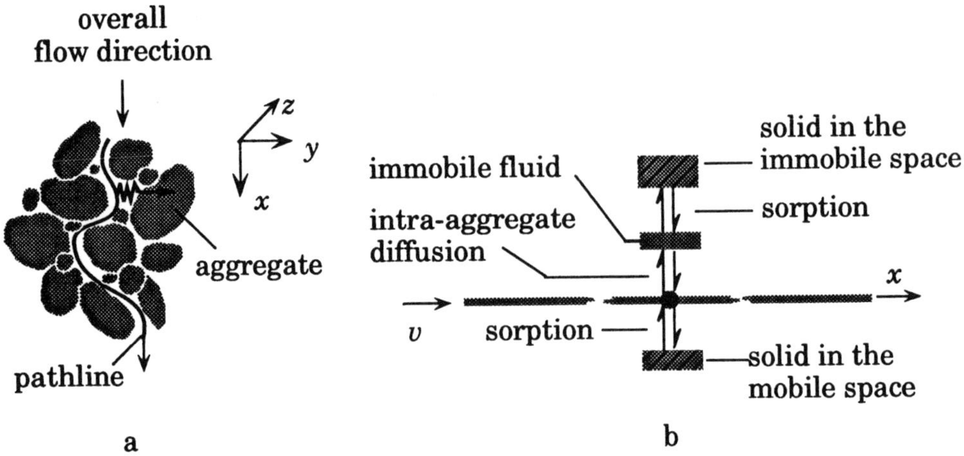

Figure 1.1. Schematic illustration of a solute pathway in an aggregated medium (a); and a conceptualization of the coupled physical/chemical processes in series (b).

dispersion and molecular diffusion temporally, the equation of mass conservation is given by

$$\frac{\partial c_m}{\partial_t} + \frac{1}{\theta_m}\frac{\partial c_a}{\partial t} + v\frac{\partial c_m}{\partial x} = 0 \tag{1.1a}$$

where t is time (T); c_m concentration in the mobile fluid (M/L^3); v the pore velocity (L/T); $c_a = f\rho S_m + (1-f)\rho S_{im} + \theta_{im} c_{im}$, the total concentration in the overall immobilized phases (M/L^3); f the fraction of sorption sites directly exposed to the mobile fluid; and ρ the medium bulk density (M/L^3). The total concentration c_a can be generally described by a convolution integral:

$$c_a(t,x) = \int_0^t c_m(t'x)\, h\,(t-t')dt' \tag{1.1b}$$

where $h(t)$ is a density function describing the process of mass withdrawal and return between the mobile fluid and the overall immobilized phases, which is flexible to be further specified as a first-order or a diffusion process in later sections. Note that the local dispersion term has been omitted in Eq. 1.1a; however, the macroscopic dispersive effect is retained through the ensemble approaches (see the next section) used to describe the random velocity field (Simmons, 1982).

The Laplace transform of Eq. 1.1 with zero initial conditions of c_m and c_a is:

$$s\hat{c}_m + \frac{1}{\theta_m}s\hat{c}_a + v\frac{d\hat{c}_m}{dx} = 0 \tag{1.2a}$$

and

$$\hat{c}_a = \hat{c}_m\,\hat{h}(s) \tag{1.2b}$$

where s is the Laplace parameter, and the symbol "$\wedge$" denotes that a variable has been Laplace-transformed. Under a Dirac injection of unit mass at $x=0$:

$$c_m(t, x=0) = \delta_+(t)$$

Eq. 1.2 has the solution:

$$\hat{c}_m = \exp[-s\tau(1 + H(s))] \tag{1.3}$$

where

$$\tau = \int_0^{x=L} \frac{d\xi}{v} \tag{1.4}$$

and

$$H(s) = \frac{1}{\theta_m}\hat{h}(s) \tag{1.5}$$

Note that τ in Eq. 1.4 is the residence time of a solute that was released at the inlet $x=0$ and reached the outlet $x=L$ with microscopic pore velocity v (L/T) under no influence of any physical/chemical processes. The pore velocity v may be assumed here, in the microscopic point of view, as a mean projection of the three-dimensionally orientated velocity along the particle trajectory onto the x coordinate (Cvetkovic and Shapiro, 1990). $H(s)$ in Eq. 1.5 is also called the transfer function for the nonequilibrium reaction processes (Villermaux, 1974). Detailed forms of $H(s)$ will be discussed and specified in later sections.

Statistical Ensemble Approaches

Consider a uniform application of a large number of solute particles that are simultaneously released over an inlet (soil) surface. Two extreme cases may influence each solute particle during its transport. In the first, each particle travels with the same original velocity as if it were trapped on a pathway during its whole trip from the inlet to the outlet. This might be the case of a bundle of isolated capillaries (Rao et al., 1976), representing situations wherein long and continuous preferential flow pathways exist. In such a system the relationship between residence time and velocity for each individual particle simply becomes $\tau=L/v$ by integrating Eq. 1.4, and thus the RTD of many particles can be directly derived from the knowledge of the velocity distribution. By knowing the RTD, an ensemble mean concentration of Eq. 1.4 is readily calculated, giving a BTC at the outlet plane. An important characteristics of this approach is that the egodic hypothesis (Beran, 1968) is not valid for the random velocity field because the probability of a particle having a given value of travel speed directly depends upon its initial location on the input surface. It has been shown that solute transport in such a system is no longer Fickian in the sense that the effect of dispersion caused by velocity fluctuations turns out to be time-dependent (Simmons, 1982; Jury and Roth, 1990).

In the second extreme case, all solute particles would not forever remain on their original pathways. Rather, each particle may remain for some time on the same pathline, and travel a certain distance with a fixed or nearly fixed velocity. But lateral diffusion (molecular or mechanical) may also interrupt and reduce its residence within that pathline, thus shifting it away to any other streamlines on the plane normal to the mean flow direction x. As it shifts among all different streamlines on the plane, each individual particle appears traveling with a random velocity along the flow direction x.

The relationship between the particle residence time τ and its travel velocity v, defined by Eq. 1.4, therefore becomes a random integral and cannot be evaluated regularly. Due to this reason, the RTD cannot be directly derived through the knowledge of the distribution of the pore velocity v as before.

Detailed knowledge of the local structure of the statistical properties of the pore velocity v in the porous medium, such as the spatial autocorrelation function of v (Beran, 1968), is needed to correctly determine the RTD. Since the autocorrelation function describes how the stochastic field of v evolves in space (or time), theoretically it should identify the transport characteristics in the preasymptotic stage (Sposito et al., 1986). Here, we simply adopt the asymptotic behavior of the solute transport in porous media based on the central limit theory; that is, all solute transport in porous media will eventually approach Gaussian whenever the transport scale is much longer than that of the velocity correlation length (Dagan, 1986). With this condition the ergodic theorem applies to the random velocity field v; i.e., the probability distribution of a single particle experiencing all possible velocities is identical to that of many particles. For the residence time τ of a particle, therefore, the positively-valued inverse Gaussian density function holds (Johnson and Kotz, 1970):

$$p_\tau(\tau; L) = \frac{\bar{\tau}^{3/2}}{\sigma_\tau\sqrt{2\pi\tau^3}} \exp\left[-\frac{\bar{\tau}(\tau-\bar{\tau})^2}{2\sigma_\tau^2\tau}\right] \tag{1.6}$$

where $p_\tau(\tau; L)$ is the RTD, the probability for a particle to have residence time between τ and $d\tau$ before exiting the system at L; σ_τ^2 is the variance of τ; and $\bar{\tau}$ is the mean residence time, which is determined by taking the expectation of Eq. 1.4:

$$\begin{aligned} \bar{\tau} &= E[\tau] = \int_0^L E\left[\frac{1}{v}\right] d\xi \\ &= \frac{L}{1/E[1/v]} = \frac{L}{U} \end{aligned} \tag{1.7}$$

where U is the harmonic mean of v; and $E[\ ]$ denotes expectation.

By knowing the distribution of residence time, $p_\tau(\tau\,;\,x)$, the expected output concentration then can be calculated through Eq. 1.3 in the Laplace domain:

$$\hat{C}_m(s, L) = E[\hat{c}_m(s)] = \int_0^\infty \hat{c}_m(s)\, p_\tau(\tau, L)\, d\tau \tag{1.8}$$

where the capital $\hat{C}_m$ denotes the ensemble mean of c_m in the Laplace domain. By substituting Eqs. 1.3 and 1.6 into 1.8, the final result is (Chen, 1994):

$$\hat{C}_m(s, L) = \exp\left\{(1/\epsilon_\tau)^2\left(1-\sqrt{1+2s\bar{\tau}\,\epsilon_\tau^2\,[1+H(s)]}\right)\right\} \tag{1.9a}$$

where $\epsilon_\tau = \sigma_\tau/\bar{\tau}$, is the coefficient of variation (CV) of residence time τ. Under a square pulse input, Eq. 1.9a becomes:

$$\hat{C}_m(s, L) = \frac{1-\exp(-sT_0)}{s} \exp\left\{(1/\epsilon_\tau)^2\left(1-\sqrt{1+2s\bar{\tau}\epsilon_\tau^2[1+H(s)]}\right)\right\} \tag{1.9b}$$

where T_0 is the length of the pulse (L).

The inverse of Eqs. 1.9a or 1.9b provides an expected concentration output across the exit plane at L under the corresponding input boundary conditions in a porous system, which is an equivalent BTC in column studies. In particular, by defining the dispersion coefficient, $D=\sigma_\tau^2 x^2/(2\bar{\tau}^3)=\sigma_\tau^2 U^2/(2\bar{\tau})$, Chen (1994) has shown that the inverse of Eq. 1.9 gives the same solution as in the CDE approach as long as $H(s)$ identifies a same nonequilibrium process (e.g., the two-site sorption model). It can also be shown that the relationship between the CV, ϵ_τ, and the column Peclet number P is: $P=2/\epsilon_\tau^2$. Hence, the CDE can be thought as a subcase of RTD theory (i.e., with a specific RTD of the inverse Gaussian probability density function (pdf) Eq. 1.6).

A Probabilistic Representation of Sorption Rate Heterogeneity and Multiprocess Nonequilibrium

Single Rate Sorption Coupled with Physical Nonequilibrium

Since a general convolution integral was used in Eq. 1.2b to represent the simultaneous physical/chemical processes, the corresponding Laplace transform, $H(s)$, remains undetermined. Consider now a first-order assumption for both sorption and intra-aggregate diffusion processes. Rewrite the total concentration in the overall immobilized phases:

$$c_a = f\rho S_m + (1-f)\rho S_{im} + \theta_{im} c_{im} \tag{1.10}$$

Note that all the mass in the immobile region (i.e., the last two terms in the right-hand side of Eq. 1.10) is subject to the mass transfer of the first-order intra-aggregate diffusion. That is, in the differential form, we have:

$$\frac{\partial c_a}{\partial t} = f\rho\frac{\partial S_m}{\partial t} + \alpha(c_m - c_{im}) \tag{1.11}$$

where α is the rate coefficient of the first-order intra-aggregate diffusion process, (T^{-1}). Similarly assuming the first-order sorption for the sorbed phases in both regions, we have:

$$\frac{\partial S_m}{\partial t} = k(K_D c_m - S_m) \tag{1.12}$$

and

$$\frac{\partial S_{im}}{\partial t} = k(K_D c_{im} - S_{im}) \tag{1.13}$$

where k is the rate coefficient for sorption (T^{-1}); K_D is the partition coefficient between fluid and solid when sorption equilibrium achieved, (L^3/M). In the Laplace domain, Eqs. 1.10, 1.11, 1.12, and 1.13 become:

$$\hat{c}_a = f\rho\,\hat{S}_m + (1-f)\rho\,\hat{S}_{im} + \theta_{im}\hat{c}_{im} \tag{1.14}$$

$$s\,\hat{c}_a = sf\rho\hat{S}_m + \alpha\,\hat{c}_m + \alpha\,\hat{c}_{im} \tag{1.15}$$

$$s\,\hat{S}_{im} = k K_D\,\hat{c}_{im} + k\hat{S}_{im} \tag{1.16}$$

$$s\,\hat{S}_m = k K_D\,\hat{c}_m + k\hat{S}_m \tag{1.17}$$

Eliminating $\hat{c}_{im}$, $\hat{S}_m$, and $\hat{S}_{im}$ in Eqs. 1.14, 1.15, 1.16, and 1.17, and rearranging give:

$$\frac{\hat{c}_a}{\hat{c}_m} = \frac{1}{1/A(t_s;\, s) + t_e s} + \frac{f\rho K_D}{1 + st_s} \tag{1.18}$$

where

$$A(t_s;\, s) = \theta_{im} + \frac{(1-f)\rho\; K_D}{1 + t_s s} \tag{1.19}$$

$t_e = 1/\alpha$, T; and $t_s = 1/k$, T. Note that defining both t_e and t_s as reciprocals of their correspondent first-order rate coefficients has the advantages of converting the large magnitudes of α and k to small values, giving a convenient representation of instantaneous sorption (i.e., α and k at infinity correspond to $t_e = t_s = 0$). The redefined coefficients t_e and t_s have the unit of time, thus are the characteristic times of either the intra-aggregate diffusion or the sorption process. By definition in Eq. 1.5, $H(s)$ therefore is expressed as:

$$H(s) = \frac{1}{\theta_m}\left[\frac{1}{1/A(t_s;\, s) + t_e s} + \frac{f\rho K_D}{1 + st_s}\right] \tag{1.20}$$

Eq. 1.20 is a first-order intra-aggregate diffusion model coupled with a first-order sorption process. In the case of no sorption, it is a unique first-order intra-aggregate diffusion model (FO).

Heterogeneous Sorption Coupled with Physical Nonequilibrium

While soil may be considered chemically homogeneous on the basis of conventional macroscopic properties, experimental evidence has shown that heterogeneity in soil physical and chemical properties are inevitable at microscopic scales (Weber et al., 1992). Variability in the microsurface properties, composition, and structure of soil particles causes considerable uncertainties in local sorption processes. As a result, the characteristic sorption time, t_s, may change in a random manner and thereby must be characterized by distribution functions. The simplest distributed sorption rate model is the two-site approach, in which the t_s takes values of either 0 (equilibrium sites) or >0 (time-dependent sites). To be general, suppose that a traveling solute particle may take a value of t_s between t_s and $t_s + dt_s$ with probability $p_s\,dt_s$ for sorption, where p_s is the probability density function (pdf) for t_s. The transfer function of Eq. 1.20 is then evaluated by taking a mathematical expectation as follows:

$$H(s) = \frac{1}{\theta_m} \int_0^\infty \left[\frac{1}{1/A(t_s;\ s) + t_e s} + \frac{f\rho K_D}{1 + st_s} \right] p_s(t_s) dt_s \tag{1.21}$$

H(s) in Eq. 1.21 represents a mean transfer function of the heterogeneous sorption processes.

Note that for simplicity, we have assumed in Eq. 1.21 an independent and constant K_D for all possible values of t_s. This means that when true sorption equilibrium is reached, all sorption sites will have a same distribution coefficient. As shown by Chen and Wagenet (1995), when no physical nonequilibrium (i.e., $f=1$, $\theta_{im}=0$, and $t_e=0$ in Eq. 1.21) involved, a constant K_D is equivalent to the mean value of a bulk soil sample by assuming a random K_D, independent of t_s.

The pdf of t_s in Eq. 1.21 takes different forms according to different interpretations for sorption. When soil is assumed to consist of *m* distinct sorption sites and each of them has different accessibility to sorbate molecules, the pdf of t_s is then discrete, and is rewritten as:

$$p_s(t_s) = \sum_{i=1}^{m} F_i \delta(t_s - t_{si}) \tag{1.22}$$

where F_i is the probability (or weight) of a solute molecular being sorbed onto site *i* with time coefficient t_{si}. Substituting Eq. 1.22 into 1.21 gives a simple algebraic sum of *H(s)*:

$$H(s) = \frac{1}{\theta_m} \sum_{i=1}^{m} F_i \left[\frac{1}{1/A(t_{si};\ s) + t_e s} + \frac{f\rho K_D}{1 + t_{si} s} \right] \tag{1.23}$$

When $\theta_{im} \neq 0$ and $t_e \neq 0$, Eq. 1.23 is a multisite sorption model coupled with an intra-aggregate diffusion process with constant time coefficient t_e. The well-recognized two-site sorption model (TS) therefore corresponds to $m=2$ in Eq. 1.23, in which one site with $t_{s1}=0$ (equilibrium), and the other $t_{s2} > 0$ (time-dependent). In this two-site case, F_1 and F_2 therefore serve as the fractions of both sites, and $F_1 = 1 - F_2$. We hereafter refer to this special case as a first-order intra-aggregate diffusion and two-site sorption (FOTS) model. When no intra-aggregate diffusion exists (i.e., $\theta_{im}=0$ and $t_e=0$), the abbreviation TS (two-site sorption) model is used.

The discrete multisite model creates two parameters (F_i and t_{si}) for each individual type of sorption sites. When there are more than two types of sorption sites, Eq. 1.23 becomes less practical because of too many estimated parameters. A continuous pdf of t_s is therefore useful since continuous distributions usually need only two parameters (e.g., the mean and variance). Assuming a generalized gamma distribution for t_s:

$$p_s(t_s) = \frac{\beta^n t_s^{n-1} \exp(-\beta\, t_s)}{\Gamma(n)} \tag{1.24}$$

where $\Gamma(n)$ is the gamma function:

$$\Gamma(n) = \int_0^\infty \xi^{n-1} \exp(-\xi) d\xi$$

In Eq. 1.24, *ß* and *n* are scale and shape parameters determining the mean and variance of the distribution, respectively:

$$\bar{t}_s = \frac{n}{\beta} \tag{1.25a}$$

$$\sigma_s^2 = \frac{n}{\beta^2} \tag{1.25b}$$

The effect of the parameter *n* on the shape of the pdf is illustrated in Figure 1.2. In the figure, all curves have the same mean of unity (*n*/*ß*=*1*). As shown in the figure, the distribution is increasingly skewed toward the origin as *n* decreases, and it becomes more symmetric as *n* increases. A special case is that when *n* = *1*, an exponential pdf results. The gamma pdf is therefore very flexible to describe a wide range of distributions from highly skewed to very symmetric.

Given the gamma pdf of t_s as Eq. 1.24, the expression of *H(s)* is then derived by substituting 1.24 into 1.21 and integrating (Gradshteyn and Ryzhik, Eq. 3.383.10, page 366, 1994):

$$H(s) = \frac{1}{\theta_m}\Big[B_1 + B_2(B_3\beta)^{n-1}\exp(B_3\beta)\Gamma(1-n, B_3\beta) + f\rho K_D\left(\frac{\beta}{s}\right)^n \exp\left(\frac{\beta}{s}\right)\Gamma\left(1-n, \frac{\beta}{s}\right)\Big] \tag{1.26}$$

where $\Gamma(\xi, z)$ is the incomplete gamma function of *z* (Abramowitz and Stegun, Eq. 6.5.3, page 260, 1972); B_1, B_2, and B_3 are constants, defined by:

$$B_1 = \frac{\theta_{im}}{1+\theta_{im} s\, t_e}$$

$$B_2 = \frac{(1-f)\beta\rho K_D}{(1+\theta_{im} s\, t_e)^2 s}$$

and

$$B_3 = \frac{1}{s} + \frac{(1-f)\rho K_D t_e}{1+\theta_{im} s t_e}$$

Eq. 1.26 is referred to as the first-order intra-aggregate diffusion and gamma sorption (FOGS) model. In lack of intra-aggregate diffusion, the gamma sorption (GS) model is then used.

Degradation

Many organic compounds in soils/aquifers are subject to degradation due to microbiological and chemical reactions. The degradation process is usually represented by a lumped first-order reaction:

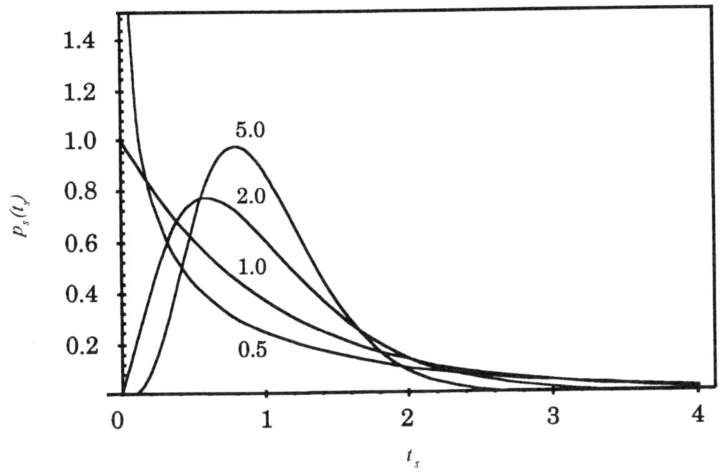

Figure 1.2. Gamma density function. The values of parameters, *n* and *ß*, are both set equal, and shown on each lines.

$$dC/dt = -\gamma C \tag{1.27}$$

where γ is the rate coefficient, (1/T); *C* is the overall concentration in both solution and sorbed phases of soil, (M/L^3). If the first-order degradation is assumed everywhere in the soil system with the same constant rate coefficient γ, it can be shown that in the Laplace domain, the transform parameter *s* in the previously developed models simply needs to be replaced by $s+\gamma$ (Sardin et al., 1991; Jury and Roth, 1990).

To summarize the developed models to be examined next, we used acrostics, TS, FOTS, GS, and FOGS, to represent Eq. 1.9 (either Eq. 1.9a or 1.9b, depending on boundary conditions for hydrodynamic transport) coupled with four combinations of the transfer function *H(s)* of nonequilibrium processes (Eqs. 1.23 and 1.26). FOTS and TS stand for a two-site sorption version of Eq. 1.23 (i.e., $m=2$) with or without (i.e., $\theta_{im}=0$ and $t_e=0$) the intra-aggregate diffusion process, respectively. Similarly, FOGS and GS denote that Eq. 1.26 is used in Eq. 1.9, with or without (i.e., $\theta_{im}=0$ and $t_e=0$) the intra-aggregate diffusion process, respectively. As all model solutions in this study were derived in the Laplace domain, the inverse transforms were necessary and all were performed numerically using the program provided of Jury and Roth (1990), which is based on an algorithm proposed by Talbot (1979). The FORTRAN-77 codes and all other details are found in Chen (1994).

Effects of Sorption Rate Heterogeneity on Transport

Sorption Nonequilibrium

For simplicity, we first consider the sorption nonequilibrium process without intra-aggregate diffusion in the porous medium (i.e., the TS and GS models). This allows focus on the effect of the variability of sorption time t_s on solute spreading during transport. To quantitatively evaluate the effect and to obtain implications, the time moment method is utilized, which was described in detail by Valocchi (1985). Time moments serve as descriptors of a concentration BTC; for example, the first four moments indicate the mean breakthrough time, the degree of spreading, the asymmetry and the peakedness of the BTC, respectively. It is expected that each higher moment provides additional information about the breakthrough.

The ith absolute moment of a concentration BTC such as defined by Eq. 1.9a under a Dirac input condition is calculated as:

$$m_i = (-1)^i \lim_{s \to 0} \left[\frac{d^i}{ds^i} \hat{C}_m(s, x) \right] \tag{1.28}$$

Given the absolute moments calculated as Eq. 1.28, the corresponding central moments can be obtained as described by Abramowitz and Stegun (Eq. 26.1.14, page 928, 1972). Before performing the detailed moment calculation however, we first reduce the previously presented models to dimensionless forms for convenience of comparison. Define the dimensionless variables as:

$$T = t/\bar{\tau} \quad at\ x \quad X = x/L$$

and correspondingly,

$$T_e = t_e/\bar{\tau}, \text{ and } T_s = t_s/\bar{\tau}$$

where L is the column length, (L); T_e and T_s are dimensionless characteristic times for sorption and intra-aggregate diffusion, respectively. The reduced model solution of Eq. 1.9a under a Dirac injection is (Chen, 1994):

$$\hat{C}(s; X) = \exp\left\{ (1/\epsilon_\tau)^2 \left[1 - \sqrt{1 + 2 \epsilon_\tau^2 \, s[1 + H(s)]} \right] \right\} \tag{1.29}$$

Eq. 1.21 in the dimensionless form is:

$$H(s) = \frac{1}{\theta_m} \int_0^\infty \left[\frac{1}{1/A(T_s;\ s) + T_e s} + \frac{f\rho K_D}{1 + sT_s} \right] p(T_s) dT_s \tag{1.30}$$

where $p(T_s)$ now denotes the pdf of the dimensionless variate T_s. If the Gamma pdf is assumed for t_s, it can be shown that T_s is also Gamma-distributed but with parameters n and $\beta\,\bar{\tau}$ (see Chen, 1994). When only sorption process considered, Eq. 1.30 is simplified as:

$$H(s) = \frac{1}{\theta} \int_0^\infty \frac{\rho K_D}{1 + t_s s} p(T_s) dT_s \tag{1.31}$$

The absolute moments of Eq. 1.29 with Eq. 1.31 then can be calculated based on Eq. 1.28. And the corresponding central moments of the first four are:

$$m_{BTC} = R \tag{1.32}$$

$$\sigma^2_{BTC} = R^2 \in^2_\tau + 2\, K \frac{\bar{t}_s}{\bar{\tau}} \tag{1.33}$$

$$\mu_3 = 3\, R^3 \in^4_\tau + 6\, K\, R \in^2_\tau \frac{\bar{t}_s}{\bar{\tau}} + 6\, K\,(1+\in^2_{ts}) \left(\frac{\bar{t}_s}{\bar{\tau}} \right)^2 \tag{1.34}$$

$$\mu_4 = 3\, R^4 (1 + 5\in^2_\tau) \in^4_\tau + 12 K R^2 \in^2_\tau (1 + 3 \in^2_\tau) \frac{\bar{t}_s}{\bar{\tau}} + 12\, K \Big[K(1+\in^2_\tau) + 2\, R \in^2_\tau (1+\in^2_{ts}) \Big] \left(\frac{\bar{t}_s}{\bar{\tau}} \right)^2 + 24\, K \frac{E[t_s^3]}{\bar{\tau}^3} \tag{1.35}$$

where $K = K_D\rho/\theta$; $R = 1 + K$, is the retardation factor; $\in_{ts} = \sigma_{ts}/\bar{t}_s$, is the CV of t_s (Eq. 1.24); $E[t_s^3]$ is the third absolute moment of the pdf of t_s (Eq. 1.24); m_{BTC} and σ^2_{BTC} are the BTC mean and variance; μ_3 and μ_4 are the BTC third and fourth central moments, respectively. Note that $\bar{T}_s = \bar{t}_s/\bar{\tau}$ in the above equations, representing the mean of the dimensionless sorption time constant T_s.

Three interesting results are demonstrated in the expressions of the first four moments for a given dispersive system (i.e., $\in_\tau$ is given). First, the mean breakthrough time (m_{BTC}) is solely determined by the equilibrium properties (R), while the spreading and tailing are described by the BTC variance (σ^2_{BTC}) and its corresponding higher moments (μ_3 and μ_4) are influenced by both equilibrium and sorption rate heterogeneity ($\in_{ts}$). Second, the influence of the mean sorption time constant occurs from the BTC variance onward, whereas the impact of sorption rate heterogeneity ($\in_{ts}$) starts from the third, indicating its important influence on BTC tailing. This also indicates that the ultimate similarity of BTCs from different nonequilibrium models can be achieved up to the third moment for any sorption rate distributions characterized by the same mean and variance. Third, the dependence of higher BTC moments on both the ratio of $\bar{t}_s/\bar{\tau}$ (the reverse also called the Damkohler number) and $\in_{ts}$, suggests that only when the variability of sorption sites is given may the Damkohler number serve as a sufficient measure of the severity of local nonequilibrium conditions.

To gain further insight into the effect of sorption rate heterogeneity on solute transport processes, we compare the two models, TS and GS, both having very different sorption rate distributions, i.e., Eq. 1.22 with *m=2*, and Eq. 1.24, respectively. These two models are both characterized by two parameters determining their means and variances. That is, the GS model has n and $ß$ and the TS has F_1 and t_{s2}. The effect of these two distributions therefore will be evaluated on the basis of having the same mean and variance. As seen in Eqs. 1.32 to 1.35, the same mean and variance of the sorption time constant (t_s) will make the first three BTC moments identical. On this basis, any discrepancy between the two model predictions is solely attributed to the different nature of the distribution of sorption rates.

Observing the moment representations of Eq. 1.32 to 1.35, the difference of the fourth moment between the two BTCs (Δ_4) is:

$$\Delta_4 = \frac{24K|M_3 - M_3'|}{\bar{\tau}^3} \tag{1.36}$$

where M_3 and M'_3 are the third moments of the two different pdfs of t_s, respectively. Assuming the difference of the fourth moment reflects the difference between the TS and GS model predictions, Eq. 1.36 then serves as a criterion for the discrepancy between the two models. A similar criterion was also used in the work by Parker and Valocchi (1986) for quantifying the deviation of the local equilibrium assumption from the first-order nonequilibrium process.

For the TS and GS pdfs (i.e., Eq. 1.22 with $m=2$, and Eq. 1.24), the equivalent mean and variance of t_s can be obtained by equating the two pairs of parameters in each pdf (Chen, 1994):

$$F_1 = 1/(1+n) \tag{1.37}$$

and

$$t_{s2} = (1+n)/\beta \tag{1.38}$$

where all parameters are defined as before. Given Eqs. 1.37 and 1.38, Eq. 1.36 becomes:

$$\Delta_4 = 24(R-1)\in_{ts}^2 (1+\in_{ts}^2)\left(\frac{\bar{t}_s}{\bar{\tau}}\right)^3 \tag{1.39}$$

Eq. 1.39 demonstrates that the deviation of the fourth moment between the TS and GS models increases rapidly with the increase of variability of the sorption time constant t_s (fourth power) and the ratio $\bar{t}_s/\bar{\tau}$, the mean characteristic time of sorption to the mean residence time of transport (third power). This indicates that difference in describing slow BTC tailing between the two models can be significant, especially when the variability of sorption rates are large or sorption are markedly slower than transport. Eq. 1.39 also indicates that the difference between the fourth moments increases linearly with the increase of the retardation coefficient R.

To further illustrate the effects of $\in_{ts}$, $\bar{t}_s/\bar{\tau}$, and R on solute transport and the discrepancy between the two model predictions, systematic simulation exercises were performed as three cases based on the parameter values given in Table 1.1. Simulation results are shown in Figures 1.3 to 1.5, where all concentrations in the figures are presented in the log scale to allow scrutiny of the BTC tails. As expected from the proceeding discussions, increasing all the three coefficients increases the difference between the two model predictions. Generally, predictions by the GS model are more spread but less peaked than the TS predictions in various cases. Since in each case both models have the same mean and variance of the sorption time constant t_s and all transport related parameters, the difference between model predictions is solely attributed to the different assumptions regarding to the sorption rate distribution. The linear decline of all TS-simulated BTC tails on the log scale characterizes the single-valued first-order rate process assumed in the model. Conversely, the continuously extended tailing generated by the GS model signifies the effect of the gamma-distributed sorption rate, leading to gradual prolongation of leaching.

Figure 1.3 illustrates the effect of the two TS and GS pdfs under different ratios of $\bar{t}_s/\bar{\tau}$. It is seen that the sharpness of BTC peaks from the TS model develops rapidly as the ratios of $\bar{t}_s/\bar{\tau}$ increases from 0.1 to 10.0 (Figures 1.3a to 1.3d), indicating high dependence of the TS model predictions upon the relative characteristic time of sorption and transport. When the ratio is high, sorption is generally slower than transport and, nonequilibrium conditions present. For the TS model this means that most sol-

Table 1.1. Parameter Values for the Simulation of Transport Under Sorption Nonequilibrium.

Cases	$\bar{t}_s/\bar{\tau}$	ϵ_{ts}	R	GS Model ß	GS Model n	TS Model F_1	TS Model t_{s2}	ϵ_{τ}
1a	0.1	2.0	3.4	0.005	0.25	0.8	250.0	0.2
1b	1.0	2.0	3.4	0.005	0.25	0.8	250.0	0.2
1c	10.0	2.0	3.4	0.005	0.25	0.8	250.0	0.2
1d	20.0	2.0	3.4	0.005	0.25	0.8	250.0	0.2
2a	2.0	0.3	3.4	0.22	11.11	0.083	54.5	0.2
2b	2.0	1.5	3.4	0.0089	0.44	0.69	162.5	0.2
2c	2.0	3.0	3.4	0.0022	0.11	0.90	500.0	0.2
2d	2.0	5.0	3.4	0.0008	0.04	0.96	1300.0	0.2
3a	5.0	2.0	1.24	0.005	0.25	0.8	250.0	0.2
3b	5.0	2.0	3.4	0.005	0.25	0.8	250.0	0.2
3c	5.0	2.0	5.8	0.005	0.25	0.8	250.0	0.2
3d	5.0	2.0	13.0	0.005	0.25	0.8	250.0	0.2

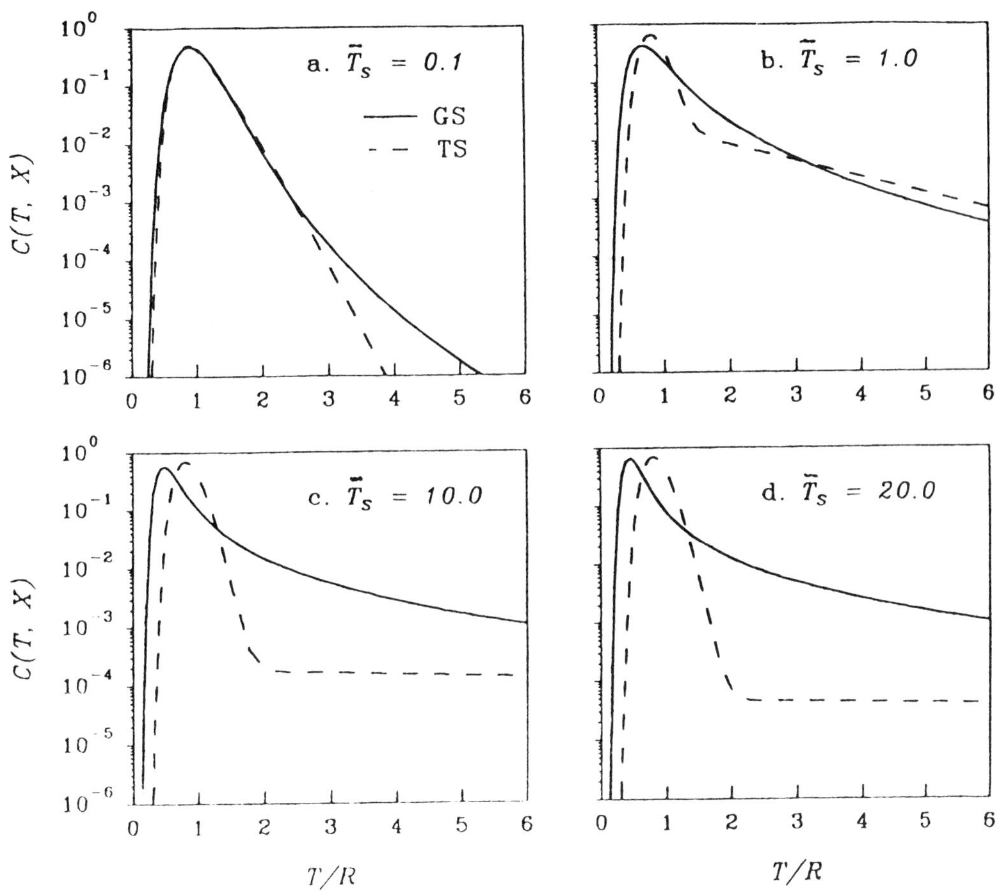

Figure 1.3. Effect of $\bar{T}_s(=\bar{t}_s/\bar{\tau})$ on concentration BTCs of two-site (dashed line) and γ-sorption (solid line) models (case 1 in Table 1.1). Values of $\bar{t}_s/\bar{\tau}$ are indicated on each plot (Chen and Wagenet, *Environ. Sci. Technol.*, 29, p. 2731, Figure 4, 1995. With permission).

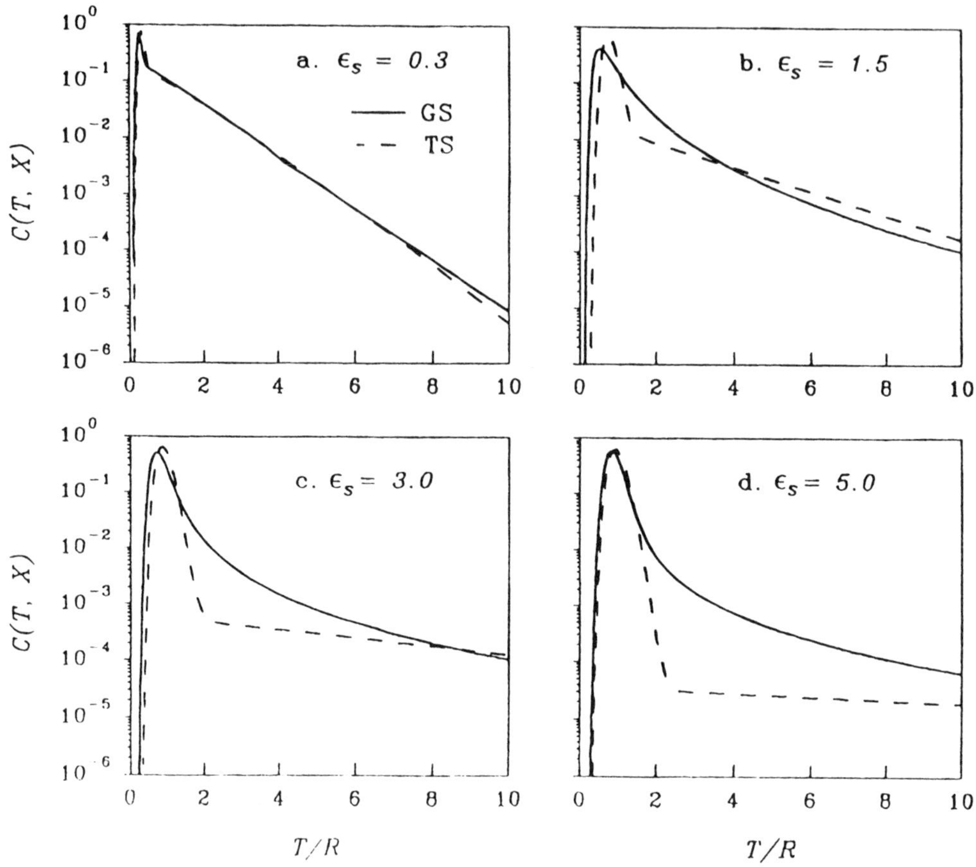

Figure 1.4. Effect of ϵ_{ts} on concentration BTCs of two-site (dashed line) and γ-sorption (solid line) models (Case 2 in Table 1.1). Values of ϵ_{ts} are indicated on each plot (Chen and Wagenet, *Environ. Sci. Technol.*, 29, p. 2731, Figure 5, 1995. With permission).

utes are likely to bypass the single-valued time-dependent sorption site, resulting in a sharp peak and a very small amount of sorbed solutes to produce flattened tails. In the low ratio case, nonequilibrium conditions are less severe, and most solutes have enough time to react with the unique time-dependent site, giving a wider peak spreading and a closer match to the GS model predictions (Figure 1.3a).

Compared to the TS model, the GS model has been less affected by this ratio, indicating the complimentary effect of a continuum of sorption sites. The high dependence of deviations between different sorption-site distribution models on the ratio $\bar{t}_s/\bar{\tau}$ was also found by Sardin et al. (1991). They compared different distribution models having the same mean t_s, and attributed the deviation between model predictions to the variance of t_s. As seen from the current simulations, where the same mean and variance were used for t_s in all comparison cases, substantial deviation between BTCs still exists, indicating the intrinsic difference between TS and GS pdfs in describing sorption rate heterogeneity. The high dependence of the TS model on the ratio also implies that this model can be highly velocity-dependent. This observation will be further verified with experimental data in the later section.

The effects of different CV values of t_s are shown in Figure 1.4. Theoretically, when the value of the CV (ϵ_{ts}) tends to zero, all model predictions will converge to the results of a one-site model, representing homogeneous sorption. This trend is shown in Fig-

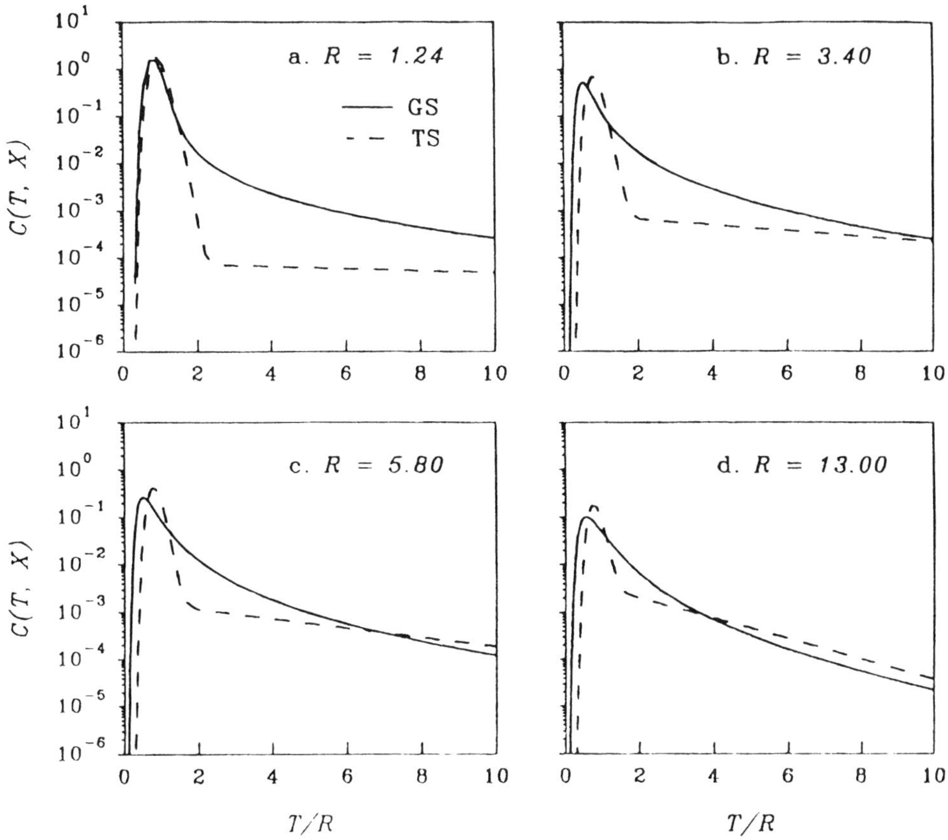

Figure 1.5. Effect of R on concentration BTCs of two-site (dashed line) and γ-sorption (solid line) models (Case 3 in Table 1.1). Values of R are indicated on each plot (Chen and Wagenet, *Environ. Sci. Technol.*, 29, p. 2732, Figure 6, 1995. With permission).

ure 1.4, where a closer match between the TS and GS model simulations was achieved when the coefficient of variance decreased from the highest 5.0 (Figure 1.4d) to the lowest 0.3 (Figure 1.4a). An interesting result shown in Figure 1.4 is that the deviation between these two models appears primarily on the tailing as the variability of sorption rates increases. Although intuitively, higher variability of sorption sites would spread the solute peak more significantly, peak spreading is limited in the TS model since the increase of the sorption rate variability means the increase of rate deviation between the two distinct sites. That is, the single-valued time-dependent site has to be very slow, and most mass will bypass the slow site, resulting in a sharpened instead of more distributed peak.

Figure 1.5 compares the influence of different retardation values (R) on the two model simulation results. For the TS model (dashed lines), an increase in R increases the peak retardation, as well as the concentration level in the BTC tail. Comparatively, the GS model simulations are less influenced. Because the TS model assumes a single-valued rate (desorption) constant (k) for the unique time-dependent site and because K_D is the ratio of the forward (sorption) to the reverse (desorption) rate constants, the increase of R (i.e., K_D increases) would only increase the forward mass transfer rate, thus resulting in enhanced retardation and tailing for this model. Contrast to the single-valued rate constant assumed in the TS model, the GS model was hypothesized to contain a distribution of t_s (or equivalently k). Thus, a change in R would not significantly change the model predictions, which indicates that the current GS model with no correlation between K_D and t_s may well approximate the cases where K_D and t_s are not strictly correlated as discussed by Chen and Wagenet (1995).

Multiprocess Nonequilibrium

When an intra-aggregate diffusion process is imposed on sorption, multiprocess nonequilibrium results. The transport pattern of solutes may change substantially, depending upon the extent to which the intra-aggregate diffusion may dominate the overall process. To illustrate, we list the first two BTC central moments according to the procedures discussed in the last section while we neglect the higher, which are somewhat lengthy. The nonequilibrium transfer function *H(s)* now includes both the first-order intra-aggregate diffusion (t_e) and sorption (t_s) (i.e., Eq. 1.30). From Eqs. 1.29 and 1.30, the first and second moments are:

$$m_{BTC} = R_m + \frac{\theta_{im}}{\theta_m} \tag{1.40}$$

$$\sigma^2_{BTC} = \left(R_m + \frac{\theta_{im}}{\theta_m} \right) \epsilon_\tau^2 + \frac{2}{\theta_m} \left\{ [(1-f)K_D\rho + \theta_{im}]^2 \frac{\bar{t}_e}{\bar{\tau}} + K_D\rho \frac{\bar{t}_s}{\bar{\tau}} \right\} \tag{1.41}$$

where $R_m = 1 + \rho K_D/\theta_m$, and $K_{im} = (1-f)\rho K_D + \theta_{im}$.

Comparing Eq. 1.40 to Eq. 1.32, the mean breakthrough time is longer under the addition of intra-aggregate diffusion, indicating that transport is prolonged due to the additional resistance. Increase in the BTC variance (Eq. 1.41) and higher moments (not shown) is also expected. A special case is that when the intra-aggregate diffusion is fast compared to transport ($\bar{t}_e/\bar{\tau} << 1$), and the term containing $\bar{t}_e/\bar{\tau}$ in Eq. 1.41 can be neglected. In this case, the retardation factor can be redefined as $R = R_m + \theta_{im}/\theta_m$, and Eq. 1.41 becomes identical to Eq. 1.33 of the non-intra-aggregate diffusion case. This means that the fast intra-aggregate diffusion may simply behave as an additional retardation factor and any nonequilibrium is primarily attributed to sorption. Such cases, for example, can be the transport of sorbing solutes in soils containing small aggregates.

To further illustrate, several simulated BTCs assuming multiprocess nonequilibrium processes (Eqs. 1.29 with 1.30) are shown in Figures 1.6 and 1.7. Parameters used were reported in Table 1.2. Note that the dispersive effect in all cases was fixed to $\epsilon_\tau = 0.2$, and the mobile water content (θ_m) was assumed to occupy 60% of the total pore volume (θ). Moreover, the fraction of sorption sites (f) that are directly exposed to the mobile fluid was set to zero in the cases of intra-aggregate diffusion, meaning that all sorbate molecules were subject to the physical process prior to sorption.

Comparison of simulated BTCs assuming multiprocess nonequilibrium to that assuming a unique sorption process is shown in Figure 1.6 (Cases 1 and 2 in Table 1.2). Compared to the BTC with the only sorption process, the addition of the intra-aggregate diffusion has accelerated the peak breakthrough due to the reduction of mobile space. The addition of the intra-aggregate diffusion also reduces the deviation between the FOTS and FOGS sorption models, indicating the control of this physical process. The controlling effect is further illustrated in Figure 1.7 (Cases 2 and 3 in Table 1.2). As seen in the figure, when the intra-aggregate diffusion process is fast compared to transport (i.e., $\bar{t}_e/\bar{\tau} = 0.08$ in the figure), the deviation between the FOTS and FOGS models increases, and both predicted BTCs were not advanced to the de-

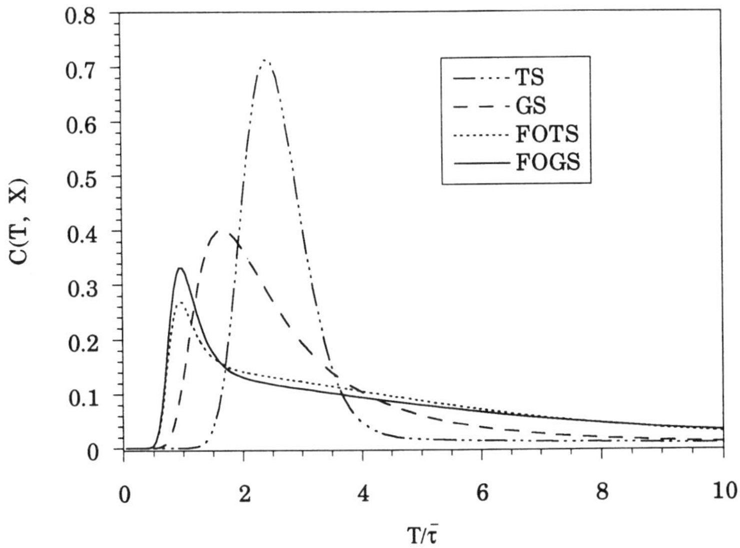

Figure 1.6. Effect of first-order intra-aggregate diffusion on sorption process. Parameters are given in Cases 1 and 2 in Table 1.2 (Chen, 1994).

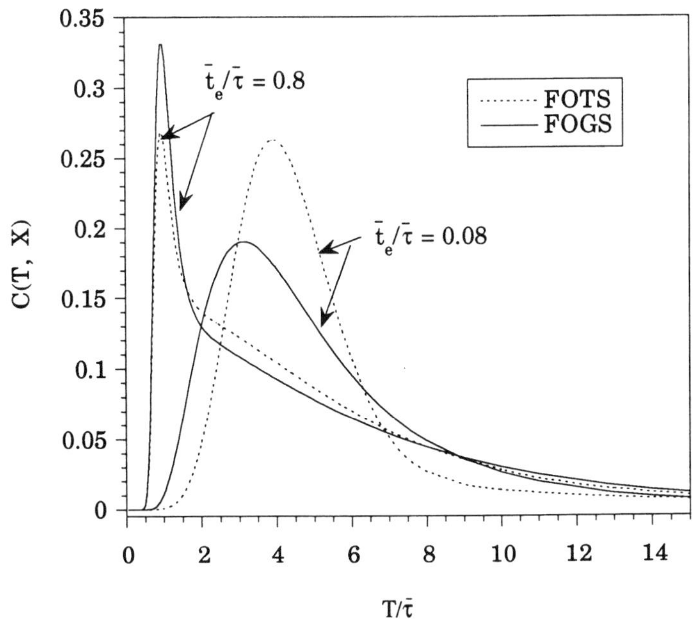

Figure 1.7. Control of intra-aggregate diffusion over sorption. Parameters are given in Table 1.2 as Cases 2 and 3 (Chen, 1994).

gree of the larger ratio case ($\bar{t}_e/\bar{\tau}$= 0.8). This means that in the case of fast intra-aggregate diffusion, the constraint of the physical diffusion process on mass transfer is minimized, and its ultimate effect appears as a retardation factor. The control of intra-aggregate diffusion should also depend upon the value of f. Smaller f indicates more sorption sites that are not directly accessible to the mobile fluid, and thus not free

Table 1.2. Parameter Values for the Simulation of Transport Under Multiprocess Nonequilibrium Conditions (Chen, 1994).

Cases		f	$\frac{\bar{t}_e}{\bar{\tau}}$	$\frac{\bar{t}_s}{\bar{\tau}}$	R	ϵ_{ts}	β	n	F_1	t_{s2}	$\frac{\theta_m}{\theta}$	ϵ_t
1a	TS	1.0	NP[a]	0.5	3.4	1.41	NP	NP	0.67	150.0	1.0	0.2
1b	GS	1.0	NP	0.5	3.4	1.41	0.01	0.5	NP	NP	1.0	0.2
2a	FOTS	0	0.08	0.5	3.4	1.41	NP	NP	0.67	150.0	0.6	0.2
2b	FOGS	0	0.08	0.5	3.4	1.41	0.01	0.5	NP	NP	0.6	0.2
3a	FOTS	0	0.8	0.5	3.4	1.41	NP	NP	0.67	150.0	0.6	0.2
3b	FOGS	0	0.8	0.5	3.4	1.41	0.01	0.5	NP	NP	0.6	0.2

[a] Not applicable.

from the physical diffusion process. For well-aggregated soils, most sorption sites may reside in the aggregate region, and intra-aggregate diffusion often dominates the consequent sorption process. Under these circumstances, sorption rate heterogeneity may be well approximated by simple distribution models (e.g., TS) or even by direct use of the local equilibrium assumption for the sorption process.

MODEL EVALUATION

In this section, we use part of the results of Chen and Wagenet (1997) and Chen (1994) to demonstrate some of the evaluations that have been done on the established sorption models (GS, FOGS, TS, and FOTS). The evaluation was based on model independent predictions; i.e., model parameters were estimated from an independent experiment and were used to predict observations under different experimental conditions. Different experimental conditions were primarily limited to physical aspects, including water flow rates, column lengths, input concentration and pulse lengths. According to the significance level of physical intra-aggregate diffusion, the results are organized into two cases below.

Case 1

In this case, sorption was primarily responsible for nonequilibrium and the physical intra-aggregate diffusion process was minor. Two sets of experiments, batch and column, were performed using ^{14}C atrazine [2-chloro-4-ethylamino-6-isopropylamino-s-triazine] and a Niagara silt loam (organic matter content, 5.96%; silt, 71.0%; clay, 18.3%; pH, 5.73; and CEC, 14.87 cmol/kg) which was air-dried and passed through a 1-mm sieve. In the batch experiments, nine atrazine sorption isotherms (concentration ranging from 0.05 to 20 μg/mL) were completed for each equilibration time of 1, 3, 18, 24, 42, 48, 94, 115, and 237 hours (Table 1.3). Detailed experimental procedures are found in Chen and Wagenet (1997).

The column experiments consisted of two glass columns of different lengths (6.0 and 15.5 cm; Table 1.4). Both columns were carefully packed with the prepared soil samples in uniform bulk density. An HPLC pump was used to saturate the columns with 0.005 M $CaSO_4$ and to establish a steady-state water flow condition under a desired flow rate. A pulse of the mixed aqueous solution of atrazine and Cl^- was applied to each column and effluent samples from each experiment were collected. The shorter column (Col. I) was further distinguished as four different velocity cases (Ia, Ib, Ic, and Id; Table 1.4). After the displacement of the first pulse of chemicals was completed

Table 1.3. Isotherm constants derived from batch experiments.[a]

Equilibration time (hours)	Nonlinear Freundlich Isotherm ($S=K_f\,C^N$) K_f (mL g^{-1} μg^{1-N})	N	r^2	Linear Freundlich Isotherm ($S=K_D C$) K_D (mL/g)	r^2
1.0	1.22	0.91	0.9992	1.04	0.9922
3.0	1.48	0.90	0.9970	1.24	0.9692
18.0	2.51	0.87	0.9988	2.01	0.9909
24.0	1.99	0.91	1.0000	1.71	0.9968
42.0	2.62	0.85	0.9990	2.06	0.9890
48.0	2.56	0.88	0.9999	2.11	0.9936
94.0	3.07	0.84	1.0000	2.49	0.9994
115.0	3.16	0.87	0.9998	2.63	0.9926
237.0	3.73	0.86	0.9989	3.02	0.9814

[a] Chen and Wagenet, Table 2; *Soil Sci. Soc. Amer. J.* (1997, with permission).

(Ia), the column was disconnected from the pump and the whole column was sterilized under a Co (γ–irradiation) source (total dose = 2.5 mega-rads). Three more velocity cases (Ib-c) were then accomplished in the sterilized column. The purpose of sterilization was to avoid potential biodegradation of atrazine during the slowest velocity run (Case Ib), which lasted about 10 days. As discussed in detail by Chen and Wagenet (1997), none of the batch or column studies had shown any significant atrazine degradation over a 10-day period.

Parameter Estimation

Cl^- BTCs from each displacement experiment were used to estimate the two transport-related parameters, pore-water velocity (v) and the CV of hydrodynamic residence time (ϵ_τ) (incorporated in the Peclet number, $P = 2/\epsilon_\tau$, Table 1.4). The retardation factor for Cl^- was fixed to 1.0 during the fitting. With the estimates of v and P, the three sorption parameters in each of the sorption models were estimated using a given atrazine BTC obtained from a specific flow rate and column length. The three sorption parameters were: K_D, F_1, and t_{s2} for the TS model, i.e., Eq. 1.9b with Eq. 1.23 ($m=2$, $\theta_{im}=0$, and $t_e=0$); K_D, β, and n for the GS model, i.e., Eq. 1.9b with Eq. 1.26 ($\theta_{im}=0$ and $t_e=0$). The K_D value obtained from batch experiments was not employed directly since nonequilibrium existed.

A nonlinear least square method (NLLS) (Marquardt, 1963) was employed to optimize model predictions to the given observed BTC. The FORTRAN 77 codes were adapted from Parker and van Genuchten (1984), with the incorporation of the nonequilibrium model (Chen, 1997). The model was solved analytically in the Laplace domain, and was numerically inverted with a program provided by Jury and Roth (1990). With all parameters estimated, model evaluation procedures were based on the extrapolations of these "best-fit" parameters to results obtained under different experimental conditions.

Results and Discussion

Sorption isotherms obtained from various equilibration times (Table 1.3) were fitted by both linear and nonlinear forms of the Freundlich equation. Six representative

Table 1.4. Experimental Conditions for the Miscible Displacement Experiments of Cl^- and Atrazine.[a]

Columns	Experimental Cases	Soil Particle Size (mm)	Atrazine C_0 (μg/mL)	L (cm)	Measured Velocity (cm/hr)	θ (cm^3/cm^3)	Pulse $=vT_0/L$	Fitted Values from Cl^- Data	
								v (cm/hr)	$P = 2/\epsilon_\tau^2$
I	Ia	<1.0	5	6	2.2	0.53	1.22	2.28	57
	Ib[b]	<1.0	5	6	0.51	0.53	1.10	0.47	57
	Ic[b]	<1.0	5	6	3.83	0.53	1.17	3.61	105
	Id[b]	<1.0	5	6	21.27	0.53	1.03	19.28	83
II	II	<1.0	2	15.5	2.43	0.49	1.75	2.59	111

[a] Chen and Wagenet, Table 1; *Soil Sci. Soc. Amer. J.* (1997, with permission).
[b] Sterilized by γ-irradiation.

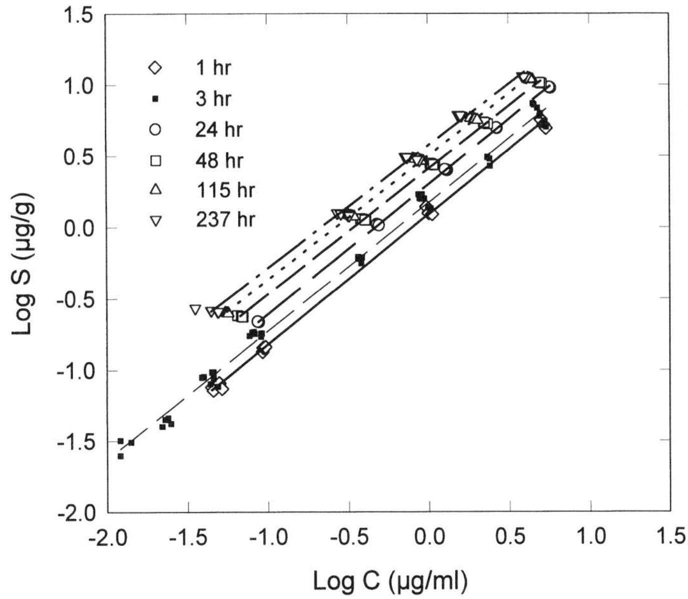

Figure 1.8. Atrazine sorption isotherms determined under six different equilibration times on a Niagara silty loam in batch experiments. Lines were fitted by the nonlinear Freundlich equation (Chen and Wagenet, Figure 1; *Soil Sci. Soc. Amer. J.*, 1997. With permission).

nonlinear isotherms (Figure 1.8) demonstrate that significant increase of atrazine sorption was observed as the equilibration time increased from 1 to 237 hr, indicating that sorption had never reached equilibrium over this range of time scales. The general increasing trend of K_D and K_f values in the respective linear and nonlinear Freundlich equations is also apparent (Table 1.3). However, the effect of the observed isotherm nonlinearity was found to be minor and thus the linear form of isotherms was used throughout the modeling exercises (Chen and Wagenet, 1997).

The overall shapes of the measured Cl^- BTCs (Figure 1.9) were generally neither skewed nor retarded to a significant degree, indicating that physical intra-aggregate diffusion was negligible in the sieved (<1 mm) soil samples. Observed atrazine BTCs indicated in all cases that atrazine was both retarded and leftward skewed (Figure 1.10). The leftward shifts of the BTCs (early breakthrough) with increasing flow velocity in the three irradiated cases (Ib–d) or decreasing column length in the two nonirradiated cases (Ia and II) are demonstrated in this figure. However, the left-hand shift of BTCs with increasing velocity was not observed between the nonirradiated (Ia) and irradiated (Ib) treatments. As shown in Figure 1.10, the BTC of nonirradiated Case Ia occurred even later than that of irradiated Case Ib regardless of its nearly five times higher velocity. Given the same hydrodynamic conditions ($P = 57$, Table 1.4), this inconsistency was likely due to potential alteration of soil properties by γ-irradiation (Chen and Wagenet, 1997).

The three velocity cases of the γ-irradiation treatment were considered first and separately. A 3-parameter fit, as described above, was performed by applying either TS or GS to the BTC data obtained from Case Ib under the slowest velocity conditions (Table 1.4). Optimized parameters are in Table 1.5 and comparisons between fitted and observed BTCs are in Figure 1.11a. With the optimized parameters, independent predictions were then made by each model for the two higher velocity cases (Ic and Id), with the corresponding results compared in Figures 1.11b and 1.11c. Excellent agreement

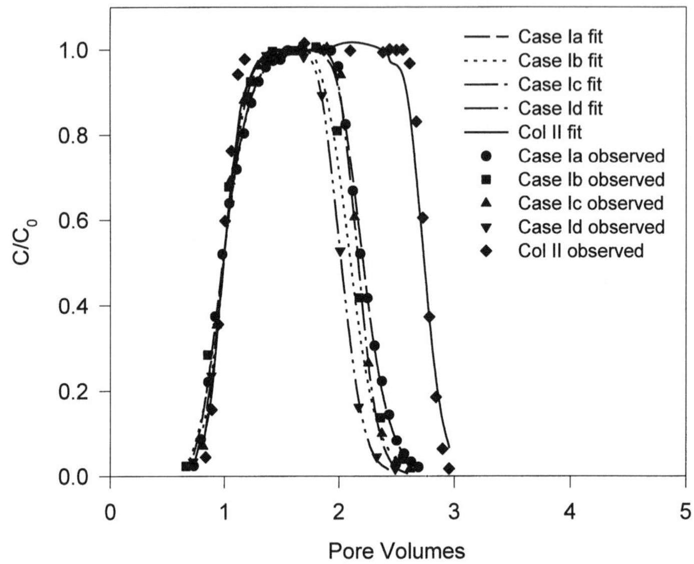

Figure 1.9. Observed and fitted Cl^- breakthrough curves (BTCs) from all column experiments. Curve-fit was performed by adjusting the CV of residence time and pore-water velocity, with the optimized values reported in Table 1.4 (Chen and Wagenet, Figure 3; *Soil Sci. Soc. Amer. J.*, 1997. With permission).

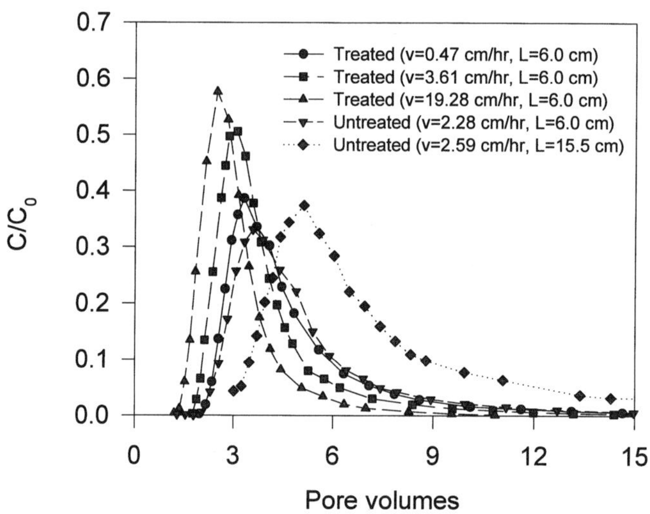

Figure 1.10. Observed atrazine breakthrough curves (BTCs) from all column experiments. Treatment indicates γ-irradiation. Col. I (L=6.0 cm, Table 1.4) was treated after its first run. Three more velocity cases were performed on the same column after irradiation. Col. II (L=15.5 cm, Table 1.4) was never treated (Chen and Wagenet, Figure 4; *Soil Sci. Soc. Amer. J.*, 1997. With permission).

between the GS model predictions and the BTC measurements under both higher velocity cases indicated a very weak dependence of the GS model on velocity, over the velocity range used in this study (0.47–19.28 cm/hour). On the other hand, predictions by the TS model were poor, with simultaneous overestimation of the BTC peaks and underestimation of the tailing.

Table 1.5. Optimized Sorption Parameters from Two Atrazine Column Experimental Treatments.[a]

Exp.	TS			GS		
Cases	K_D (mL/g)	F_1	t_{s2} (hr)	K_D (mL/g)	β (hr^{-1})	n
Ia	1.81	0.48	5.38	4.58	0.000120	0.191
Ib[b]	1.42	0.57	19.58	2.46	0.000121	0.152

[a] Chen and Wagenet, Table 4; *Soil Sci. Soc. Amer. J.*, 1997. With permission.
[b] Sterilized by γ-irradiation.

The TS model overpredicted the retardation of BTCs for the two higher velocity cases (Figures 1.11b and 1.11c), despite having the fitted K_D value (1.42 mL/g) lower than the GS-fitted (2.46 mL/g) (Table 1.5). The conflict arose from the fact that the observed leftward shift of the BTC with increasing velocity required the adjustment of both K_D (to reduce retardation) and the single rate parameter (t_{s2}) (to maintain tailing) under the assumption of the two distinct sorption sites in the TS model. This does not conflict, however, with the definition of the K_D being "constant" either in the model or in "real-life." It means that because of the limitation of the "two-site" assumption, the K_D value has to be "apparently" adjusted with the rate parameter to allow flexibility to describe different experimental conditions. However, this is not the case for the GS model where a distribution of various sorption rate constants is assumed. And more rigorously, there is no absolute "equilibrium site" in the GS model that is solely determined by its equilibrium partitioning coefficient. Presumably, if a specific sorption process dominates under a given velocity condition, the GS model would have a corresponding segment in its distribution to reflect this dominating sorption process, without invoking apparent model parameter adjustment such as the K_D. The significant leftward shift of BTCs with increasing velocity, as well as the persistent tailing, was also observed for a number of sorbing organic compounds by Kay and Elrick (1967); Davidson and McDougal (1973); Hutzler et al. (1986); Lee et al. (1988); among others.

None of the two models, however, were be able to predict atrazine BTCs obtained from the nonirradiated Case Ia and Col II using the same parameters determined from the irradiated Case Ib (Chen and Wagenet, 1997). A new set of sorption parameters for each model, therefore, was estimated from a nonirradiated case and was evaluated similarly independently. Case Ia (Table 1.4) was chosen for the new parameter estimation (results in Table 1.5). The curve fitting results and the independent predictions are shown in Figures 1.12a and 1.12b, respectively. Although the GS independent predictions demonstrated an overall better performance than TS in Figure 1.12b, discrimination between the two models was not as obvious as in the different velocity cases (i.e., Cases Ic and Id). This was attributed to the smaller difference of time-scales (mean travel time) between Case Ia and Col II. The calculated mean travel time ($=L/v_{\text{fitted}}$, based on Table 1.4) for Col II was about 2.3 times longer than for Case Ia, but the difference was much larger among the three previously reported cases (travel time in Case Ib was 8 to 41 times longer than in the fast Cases Ic and Id).

The independent predictions by both TS and GS models using the nonirradiated column-derived parameters were further tested by the batch experimental results where none of the batch soil samples received γ-irradiation. As shown in Figure 1.13, the batch isotherm-derived K_D values obtained from different equilibration time treatments (Table 1.3) were plotted against equilibrium time and compared to the values predicted by the two nonequilibrium sorption models. The model simulations were accomplished with the two individual sorption submodels in their forms of transfer

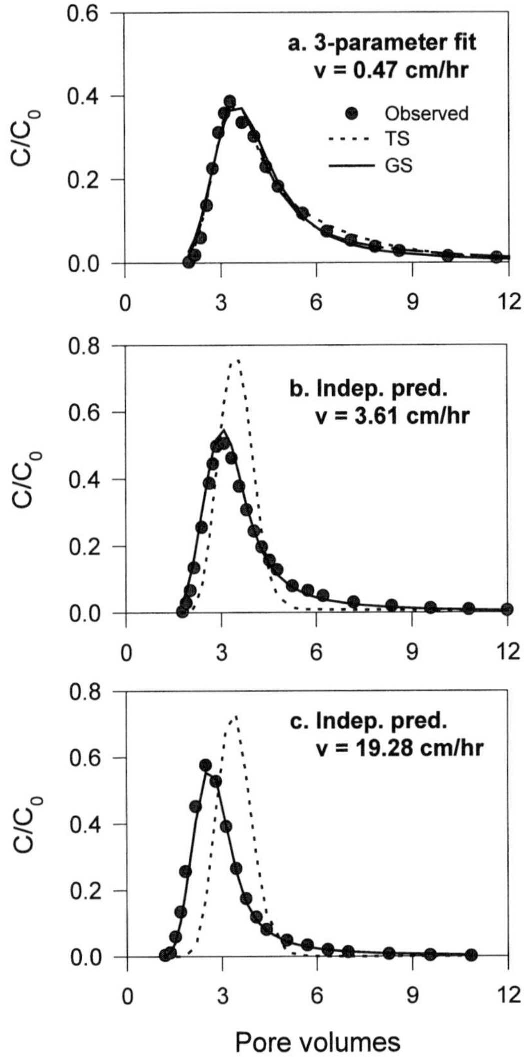

Figure 1.11. Results of a 3-parameter fit by each model to γ-irradiated Case Ib (a) and the corresponding independent predictions of the same γ-irradiated column, but under different velocities (b and c) (Chen and Wagenet, Figure 5; *Soil Sci. Soc. Amer. J.*, 1997. With permission).

functions (i.e., Eqs. 1.23 and 1.26, respectively). The simulations were accomplished with parameters identified from Case Ia (Table 1.5) under a continuous input (step function) of a solution concentration (1 μg/mL) starting from 0 initial concentrations. The corresponding output was the amount of sorption (S, μg/g) changing with time. The simulated K_D value (μL/g) at a given time was then calculated as the ratio of S to the solution concentration at the same time.

Figure 1.13 demonstrates that the asymptotic (equilibrium) K_D values were approached much faster by TS, whereas the GS model was obviously slower and in better agreement with the trend shown by the batch isotherm-derived values during the overall course of approaching equilibrium. The TS model, for example, approached its asymptote after about 24 hr, whereas the GS model achieved only 46% of mass sorbed at equilibrium after 48 hr. Extending the GS simulation further (not shown in Figure 1.13) suggested that 3900 hr were required to reach 90% of equilibrium sorption. The

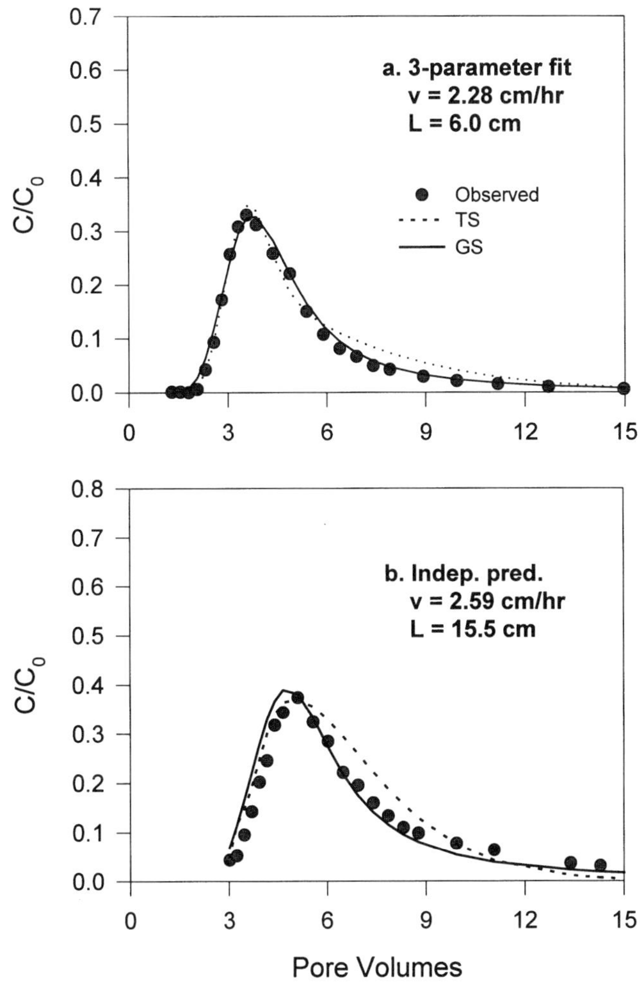

Figure 1.12. Results of a 3-parameter fit by each model to nonirradiated Case Ia (a) and the corresponding independent predictions of nonirradiated Col II (b) (Chen and Wagenet, Figure 7; *Soil Sci. Soc. Amer. J.*, 1997. With permission).

continuously decreasing sorption rate after a fast initial uptake during the first 24 hr was taken as evidence that more resistant sorption sites (domains) must have been involved in the process as time increased. This suggests that the two distinct sorbing fractions assumed in the TS model may not be sufficient to describe sorption over a variety of time-scales caused by a wide range of heterogeneous sorbing components in soil.

Case 2

A set of Cl^- and atrazine BTC data from Gamerdinger et al. (1991) was selected by Chen (1994) as a case for evaluating FOTS and FOGS under multiprocess nonequilibrium conditions. In these experiments, asymmetry in both Cl^- and atrazine BTCs was observed from the same column experiment, indicating that intra-aggregate diffusion and sorption occurred simultaneously during transport (Figures 1.14, 1.15, and 1.16). The column experiments were performed using a Rinebeck silty clay loam (fine, illitic,

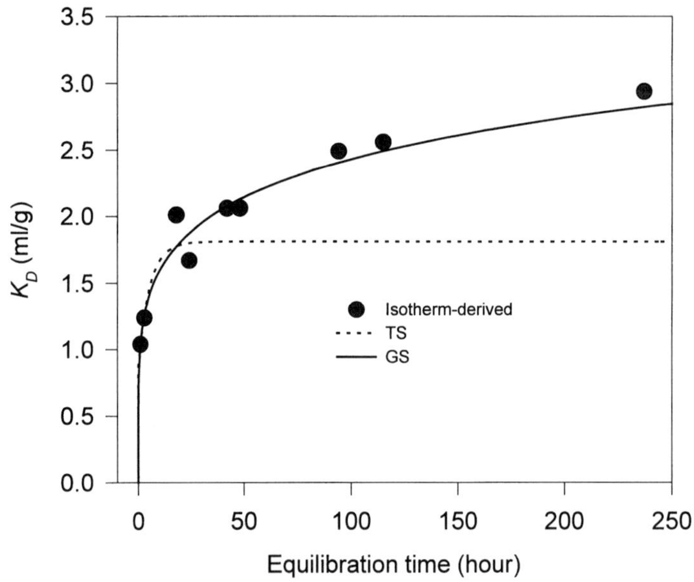

Figure 1.13. Increase of isotherm-derived K_D values with equilibration time in batch experiments, compared to each model-predicted sorption behavior using parameters obtained from the 3-parameter fit of untreated Case Ia (Table 1.5) (Chen and Wagenet, Figure 8; *Soil Sci. Soc. Amer. J.*, 1997. With permission).

mesic Aeric Ochraqualfs) passed through a 2-mm sieve. Relevant soil physical and chemical properties were reported by Gamerdinger et al. (1991).

Parameters for transport and intra-aggregate diffusion were obtained by fitting the first-order intra-aggregate diffusion model (FO) to the measured Cl^- BTCs (Figure 1.14). The four parameters estimated are v, ϵ_τ, t_e, and θ_m, where t_e and θ_m are the characteristic reaction time of the first-order intra-aggregate diffusion (hr) and the mobile water content (cm^3/cm^3), respectively.

The fitted transport and physical process-related parameters were then applied to the multiprocess model (FOTS or FOGS) for describing atrazine BTCs. While it is recognized that the magnitude of t_e may be larger for atrazine than for Cl^- due to the smaller diffusivity of atrazine in water (Brusseau and Rao, 1989), the correction for this was not found feasible (Chen, 1994). The direct estimate of t_e from Cl^- therefore was used for atrazine. The K_D value was determined as 3.16 (L/kg) from batch studies (Gamerdinger et al., 1991), and degradation was found only in the slower column (Figure 1.15) with the first-order rate coefficient γ as 0.0008 h^{-1} (Gamerdinger et al., 1991). The fraction of sorption sites exposed directly to the mobile fluid (f) was approximated as the ratio of θ_m/θ, according to Nkedi-Kizza et al. (1983). With all these parameters fixed, the remaining sorption parameters are t_{s2} and F_1 for TS, are n and β for GS, which were further fitted from the lower velocity case (Figure 1.15).

Using parameters determined above, FOTS and FOGS were then applied to predicting the measured atrazine BTC obtained from the higher velocity case (Figure 1.16). It is clear that although the two models were both able to fit the BTC similarly well (Figure 1.15), the independent predictions by FOGS for the higher velocity case was much better than that by FOTS (Figure 1.16). The sharp and over-retarded peak and rapidly flattened tail predicted by the FOTS model (dotted line) once again indicate that this two-site model was not able to completely identify all possible sorption

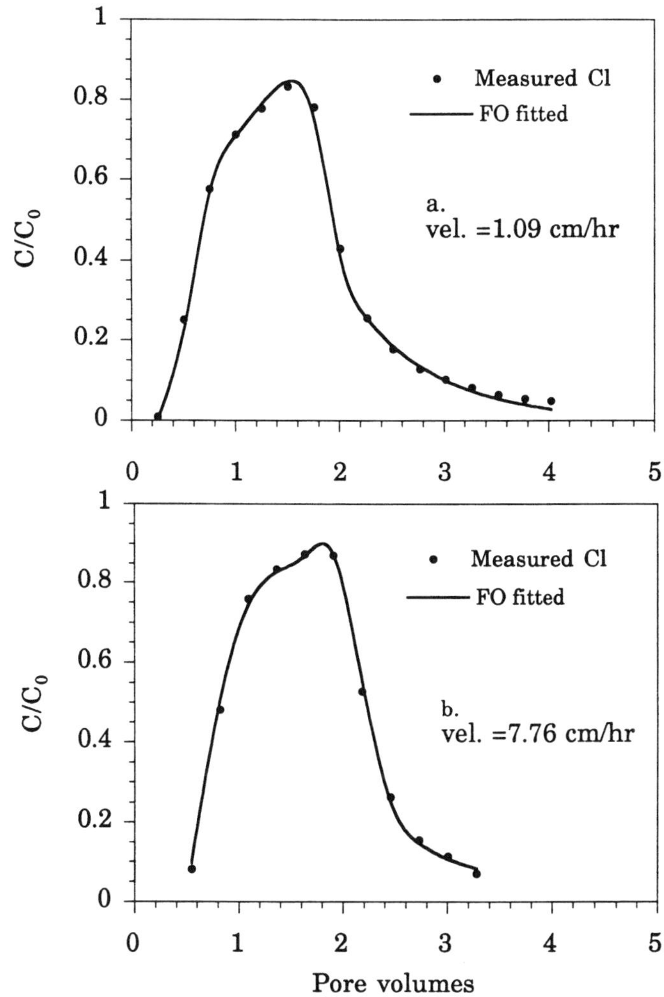

Figure 1.14. Fitted and measured Cl^- breakthrough curves. Parameters obtained using the first-order intra-aggregate diffusion model (FO) were: for the low velocity case (a) $v = 1.14$ cm/hr, $\epsilon_\tau = 0.27$, $t_e = 57.7$ hr, $\theta_m/\theta = 0.624$, (measured $\theta = 0.513$); for the high velocity case (b), $v = 7.21$ cm/hr, $\epsilon_\tau = 0.27$, $t_e = 18.74$ hr, $\theta_m/\theta = 0.725$ (measured $\theta = 0.498$) (Chen, 1994).

processes that were operating but to a different dominating degree in different flow rate conditions.

SUMMARY

Solute nonequilibrium transport in soil was analyzed separately through two general types of processes: hydrodynamic and nonequilibrium. The hydrodynamic process was described based on residence time distribution theory, while the physical/chemical nonequilibria were represented by distributed sorption rate heterogeneity models, representing variability in soil microscopic surface properties and composition. The residence time distribution approach connects solute transport to the variability of pore water velocity, which is further related to soil hydraulic properties.

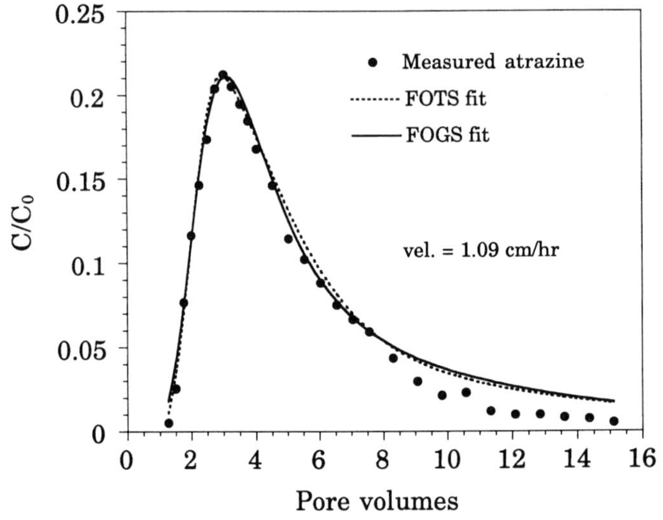

Figure 1.15. Measured and fitted atrazine breakthrough curves from the lower velocity case. The fitted parameters were: for FOTS, $F_I = 0.48$, $t_{s2} = 15.7$ hr; for FOGS, $ß = 0.0159$, $n = 0.346$ (Chen, 1994).

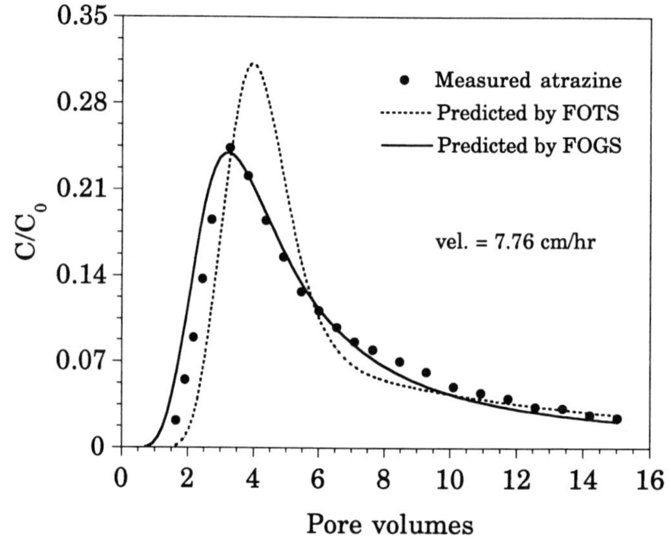

Figure 1.16. Independent predictions of atrazine breakthrough in the higher velocity case by the two multiprocess models, FOTS and FOGS. Sorption-related parameters were obtained from the lower velocity case (Figure 1.15) (Chen, 1994).

Variability of the sorption characteristic time demonstrated to influence the higher moments (third or higher) of concentration BTCs. Two typical nonequilibrium models that represent sorption rate heterogeneity as either two-site or more generally distributed (γ-pdf) were investigated. Moment analysis indicated that even when the first three BTC moments of the two models were matched, substantial deviation between both models still existed. Nonequilibrium conditions in a given dispersive flow system depend not only on the ratio of the characteristic times for sorption and transport, but also on the variability of sorption rate heterogeneity and the sorption equilibrium properties of the porous medium. Generally, the two-site model is limited in its ability to incorporate a wide range of sorption sites having different sorption rates, while the γ-model is more flexible (less velocity-dependent) and better able to provide smooth and gradually flattened BTC tailing.

The two-site sorption model (TS/FOTS) and the γ-distributed sorption rate model (GS/FOGS) were evaluated through pesticide miscible displacement and batch experiments. Generally, under all experimental conditions involved in the study, the GS/FOGS models were superior to TS and FOTS. Successful agreement between GS/FOGS model predictions and the measured pesticide BTCs indicated a very weak velocity dependency of this model over the velocity range used (from 0.47 to 19.28 cm/hr). For the same velocity range, TS/FOTS were likely to require different sets of parameters in order to sufficiently describe the observed BTCs. In batch sorption experiments, the increasing trend of the isotherm-derived partitioning coefficient (K_D) with equilibration time (ranging from 1 to 237 hr) was well-predicted independently by the GS model. The TS model was not able to reconcile the entire range of change in K_D. The weak time-scale dependency of GS found in both column and batch experiments therefore supports the assumption that a wide range of heterogeneous sorption sites/domains were involved in the soil sorption process, but to a variable degree as time elapsed.

REFERENCES

Abramowitz, M. and I.A. Stegun. *Handbook of Mathematical Functions with Formulas, Graphs, and Mathematical Tables,* Dover Publications, Inc., New York, 1972, p. 260.

Aris, R. *The Scope of R. T. D. Theory, Residence Time Distribution Theory in Chemical Engineering,* Pethö, A. and R.D. Noble (Eds.), Verlag Chemie GmbH, D-6940 Weinheim, 1982, pp. 1–21.

Bachmat, Y. and J. Bear. Macroscopic modeling of transport phenomena in porous media: 1. The continuum approach, *Trans. Porous Media,* 1: 213–240, 1986.

Ball, W.P. and P.V. Roberts. Long-term sorption of halogenated organic chemicals by aquifer material: 2. Intraparticle diffusion, *Environ. Sci. Technol.,* 25:1237–1249, 1991.

Bear, J. *Dynamics of Fluids in Porous Media,* Dover Publications, Inc. Mineola, NY, 1972, p. 15.

Beran, M.J. *Statistical Continuum Theories,* Wiley Interscience, New York, 1968.

Brusseau, M.L. and P.S.C. Rao. The influence of sorbate-organic matter interactions on sorption nonequilibrium. *Chemosphere,* 18:1691–1706, 1989.

Brusseau, M.L., R.E. Jessup, and P.S.C. Rao. Modeling the transport of solutes influenced by multiprocess nonequilibrium, *Water Resour. Res.,* 25:1971–1988, 1989.

Brusseau, M.L. Nonequilibrium transport of organic chemicals: The impact of pore-water velocity, *J. Contam. Hydrol.,* 5:215–234, 1992.

Cameron, D.A. and A. Klute. Convective-dispersive solute transport with a combined equilibrium and kinetic adsorption model, *Water Resour. Res.,* 13:183–188, 1977.

Chen, C. and R.J. Wagenet. Simulation of water and chemicals in macropore soils: Part 1. Representation of the equivalent macropore influence and its effect on soil-water flow, *J. Hydrol.,* 130:105–126, 1992.

Chen, W. Solute Nonequilibrium Transport in Heterogeneous Porous Media: A Unified Probability Distribution Approach. PhD dissertation, Cornell University, 1994.

Chen, W. and R.J. Wagenet. Solute transport in porous media with sorption-site heterogeneity, *Environ. Sci. Tech.,* 29:2725–2734, 1995.

Chen, W. and R.J. Wagenet. Description of atrazine transport in soil with heterogeneous nonequilibrium sorption, *Soil Sci. Soc. Amer. J.,* 61:360–371, 1997.

Coats, K.H. and B.D. Smith. Dead-end pore volume and dispersion in porous media, *Soc. Pet. Eng. J.,* 4:73–80, 1964.

Connaughton, D.F, J.R. Stedinger, L.W. Lion, and M.L. Shuler. Description of time varying desorption kinetics: Release of naphthalene from contaminated soils, *Environ. Sci. Technol.,* 27:2397–2403, 1993.

Cvetkovic, V.D. and A.M. Shapiro. Mass arrival of sorptive solute in heterogeneous porous media, *Water Resour. Res.,* 26:2057–2067, 1990.

Dagan, G. Solute transport in heterogeneous porous formations, *J. Fluid Mech.,* 145:151–177, 1984.

Dagan, G. Stochastic theory of groundwater flow and transport: Pore to laboratory, laboratory to formation, and formation to regional scale, *Water Resour. Res.,* 22:120s–134s, 1986.

Dao, T.H. and T.L. Lavy. Atrazine adsorption on soil as influenced by temperature, moisture content and electrolyte concentration, *Weed Sci.,* 26:303–308, 1978.

Davidson, J.M., C.M. Rieck, and P.W. Santelmann. Influence of water flux and porous material on the movement of selected herbicides, *Soil Sci. Soc. Am. Proc.,* 32:629–633, 1968.

Davidson, J.M. and J.R. McDougal. Experimental and predicted movement of three herbicides in water-saturated soil, *J. Environ. Qual.,* 2:428–433, 1973.

Gamerdinger, A.P., A.P. Lemley, and R.J. Wagenet. Nonequilibrium sorption and degradation of three 2-chloro-s-triazine herbicides in soil-water systems, *J. Environ. Qual.,* 20:815–822, 1991.

Gamerdinger, A.P., R.J. Wagenet, and M.Th. van Genuchten. Application of two-site/two-region models for studying simultaneous nonequilibrium transport and degradation of pesticides, *Soil Sci. Soc. Am. J.,* 54:957–963, 1990.

Garbarini, D.R. and L.W. Lion. Influence of the nature of soil organics on the sorption of toluene and trichloroethylene, *Environ. Sci. Technol.,* 20:1263–1269, 1986.

Gelhar, L.W. Stochastic subsurface hydrology from theory to application, *Water Resour. Res.,* 22:135s–145s, 1986.

Gelhar, L.W., A.L. Gutjahr, and R.L. Naff. Stochastic analysis of macro-dispersion in a stratified aquifer, *Water Resour. Res.,* 15:1387–1397, 1979.

Gerke, H.H. and M.T. van Genuchten. A dual-porosity model for simulating the preferential movement of water and solutes in structured porous media, *Water Resour. Res.,* 29:305–319, 1993.

Germann, P.F. and K. Beven. Kinematic wave approximation to infiltration into soils with sorbing macropores, *Water Resour. Res.,* 21:990–996, 1985.

Germann, P.F. Preferential flow and the generation of runoff: 1. Boundary layer flow theory, *Water Resour. Res.,* 26:3055–3063, 1990.

Gradshteyn, I.S. and I.M. Ryzhik. *Table of Integrals, Series, and Products,* Academic Press, Inc., New York, 1994, p. 319.

Huang, P.M., R. Grover, and R.B. Mckercher. Components and particle size fractions involved in atrazine adsorption by soil, *Soil Sci.,* 138:20–24, 1984.

Hutson, J.L. and R.J. Wagenet. A multi region model describing water flow and solute transport in heterogeneous soils. *Soil Sci. Soc. Am. J.,* 59:743–751, 1995.

Hutzler, N.J., J.C. Crittenden, and J.S. Gierke. Transport of organic compounds with saturated groundwater flow: Experimental results. *Water Resour. Res.,* 22:285–295, 1986.

Johnson, N.L. and S. Kotz. *Distribution in Statistics: Continuous Univariate Distributions,* John Wiley & Sons, 1970, pp. 137–153.

Jury, W.A. and K. Roth. *Transfer Functions and Solute Movement through Soil: Theory and Applications,* Birkhäuser Verlag Basel, 1990.

Karickhoff, S.W. and K.R. Morris. Sorption dynamics of hydrophobic pollutants in sediment suspensions, *Environ. Toxicol. Chem.,* 4:469–479, 1985.

Kay, B.D. and D.E. Elrick. Adsorption and movement of lindane in soils, *Soil Sci.,* 104:314–322, 1967.

Kookana, R.S., L.A.G. Aylmore, and R.G. Gerriste. Time-dependent sorption of pesticides during transport in soils, *Soil Sci.,* 154:214–225, 1992.

Lafleur, K.S. Sorption of pesticides by model soil and agronomic soil: Rates and equilibria, *Soil Sci.,* 127:94–101, 1979.

Lavy, T.L. Micromovement mechanisms of *s*-triazines in soil, *Soil Sci. Soc. Am. Proc.,* 32:377–380, 1968.

Lee, L.S., P.S.C. Rao, M.L. Brusseau, and R.A. Ogwada. Nonequilibrium sorption of organic contaminants during flow through columns of aquifer materials, *Environ. Toxicol. Chem.,* 7:779–793, 1988.

Marquardt, D.W. An algorithm for least-squares estimation of nonlinear parameters, *J. Soc. Ind. Appl. Math.,* 11:431–441, 1963.

Nauman, E.B. and B.A. Buffham. Mixing in Continuous Flow Systems, John Wiley & Sons, Inc., New York, 1983.

Nielsen, D.R., M.Th. van Genuchten, and J.W. Biggar. Water flow and solute transport processes in the unsaturated zone, *Water Resour. Res.,* 22:89S–108S, 1986.

Nkedi-Kizza, P., J.W. Bigger, M.Th. van Genuchten, P.J. Wierenga, H.M. Selim, J.M. Davidson, and D.R. Nielsen. Modeling tritium and chloride 36 transport through an aggregated oxisol, *Water Resour. Res.,* 19:691–700, 1983.

Parker, J.C. and A.J. Valocchi. Constraints on the validity of equilibrium and first-order kinetic transport models in structural soils, *Water Resour. Res.,* 22:399–407, 1986.

Parker, J.C. and M.Th. van Genuchten. Determining Transport Parameters from Laboratory and Field Tracer Experiments. Bull. 84-3, Va. Agric. Exp. Stn., Blacksburg, 1984.

Passioura, J.B. Hydrodynamic dispersion in aggregated media, *Soil Sci.,* 111:339–344, 1971.

Pignatello, J.J. Sorption dynamics of organic compounds in soils and sediments, In Reactions and Movement of Organic Chemicals in Soil, SSSA Special Publication No. 22, 1989, pp. 45–80.

Pignatello, J.J., F.J. Ferrandino, and L.Q. Huang. Elution of aged and freshly added herbicides from a soil, *Environ. Sci. Technol.,* 27:1563–1571, 1993.

Rao, P.S.C., J.M. Davidson, R.E. Jessup, and H.M. Selim. Evaluation of conceptual models for describing nonequilibrium adsorption-desorption of pesticides during steady-flow in soils, *Soil Sci. Soc. Am. J.,* 43:22–28, 1979.

Rao, P.S.C., D.E. Rolston, R.E. Jessup, and J.M. Davidson. Solute transport in aggregated porous media: Theoretical and experimental evaluation, *Soil Sci. Soc. Am. J.,* 44:1139–1146, 1980a.

Rao, P.S.C., R.E. Jessup, D.E. Rolston, J.M. Davidson, and D.P. Kilcrease. Experimental and mathematical description of nonadsorbed solute transfer by diffusion in spherical aggregates, *Soil Sci. Soc. Am. J.,* 44:684–688, 1980b.

Rao, P.S.C., R.E. Green, L.R. Ahuja, and J.M. Davidson. Evaluation of a capillary bundle model for describing solute dispersion in aggregated soil, *Soil Sci. Soc. Am. J.,* 40:815–819, 1976.

Rasmuson, A. The effect of particles of variable size, shape and properties on the dynamics of fixed beds, *Chem. Eng. Sci.,* 40:621–629, 1985.

Rutherford, D.W., C.T. Chiou, and D.E. Kile. Influence of soil organic matter composition on the partition of organic compounds, *Environ. Sci. Technol.,* 26:336–340, 1992.

Ruthven, D.M. and K.F. Loughlin. The effect of crystallite shape and size distribution on diffusion measurements in molecular sieves, *Chem. Eng. Sci.,* 26:577–584, 1971.

Sardin, M., D. Schweich, F.J. Leij, and M.Th. van Genuchten. Modeling the nonequilibrium transport of linearly interacting solutes in porous media: A review, *Water Resour. Res.,* 27:2287–2307, 1991.

Selim, H.M., J.M. Davidson, and R.S. Mansell. Evaluation of a two-site adsorption-desorption model for describing solute transport in soils, in *Proc. Summer Simul. Conf.,* Washington, DC, 12–14 July, 1976, pp. 444–448.

Simmons, C.S. A stochastic-convective transport representation of dispersion in one-dimensional porous media, *Water Resour. Res.*, 18:1193–1214, 1982.

Sposito, G., W.A. Jury, and V.K. Gupta. Fundamental problems in the stochastic convection-dispersion model of solute transport in aquifers and field soils, *Water Resour. Res.*, 22:77–88, 1986.

Talbot, A. The accurate numerical inversion of Laplace transforms, *J. Inst. Math. Appl.*, 23:97–120, 1979.

Valocchi, A.J. Validity of the local equilibrium assumption for modeling sorbing solute transport through homogeneous soils. *Water Resour. Res.*, 21:808–820, 1985.

van Genuchten, M.Th. and F.N. Dalton. Models for simulating salt movement in aggregated field soils, *Geoderma*, 38:165–183, 1986.

van Genuchten, M.Th. and P.J. Wierenga. Mass transfer studies in sorbing porous media, I. Analytical solutions, *Soil Sci. Soc. Am. Proc.*, 40:473–480, 1976.

van Genuchten, M.Th. and R.J. Wagenet. Two-site/two-region models for pesticide transport and degradation: Theoretical development and analytical solutions, *Soil Sci. Soc. Am. J.*, 53:1303–1310, 1989.

Villermaux, J. Deformation of chromatographic peaks under the influence of mass transfer phenomena, *J. Chromatogr. Sci.*, 12:822–831, 1974.

Villermaux, J. Theory of linear chromatography, in Percolation Processes, Theory and Applications, Rodrigues, A.E. and D. Tongeur (Eds.), NATO ASI Series, Series E, Vol. 33, Sijthoff and Noordhoff, Rockville, MA, 1981, pp. 83–140.

Wagenet, R.J. Measurement and interpretation of spatially variable soil leaching processes, in *Soil Spatial Variability*, Nielsen, D.R. and J. Bouma (Eds.), Centre for Agricultural Publishing and Documentation (PUDOC), Wageningen, The Netherlands, 1985, pp. 209–230.

Wagenet, R.J. and P.F. Germann. Concepts and Models of Water Flow in Macropore Soils, The Connecticut Agricultural Experiment Station, New Haven, Bulletin 876, December 1989.

Wagenet, R.J. and P.S.C. Rao. Modeling Pesticide Fate in Soil, in *Pesticides in the Soil Environment*, Cheng, H.H. (Ed.), SSSA Book Series, No. 2, Soil Science Society of America, Madison, WI, 1990, pp. 351–399 (Ch. 10).

Weber, W.J., Jr., P.M. McGinley, and L.E. Katz. A distributed reactivity model for sorption by soils and sediments, 1. Conceptual basis and equilibrium assessments, *Environ. Sci. Technol.*, 26:1955–1962, 1992.

Wu, S.C. and P.M. Gschwend. Sorption kinetics of hydrophobic organic compounds to natural sediments and soils, *Environ. Sci. Technol.*, 20:717–725, 1986.

CHAPTER TWO

Hydraulic and Physical Nonequilibrium Effects on Multiregion Flow

G.V. Wilson, J.P. Gwo, P.M. Jardine, and R.J. Luxmoore

It is well accepted that soil consists of a continuous distribution of particle sizes which are arbitrarily and inconsistently segregated into particle size classes of sand, silt, and clay. In an analogous manner, soil consists of a continuous distribution of pore sizes which may be segregated into pore size classes of macro-, meso-, and micropore. Since a secondary pore class, not considered macropore or micropores, significantly contributes to preferential flow under variably saturated conditions, a multiregion flow approach is needed. This chapter will show that a key mechanism to solute transport through such a triple-porosity, triple-permeability media under transient flow conditions is the mass transfer among regions. Preferential flow in unsaturated soil invariably results in hydraulic gradients between the preferential flow region and the less mobile regions. Under such hydraulic nonequilibrium conditions (HNE), advection will cause mass transfer of solutes between regions. Preferential flow also results in physical nonequilibrium (PNE); i.e., concentration gradients, between flow regions. Thus, mechanistically rigorous transport modeling of multiregion flow must incorporate both advective and diffusive mass transfer. This chapter presents multiregion flow and transport concepts and justification, experimental assessment of intra-region transport properties and inter-region mass transfer processes, and multiregion flow and transport modeling.

MULTIREGION FLOW AND TRANSPORT JUSTIFICATION

An established concept in soil science is that soils consist of a continuous distribution of particle sizes that can be segregated into selected particle size classes from which soil texture may be classified. Three size classes entrenched in the literature are sand, silt, and clay; however, these particle sizes are arbitrarily and inconsistently defined (Figure 2.1). The United States department of Agriculture (USDA), American Society for Testing and Materials (ASTM), International Soil Science Society (ISSS), and other agencies have differing classification schemes. Analogously, soils also consist of a continuous distribution of pore sizes. Luxmoore et al. (1990) reviewed the pore size schemes that have been proposed in the literature and found that numerous schemes have been used. Thus, equivalent pore sizes have, since 1951, been arbitrarily and inconsistently segregated into pore size classes (Figure 2.2). Luxmoore (1981) promoted the

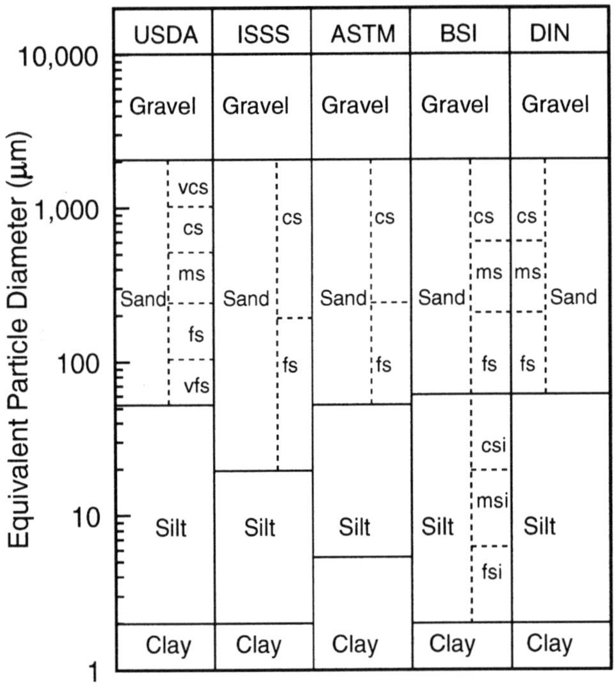

Figure 2.1. Classification schemes: U.S. Department of Agriculture (USDA), International Soil Science Society (ISSS), American Society for Testing and Materials (ASTM), British Standards Institute (BSI), and German Standards (DIN).

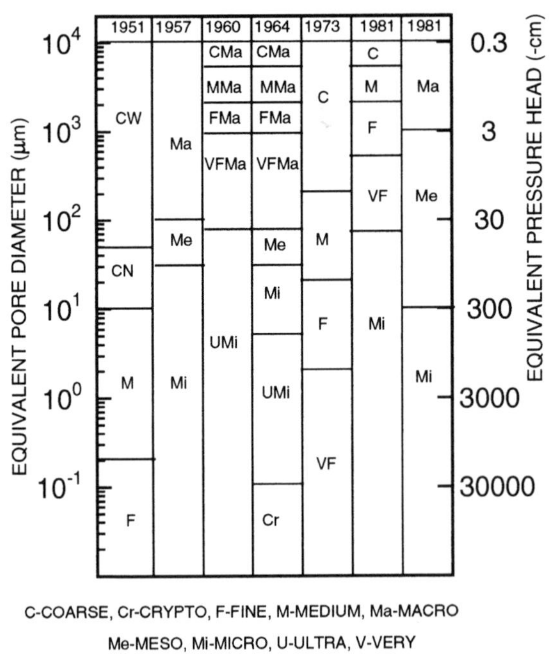

Figure 2.2. Pore size classification schemes from Luxmoore et al. (1990). With permission.

designation of macro-, meso-, and micropore just as particles are designated sand, silt, and clay. It can be questioned whether there is a distinct change of process for these three pore classes (Bouma, 1981a). However, the same question can be made of three particle size classes. There is no chemical process distinction between sand and silt size classes, yet they have proved extremely useful. Many soil physical processes are quantified in terms of the distribution of these arbitrary particle classes. The macropore class has become well accepted for its importance to preferential flow (Beven and Germann, 1982). So the question arises as to whether the segregation of preferential flow paths into two pore size classes is warranted.

Dye infiltration experiments by Omoti and Wild (1979) revealed rapid transport primarily through pores too small to be considered macropores. It is common for dye staining patterns to reveal preferential flow paths that are not morphologically distinguishable from the matrix (Hornberger et al., 1991). Infiltration—drainage experiments performed by Germann and Beven (1981) revealed a secondary pore system, not considered macropore or micropores, that significantly contributed to preferential flow under variably saturated-unsaturated conditions. Roth et al. (1991) monitored the rainfall-driven movement of a surface-applied chloride tracer along a 12 m long transect to a depth of 2.4 m with 110 suction samplers. They found that the single tracer pulse separated into a slowly moving pulse through the soil matrix and a series of fast preferential flow pulses. The sporadic preferential flow pulses resulted in rapid solute transport to 220 cm and accounted for 58% of the total mass. However, flow through the secondary pore system within the soil matrix accounted for the remaining 42% and moved solutes to 84 cm.

For structured soil, the dual porosity or mobile—immobile flow regions approach is widely accepted. Since a secondary flow (i.e., mesopores) system exists, then is a mobile-immobile approach warranted? Jury and Flühler (1992) noted that "substantial water flow may be occurring in the surrounding matrix as well as in the preferential flow region. Solutes partition into a rapid or preferential flow region and a slower but still mobile matrix flow region, each of which may embody a smaller but significant degree of water flow." They concluded that "the (dual porosity) model is unsatisfactory for representing media wherein transport occurs both in preferential flow region and in the bulk matrix." Thus, a multiregion, or at least a three-region approach is justified for certain media.

In a simplistic sense, multiregion flow and transport is an extension of the dual porosity or mobile-immobile concept whereby porous media is considered to possess multiple distinct porosity-permeability classes. The term *region* is used to convey a macroscopic description as opposed to a discrete description; e.g., fracture networks. Bai et al. (1993) presented the following combinations of porosity-permeability classes: (i) single-single, (ii) dual-single, (iii) dual-dual, (iv) triple-dual, and (v) triple-triple. The single porosity-single permeability model was the historical view of soil for which many of our theories on water and solute movement originated. It may be satisfactory for homogeneous soils such as structureless sands although recent work (Glass and Steenhuis, 1984; Hillel, 1993; Kinsall et al., 1997) has shown that preferential flow via fingering may be significant even for this condition. Bai et al. (1993) presented the dual porosity-single permeability model as a representation of a fractured reservoir of high water storage but low permeability due to the discontinuity of pores. Dual porosity-dual permeability models became popular in the 1970s for representing fractured reservoirs as well as vadose zone soils with structural cracks or biological channels. Despite the numerous inferences to the triple porosity model, it has not found acceptance until recently. The scheme proposed by Luxmoore (1981) of a triple porosity-dual permeability model views the micropore region as immobile but available for plant extraction. A partially weathered shale soil, such as at Melton Branch watershed on the Oak Ridge Reservation (Wilson et al., 1993) is a good example. This subsoil consists of a highly fractured saprolite. The rock-like saprolite appears as a single unit *in situ;* however, upon removal it is easily dissected into smaller units due to microfractures within each unit. These microfractures are partially filled with translocated clay and coated with Fe and Mn oxides indicating that they are hydrologically active preferential flow paths. The remaining saprolite has a high porosity but low permeability. A more generic description is one in which soils are highly structured such that macropores occur as voids between macroaggregates (peds) which themselves are highly permeable due to smaller voids between microaggregates. Radulovich et al. (1992) called these inner macroaggregate voids *interpedal pores.*

This chapter presents the concept of multiregion media consisting of three pore size classes. Under our multiregion flow concepts, preferential flow is considered to include a secondary pore class; e.g., mesopores, that is capable of high flow for extended periods following drainage events. This secondary pore class may consist of microfractures, interpedal pores, or biological features. Thus movement within the soil matrix or macroaggregates is included such that three velocity profiles exist. Conceptually, a representative elemental volume, REV, at any point in the soil consists of three regions, Figure 2.3a, with its own flow and transport parameters.

For explanation purposes, consider a soil profile that is initially saturated, pressure head ≥ 0, and allowed to drain. Initially, flow would occur through all three regions, but most rapidly through macropores, with inter-region mass transfer between all regions (Figure 2.3a). Due to the high pore water velocity of macropores, slightly negative soil water pressures develop rather quickly and would result in large pores emptying. Conditions below saturation (i.e., macropores empty) and above field capacity generally persist for several days. Flow during this period occurs predominately, maybe exclusively, through mesopores; however, soil water in micropores is significant to the inter-region mass transfer that would occur between these two regions (Figure 2.3a). After soil water pressures decrease below field-capacity, flow would occur only through micropores with no mass transfer between regions. According to de Marsily (1986), the water thickness adjacent to particle surface must be >1 μm to be mobile due to viscous drag forces acting upon the water molecules near the surface. Thus, flow through micropores may be negligible; however, they would still serve as a source/sink of solutes for the other mobile regions as well as for biological uptake.

Given the concepts presented above, one may misinterpret that macropores are only active under saturated conditions and that only diffusion is responsible for mass transfer between regions. The idea that macropores must be open to the surface, soil fully saturated, and the surface ponded for macropores to be conductive was prevalent in past research (Bouma, 1981b; Beven and Germann, 1982). However, many have noted that macropore flow may occur under unsaturated conditions (De Vries and Chow, 1978; Hammermeister, et al., 1982; Radulovich et al., 1992; Sollins and Radulovich, 1988; Wilson et al., 1990, 1991a,b). It should be noted that the previous discussion was restricted to drainage from an initially saturated soil.

To better understand the concepts of unsaturated multiregion flow, consider the scenario in which the upper layer of the vadose zone is contaminant rich and the surface receives rainfall when the soil profile is below field-capacity. This situation is indicative of agricultural soils in which the surface layer is rich in nutrients and pesticides from previous applications. Wilson and Luxmoore (1988) suggested that due to their high flow capacity, mesopores are capable of infiltrating most rainfall events without macropores contributing. Since soil contains orders of magnitude more mesopores than macropores (Luxmoore et al., 1990), mesopores can converge sufficient flow into the

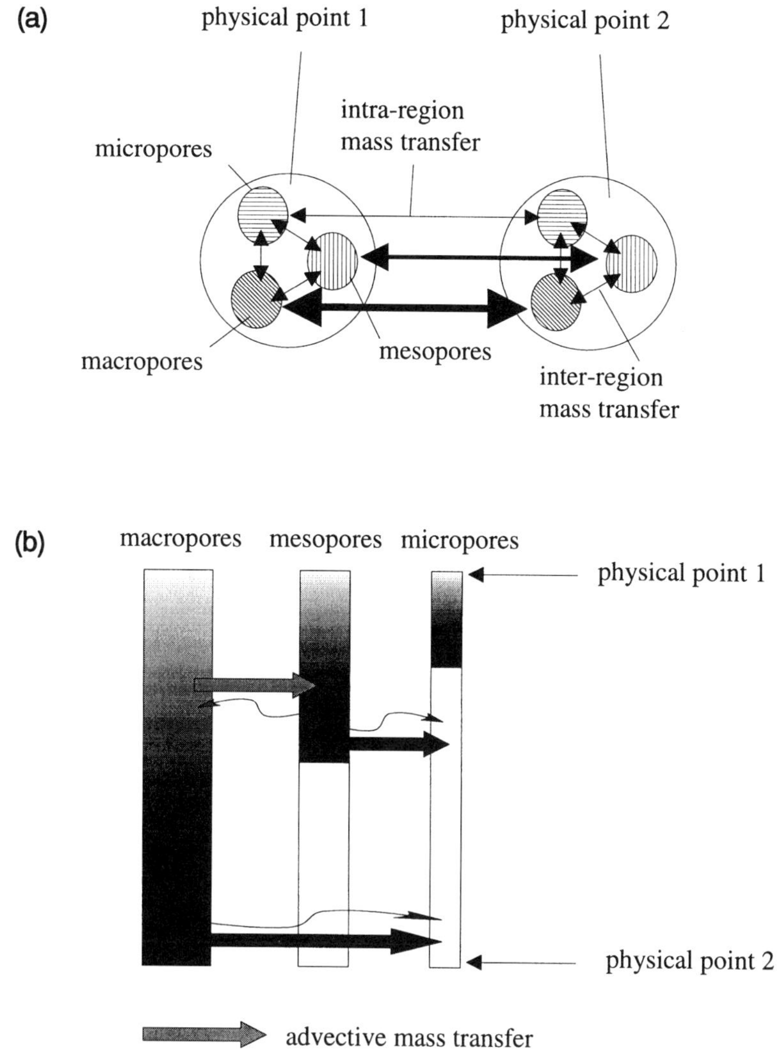

Figure 2.3. Triple-porosity, triple-permeability example of multiregion flow and transport concepts. (A) The Rev (large circle) at two physical points consist of three pore regions. Intra-region flow and transport indicated by lines between large circles (Rev's) and inter-region transfer indicated by lines within each large circle (Rev). (B) Both advective and diffusive mass transfer may occur in parallel or counter to each other. (Figure from Gwo et al., 1996). With permission.

less frequent macropores to initiate macropore flow (Wilson et al., 1991a; Radulovich et al., 1992) even under unsaturated conditions (Figure 2.3b). Once macropore flow is initiated, the unsaturated macroaggregates (mesopores and micropores) at subsequent depths/distances are bypassed due to the high velocity in the macropore region. This variably saturated preferential flow scenario explicitly means that hydraulic nonequilibrium exists; i.e., hydraulic gradients exist between flow regions. Clean rainwater infiltrating into mesopores and converging into macropores would bypass the contaminant-rich micropore water. Thus, physical nonequilibrium, i.e., concentration gradients, would also exist between regions and cause diffusive transfer. Diffusion from the micropore region would serve as a source to supply contaminants to the mobile regions. However, pressure gradients would exist initially from the mesopore and macropore regions toward the micropore region. Thus, diffusion of contaminants from the micropore region would initially be counter to the inter-region advective transfer, Figure 2.3b. The resulting export of contaminants out the profile would be slow, followed by a gradual increase as advective mass transfer between regions subsides. Here, macropores would serve to provide relatively clean water to aquifers and streams. In the absence of macropores, the initiation of outflow may require a longer period; however, the eventual export of contaminants would be much greater due to the increased displacement of this contaminant-rich micropore water. It is evident from this example that advection as well as diffusion of solutes between pore regions are relevant processes in solute transport under field conditions.

MULTIREGION FLOW AND TRANSPORT MEASUREMENT

As we have shown, the literature is well represented by investigations affirming the significance of multiple flow regions and the need for experimental and numerical assessment of multiregion flow and transport processes. However, measurement of the flow and transport properties of each pore class is an enormously difficult problem. A major difficulty is the sampling of one pore class without creating or enhancing its flow while excluding the other pore classes from contributing to the sample.

Flow Measurement

For multiregion flow measurement, the tension infiltrometer developed by Perroux and White (1988) has been successful. This device allows water to be infiltrated into the soil under a tension, thereby excluding the larger pores from contributing to the infiltration process. By determining the infiltration rate under positive pressure (saturated) and selected negative pressures successively, the hydrologic contribution of corresponding pore classes may be quantified. The capillary equation relates the equivalent pore diameter, r (cm), to the tension, h (cm) applied by the infiltrometer by

$$r = -2\sigma \cos\acute{\alpha} / \rho g h = -0.15/h \tag{2.1}$$

where σ is the surface tension, $\acute{\alpha}$ is the contact angle, ρ is the water density, and g is the gravitational constant. Tensions of 2, 5, and 14-cm, as used by Watson and Luxmoore (1986) and Wilson and Luxmoore (1988), excluded pores larger than 0.75, 0.3, and 0.11 mm radius, respectively, from contributing to the infiltration. The difference between the infiltration rates at two consecutive tensions, I, is the contribution of that pore class to flow. These authors found, for the Walker and Melton Branch subcatchments, that macropores (pores $\geq$ 1.5 mm diam.) accounted for 85 and 73% of the saturated flow, respectively. Not only can the velocity contribution of each flow region be measured with this technique, but the hydrologically active porosity, f, number of contributing pores per unit volume, N, and pore surface area of each region can also be estimated by

$$f = 8\nu I/\rho g r^2, \text{ where } f = N\nu r^2 \tag{2.2}$$

and ν is the viscosity of water. Luxmoore et al. (1990) used Eq. 2.2 with data from Wilson and Luxmoore (1988) and estimated three orders of magnitude increase in the number of pores contributing to flow as the pore size class decreased from > 1.5 mm diameter to between 0.6 and 0.22 mm diameter (Table 2.1).

Table 2.1. Ponded (0 Tension) and Tension Infiltration Data and Calculated Attributes for Three Pore Classes Estimated from Capillary-Surface Tension and Poiseuille's Flow Equations for Walker Branch Watershed.[a]

Infiltration Tension Head (cm)	Mean Flux $x10^{-5}$ ms^{-1}	Pore Class Diameter mm	Porosity $x10^{-3}$ m^3m^{-3}	Number of Pores $x10^6$ m^{-3}	Surface Area m^2m^{-3}
0	11.5				
2	1.7	>1.5	0.32	0.12	0.86
5	1.1	1.5 to 0.6	0.21	1.27	1.43
14	0.1	0.6 to 0.2	1.18	141.00	21.40

[a] Wilson and Luxmoore, 1988; Luxmoore et al., 1990. With permission.

Modeling the transient flow behavior of variably saturated multiregion systems requires knowledge of the functional relationships among water content (θ), pressure head (h), and hydraulic conductivity (K) for each flow region. Smettem et al. (1991) proposed a method for modeling θ and K as a function of h for a macropore-micropore system. They used the van Genuchten (1980) model for θ(h) of the macropore (inter-aggregate) region such that

$$\theta = \theta_j + [\theta_s - \theta_j]/[1 + (\alpha h)^n]^m, \text{ for } h>h_j \tag{2.3}$$

where subscripts s and j designate the saturated value and junction point between macropore and micropore regions, respectively. The θ(h) relationship for the micropore region was described for $h < h_j$ by

$$\theta = \theta_r+[\theta_j-\theta_r]/[1+(\alpha' h_t)^{n'}]^{m'} \tag{2.4}$$

where h was transformed such that $h_t=h-h_j$. The θ_r, α, and n were fitted parameters with m=1–1/n. This analysis was extended by Wilson et al. (1992) to a three-region system. Phillips et al. (1989) demonstrated that drainage vertically downward through artificial macropores occurs under slightly (–10 cm) negative pressure heads, while Scotter (1978) suggested that macropores drain until around –15 cm. The demarcation between the macropore and mesopore regions was assumed by Wilson et al. (1992) to occur at h = –10 cm, and between the mesopore and micropore regions was at –250 cm. Schuh and Cline (1990) reported that the K(h) versus h relationship is typically assumed as constant or linear for the region close to saturation. Jarvis et al. (1991) assumed a linear θ(h) function and a simple power function for K(θ) for the macropore region. However, Gwo et al., 1991 (unpublished data) found that the linear θ(h) relationship failed to produce convergence when utilized in a 3-dimensional multiregional flow model called MURF. Wilson et al. (1992) used a Fermi function for θ(h) and K(θ) of the macropore region to avoid the numerical problems caused by the discontinuous derivative of the linear θ(h) function at h=0. The Fermi functions for θ(h) and K(θ) asymptotically approach θ_s and K_s, respectively, as the pressure head increases. The Fermi function for θ(h) was

$$\theta = (\theta_{macro} - \theta_{meso}) / \{1 + \exp[-\gamma(h - h_\theta)]\} + \theta_{meso} \tag{2.5}$$

and for K(θ) was

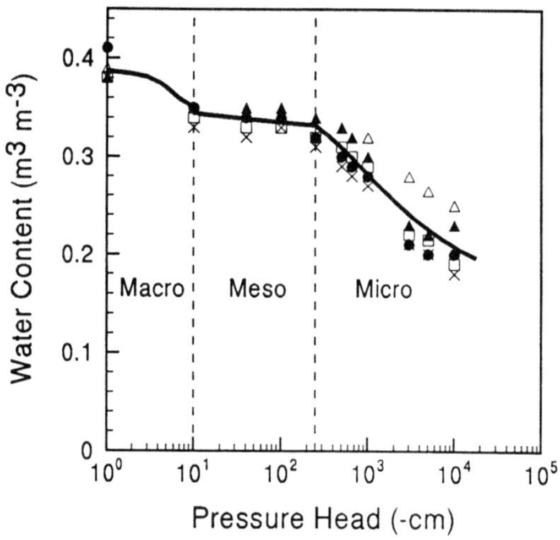

Figure 2.4. Water retention data and functions for three-region media from Wilson et al. (1992). With permission.

$$\log_{10}(K / K_{macro}) = \in / \{1 + \exp[-\beta(h - h_k)]\} - \in \quad (2.6)$$

where γ, β, and $\in$ are fitted parameters with h_θ and h_k set at –5 and –10 cm, respectively.

Wilson et al. (1992) successfully modeled the $\theta(h)$, Figure 2.4, and K(h) for a three-region soil using the Fermi function for macropores combined with the van Genuchten model for mesopore and micropore regions. They obtained continuity in $\theta(h)$ and K(h) between regions and a good agreement between predicted and measured values.

Transport Measurement

Measurement of the transport through individual pore regions is more complicated than the flow analysis, and the measurement of transfer between regions is particularly difficult. The transport parameters controlling the movement of solutes through the three-region saprolitic subsoil was first studied by Jardine and colleagues utilizing conventional miscible displacement techniques adapted for single-region media. Using undisturbed soil columns, Jardine et al. (1988) found that batch adsorption isotherms generated on disturbed samples, over predicted the solute reactivity (Figure 2.5a) and thereby underpredicted the potential transport due to preferential flow through the saturated undisturbed media. They devised a dynamic adsorption isotherm method using undisturbed columns to independently determine the reactivity that accurately predicted the preferential transport through saturated undisturbed soil columns (Figure 2.5b). This approach lumped the contribution of pore regions and did not provide a distinction of the effect of the macropore region to transport. Jardine et al. (1993) revised the miscible displacement experiment by establishing flow through undisturbed soil columns from the same hillslope under negative pressure heads. These columns were equipped with fritted glass end-plates. Steady state flow was established under a unit hydraulic gradient by maintaining the inlet and outlet under the same negative pressure head. Miscible displacements were conducted under positive pressure heads for saturated flow, followed by negative pressure heads for unsaturated conditions to exclude macropore flow. They found that the nonreactive solute breakthrough curves (BTC) were nonsymmetric, Figure 2.6, for saturated conditions due to

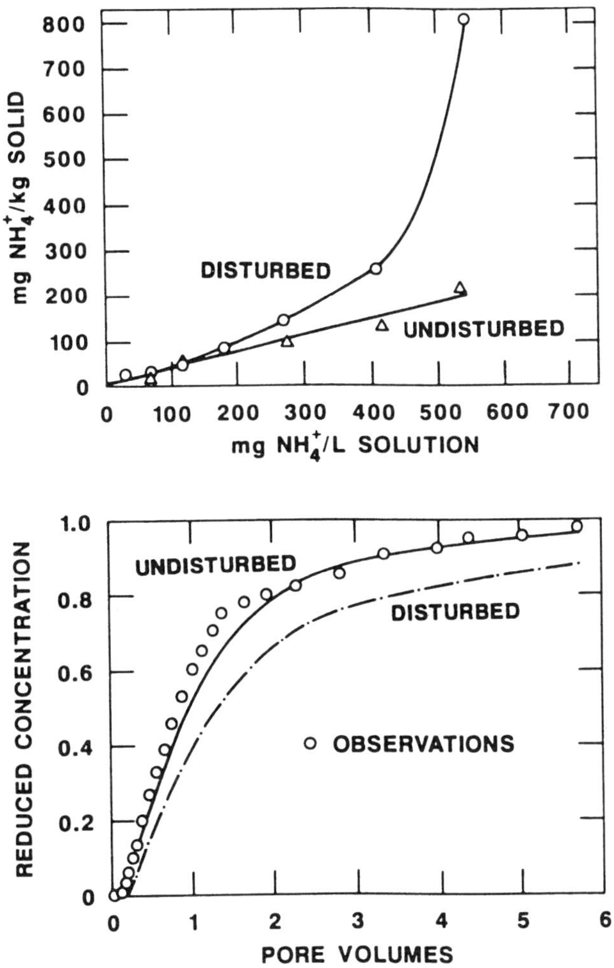

Figure 2.5. (a) Ammonium adsorption isotherms for disturbed and undisturbed soil and (b) observed BTC (open circles) and BTCs predicted for disturbed and undisturbed adsorption parameters. (Figures from Jardine et al., 1988). With permission.

preferential flow. One of the more interesting findings was that the pore water velocity decreased by 5-, and 40-fold from the saturated rate due to respective pressure heads of –10, and –15 cm, while the decrease in water content was minimal (0.55 m^3m^3 at h=0 to 0.51 m^3m^3 at h=–15cm). This is a clear example of the significance of channelized flow through pores that hold water at > –10 cm pressure head even though their porosity is extremely small (Wilson and Luxmoore, 1988). The reactivity of Sr^{2+} and Co^{2+} increased dramatically with the slight decrease in pressure head from 0 to –10 cm. This was believed to be due to the decrease in preferential bypass when macropores empty, and due to the greater surface area of mesopores (Table 2.1).

The significance of these lab-scale steady-state flow studies to field-scale transient flow transport processes requires elaborate research facilities to test their validity. The Walker Branch and Melton Branch Subsurface Transport Facilities on the Oak Ridge Reservation were developed for just such purposes. These facilities consisted of lateral flow collection pans placed against the face of a subsurface trench excavated across the outflow region of each subcatchment (Figure 2.7). Lateral subsurface flow from the

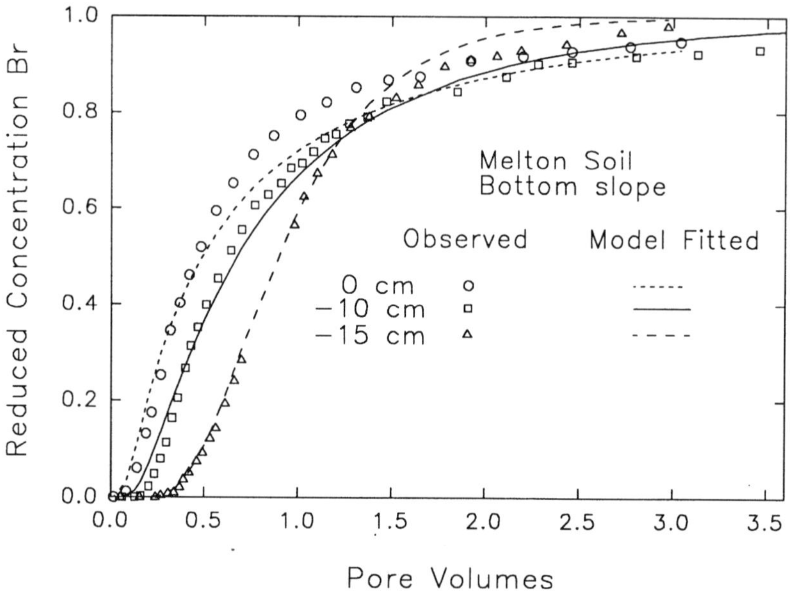

Figure 2.6. Bromide BTCs for three steady-state pressure heads and the single region ADE fitted models. (Figure from Jardine et al., 1993). With permission.

subcatchment was collected by six pans pressed against the vertical face of a 2 m deep and 16 m long trench. Flow collected by these pans and flow from the trench floor that entered the trench below the pans was monitored continuously by a series of tipping bucket rain gauges and H-flumes equipped for computer-based data acquisition. It was assumed that the pans collected free drainage through macropores due to the requirement for positive pressures to produce outflow into the trench collectors. Instrumentation at Walker Branch included: 15 shallow perched water table wells with pressure transducers interfaced to a data logger, 19 ceramic cup solution samplers under falling vacuum of 500 cm, and 48 tensiometers connected to a single pressure transducer interfaced to a data logger; whereas, Melton Branch included: 30 shallow perched water table wells, 45 ceramic cup solution samplers under continuous vacuum of 500 cm, and 75 tensiometers. It was further assumed that solution samplers would sample movement through mesopores due to the sampler cup installed into a slurry of excavated material that precluded macropores from directly contacting the cup.

Wilson et al. (1990, 1991a,b) monitored water and natural solute transport through the Walker Branch hillslope during storm events. They observed a general increase in solute concentrations as lateral flow rates increased, followed by a decrease in concentration that was generally slower than the decrease in flow rate. They reasoned that micropores provided an unlimited supply of solutes that flushed out of the matrix through mesopores into the highly conductive macropore system. To further quantify these multiregion processes under field-scale transient flow conditions, a tracer release was made at the ridge top of the Melton Branch hillslope by Wilson et al. (1993) during a storm event. They observed extremely rapid lateral transport through the forested hillslope with Br^- breakthrough at the trench pans 70 m from the line source within 3 hr, Figure 2.8b. Sampling of the movement through the matrix using suction cup solution samplers suggested that Br^- only moved 3 m downslope of the line source within this first 3 hr, Figure 2.8a. This supported the inference of a slower velocity

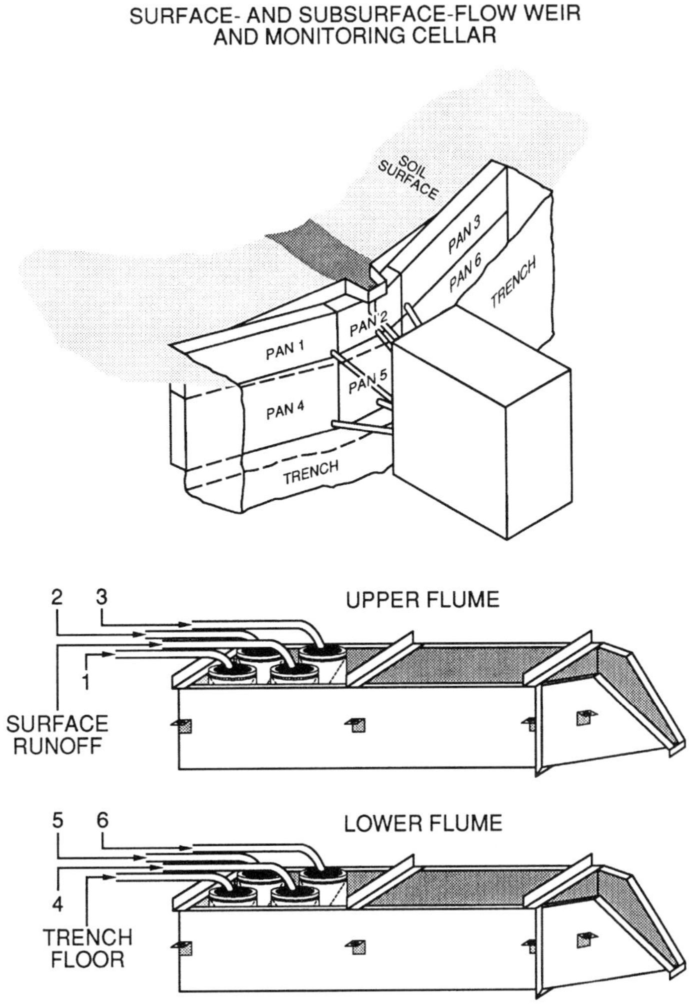

Figure 2.7. Top illustration depicts the stainless steel pans pressed against the trench face for collecting free-lateral flow, and lower part illustrates the two H-flumes that contain instruments for continuous monitoring of subsurface flow (Wilson et al., 1991a. With permission.)

front moving through the microfracture network. However, further sampling revealed that the tracer plume was refracted in the direction of the dip in the fractured saprolitic subsoil and skirted around the remaining solution samplers. While the tracer release revealed a very rapid transport through the fracture region, the mass transfer into the low permeability saprolite matrix must have been significant since 50% of the applied tracer was still "immobilized" some six months and numerous storm events later. This suggests that the mass transfer rate was large. However, storm events following the tracer release, in which the tracer must be transferred from the saprolite matrix into the fracture system to be exported, revealed substantial delay in transport relative to the subsurface flow hydrograph (Figure 2.9). Stable isotope and solute chemistry analysis revealed that subsurface flow was predominantly new water at peak flow and almost exclusively old water during the recession limb of the subsurface hydrograph. This difference between rapid transfer during tracer application and slower transfer during tracer flushing could be due to either a hysteretic effect on the transfer rate coeffi-

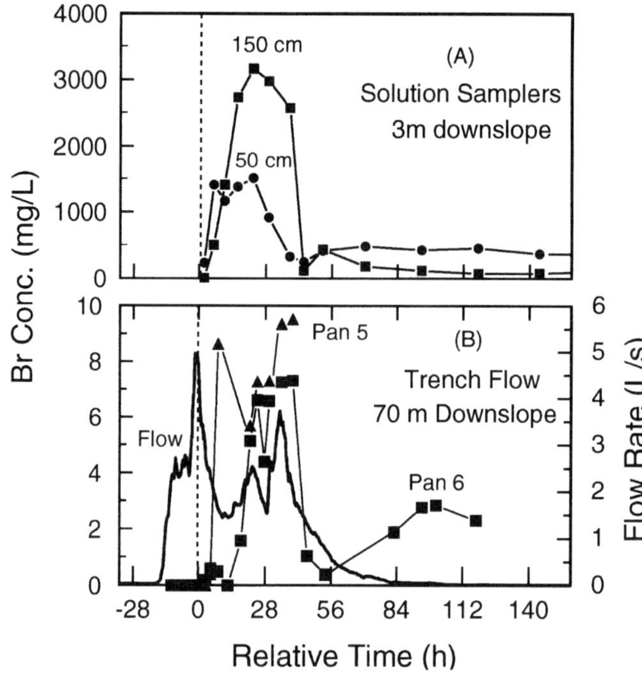

Figure 2.8. Bromide dynamics observed in (A) solution samplers 3 m downslope of the line source and (B) subsurface flow hydrograph (solid line) for lower pans and Br concentration in flow through pans 5 (triangles) and 6 (squares) from Wilson et al. (1993). With permission.

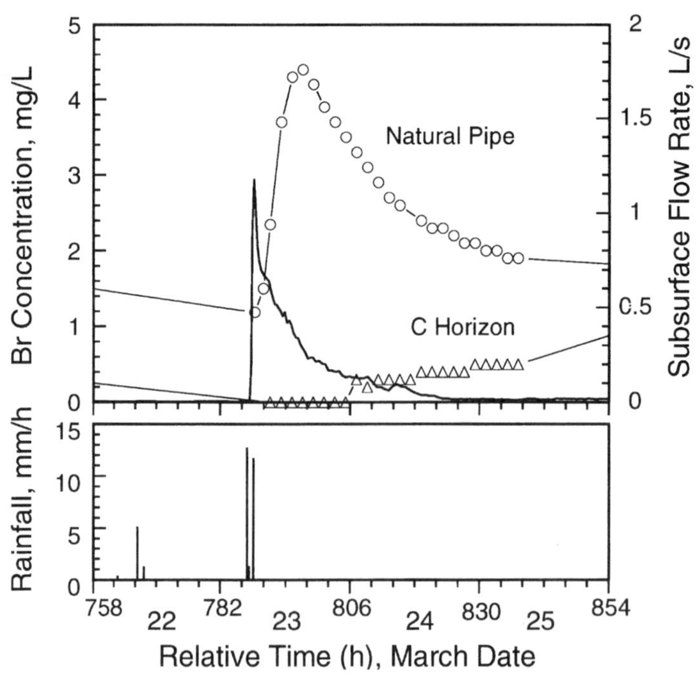

Figure 2.9. Subsurface flow hydrograph (solid line), bromide concentration from a natural pipe (open circles) and the lower pans (open triangles), and rainfall rate during the March 1991 storm event (Wilson et al., 1993, with permission).

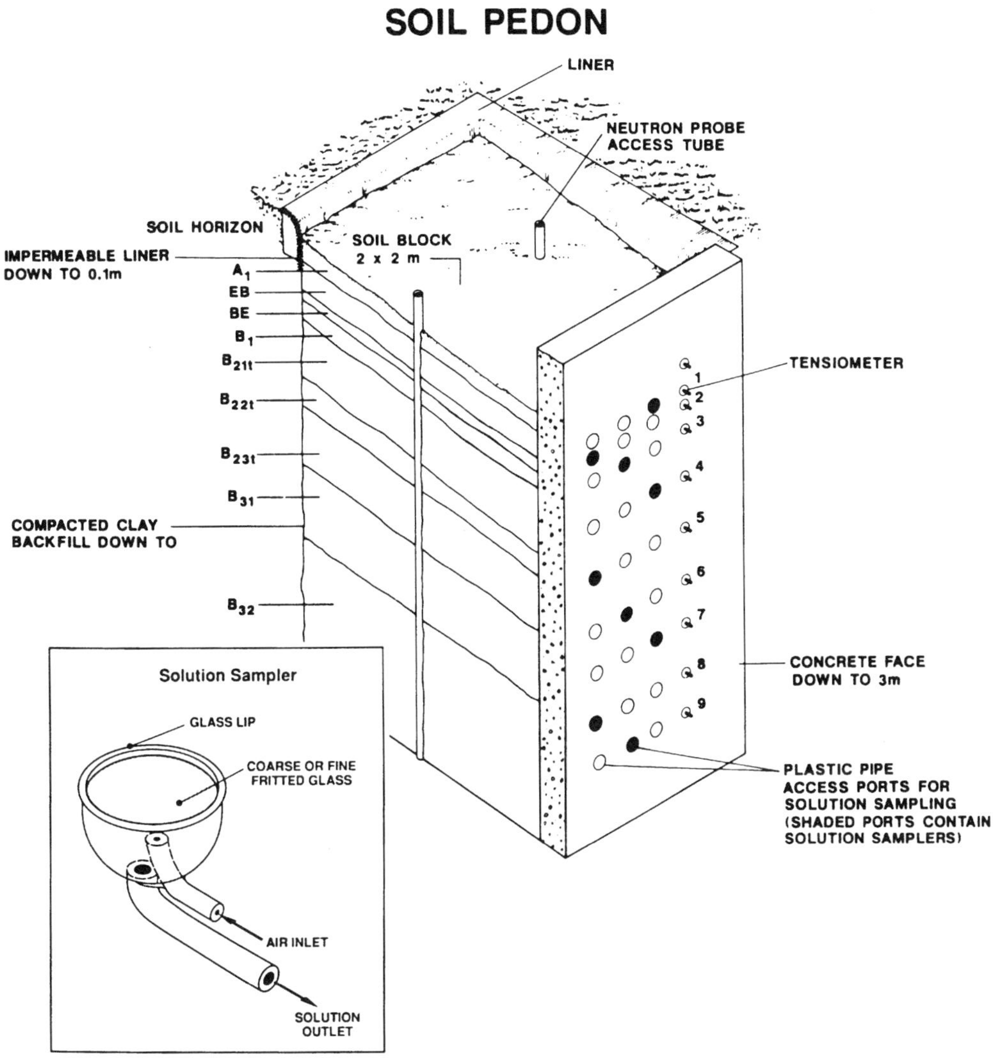

Figure 2.10. Diagram of the soil block facility for field-scale one-dimensional transport studies. The insert illustrates the fritted-glass plate samplers installed in the block at locations indicated by the solid circles (Jardine et al., 1990, with permission).

cients or due to advective and diffusive transfer working in parallel during release and counter during flushing.

Jardine et al. (1990) also attempted to distinguish the contribution of pore regions to transport under field-scale transient flow conditions using the 2 m × 2 m by 3 m deep *in situ* soil block at the Walker Branch Subsurface Transport Facility. They instrumented the soil block with 5.7 cm diameter fritted-glass plate lysimeters for sampling two pore classes, Figure 2.10. Rapid flow through large pores was sampled with high conductivity porous glass plates with a bubbling pressure of 20 cm. Movement through smaller pores within the soil matrix was sampled with low conductivity porous glass with a bubbling pressure of 300 cm. All plate lysimeters were pressed into firm contact with the upper surface of auger holes bored horizontally into the soil. The excavated soil material was packed on top of the fine porosity plates to ensure pore continuity while excluding macropores from direct contact with the sampler. It was assumed that macropore flow above the sampler area would bypass the sampler. While the assumption of pore classes sampled may be questioned, the simultaneous transport through both a fast flow region, and a slower flow region under intermittent natural rainfall conditions was effectively demonstrated. They observed significant movement through the small-pore region that was more continuous than the transport through larger pores which responded to storm events much like discrete flow pulses, Figure 2.11. However, due to the transient variably saturated flow conditions, mass transfer between pore classes complicated the findings. They reasoned that hydraulic and physical nonequilibrium between pore classes near the soil surface shortly after tracer application caused advection and diffusion, respectively, from the solute rich small pores into the larger pores. This resulted in rapid transport to deeper depths where the mass transfer from the large to smaller pores occurs by the same mechanisms. With time, the movement within the smaller pore region resulted in higher concentrations in the small pore class at these deeper depths (Figure 2.11).

Due to the complication of mass transfer resulting from the hydraulic and physical nonequilibrium induced by preferential flow under transient variably saturated flow conditions, an extensive steady-state flow tracer study was conducted on the soil block at Melton Branch. The soil profile was instrumented similarly to the Walker Branch soil block but with free drainage (fracture flow) collected in 5 cm × 30 cm long fritted-glass plate lysimeters, while flow was also collected under 30 cm tension from a 5 cm × 30 long plate lysimeter and a 5.7 cm diameter plate lysimeter under 500 cm tension. Steady-state flow was established under three successive infiltration rates (330, 30, and 3 cm/d) with Br applied as a pulse. The objective was to determine the transport parameters for three different flow regions while minimizing the complication of mass transfer between regions. The free-drainage samples produced flow almost exclusively under the highest infiltration rate only. Thus, comparison among three flow regions was possible from this experiment. Three distinct breakthrough curves were observed, Figure 2.12, enabling the quantification of pore flow velocity and dispersion parameters as a function of depth and flow region. However, the diminutive differences in BTCs under the highest infiltration rate and the substantial differences in the transport parameters for the smallest pore class during the low infiltration rate experiment suggest that the rapid fracture flow was not being excluded from lysimeters extracting water under high tensions during the high infiltration experiment. Thus, inter-region transport parameters remained incompletely quantified even with these elaborate research facilities.

Inter-Region Mass Transfer:

For any type of multiregion flow model, the inter-region transfer is a crucial component (Skopp and Gardner, 1992). As stated earlier, this is the most difficult aspect of multiregion flow and transport to measure. With the traditional dual porosity (mobile-immobile) approach, solutes are transported rapidly through a single preferential flow region with mass transferred only by diffusion between this region and the stagnant water within the soil matrix. Most of the research to date on preferential flow has considered only diffusive transfer (Brusseau et al., 1989a,b; Leij et al., 1993; Maloszewski and Zuber, 1993; van Genuchten and Wierenga, 1976).

The chapter by Jardine et al. in this book reviews methods for quantifying the transfer due to physical nonequilibrium, therefore this chapter will not repeat this review except to briefly mention the flow-interrupt method. The technique involves displacement of a nonreactive tracer, inhibiting displacement for a designated time period (interruption), and then restarting the displacement. If molecular diffusion

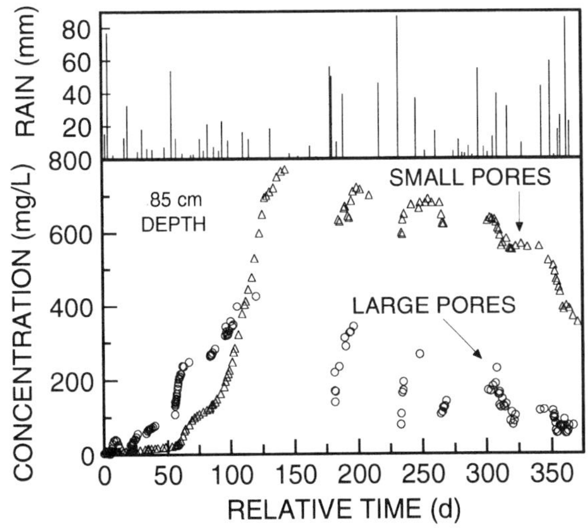

Figure 2.11. Breakthrough of bromide through large pores (circles) and small pores (triangles) at the 85 cm depth of the Walker Branch soil block (replotted from Jardine et al., 1990, with permission).

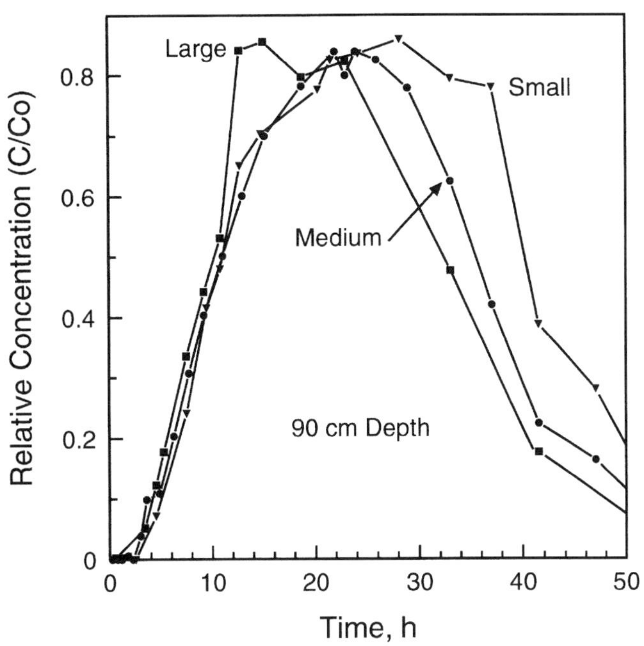

Figure 2.12. Breakthrough of bromide through three pore regions at the 90 cm depth of the Melton Branch soil block during the 300 cm/d infiltration experiment.

into the matrix of porous media is significant, the tracer concentration will be less upon flow restart than prior to flow interruption for an interrupt during the tracer injection while the opposite will occur for interrupt during the tracer flushing. The technique will work only when the pore water velocity in the preferential flow path is greater than the rate of diffusion into the matrix. Reedy et al. (1996) employed the flow interruption method on undisturbed columns from the Melton Branch Subsur-

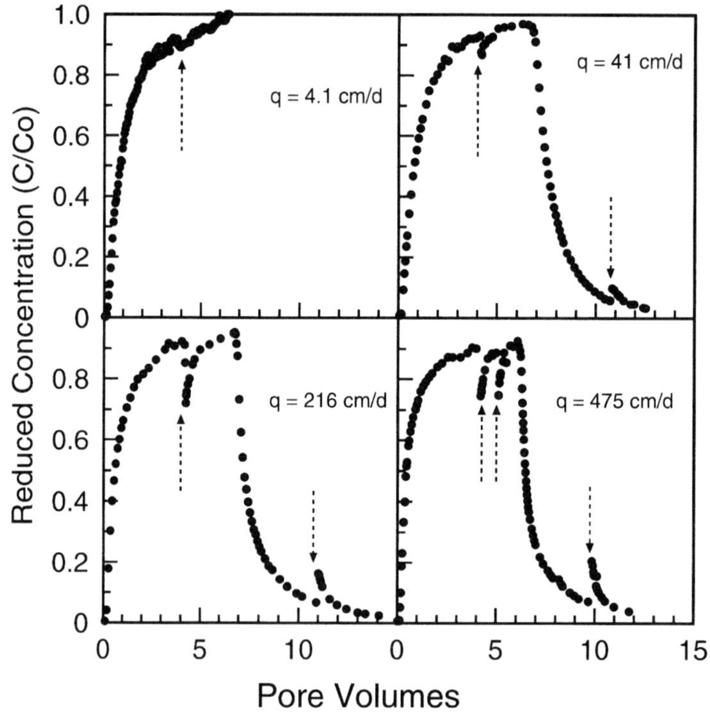

Figure 2.13. Breakthrough of bromide for undisturbed Melton Branch soil columns at four flow rates with an interrupt (indicated by arrows) duration of 7 d (replotted from Reedy et al., 1996, with permission).

face Transport Facility. They quantified the effect of flow rate on the physical nonequilibrium by conducting miscible displacement experiments at four flow rates covering 2 orders of magnitude, Figure 2.13. They also investigated the effect of interrupt period on physical nonequilibrium by varying the interrupt period at an intermediate flow rate, Figure 2.14. They found that diffusive transfer, indicated by the degree of perturbation in effluent concentration during the interrupt, was minimal if the interrupt period was short (6 h). This does not mean that physical nonequilibrium was nonexistent as evidenced by the fact that the solute perturbation was substantial at this flow rate when interrupt time was extended. The physical nonequilibrium did appear to be minimal when the flow rate was low (4.1 cm/day) as would be expected. However, diffusive transfer due to physical nonequilibrium in response to rapid interregion advective transport was significant over a large range of fluxes above 4 cm/day and time periods above 6 h. More recent displacement experiments through this triple-porosity media using multiple nonreactive tracers with different diffusion coefficients, indicated that the process of diffusion between different pore-regions occurred throughout all stages of solute transport and breakthrough (O'Brien et al., 1997).

Recent research, however, has questioned the all-encompassing validity of the diffusion-only assumption. Luxmoore and Ferrand (1992) used percolation theory to model a mobile-immobile system at the pore-scale. In this model, each pore in a lattice network of pores was considered a unique radius which was connected to neighbors by tubes of defined radii. A cluster of pores that are interconnected from the top to the bottom of the network is known as the backbone. Connected to the backbone (i.e., preferential flow path) are fluid-filled pores called backwater that do not have any

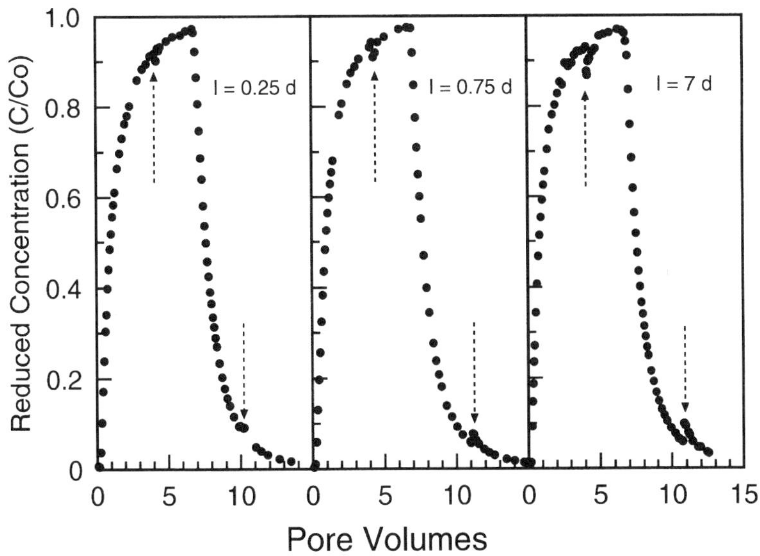

Figure 2.14. Breakthrough of bromide for undisturbed Melton Branch soil columns at three interrupt (indicated by arrows) durations under the 41 cm/d flow rate (replotted from Reedy et al., 1996, with permission).

flow-through connections but do have diffusive exchange of solutes with fluid in the backbone. Poiseuille flow, depending on the fourth power of the tube radius, was used to calculate flow in the tubes. The pathlength-supply process proposed by Luxmoore et al., 1990, viewed rising flow rates to be composed of longer flow-paths through the landscape which allowed enhanced diffusive transfer of solute from backwater pores, thus increasing concentrations. Luxmoore and Ferrand (1992) evaluated this proposed mechanism and estimated that the surface area of interaction between backwater and backbone was insufficient to account by diffusion alone for rising solute concentrations with rising flow rates.

It should be remembered that preferential flow in unsaturated soil is also a hydraulic nonequilibrium process in that hydraulic gradients exist between flow regions. Under such conditions, advection, which is orders of magnitude more rapid than diffusion, will cause mass transfer between regions. Zimmerman et al. (1993) incorporated a semianalytical exchange equation to simulate advective transfer between flow regions. For characterization of a mechanistic advective transfer, one would need to measure the pressure head in each region, the transfer coefficient, and the length scale. Techniques for measurement of these advective transfer parameters are almost nonexistent. Even if techniques could be developed for these measurements, the spatial variability associated with such microscale properties would render field-scale application impractical. A manageable proposition is to measure the macroscopic effect of advective transfer analogous to how diffusive transfer is parameterized.

A technique that shows promise for separating the advective from diffusive transfer is the use of multiple tracers. Maloszewski and Zuber (1993) proposed that double tracing; i.e., two solutes of different diffusion coefficients, be used as a rule for determining the importance of diffusion from the preferential flow region into the matrix. McKay et al. (1993) used bacteriophage to improve the characterization of fracture flow due to their exclusion from the matrix. We propose that this be extended to include two sizes of colloids along with a nonreactive solute. One size colloid, $d \geq 1$ mm, would be excluded from mesopores while a secondary size colloid, 1 mm > d > 0.01 mm, could enter mesopores but be excluded from micropores. The transport experiments should be run first in a steady-state flow condition whereby advective transfer is excluded, followed by transient flow transport experiments whereby advective and diffusive transfer are active.

MULTIREGION FLOW AND TRANSPORT MODELING

The question then arises as to how to model multiregion flow and transport. There have been many approaches presented recently. Skopp and Gardner (1992) proposed a multidomain model in which more than one pore class contributed to water and solute movement. Steenhuis and Parlange (1988) proposed a model that divides soil water into a number of pore groups in which water and solutes move at different rates with mixing between pore classes. Jarvis et al. (1991) developed MACRO, a two-region transient flow and transport model that requires functional hydraulic properties partitioned into macropore-micropore regions. Flow through the micropore region was modeled by Richard's Equation. Beven and Germann (1982) concluded that the goal in modeling a preferential flow system is to integrate the various components (i.e., flow regions) into one flow theory that reduces to a Darcy-type model in the absence of macropores. Due to possibilities of turbulent flow in the macropore region, a Darcy-type model may not be applicable for this region. One may need to incorporate a stochastic-type model for the macropore region with Darcy-type models for the other regions. The difficulty is in mechanistically integrating these contrasting model approaches. An alternative approach is to apply a Darcy-type model macroscopically to all regions with a variable hydraulic conductivity function for the macropore region.

Gwo et al. (1991) developed a model, MURF, which allows the soil to be modeled as a porous media composed of multiple flow regions. Intra-region flow; i.e., through each region, is modeled by Richard's equation while intra-region transport, through each region, is modeled in MURT as an advective dispersive process. Thus, the lumped parameter description of dispersion for a single region ADE is segregated into distinct regions. The unique feature of MURF and MURT is that first-order, steady, finite-volume, advective and diffusive transfer schemes were incorporated into these codes to allow inter-region transfer; i.e., between regions, by fluid flow and solute movement, respectively. The mathematical formulism of these codes will not be discussed here but can be found elsewhere (Gwo et al., 1994, 1995a,b). Selected simulations from Gwo et al. (1996) are presented to demonstrate the capabilities of these codes in describing the flow and transport processes of a triple-porosity, triple-permeability variably saturated flow system. A 200 cm long soil column with an initial pressure head of –90 cm at the top was modeled with the bottom maintained at 0 cm pressure head. A 5 h fluid pulse at a rate of 9 cm h^{-1} was applied to the top and allowed to drain for 10 h. The fluid applied during the first two hours of the pulse contained a nonreactive tracer. Parameters for the hydraulic and transport properties were obtained from Wilson et al. (1992) and Jardine et al. (1993). Given that the pressure head varies with time within each region, the transfer coefficients for each region may also vary with time. However, the advective and diffusive transfer coefficients were held constant with time but were varied by five and seven orders of magnitude, respectively, for each simulation.

Hydraulic nonequilibrium among flow regions diminished with proximity to the bottom boundary condition (h = 0 cm). However, at just 95 cm above the bottom boundary the effect of multiple permeabilities is instantaneously evident upon pulse application (Figure 2.15). The pressure head in each pore region at this depth, Figure 2.15a, increased but to different degrees. Thus, inter-region fluid was transferred by

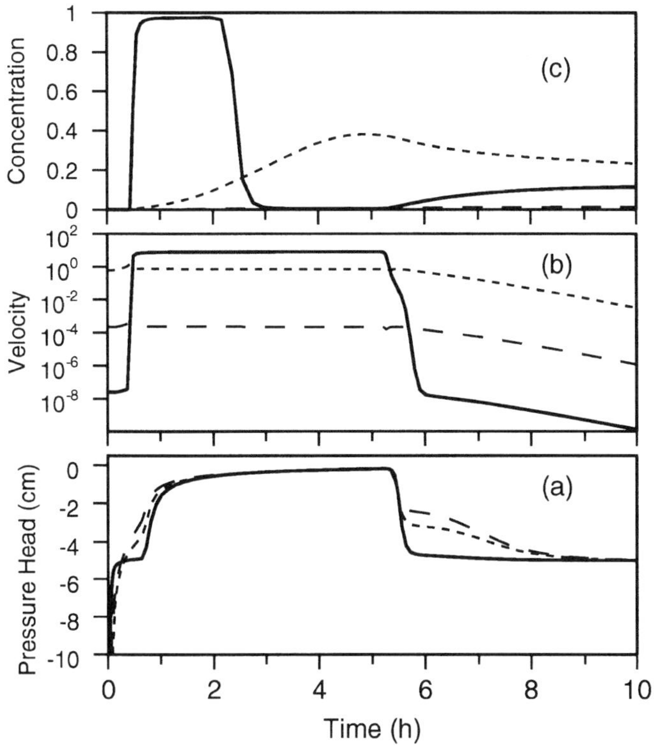

Figure 2.15. (a) Pressure head, (b) water velocity, and (c) solute BTC dynamics for the macropore (solid), mesopore (small dashes), and micropore (large dashes) regions at the 95 cm column height (Gwo et al., 1996, with permission).

advection due to the hydraulic nonequilibrium (pressure gradients). The inter-region advective transfer eventually, after a time of 2 h to 5 h, led to equilibrium among regions. It should be noted that even though pressure heads were in equilibrium among pore regions during this time, distinct pore region conductivities resulted in pore water velocities that varied by several orders of magnitude (Figure 2.15b). Thus, when the inter-region advective transfer phased out, the intra-region velocity differences remained. When the pulse application was terminated at t=5 h, the perturbation caused nonequilibrium conditions to quickly redevelop within the system. Due to its high conductivity, the macropore region exhibited the most rapid decline in velocity, Figure 2.15b, and establishment of its steady-state pressure head, Figure 2.15a. Note in Figure 2.15a the hysteretic response in the establishment of inter-region equilibrium. Pressure gradients among pore regions were more dramatic and the time required to reach equilibrium among flow region was greater after the pulse was terminated at t> 5h, than at initiation, t=0.

The extent of the pressure gradient that developed between flow regions was dependent upon the inter-region transfer coefficients for these regions, Figure 2.16. The smaller the transfer coefficient, the larger the nonequilibrium was between regions. Additionally, the smaller the transfer coefficient, the longer the time necessary to establish inter-region equilibrium. The fluid velocity within each flow region was directly controlled by its intra-region pressure gradient; however, it was also indirectly controlled by the inter-region pressure gradients and their transfer coefficients. For example, the macropore region had the highest hydraulic conductivity and the

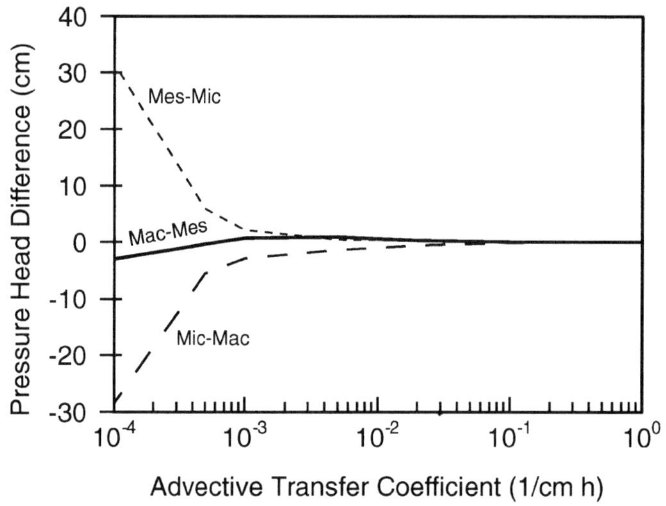

Figure 2.16. Effects of advective transfer coefficients on the pressure head gradients between the macro-mesopore (solid), meso-micropore (small dashes), and micro-macropore (large dashes) regions (Gwo et al., 1996. With permission.)

smallest porosity, thus it was most sensitive to water content and pressure head variations. Gwo et al. (1996) showed that high advective transfer coefficients allowed more water to be transferred from mesopore and micropore regions into the macropore region, thereby increasing the hydraulic conductivity of this region before the fluid front arrived. They further showed that low advective transfer coefficients result in the isolation of flow regions. As discussed earlier, experimental quantification of the advective transfer coefficient is problematic. The degree to which the pressure gradients depicted in Figure 2.16 and Figure 2.15a develop under field conditions is uncertain due to our inability to measure the water pressure in one region while excluding the other. It could explain why De Vries and Chow (1978) observed macropore flow in the direction opposite the hydraulic gradient as indicated by tensions. If the tensiometers are within the soil aggregate (matrix), they would not reflect the pressure gradient of the macropore region.

Despite solute concentrations at the upper boundary being equal among flow regions at the initiation of flow and solute injection, multiple solute fronts quickly develop within the system due to the multiple velocity fields, Figure 2.15b. Due to the multiple solute fronts, not only does local hydraulic nonequilibrium develop, but physical nonequilibrium develops as well. Inter-region diffusive transfer serves to move solutes from high to low concentrations while pressure gradients may move solutes in either direction. The end result is a system of multiple sources and sinks for the solute. This complex transport process can be seen from this simple scenario simulated by Gwo et al. (1996). The breakthrough of the nonreactive tracer at 95 cm from the column bottom, Figure 2.15c, was first observed in the macropore region. A delayed breakthrough was observed for the mesopore region while breakthrough from the micropore region was not observed. The flux averaged concentration would appear as a dual peak BTC common for dual porosity-dual permeability media (Hornberger et al., 1991).

As shown earlier, advective inter-region transfer was occurring during the initial stage of flow and tracer injection due to hydraulic nonequilibrium between regions. During the period that the solute in the macropore region was exiting the column, t=

2 h, the solute concentration in the mesopore region at the exit was increasing due to this initial advective transfer near the surface and due to diffusive transfer along the column length where the solute fronts had separated. Due to the larger storage capacity of the micropore region, inter-region transfer did not result in a noticeable solute concentration increase in this region.

At the time that the solute pulse was terminated, $t = 2$ h, hydraulic equilibrium existed throughout the column even though the water application was continued under a steady rate until $t = 5$ h. After solute termination, $t > 2$ h, the solutes in the macropore region were quickly exhausted due to their high water velocity. Additionally, solutes were being transferred by diffusion from the mesopore to macropore region near the surface and from the macropore to mesopore region near the bottom. After the water application was terminated, $t = 5$ h, the velocity of the macropore region quickly diminished (Figure 2.15b), and the major contributor to solute export was the mesopore region. As the solute front in the mesopore region exited the column, the concentration in the macropore region increased due to inter-region transfer by diffusion and advection from the mesopore to macropore region. Advection was also occurring from the micropore to macropore region at this depth and time; however, this inter-region transfer had the opposite effect, serving to dilute the macropore region. Due to the order of magnitude lower rate of transfer for the micropore region, dilution of macropore water from micropore water was minimal. At this depth and time, solute was diffusing from the mesopore to the micropore region counter to the advection. Both transfer processes, though opposite in direction, were causing a decrease concentration in the mesopore region. Thus, the secondary peak in the BTC was not as large as would be observed if regions were isolated. The bottom line is that not only must flow velocities through each region be incorporated, but advective and diffusive transfer must also be included in the multiregion flow and transport model.

While a flux averaged measurement would exhibit a bimodal BTC, this approach would fail to represent the HNE and PNE processes causing the dual peaks. This is clearly seen in the solute concentration profiles in the column, Figure 2.17. At 0.32 h after solute injection, the solute had moved more than 60 cm through the macropore region, and to a lesser degree in the mesopore region, but had not entered the micropores. Due to the extremely small porosity of the macropore region, soil sampling, which produces a volume averaged resident concentration, would conclude limited solute movement in the upper 50 cm and no movement below this depth. However, a flux sampler which can collect drainage from macropores, such as a tension-free pan lysimeter, would have detected significant movement at the 50 cm depth. This is exactly what Essington et al. (1995) observed in their study on pesticide movement in long-term tillage plots. Depth-incremented soil sampling in the no-till soil indicated that the pesticide had not moved below the 60 cm depth. Traditional pesticide analysis would assume that all loss of mass above the depth of penetration was due to degradation. However, tension-free pan lysimeters at the 90 cm depth, which collect drainage through macro- and mesopores, revealed as much as 53% of the applied pesticide had been transported out of the root zone and was not detected by soil sampling. They also found that between 68 and 100% of the pesticide transport was during the first two storm events following application. They concluded that rainfall timing relative to application was the key factor to pesticide mobility. This is likely a result of the contribution of the mass transfer processes and multireactive adsorption processes (Wilson et al., 1998) to transport. However, ignoring these HNE and PNE processes would result in an extreme overestimation of pesticide degradation and an extreme underestimation of leaching potential.

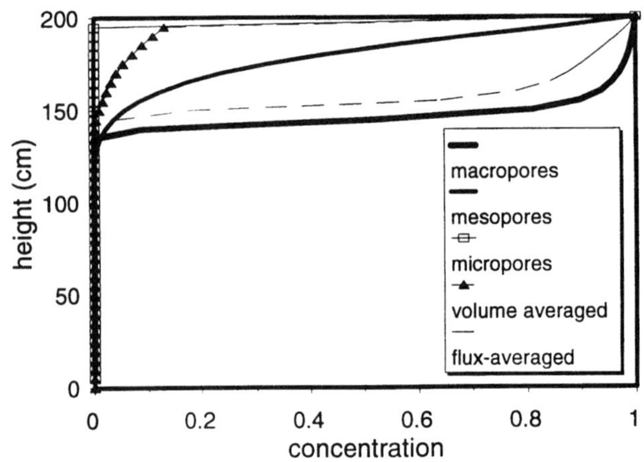

Figure 2.17. Solute concentration profile at time = 0.32 h for each pore region along with the volume averaged resident concentration (solid triangles) and the flux averaged solution concentration (dashed line). Unpublished data from the simulations of Gwo et al., 1996. With permission.

SUMMARY

Despite the arbitrary criteria for segregating soil into macro-, meso-, and micropore regions, analogous to the arbitrary segregation of particles into sand, silt, and clay, the physical significance of this segregation was demonstrated. The importance of macropore flow and transport has been widely accepted. This chapter documented from the literature the significance of mesopore flow and demonstrated conceptually and through numerical simulations its importance to transport. Gwo et al. (1996) concluded that intra-region advective-dispersive transport was negligible for the micropore region as it was virtually stagnant for their simulations. However, the importance of inter-region transfer with this region was demonstrated as it served as a source/sink for the other regions. The importance of including advection in addition to diffusion in the inter-region transfer was also demonstrated.

This chapter showed how macropore flow may function as a serious threat to groundwater supplies due to the bypass of the chemically reactive and biologically active intra-aggregate regions (meso- and micropores). However, it should be kept in mind that the conclusions from the work of Gwo et al. (1996) with regard to the insignificance of movement within the micropore region were based on simulations of a single rainfall event of less than one day duration. They did not consider the significance of this slowly-moving micropore front with its high solute concentrations over the time frames relevant to remediation scenarios. As McKay et al. (1993) pointed out, for a scenario where the contaminant resides in the matrix and is exported from the system as the solute is transferred from the matrix into the more mobile regions, "it may take many years...to flush out the contaminants." Macropore and mesopore transport would result in sporadic flushing of contaminants during individual storms. While this is occurring, the contaminant may be slowly migrating through the micropore region. Low inter-region transfer rates with the micropore region would render pump and treat remediation strategies to be of limited value. To be effective, the inter-region transfer coefficients for the micropore region would need to be appreciable. Since this contaminant-rich water remains somewhat isolated in the micropore region, it may exhibit negligible movement over timescales of storm events but may directly contribute to export into stream or groundwater supplies over timescales of decades.

In addition, the conclusions by Gwo et al. (1996) that characterization efforts should focus more on diffusive transfer than advective transfer may be an artifact of simulating conditions in which the contaminant is applied as an input flux to a clean system. A more common scenario that has received much less attention in transport studies is the condition in which the contaminant resides predominantly in the less mobile regions between storm events with transfer occurring into the more mobile regions during an event. For such conditions, and particularly for the extremely dynamic transient flow conditions experimentally assessed by Jardine et al. (1990) and Wilson et al. (1993), it is possible that advective transfer is more important than the model simulations of Gwo et al. (1996) suggest.

Hydraulic and physical nonequilibrium among flow regions can be important mechanisms for mass transfer between regions. Quantification of these transport processes is critical to water quality preservation. For this reason, it is imperative that we apply multiregion flow and transport approaches to develop technologies capable of characterizing and manipulating these processes for remediation.

REFERENCES

Bai, M., D. Elsworth, and J.C. Roegiers. Multiporosity/multipermeability approach to the simulation of naturally fractured reservoirs. *Water Resour. Res.,* 29:1621–1633, 1993.

Beven, K. and P. Germann. Macropores and water flow in soils. *Water Resour. Res.,* 18:1311–1325, 1982.

Bouma, J. Comment on 'Micro-, meso-, and macroporosity of soil.' *Soil Sci. Soc. Am. J.,* 45:1244–1245, 1981a.

Bouma, J. Soil morphology and preferential flow along macropores. *Agric. Water Manage.,* 3:235–250, 1981b.

Brusseau, M.L., R.E. Jessup, and P.S.C. Rao. Modeling the transport of solutes influenced by multiprocess equilibrium. *Water Resour. Res.,* 25:1971–1988, 1989a.

Brusseau, M.L., P.S. C. Rao, R.E. Jessup, and J.M. Davidson. Flow interruption: A method for investigating sorption nonequilibrium. *J. Contam. Hydrol.,* 4:223–240, 1989b.

de Marsily, G. *Quantitative Hydrogeology.* Academic Press, Inc., New York, 1986, p. 23.

De Vries, J. and T.L. Chow. Hydrologic behavior of a forested mountain soil in Coastal British Columbia. *Water Resour. Res.,* 14:935–942, 1978.

Essington, M.E., D.D. Tyler, and G.V. Wilson. Fluometuron behavior in long-term tillage plots. *Soil Sci.* 160:405–414, 1995.

Germann, P.F. and K. Beven. Water flow in soil macropores 1. An experimental approach. *J. Soil Sci.,* 32:1–13, 1981.

Glass, R.J. and T.S. Steenhuis. Factors influencing infiltration flow instability and movement of toxics in layered sandy soil. American Society of Agricultural Engineers, ASAE Technical Paper 84-2508. ASAE, St. Joseph, MI, 1984.

Gwo, J.P., G.T. Yeh, and G.V. Wilson. Proceedings of the International Conference on Transport and Mass Exchange Processes in Sand and Gravel Aquifers: Field and Modeling Studies, Ottawa, Canada, 2, 1991, pp. 578–589.

Gwo, J.P., P.M. Jardine, G.T. Yeh, and G.V. Wilson. MURF user's guide: A finite element model of multiple-pore-region flow through variably saturated subsurface media. ORNL/GWPO-011. Oak Ridge National Laboratory, TN, 1994.

Gwo, J.P., P.M. Jardine, G.T. Yeh, and G.V. Wilson. MURT user's guide: A hybrid Lagrangian-Eulerian finite element model of multiple-pore-region solute transport through subsurface media. ORNL/GWPO-015. Oak Ridge National Laboratory, TN, 1995a.

Gwo, J.P., P.M. Jardine, G.V. Wilson, and G.T. Yeh. A multi-pore-region concept to modeling mass transfer in subsurface media. *J. Hydrol.,* 164, 217–237, 1995b.

Gwo, J.P., P.M. Jardine, G.V. Wilson, and G.T. Yeh. Using a multiregion model to study the effects of advective and diffusive mass transfer on local physical nonequilibrium and solute mobility in a structured soil. *Water Resour. Res.,* 32:561–570, 1996.

Hammermeister, D.P., G.F. Kling, and J.A. Vomocil. Perched water tables on hillsides in western Oregon: I. Some factors affecting their development and longevity. *Soil Sci. Soc. Am. J.,* 46:812–818, 1982.

Hillel, D. Unstable flow: A potentially significant mechanism of water and solute transport to groundwater, in *Water Flow and Solute Transport in Soils,* Russo, D. and G. Dagan (Eds.), Springer-Verlag, New York, 1993, pp. 123–134.

Hornberger, G.M., P.F. Germann, and K.J. Beven. Throughflow and solute transport in an isolated sloping soil block in a forested catchment. *J. Hydrol.,* 124:81–99, 1991.

Jardine, P.M., G.V. Wilson, and R.J. Luxmoore. Modeling the transport of inorganic ions through undisturbed soil columns from two contrasting watersheds. *Soil Sci. Soc. Am. J.,* 52:1252–1259, 1988.

Jardine, P.M., G.V. Wilson, and R.J. Luxmoore. Mechanisms of unsaturated solute mobility during storm events. *Geoderma,* 46:103–118, 1990.

Jardine, P.M., G.K. Jacobs, and G.V. Wilson. Unsaturated transport processes in undisturbed heterogeneous porous media I. Inorganic contaminants. *Soil Sci. Soc. Am. J.,* 57:945–954, 1993.

Jarvis, N.J., P.-E. Jensson, P.E. Dik, and I. Messing. Modeling water and solute movement in macroporosity soil. I. Model description and sensitivity analysis. *J. Soil Sci.,* 42:59–70, 1991.

Jury, W.A. and H. Flühler. Transport of chemicals through soil: Mechanisms, models, and field applications. In *Advances in Agronomy.* Sparks, D.L. (Ed.), Academic Press, Inc., 1992, pp. 141–201.

Kinsall, B.L., G.V. Wilson, A.V. Palumbo, and T.J. Phelps. Soil property influences on heterogeneity in unsaturated flow and microbial transport through undisturbed soil blocks. *Soil Sci. Soc. Am. J.* (in review), 1997.

Leij, F.J., N. Toride, and M.T. van Genuchten. Analytical solutions for non-equilibrium solute transport in three-dimensional porous media. *J. Hydrol.,* 151:193–228, 1993.

Luxmoore, R.J. and L.A. Ferrand. *Water Flow and Solute Transport in Soils: Modeling and Applications.* Russo, D. and G. Dagan (Eds.), Springer-Verlag, New York, 1992, pp. 45–60.

Luxmoore, R.J., P.M. Jardine, G.V. Wilson, J.R. Jones, and L.W. Zelazny. Physical and chemical controls of preferred path flow through a forested hillslope. *Geoderma,* 46:139–154, 1990.

Luxmoore, R.J. Micro-, meso-, and macroporosity of soil. *Soil Sci. Soc. Am. J.,* 45:671–672, 1981.

Maloszewski, P. and A. Zuber. Tracer experiments in fractured rocks: Matrix diffusion and the validity of models. *Water Resour. Res.,* 29:2723–2735, 1993.

McKay, L.D., R.W. Gillham, and J.A. Cherry. Field experiments in a fractured clay till 2. Solute and colloid transport. *Water Resour. Res.,* 29:3879–3890, 1993.

O'Brien, R., P.M. Jardine, J.P. Gwo, L.D. McKay, and A. Harton. Experimental and numerical evaluation of solute transport processes in fractured saprolites. *Water Resour. Res.,* (in review), 1997.

Omoti, U. and A. Wild. Use of fluorescent dyes to mark the pathways of solute movement through soils under leaching conditions, 2. Field experiments. *Soil Sci.,* 128:98–104, 1979.

Phillips, R.E., V.L. Quisenberry, J.M. Zeleznik, and G.H. Dunn. Mechanism of water entry into simulated macropores. *Soil Sci. Soc. Am. J.,* 53:1629–1635, 1989.

Perroux, K.M. and I. White. 1988. Designs for disc permeameters. *Soil Sci. Soc. Am. J.,* 52:1205–1215, 1988.

Radulovich, R., P. Sollins, P. Baveye, and E. Solórzano. Bypass water flow through unsaturated microaggregated tropical soils. *Soil Sci. Soc. Am. J.,* 56:721–726, 1992.

Reedy, O.C., P.M. Jardine, H.M. Selim, and G.V. Wilson. Quantifying the diffusive mass transfer of non-reactive solutes in undisturbed subsurface columns using flow interruption. *Soil Sci. Soc. Am. J.,* 60:1376–1384, 1996.

Roth, K., W.A. Jury, H. Flühler, and W. Attinger. Transport of chloride through an unsaturated field soil. *Water Resour. Res.,* 27:2533–2541, 1991.

Schuh, W.M. and R.L. Cline. Effect of soil properties on unsaturated hydraulic conductivity pore interaction factors. *Soil Sci. Soc. Am. J.,* 54:1509–1519, 1990.

Scotter, D.R. Preferential solute movement through large soil voids. 1: Some computations using simple theory. *Aust. J. Soil Res.,* 16:255–267, 1978.

Skopp, J. and W.R. Gardner. Miscible displacement: an interacting flow region model. *Soil Sci. Soc. Am. J.,* 56:1680–1686, 1992.

Smettem, K.R.J., D.J. Chittleborough, B.G. Richards, and F.W. Leaney. The influence of macropores on runoff generation from a hillslope soil with a contrasting textural class. *J. Hydrol.,* 122:235–252, 1991.

Sollins, P. and R. Radulovich. Effects of soil physical structure on solute transport in a weathered tropical soil. *Soil Sci. Soc. Am. J.,* 52:1168–1173, 1988.

Steenhuis, T.S. and J.Y. Parlange. In *Proceedings of the International Conference on Validation of Flow and Transport Models for the Unsaturated Zone,* Wierenga, P.J. and D. Bachelet (Eds.), 88-SS-04, New Mexico State Univ., Las Cruces, NM, pp. 381–391, 1988.

van Genuchten, M.T., A closed-form equation for predicting the hydraulic conductivity of unsaturated soils. *Soil Sci. Soc. Am. J.,* 44:892–898, 1980.

van Genuchten, M.T. and P.J. Wierenga. Mass transfer studies in sorbing porous media I. Analytical solutions. *Soil Sci. Soc. Am. J.,* 40:473–480, 1976.

Watson, K.W. and R.J. Luxmoore. Estimating macroporosity in a forest watershed by use of a tension infiltrometer. *Soil Sci. Soc. Am. J.,* 50:578–582, 1986.

Wilson, G. V. and R. J. Luxmoore. Infiltration, macroporosity, and mesoporosity distributions on two forested watersheds. *Soil Sci. Soc. Am. J.,* 52:329–335, 1988.

Wilson, G.V., P.M. Jardine, R.J. Luxmoore, and J.R. Jones. Hydrology of a forested watershed during storm events. *Geoderma,* 46:119–138, 1990.

Wilson, G.V., P.M. Jardine, R.J. Luxmoore, L.W. Zelazny, D.A. Lietzke, and D.E. Todd. Hydrogeochemical processes controlling subsurface transport from an upper subcatchment of Walker Branch Watershed during storm events: 1. Hydrologic transport process. *J. Hydrol.,* 123:297–316, 1991a.

Wilson, G.V., P.M. Jardine, R.J. Luxmoore, L.W. Zelazny, D.E. Todd, and D.A. Lietzke. Hydrogeochemical processes controlling subsurface transport from an upper subcatchment of Walker Branch Watershed during storm events: 2. Solute transport process. *J. Hydrol.,* 123:317–336, 1991b.

Wilson, G.V., P.M. Jardine, and J. Gwo. Measurement and modeling the hydraulic properties of a multiregion soil. *Soil Sci. Soc. Am. J.* 56:1731–1737, 1992.

Wilson, G.V., P.M. Jardine, J.D. O'Dell, and M. Collineau. Field-scale transport from a buried line source in unsaturated soil. *J. Hydrol.* 145:111–123, 1993.

Wilson, G.V., L. Yunsheng, H.M. Selim, M.E. Essington, and D.D. Tyler. Tillage and cover crop effects on flow and multireactive transport of Fluometuron under saturated and unsaturated conditions. *Soil Sci. Soc. Am. J.,* (in review), 1997.

Zimmerman, R.W., G. Chen, T. Hadgu, and G.S. Bodvarsson. A numerical dual-porosity model with semianalytical treatment of fracture/matrix flow. *Water Resour. Res.,* 29, 2127–2137, 1993.

Chapter Three

Multiprocess Nonequilibrium and Nonideal Transport of Solutes in Porous Media

M.L. Brusseau

INTRODUCTION

The impact of porous-medium heterogeneity on transport of solutes or contaminants in the subsurface has been of interest for many years. It is now widely recognized that factors and processes related to physical heterogeneity can cause "nonideal" transport. For example, breakthrough curves obtained for transport of solute in a structured soil can exhibit skewness, with early breakthrough and tailing. As indicated by the title, physical heterogeneity is the focus of the book in which this chapter appears. However, it is important to recognize that factors and processes other than physical heterogeneity can cause nonideal transport. In addition, it is possible that solute transport may be influenced simultaneously by multiple factors and processes, each contributing to observed nonideal transport. To accurately simulate transport in such cases, it is necessary to understand the contribution of each factor and process, and to recognize the possibility of interactions between the factors and processes.

The purpose of this chapter is to discuss the nonideal transport of solute influenced by multiple, coupled processes. Examples from specific cases will be used to illustrate the impact of individual and combinations of processes on transport. Methods for identifying which factors or processes may control transport will be discussed, as will the development of mathematical models designed to account for multiple nonideality factors. This chapter is based in part on material presented in recent articles (Brusseau, 1994, 1997; Brusseau et al., 1997).

IMPACT OF NONIDEALITY FACTORS ON TRANSPORT

Ideal Versus Nonideal Transport

Four processes that control the movement of solutes in porous media are advection, dispersion, interphase mass transfer, and transformation reactions. Advection, also referred to as convection, is the transport of dissolved matter (solute) by the movement of a fluid responding to a gradient of fluid potential. Dispersion represents spreading of solute about a mean position, such as the center of mass. Phase transfers, such as sorption, liquid-liquid partitioning, and volatilization, involve the transfer of matter in response to gradients of chemical potential. Transformation reactions include any process by which the physicochemical nature of a solute is altered. Examples include biotransformation, radioactive decay, and hydrolysis.

The initial paradigm for transport of solutes in porous media was based on assumptions that the porous medium was homogeneous and that inter-phase mass transfers and transformation reactions were linear and essentially instantaneous. For discussion purposes, transport that follows this paradigm is considered to be ideal. However, it is well known that the subsurface is, in fact, heterogeneous, and that many phase transfers and reactions are not instantaneous. Thus, solute transport usually deviates from that which is expected based on the original paradigm, especially at the field scale. Such transport can be considered as nonideal. Solute transport can be investigated by use of breakthrough curves or by evaluating the spatial distribution of solute plumes. The nature of the transport nonideality displayed will be dependent, in part, upon which of these two approaches is employed.

The arrival of a solute mass at a point downgradient from where it is first introduced can be described by a breakthrough curve. The measurement and analysis of breakthrough curves is a widely used means of investigating solute transport in porous media. An example of a breakthrough curve generated for ideal conditions (e.g., homogeneous porous medium, instantaneous sorption) is shown in Figure 3.1. An illustration of a breakthrough curve exhibiting nonideal transport is also shown in Figure 3.1. This curve is clearly asymmetrical, and exhibits early breakthrough and tailing compared to the sharp, symmetrical curve obtained for ideal conditions. The existence of nonideal transport can thus be evaluated by examining the disposition of breakthrough curves.

The movement and evolution of the spatial distribution of dissolved matter (solute plume) in response to a fluid-potential gradient is also of interest. This is especially true for field-scale problems. Solute movement at the field scale is often examined using spatial moments calculated from the plume. The potential for nonideal transport can then be examined by evaluating the moments determined for the plume at different times. For example, the rate of displacement of the plume center of mass is expected to be uniform for ideal conditions. Conversely, several nonideality factors can cause the rate of displacement to be nonuniform. This is illustrated in Figure 3.2, where plume-displacement data obtained from a natural-gradient field experiment are shown. Clearly, the velocities of the nonreactive tracer plumes were essentially constant during the experiment, while those of the reactive organic compounds were not.

Most of the work reported to date for nonideal transport has involved analyses of the second temporal moment (i.e., spreading or dispersion of a breakthrough curve) and the second spatial moment (i.e., spreading of a plume). Under "ideal" conditions (e.g., homogeneous porous medium, instantaneous mass transfer) solute transport will exhibit minimal spreading. The existence of nonideal transport would, therefore, be indicated by the observation of increased spreading (larger second moment). However, it is also informative to examine the influence of nonideality factors on first and third spatial and temporal moments. Four major factors responsible for nonideal transport are briefly discussed below.

Nonlinear Sorption

Most solute transport models, especially for field-scale applications, include the assumption that the distribution of solute between liquid and solid phases can be described as a linear process. Such a condition would be reflected, for example, in

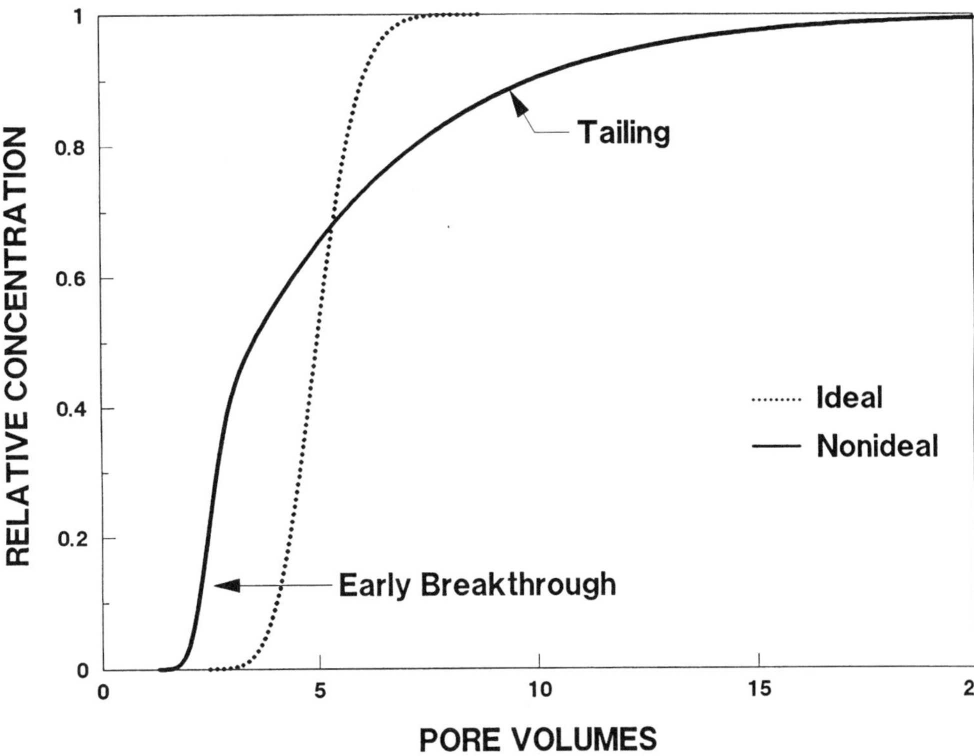

Figure 3.1. Breakthrough curves for ideal and nonideal transport. Figure from Brusseau (1994); copyright 1994, American Geophysical Union. With permission.

linear sorption isotherms. Numerous laboratory experiments have shown that sorption is often linear for low-polarity organic compounds. However, this assumption should be evaluated for each case because nonlinear isotherms have been reported for organic contaminants of interest, especially at higher concentrations. Nonlinear isotherms are typical for many polar organic, as well as for inorganic, chemicals.

It is well known that nonlinear sorption can cause asymmetrical breakthrough curves and concentration-dependent retardation (Crittenden et al., 1986; Brusseau and Rao, 1989). The concentration-dependent retardation produces a nonuniform rate of plume displacement and skewed spatial distributions (e.g., Srivastava and Brusseau, 1996). The impact of nonlinear sorption on transport is dependent in part on the degree of spreading experienced by the plume. Therefore, nonlinear sorption has a greater potential for impact when coupled with factors that cause a large degree of spreading (e.g., physical/chemical heterogeneity).

Rate-Limited Sorption

Most field-scale solute transport models include the so-called local equilibrium assumption, which specifies that interactions between the solute and the sorbent are so rapid in comparison to hydrodynamic residence time that the interactions can be considered instantaneous. However, based on laboratory experiments, it has long been known that sorption/desorption of many organic compounds by soils can be significantly rate limited. Research has shown that sorption/desorption of both organic and

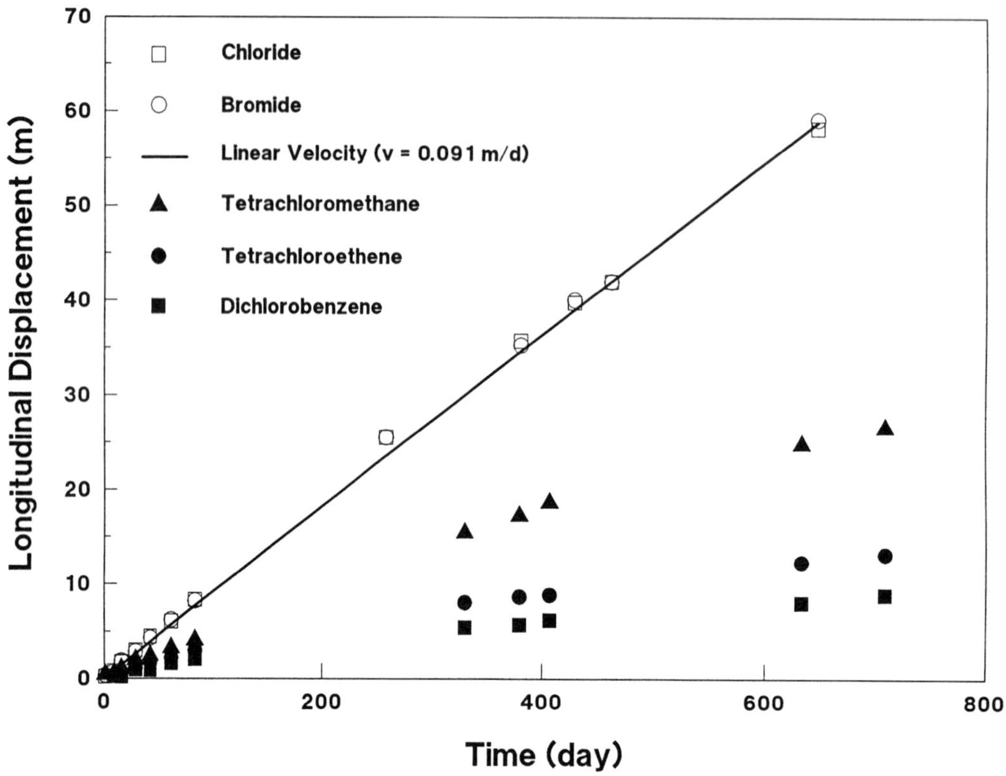

Figure 3.2. Examples of uniform (chloride and bromide) and nonuniform (organic solutes) rates of plume displacement. Data from Roberts and Mackay, 1986. With permission.

inorganic chemicals by aquifer materials can also be significantly rate limited. The question to be addressed is: how important is rate-limited sorption at the field scale?

Numerous experimental and theoretical studies have shown that rate-limited sorption/desorption can cause nonideal transport. This nonideality can take the form of asymmetrical breakthrough curves exhibiting early breakthrough and tailing, as well as skewed, decelerating plumes. The impact of rate-limited sorption/desorption on transport is mediated by the relative magnitudes of the characteristic "reaction" time and the contaminant residence time.

Structured/Locally-Heterogeneous Porous Media (Physical Nonequilibrium)

Nonideal transport, often referred to as physical nonequilibrium, can result from a structured or heterogeneous flow domain at the macroscopic scale (10^{-3} to 10^{-1} m). The existence of regions of smaller hydraulic conductivity within the flow domain creates a spatially variable velocity field, with minimal flow and advection occurring through the small-conductivity domains. Due to the small advective flux, these domains act as sink/source components, with rate-limited diffusional mass transfer between the advective and nonadvective domains causing enhanced spreading of the solute front. These sink/source regions can take various forms, including the internal porosity of aggregates, dead-end pores, the bulk matrix of fractured media, and the small hydraulic conductivity microlayers or laminae typically found in aquifers of sedimentary origin.

The possibility that structured or locally heterogeneous porous media may influence flow and transport has long been recognized (see Brusseau and Rao, 1990 for a recent review). Miscible displacement studies performed with several types of structured media have clearly demonstrated that these structures can cause asymmetrical breakthrough curves, with early breakthrough and tailing. The impact of structured media on solute transport will depend on the nature of the medium, the characteristics of the solute, and the residence time.

Field-Scale Heterogeneity

The influence of spatially variable hydraulic conductivity on water flow and solute transport at the "field" scale has been a major research topic for many years. The impact of hydraulic-conductivity variability on transport is often discussed in terms of the "scale effect," wherein apparent longitudinal dispersivity values measured for field-scale transport are usually much larger than those measured for transport in packed columns. A description of how spatially variable hydraulic conductivities could cause enhanced solute dispersion was given in the early 1960s (Theis, 1963; Warren and Skiba, 1964). Since then, a tremendous amount of research has demonstrated that the additional dispersion, often called macro- or full-aquifer dispersion, observed for field-scale transport of nonreactive solute is primarily a result of field-scale hydraulic-conductivity variability (Schwartz, 1977; Gelhar et al., 1979; Molz et al., 1983).

Due to the heterogeneity of subsurface systems, it is logical to expect sorption to be spatially variable. Several field-scale investigations have shown that this is indeed the case. Depending on the form of the spatial variability, nonuniform sorption may cause nonideal transport, as demonstrated by several investigations cited in Table 3.1. In real situations, spatial variations in both hydraulic conductivity and sorption would be expected. The existence and nature of correlations between the two properties would be important for such cases.

Multiple, Coupled Processes

Given the nature of natural subsurface systems, it would be expected that more than one of the factors discussed above may contribute to nonideal solute transport. Understanding the influence of multiple, coupled processes on solute transport is critical to our ability to accurately simulate transport. This topic has received relatively limited attention to date. Examples of experiments wherein nonideal solute transport was caused by more than one factor are presented below.

CASE STUDIES OF MULTIFACTOR, MULTIPROCESS NONIDEAL TRANSPORT

Laboratory Experiments

The transport, in heterogeneous porous media, of reactive contaminants undergoing rate-limited sorption is one example of a coupled-process system. In this case, it is of interest to understand the relative contributions of rate-limited sorption and physical heterogeneity to observed nonideal transport (e.g., early breakthrough and tailing). Brusseau and colleagues (Brusseau et al., 1989a; Brusseau, 1991; Brusseau and Zachara, 1993; Hu and Brusseau, 1996) have investigated a number of systems wherein transport was influenced by rate-limited sorption and physical heterogeneity, and concluded that the relative contribution of these two factors to nonideal transport is controlled, in part, by contaminant distribution potential and characteristic mass-transfer times.

Table 3.1. Multifactor Analyses of Nonideal Transport of Reactive Solutes in Heterogeneous Porous Media.[a]

Author(s)	Factors[b]	Form[c]
Smith & Schwartz (1981)	$\Delta K + \Delta K_d$	SM2
Jury (1983)	$\Delta K + \Delta K_d$	SM2
Bahr (1986)	$\Delta K + \Delta K_d$ + RLS	MF/BTC
van der Zee & van Riemsdijk (1987)	$\Delta K + \Delta K_d$ + NLS	SM2
Garabedian et al. (1988)	$\Delta K + \Delta K_d$	SM2
Valocchi (1988)	ΔK + RLS	SM0-2
Brusseau (1989)	ΔK + MT + RLS	SM2
Brusseau et al. (1989)	ΔK + MT + RLS	MF/BTC
Dagan (1989)	$\Delta K + \Delta K_d$	SM1,2
Valocchi (1989)	$\Delta K + \Delta K_d$ + RLS	SM0-2
Cvetkovic & Shapiro (1990)	$\Delta K + \Delta K_d$ + RLS	MF/BTC
Andricevic & F.-Georgiou (1991)	$\Delta K + \Delta K_d$ + RLS	SM2
Brusseau (1991a)	$\Delta K + \Delta K_d$ + MT + RLS	MF/BTC
Destouni & Cvetkovic (1991)	$\Delta K + \Delta K_d$ + RLS	MF/BTC
Kabala & Sposito (1991)	$\Delta K + \Delta K_d$	SM1,2
Brusseau (1992)	$\Delta K + \Delta K_d$ + MT + RLS	MF/BTC
Brusseau & Zachara (1992)	$\Delta K + \Delta K_d$ + MT + RLS	MF/BTC
Schafer & Kinzelbach (1992)	$\Delta K + \Delta K_d$	MF/BTC
Selroos & Cvetkovic (1992)	ΔK + RLS	MF/BTC
Bosma et al. (1993)	$\Delta K + \Delta K_d$	SM2
Dagan & Cvetkovic (1993)	ΔK + RLS	SM0-3
Tompson (1993)	$\Delta K + \Delta K_d$ + NLS	SM1-3 + MF/BTC
Quinodoz & Valocchi (1993)	ΔK + RLS	SM1,2
Bosma et al. (1994)	$\Delta K + \Delta K_d$ + NLS	SM1,2
Burr et al. (1994)	$\Delta K + \Delta K_d$ + MT/RLS[d]	SM1,2
Cvetkovic & Dagan (1994a)	ΔK + NLS	SM1,2
Cvetkovic & Dagan (1994b)	ΔK + RLS	SM0,1
Rabideau & Miller (1994)	ΔK + RLS + NLS[e]	MF/BTC
Srivastava & Brusseau (1994)	$\Delta K + \Delta K_d$	MF/BTC
Bellin & Rinaldo (1995)	$\Delta K + \Delta K_d$	SM1,2
Selroos (1995)	ΔK + RLS	MF/BTC
Bosma & van der Zee (1995)	$\Delta K/\Delta K_d$[e] + NLS	SM2
Berglund (1995)	ΔK + NLS	MF/BTC
Hu et al. (1995)	$\Delta K + \Delta K_d$ + RLS	SM1-3
Harvey & Gorelick (1995)	$\Delta K + \Delta K_d$ + MT	MF/BTC
Berglund & Cvetkovic (1996)	ΔK + NLS	MF/BTC
Kong & Harmon (1996)	$\Delta K + \Delta K_d$ + MT	SM2
Miralles-Wilhelm & Gelhar (1996)	$\Delta K + \Delta K_d$	SM0-3
Bosma et al. (1996)	$\Delta K + \Delta K_d$ + NLS	SM1,2
Xu & Brusseau (1996)	ΔK_d + RLS + MT	SM2 + MF/BTC
Srivastava & Brusseau (1996)	$\Delta K + \Delta K_d$ + MT + RLS + NLS	SM1-3 + MF/BTC
Hu & Brusseau (1996)	$\Delta K + \Delta K_d$ + MT + RLS	MF/BTC
Brusseau & Srivastava (1997a,b)	$\Delta K + \Delta K_d$ + MT + RLS + NLS	SM1,2

[a] ΔK = Spatially variable hydraulic conductivity; ΔK_d spatially variable sorption; RLS = rate-limited sorption; MT = smaller-scale heterogeneity and associated mass transfer; NLS = nonlinear sorption.

[b] Focus of analysis: MF/BTC = mass flux (e.g., breakthrough curves); SM*n* = spatial moments, where *n* = moment number.

[c] Simulations were presented for either rate-limited sorption or mass transfer (not both simultaneously).

[d] Spatially variable sorption (with spatially variable conductivity and linear, instantaneous sorption) considered separately.

[e] Spatially variable hydraulic conductivity and spatially variable sorption considered separately.

[f] From Brusseau (1997).

One such study was reported by Brusseau and Zachara (1993), who investigated the transport of Co^{2+} in a column packed with layers of two media of differing hydraulic conductivities and sorption capacities. The sorption capacities of the two media differed by about a factor of 3. The asymmetrical breakthrough curve obtained for transport of a nonreactive tracer through the column demonstrated the effect of the physical heterogeneity and associated mass transfer on transport (see Figure 3.3a). The breakthrough curve obtained for transport of Co^{2+} exhibited tailing and was shifted to the left of a simulated curve obtained for ideal conditions (homogeneous porous media, instantaneous sorption), as shown in Figure 3.3b. The comparison reveals that transport of Co^{2+} through the heterogeneous porous medium was significantly nonideal.

The optimized curve obtained for the case of hydraulic-conductivity variability, sorption-capacity variability, rate-limited mass transfer between the two layers, and rate-limited sorption/desorption matched the experimental data quite well. In contrast, a simulated curve for the case of homogeneous sorption, hydraulic-conductivity variability, and rate-limited mass transfer and sorption did not match the data well. The influence of sorption variability had to be considered to accurately simulate the transport of Co^{2+}. This was also true for each of the other processes. Accurate simulation of the data could only be obtained when all of the nonideality factors were incorporated into the model.

Borden Field Experiment

An extensive field experiment was conducted at the Canadian Air Forces Base, Borden, Ontario, to investigate the transport of nonreactive tracers and sorbing organic solutes in a shallow, unconfined sand aquifer under natural-gradient conditions (Mackay et al., 1986a; Freyberg, 1986; Curtis et al., 1986; Roberts et al., 1986; Sudicky, 1986). Relatively well-defined initial conditions were achieved by the injection of a pulse of solution (12 m^3) containing two inorganic tracers (Cl^- and Br^-) and five organic chemicals (bromoform, tetrachloromethane, tetrachloroethene, 1,2-dichlorobenzene, and hexachloroethane). A dense, three-dimensional array of sampling points was used to obtain time-series data at selected points (i.e., breakthrough curves) as well as synoptic data on plume movement. During the course of the 2-year experiment the plumes of the various solutes traveled 10 to 60 m.

Significant nonideal transport was observed for the organic solutes. First, the velocities of the centers of mass of the plumes were temporally nonuniform in that the plumes decelerated with time (see Figure 3.2). This behavior is reflected in a temporal increase in effective retardation. Second, the longitudinal spreading observed for the organic-solute plumes was about three times larger than that of the nonreactive tracers, for an equivalent travel distance. Third, the breakthrough curves measured at selected monitoring points were asymmetrical. This is illustrated by the breakthrough curve measured for tetrachloroethene at a point 5 meters downgradient from the injection zone (see Figure 3.4a). A breakthrough curve representing ideal transport (homogeneous porous media; linear, instantaneous mass transfer) is shown for comparison. The degree of asymmetry was larger for the organic solutes than for the nonreactive tracers (compare Figures 3.4b and 3.4a).

The nonideal transport of the organic solutes has been attributed primarily to rate-limited sorption/mass transfer (Roberts et al., 1986; Goltz and Roberts, 1986, 1988; Ball et al., 1990; Quinodoz and Valocchi, 1993; Cvetkovic and Dagan, 1994b; Thorbjarnarson and Mackay, 1994). Sorption of organic compounds by the Borden aquifer material has been shown to be rate-limited (Curtis et al., 1986; Lee et al., 1988; Ball and Roberts; 1991a; Brusseau and Reid, 1991; Harmon and Roberts, 1994). How-

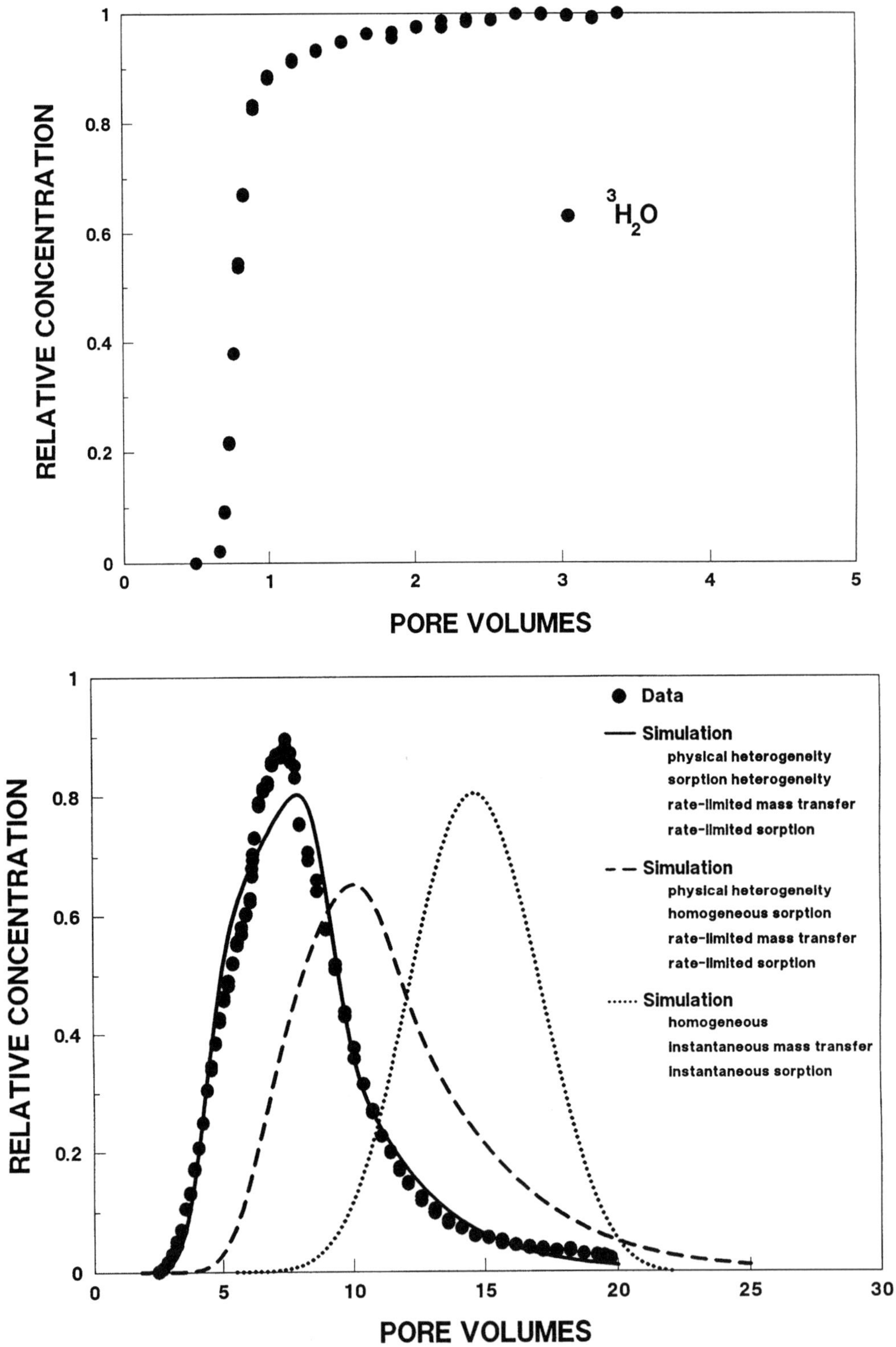

Figure 3.3. Transport of nonreactive and reactive solutes in a physically and chemically heterogeneous porous medium; (a) Tritiated water (data from Brusseau and Zachara, 1993), (b) Cobalt. Figure from Brusseau and Zachara (1993); copyright 1993, American Chemical Society. With permission.

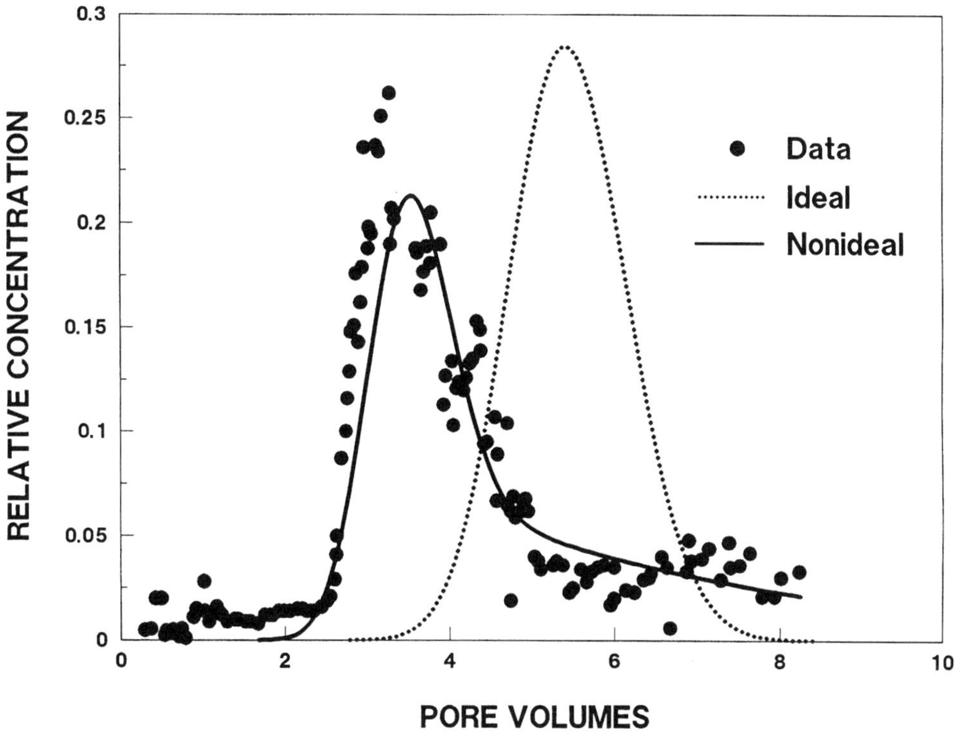

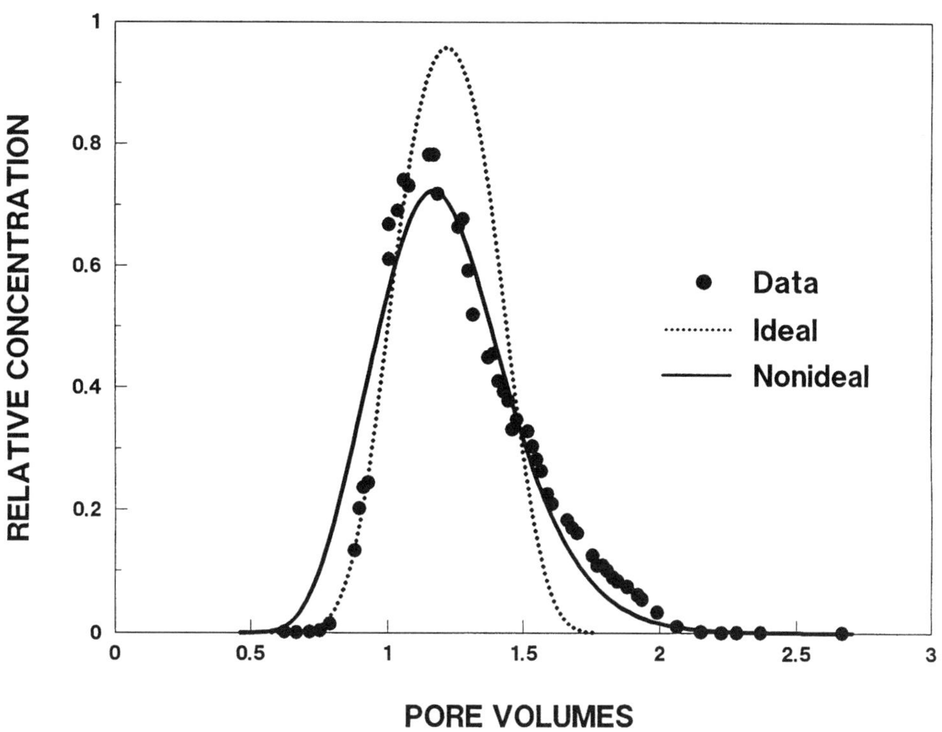

Figure 3.4. Breakthrough curves for transport of solutes during a natural-gradient field experiment; (a) Tetrachloroethene, (b) Chloride. Data from Roberts and Mackay, 1986, simulations from Brusseau, 1992, figure from Brusseau (1994); copyright 1994, American Geophysical Union. With permission.

ever, attempts to predict the field data using models based on rate-limited mass transfer alone, or in combination with hydraulic conductivity variability, have not been successful (Goltz and Roberts, 1988; Ball et al., 1990; Quinodoz and Valocchi, 1993).

It has also been shown that sorption of compounds such as tetrachloroethene for the Borden system is nonlinear (Ball and Roberts, 1991b) and spatially variable (Mackay et al., 1986b; Durant, 1986). Brusseau and Srivastava (1997a) were able to successfully predict the nonideal displacement and spreading behavior of the organic compounds measured during the Borden experiment by accounting for physical heterogeneity, sorption heterogeneity, and nonlinear sorption (see Figure 3.5). It was shown that rate-limited sorption/mass transfer had a minor impact on transport.

Cape Cod Field Experiment

A natural-gradient tracer experiment was performed in a shallow, unconfined sand and gravel aquifer located in Cape Cod, Massachusetts (Garabedian et al., 1988; Garabedian et al., 1991; LeBlanc et al., 1991). The transport of bromide, lithium, and molybdate was monitored with a dense array of multilevel samplers. The bromide solute plume had moved approximately 200 m by the 461-day sampling round, whereas the plumes for the reactive solutes had moved about half that distance.

Analysis of the field data revealed that the rate of displacement of the lithium plume decreased with time. Additionally, the longitudinal dispersion observed for lithium was more than 10 times larger than that determined for bromide (Garabedian et al., 1988). This nonideal transport was attributed to physical and geochemical heterogeneity, specifically a negative correlation between hydraulic conductivity and sorption capacity. The horizontal hydraulic conductivity is reported to vary by about one order of magnitude in the vertical direction. The sorption coefficient for lithium was observed to vary in the vertical by about a factor of two (Garabedian et al., 1988). Thus, it is quite possible that combined chemical and physical heterogeneity influenced lithium transport.

The possibility that rate-limited sorption contributed to the nonideal transport is supported by laboratory experiments reported by Wood et al. (1990). Their data showed that sorption of lithium by Cape Cod aquifer materials was significantly rate-limited. They suggested that rate-limited sorption was responsible for the nonideal transport observed during the field experiment. It is more likely, however, that lithium transport was influenced by several nonideality factors, including hydraulic-conductivity variability, sorption variability, rate-limited sorption, and nonlinear sorption (Brusseau and Srivastava, 1997b).

IDENTIFYING NONIDEAL TRANSPORT FACTORS

The transport and fate of many contaminants in subsurface systems can be influenced by several factors and processes, as illustrated above. Identification of the controlling factor/process in such systems is often difficult. Discrimination among various rate-limited and nonlinear processes is often based on inspection of the shape of breakthrough curves. However, this approach is confounded by the fact that similar effects are caused by several different factors and processes. This is illustrated in Figure 3.6, wherein similar behavior (asymmetrical breakthrough curves) is observed for solute transport controlled by different factors. The use of characterization experiments to help identify the process/factor controlling transport in a specific system will be discussed briefly below.

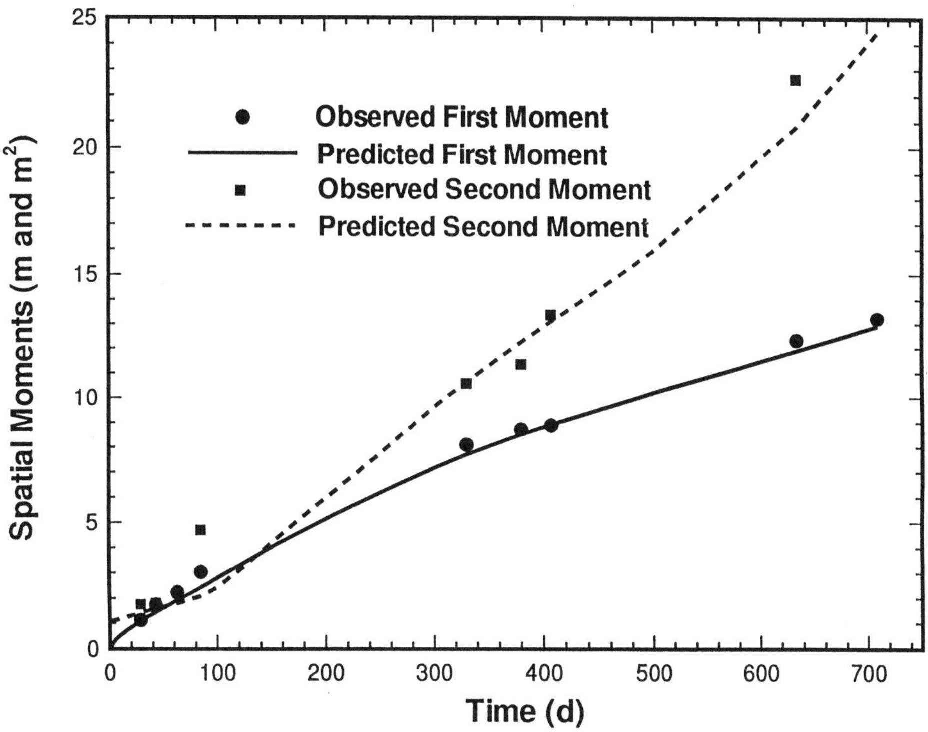

Figure 3.5. Predicted and measured plume displacement and spreading of tetrachloroethene for the Borden natural-gradient field experiment. Data from Roberts and Mackay, 1986; simulations and Figure from Srivastava and Brusseau, 1996. With permission.

Physical Heterogeneity and Physical Nonequilibrium

The use of nonreactive tracers to characterize the hydrodynamic characteristics of a porous medium is an ubiquitous component of solute transport experiments. The observation of an asymmetrical breakthrough curve for a nonreactive tracer is an indication that the flow field is not uniform, most likely due to a heterogeneous porous medium. In such systems, the possible existence of "immobile" water is of interest. Because retention and release of solute by immobile water is mediated by diffusive mass transfer, experiments based on perturbing the solute residence time or the characteristic diffusion time can be used to identify the existence of immobile water.

One often-used method to alter residence time is to conduct experiments at different pore-water velocities. Breakthrough curves for a nonreactive tracer will be influenced by changes in velocity only if transport is measurably influenced by zones of immobile water (assuming longitudinal diffusion is negligible). The impact of velocity on transport of a nonreactive tracer in an aggregated soil is illustrated in Figure 3.7, where early breakthrough is seen to increase with an increase in velocity. The observation of similar breakthrough curves for widely different velocities would indicate the absence of measurable amounts (with respect to influencing transport) of immobile water.

Recently, Brusseau (1993) proposed the use of different-sized solutes (diffusive tracers) as another approach for characterizing heterogeneous porous media. Solutes with different diffusion coefficients will experience differential rates of mass transfer and, therefore, should exhibit dissimilar transport (e.g., different degrees of asymmetry)

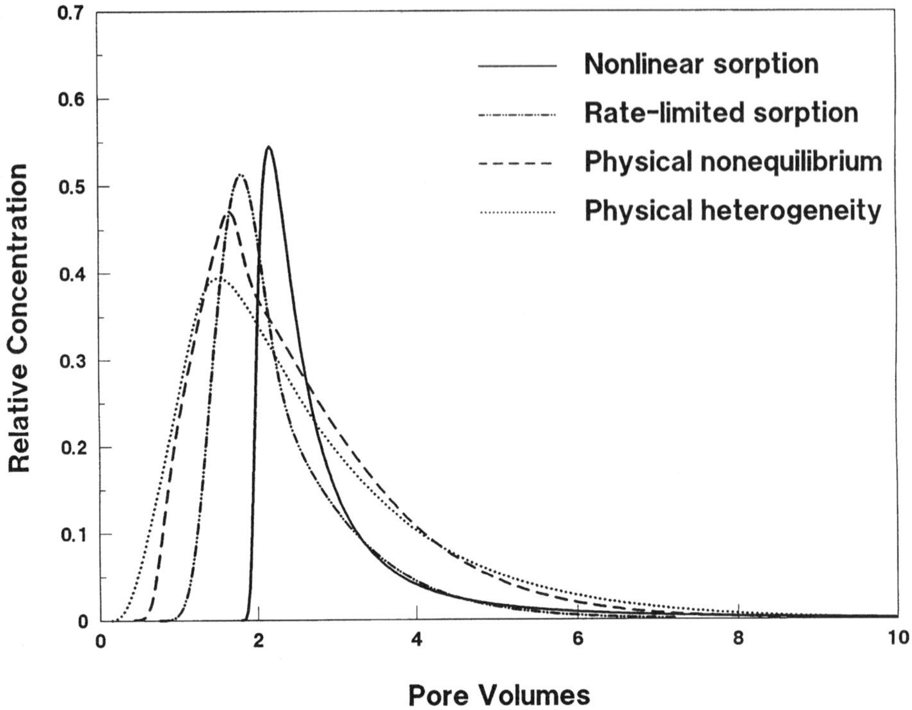

Figure 3.6. Breakthrough curves for transport influenced by different factors.

for systems influenced by diffusive mass transfer. This is illustrated in Figure 3.8a, where the breakthrough curve for the larger solute (pfba) exhibits greater tailing than does that of the smaller solute for transport in an aggregated soil. Conversely, breakthrough curves for the two solutes are identical for transport in a homogeneous soil (Figure 3.8b).

A third approach for identifying the existence of immobile water is based on the flow-interruption method of Brusseau et al. (1989b; 1997). For typical conditions, flow interruption will not have a discernible influence on transport in heterogeneous porous media as long as all fluid in the system is participating significantly in flow. Conversely, a flow interruption (or change in flow velocity) will result in an effluent-concentration perturbation when transport is influenced by immobile water. This is illustrated in Figure 3.9, wherein is presented a measured breakthrough curve for transport of ^3H_2O through an aggregated medium. A flow interruption of 1 hour caused a measurable perturbation in the effluent concentration profile.

In summary, nonideal transport (enhanced spreading, asymmetrical breakthrough curves) can be caused by both physical heterogeneity and physical nonequilibrium. In many cases, the shapes of breakthrough curves and magnitudes of spreading may be similar, which complicates differentiation between the two factors. In reality, physical nonequilibrium and physical heterogeneity are limiting cases of a continuum, with the magnitude of nonflowing fluid as the primary criterion of differentiation. Furthermore, it is well known that the definition of "nonflowing" fluid is system- and condition-dependent. Thus, differentiation between physical nonequilibrium and physical heterogeneity is an artificial construct used to aid description of solute transport. However, as shown by the examples presented in this section, it is possible for practical purposes to use characterization methods to identify systems in which significant

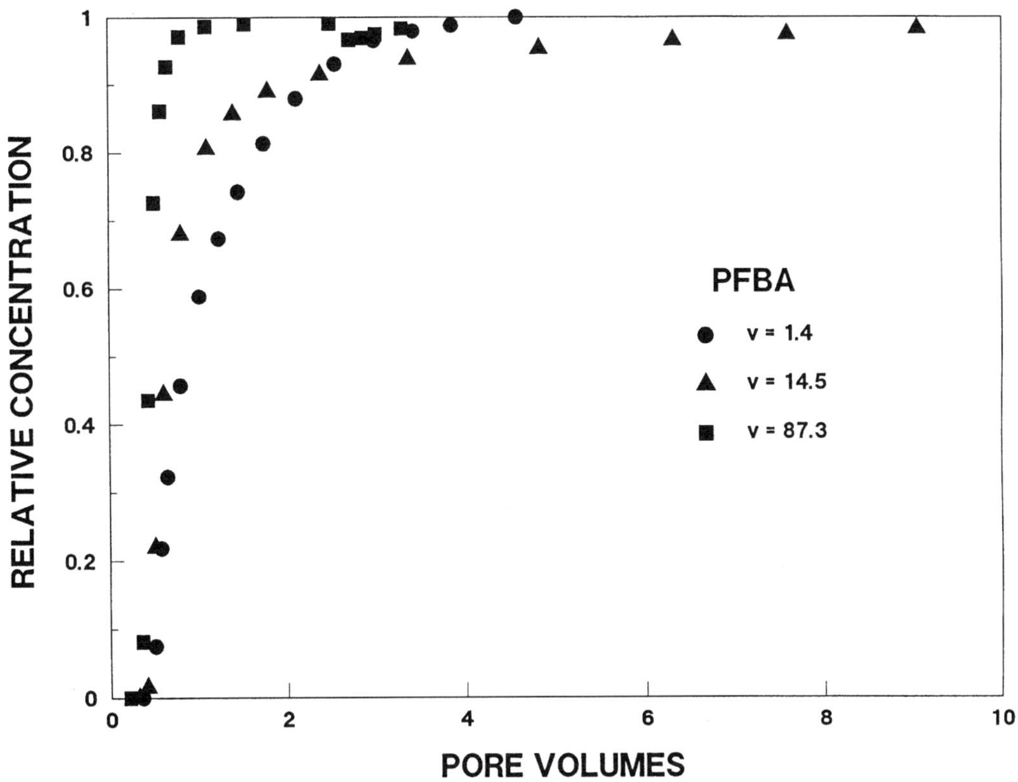

Figure 3.7. The transport of a nonreactive tracer in an aggregated soil at three pore-water velocities. Data from Brusseau et al., 1994. With permission.

volumes of nonflowing fluid exist and, concomitantly, for which diffusive mass transfer may be important.

Reactive Transport Processes

Characterization experiments similar to those discussed above can be conducted to examine the processes influencing the transport of reactive (sorbing) solutes. For example, residence-time related experiments (multiple velocities, flow interruption) can be used to discriminate between the impacts of nonlinear and rate-limited sorption. Flow interruption can also be used to differentiate between transformation reactions and irreversible sorption reactions (Brusseau et al., 1997).

SIMULATING MULTIFACTOR NONIDEAL TRANSPORT

During the past 20 years, mathematical models have been developed to account for nonideal transport. For example, models that account for spatially variable hydraulic conductivity are now common and form the basis for most advanced modeling efforts. However, despite an awareness that transport of reactive solutes is influenced by many processes and factors, model development has until recently followed a reductionist approach wherein the influence of a single factor is studied in isolation. Considering the complexity of field-scale systems, it is clear that mathematical mod-

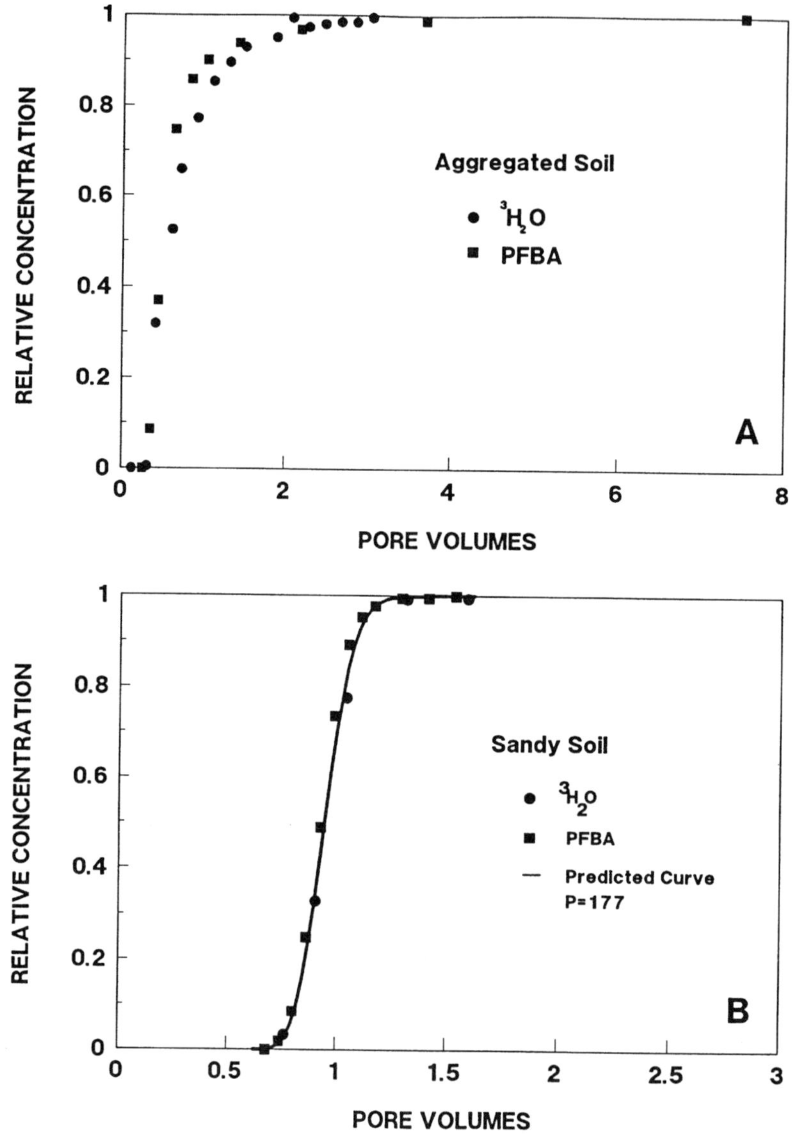

Figure 3.8. The influence of solute size on solute transport; (a) Aggregated soil, (b) Sandy (homogeneous) soil. Figures from Brusseau, 1993; copyright 1993, American Geophysical Union. With permission.

els incorporating multiple nonideality factors are needed to simulate transport of reactive solutes.

The development of mathematical models that incorporate coupled processes is integral to advancing the field. The application of models that account for only a single source of nonideality to systems affected by more than one factor yields lumped parameters. Values of these lumped parameters can usually be obtained only by calibration, and will be valid only for the specific set of conditions for which they were obtained. In addition, these lumped parameters can not supply process-discrete information and, thus, are useless for elucidating the relative contributions of various nonideality factors to total nonideality. Such informa-

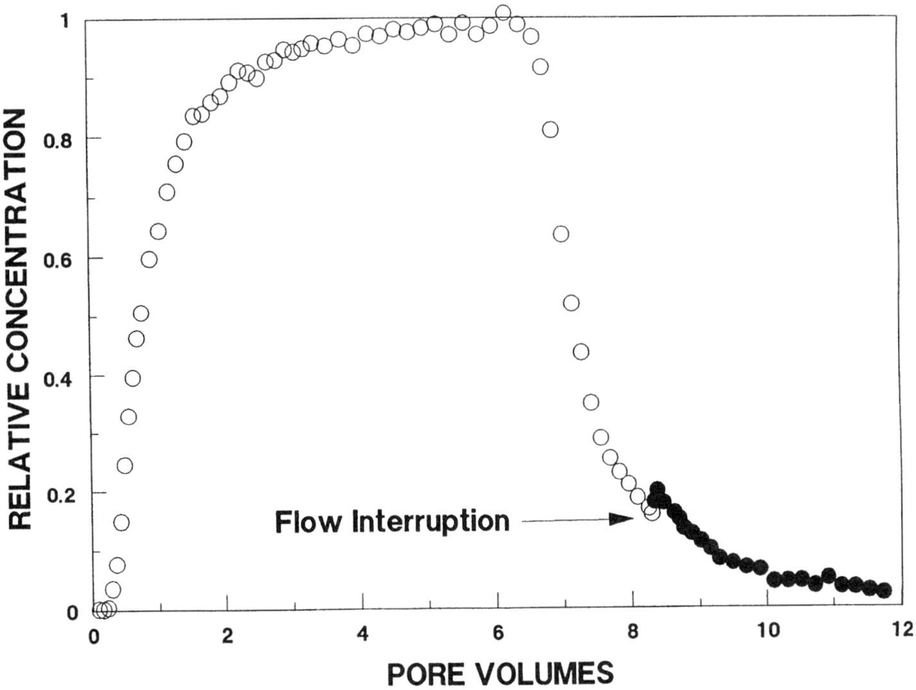

Figure 3.9. The influence of flow interruption on solute transport in an aggregated soil. Figure from Brusseau et al., 1997. With permission.

tion can only be obtained with the use of a model that accounts explicitly for the existence of multiple nonideality factors.

The recognition of the need for multifactor models has led to the development of several models that account explicitly for two or more nonideality factors. A summary of previous "multifactor" analyses of reactive-solute transport in heterogeneous porous media is provided in Table 3.1 (note that multifactor in this case means two or more factors). The influence of spatially variable hydraulic conductivity coupled with spatially variable sorption has received the most attention to date, followed by coupled hydraulic conductivity variability and rate-limited sorption. Until recently (e.g., Brusseau, 1991; Srivastava and Brusseau, 1996; Brusseau and Srivastava, 1997a,b), at most three factors were considered simultaneously in the analyses of nonideal transport. It is expected that additional multifactor transport models will be developed in the coming years.

Only a very few of the models listed in Table 3.1 have been evaluated by using them to simulate measured data. This step of "ground-truthing" a mathematical model has never been more critical or more difficult as we move to complex, multifactor models. Furthermore, the usual practice of fitting or calibrating a model to a set of measured data is becoming more and more uncertain as we develop models with more and more parameters. In such cases, the only truly valid way of evaluating a model's performance is to attempt to predict the measured data, with values for all parameters obtained independently.

Successful multifactor model evaluations based on independent predictions of measured reactive-solute transport data have been reported by Brusseau and colleagues. Brusseau et al. (1989a) and Hu and Brusseau (1996) used a one-dimensional multifactor model to predict the transport of reactive solutes through columns packed with structured porous media. The field-scale transport of several reactive solutes was successfully predicted by Brusseau (1992) and Brusseau and Srivastava (1997a), who used one-, two-, and three-dimensional multifactor nonideality models to simulate breakthrough curves and plume displacement (e.g., see Figures 3.4a and 3.5).

SUMMARY

Sufficient laboratory and field data exist to conclude that the transport of reactive solutes in the subsurface will usually be nonideal, and that the nonideal behavior may often be caused by more than one factor or process. In such cases, an accurate representation of solute transport is contingent upon understanding the contribution of each factor and process to transport. This information can be obtained by implementing advanced characterization experiments, as discussed herein. Accurate simulation of solute transport in these situations requires the use of a mathematical model that accounts for each factor and process influencing transport. A critical aspect of future research in this field is the testing of these complex models using field data, and using the models to help interpret transport and fate at the field scale.

REFERENCES

Andricevic, R. and E. Foufoula-Georgiou. Modeling kinetic non-equilibrium using the first two moments of the residence time distribution, *Stochastic Hydrol. Hydraul.*, 5:155–171, 1991.

Bahr, J.M. Applicability of the Local Equilibrium Assumption in Mathematical Models for Groundwater Transport of Reacting Solutes, Ph.D. dissertation, Stanford University, Palo Alto, CA, 1986.

Ball, W.P. and P.V. Roberts. Long-term sorption of halogenated organic chemicals by aquifer material. 1. Equilibrium, *Environ. Sci. Technol.*, 25(7):1223–1236, 1991a.

Ball, W.P. and P.V. Roberts. Long-term sorption of halogenated organic chemicals by aquifer material. 2. Intraparticle diffusion, *Environ. Sci. Technol.*, 25(7):1237–1249, 1991b.

Ball, W.P., M.N. Goltz, and P.V. Roberts. Sorption rate studies with halogenated organic chemicals and sandy aquifer materials—Implications for solute transport and groundwater remediation, in *Proceedings of the 1990 Environmental Engineering Conference, American Society of Civil Engineers,* New York, 1990, pp. 307–313.

Bellin, A. and A. Rinaldo. Analytical solutions for transport of linearly adsorbing solutes in heterogeneous formations, *Water Resour. Res.*, 31(6):1505–1511, 1995.

Berglund, S. The effect of langmuir sorption on pump-and-treat remediation of a stratified aquifer, *J. Contam. Hydrol.*, 18(3):199–218, 1995.

Berglund, S. and V. Cvetkovic. Contaminant displacement in aquifers: Coupled effects of flow heterogeneity and nonlinear sorption, *Water Resour. Res.*, 32(1):23–32, 1996.

Bosma, W.J.O. and S. van der Zee. Dispersion of a continuously injected, nonlinearly adsorbing solute in chemically or physically heterogeneous porous formations, *J. Contam. Hydrol.*, 18(3):181–196, 1995.

Bosma, W.J.P., A. Bellin, S. van der Zee, and A. Rinaldo. Linear equilibrium adsorbing transport in physically and chemically heterogeneous porous formations: 2. Numerical results, *Water Resour. Res.*, 29(12):4031–4043, 1993.

Bosma, W.J.P., S. van der Zee, A. Bellin, and A. Rinaldo. Instantaneous injection of a nonlinearly adsorbing solute in a heterogeneous aquifer, in *Transport and Reactive Processes in Aquifers,* Dracos, T.H. and F. Staufer (Eds.), Proc. IAHR/AIRH Symp., Zurich, Switzer., 11–15 April, 1994, A.A. Balkema, Rotterdam, 1994, pp. 411–417.

Bosma, W.J.P., S. van der Zee, and C.J. van Duijin. Plume development of a nonlinearly adsorbing solute in heterogeneous porous formations, *Water Resour. Res.,* 32(6):1569–1584, 1996.

Brusseau, M.L. Nonequilibrium Sorption: Investigation at Microscopic to Megascopic Scales, Ph.D. Dissertation, University of Florida, Gainesville, FL, 1989.

Brusseau, M.L. Application of a multi-process nonequilibrium-sorption model to solute transport in a stratified porous medium, *Water Resour. Res.,* 27(4):589–595, 1991.

Brusseau, M.L. Transport of rate-limited sorbing solutes in heterogeneous porous media: Application of a one-dimensional multi-factor nonideality model to field data, *Water Resour. Res.,* 28(9):2485–2497, 1992.

Brusseau, M.L. The influence of solute size, pore water velocity, and intraparticle porosity on solute dispersion and transport in soil, *Water Resour. Res.,* 29(4):1071–1080, 1993.

Brusseau, M.L. Transport of reactive contaminants in heterogeneous porous media, *Rev. Geophysics,* 32(3):285–314, 1994.

Brusseau, M.L. Nonideal transport of reactive solutes in heterogeneous porous media: Analyzing field data with mathematical models, *J. Hydrol.,* (in review), 1997.

Brusseau, M.L. and P.S.C. Rao. Sorption nonideality during organic contaminant transport in porous media, *CRC Crit. Rev. Environ. Control,* 19:33–99, 1989.

Brusseau, M.L. and P.S.C. Rao. Modeling solute transport in structured soils: A review, *Geoderma,* 46:169–192, 1990.

Brusseau, M.L. and M.E. Reid. Nonequilibrium sorption of organic chemicals by low-organic-carbon aquifer materials, *Chemosphere,* 22(3–4):341–350, 1991.

Brusseau, M.L. and J. Zachara. Transport of Co^{2+} in a physically and chemically heterogeneous porous medium, *Environ. Science Technol.,* 27(9):1937–1939, 1993.

Brusseau, M.L. and R. Srivastava. Nonideal transport of reactive solutes in heterogeneous porous media: 2. Quantitative analysis of the Borden natural-gradient experiment, *J. Contam. Hydrol.,* 1997a (in press).

Brusseau, M.L. and R. Srivastava. Nonideal transport of reactive solutes in heterogeneous porous media: Analysis of Cape Cod natural-gradient experiment, 1997b (in review).

Brusseau, M.L., R.E. Jessup, and P.S.C. Rao. Modeling the transport of solutes influenced by multi-process nonequilibrium, *Water Resour. Res.,* 25(9):1971–1988, 1989a.

Brusseau, M.L., P.S.C. Rao, R.E. Jessup, and J.M. Davidson. Flow interruption: A method for investigating sorption nonequilibrium. *J. Contam. Hydrol.,* 4(3):223–240, 1989b.

Brusseau, M.L., Z. Gerstl, D. Augustijn, and P.S.C. Rao. Simulating solute transport in an aggregated soil with the dual-porosity model: Measured and optimized parameter values, *J. Hydrol.,* 163:187–193, 1994.

Brusseau, M.L., M. Hu, and R. Srivastava. Using flow interruption to identify factors causing nonideal contaminant transport, *J. Contam. Hydrol.,* 24(3/4):205–219, 1997.

Burr, D.T., E.A. Sudicky, and R.L. Naff. Nonreactive and reactive solute transport in three-dimensional heterogeneous porous media: Mean displacement, plume spreading, and uncertainty, *Water Resour. Res.,* 30(3):791–815, 1994.

Crittenden, J.C., N.J. Hutzler, D.G. Geyer, J.L. Oravitz, and G. Friedman. Transport of organic compounds with saturated groundwater flow: Model development and parameter sensitivity, *Water Resour. Res.,* 22(3):271–284, 1986.

Curtis, G.P., P.V. Roberts, and M. Reinhard. A natural gradient experiment on solute transport in a sand aquifer. 4. Sorption of organic solutes and its influence on mobility, *Water Resour. Res.,* 22(12):2059–2067, 1986.

Cvetkovic, V.D. and A.M. Shapiro. Mass arrival of sorptive solute in heterogeneous porous media, *Water Resour. Res.,* 26(9):2057–2067, 1990.

Cvetkovic, V.D. and G. Dagan. Effect of nonlinear sorption on contamination/remediation of heterogeneous aquifers, in *Transport and Reactive Processes in Aquifers,* Dracos, T.H. and F. Staufer (Eds.), Proc. IAHR/AIRH Symp., Zurich, Switzer., 11–15 April, 1994, A.A. Balkema, Rotterdam, 1994a, pp. 411–417.

Cvetkovic, V.D. and G. Dagan. Transport of kinetically sorbing solute by steady random velocity in heterogeneous porous formations, *J. Fluid Mech.,* 265:189–215, 1994b.

Dagan, G. *Flow and Transport in Porous Formations,* Springer-Verlag, New York, 1989.

Dagan, G. and V. Cvetkovic. Spatial moments of a kinetically sorbing solute plume in a heterogeneous aquifer, *Water Resour. Res.,* 29(12):4053–4061, 1993.

Destouni, G. and V. Cvetkovic. Field scale mass arrival of sorptive solute into the groundwater, *Water Resour. Res.,* 27(6):1315–1325, 1991.

Durant, M.G. Sorption of Organic Solutes in the Borden Aquifer: Evidence of Spatial Variability. M.S. Thesis, Stanford University, Palo Alto, CA, 1986.

Freyberg, D.L. A natural gradient experiment on solute transport in a sand aquifer. 2. Spatial moments and the advection and dispersion of nonreactive tracers, *Water Resour. Res.,* 22(12):2031–2046, 1986.

Garabedian, S.P., L.W. Gelhar, and M.A. Celia. Large-scale dispersive transport in aquifers: Field experiments and reactive transport theory, Report No. 315, Dept. of Civil Engin., Mass. Inst. Technol., Cambridge, MA, 1988.

Garabedian, S.P., D.R. LeBlanc, L.W. Gelhar, and M.A. Celia. Large-scale natural-gradient tracer test in sand and gravel, 2, analysis of spatial moments for a nonreactive tracer, *Water Resour. Res.,* 27(5):911–924, 1991.

Gelhar, L.W., A.L. Gutjahr, and R.L. Naff. Stochastic analysis of macrodispersion in a stratified aquifer, *Water Resour. Res.,* 15(6):1387–1397, 1979.

Goltz, M.N. and P.V. Roberts. Interpreting organic solute data from a field experiment using physical nonequilibrium models, *J. Contam. Hydrol.,* 1(1/2):77–93, 1986.

Goltz, M.N. and P.V. Roberts. Simulations of physical nonequilibrium solute transport models: Application to a large-scale field experiment, *J. Contam. Hydrol.,* 3(1):37–63, 1988.

Harmon, T.C. and P.V. Roberts. Comparison of intraparticle sorption and desorption rates for a halogenated alkene in a sandy aquifer material, *Environ. Sci. Technol.,* 28(9):1650–1660, 1994.

Harvey, C.F. and S.M. Gorelick. Temporal moment-generating equations: Modeling transport and mass transfer in heterogeneous aquifers, *Water Resour. Res.,* 31(8):1895–1912, 1995.

Hu, B.X., F. Deng, and J.H. Cushman. Nonlocal reactive transport with physical and chemical heterogeneity: Linear nonequilibrium sorption with random K_d, *Water Resour. Res.,* 31(9):2239–21252, 1995.

Hu, Q. and M.L. Brusseau. Transport of rate-limited sorbing solutes in an aggregated porous medium: A multiprocess non-ideality approach, *J. Contamin. Hydrol.,* 24(1):53–73, 1996.

Jury, W.A. Chemical transport modeling: Current approaches and unresolved issues, Chap. 4 in *Chemical Mobility and Reactivity in Soil Systems,* Soil Science Soc. America Spec. Pub. 11, Madison, WI, 1983.

Kabala, Z.J. and G. Sposito. A stochastic model of reactive solute transport with time-varying velocity in a heterogeneous aquifer, *Water Resour Res.,* 27(3):341–350, 1991.

Kong, D. and T.C. Harmon. Using the multiple cell balance method to solve the problem of two-dimensional groundwater flow and contaminant transport with nonequilibrium sorption, *J. Contam. Hydrol.,* 23(4):285–300, 1996.

LeBlanc, D.R., S.P. Garabedian, K.M. Hess, L.W. Gelhar, R.D. Quadri, K.G. Stollenwerk, and W.W. Wood. Large-scale natural-gradient tracer test in sand and gravel, 1, experimental design and observed tracer movement, *Water Resour. Res.,* 27(5):895–910, 1991.

Lee, L.S., P.S.C. Rao, M.L. Brusseau, and R.A. Ogwada. Nonequilibrium sorption of organic contaminants during flow through columns of aquifer materials, *Environ. Toxic. Chem.,* 7:779–793, 1988.

Mackay, D.M., D.L. Freyberg, P.V. Roberts, and J.A. Cherry. A natural gradient experiment on solute transport in a sand aquifer. 1. Approach and overview of plume movement, *Water Resour. Res.,* 22(12):2017–2029, 1986a.

Mackay, D.M., W.P. Ball, and M.G. Durant. Variability of aquifer sorption properties in a field experiment on groundwater transport of organic solutes: methods and preliminary results, *J. Contam. Hydrol.,* 1(1/2):119–132, 1986b.

Miralles-Wilhelm, F. and L.W. Gelhar. Stochastic analysis of sorption macrokinetics in heterogeneous aquifers, *Water Resour. Res.,* 32(6):1541–1549, 1996.

Molz, F.J., O. Guven, and J.G. Melville. An examination of scale-dependent dispersion coefficients, *Groundwater,* 21(6):715–725, 1983.

Quinodoz, H.A. and A.J. Valocchi. Stochastic analysis of the transport of kinetically sorbing solutes in aquifers with randomly heterogeneous hydraulic conductivity, *Water Resour. Res.,* 29(9):3227–3240, 1993.

Rabideau, A.J. and C.T. Miller. Two-dimensional modeling of aquifer remediation influenced by sorption nonequilibrium and hydraulic conductivity heterogeneity, *Water Resour. Res.,* 30(5):1457–1470, 1994.

Roberts, P.V. and D.M. Mackay. A natural gradient experiment on solute transport in a sand aquifer. Technical Report No. 292, Dept. Civil Engin., Stanford Univ., Palo Alto, CA, 1986.

Roberts, P.V., M.N. Goltz, and D.M. Mackay. A natural gradient experiment on solute transport in a sand aquifer. 3. Retardation estimates and mass balances for organic solutes, *Water Resour. Res.,* 22(12):2047–2058, 1986.

Schafer, W. and W. Kinzelbach. Stochastic modeling of in situ bioremediation in heterogeneous aquifers, *J. Contam. Hydrol.,* 10(1):47–73, 1992.

Schwartz, F.W. Macroscopic dispersion in porous media: Controlling factors, *Water Resour. Res.,* 13(4)743–752, 1977.

Selroos, J. Temporal moments for nonergodic solute transport in heterogeneous aquifers, *Water Resour. Res.,* 31(7):1705–1712, 1995.

Selroos, J. and V. Cvetkovic. Modeling solute advection coupled with sorption kinetics in heterogeneous formations, *Water Resour. Res.,* 28(5):1271–1278, 1992.

Smith, L, and F.W. Schwartz. Mass Transport 2. Analysis of uncertainty in prediction, *Water Resour. Res.,* 17(2):351–369, 1981.

Srivastava, R. and M.L. Brusseau. Effect of physical and chemical heterogeneities on aquifer remediation by pump and treat, in Proceedings of the American Chemical Society National Meetings, Environmental Chemistry Division, San Diego, CA, March 13–18, 1994. Vol. 34(1), American Chemical Society, Washington, DC, 1994, pp. 262–265.

Srivastava, R. and M.L. Brusseau. Nonideal transport of reactive solutes in heterogeneous porous media: 1. Numerical model development and moments analysis, *J. Contam. Hydrol.,* 24(2):117–142, 1996.

Sudicky, E.A. A natural gradient experiment on solute transport in a sand aquifer. Spatial variability of hydraulic conductivity and its role in the dispersion process, *Water Resour. Res.,* 22(12), 2069–2082, 1986.

Theis, C.V. Hydrologic phenomena affecting the use of tracers in timing groundwater flow, in Radioisotopes in Hydrology, Inter. Atomic Energy Agency, Vienna, 1963, pp. 193–207.

Thorbjarnarson, K.W. and D.M. Mackay. A forced-gradient experiment on solute transport in the Borden aquifer 3. Nonequilibrium transport of the sorbing organic compounds, *Water Resour. Res.,* 30(2):401–419, 1994.

Tompson, A.F.B. Numerical simulation of chemical migration in physically and chemically heterogeneous porous media, *Water Resour. Res.,* 29(11):3709–3726, 1993.

Valocchi, A.J. Theoretical analysis of deviations from local equilibrium during sorbing solute transport through idealized stratified aquifers, *J. Contam. Hydrol.,* 2(3):191–207, 1988.

Valocchi, A.J. Spatial moment analysis of the transport of kinetically adsorbing solutes through stratified aquifers, *Water Resour. Res.,* 25(2):273–279, 1989.

van der Zee, S. and W.H. van Riemsdijk. Transport of reactive solute in spatially variable soil systems, *Water Resour. Res.,* 23(11):2059–2069, 1987.

Warren, J.E. and F.F. Skiba. Macroscopic dispersion, *Soc. Pet. Eng. J.*, 215–230, 1964.

Wood, W.W., T.F. Kraemer, and P.P. Hearn. Intragranular diffusion: An important mechanism influencing solute transport in clastic aquifers? *Science*, 247:1569–1572, 1990.

Xu, L. and M.L. Brusseau. Semianalytical solution for solute transport in porous media with multiple spatially variable reaction processes, *Water Resour. Res.*, 32(7):1985–1991, 1996.

CHAPTER FOUR

Coupling of Retention Approaches to Physical Nonequilibrium Models

L. Ma and H.M. Selim

INTRODUCTION

With the increasing need in land reclamation and alternative human activities benign to our environment, it is essential to quantitatively understand the behavior of environmental pollutants (e.g., heavy metals and pesticides) in soils. The study of environmental contaminants in soils is complicated by soil heterogeneity, however. One type of heterogeneity is chemical and is due to nonuniform composition of soil constituents. Another type of heterogeneity is physical and is caused by nonuniform geometry of the soil pore space.

Although physical and chemical nonequilibrium phenomena have been investigated for the last 30 years, quantitative description of these phenomena is less successful due to the complexity of the soil system. Most mathematical models apply basic physical and chemical principles without justifying their applicability. One of the earliest chemical nonequilibrium models is the kinetic one-adsorption site model (van Genuchten et al., 1974). Later, Selim et al. (1976) developed a two-adsorption site model based on the affinity of soil components to the solutes. A multireaction retention model was further developed and applied to heavy metal transport (Amacher et al., 1988; Selim et al., 1989). A chemical nonequilibrium model accounting for maximum adsorption capacity (second-order reaction model) was also proposed by Selim and Amacher (1988) and was later modified by Ma and Selim (1994a).

Several physical nonequilibrium models were based on the assumption of simplified pore size distributions. The simplest classification of soil pores is to divide soil pores as mobile and immobile regions, which in turn divides water into mobile and immobile phases. Such a mobile-immobile concept was first applied by Deans (1963), Gottschlich (1963), and Coats and Smith (1964), and was later developed by Skopp and Warrick (1974) and van Genuchten and Wierenga (1976) in soil science. The model of van Genuchten and Wierenga (1976) has been widely cited in the literature. However, coupling of physical and chemical nonequilibrium was not introduced until recently. Selim and Amacher (1988) coupled a one-site kinetic second-order approach with the mobile-immobile model. Brusseau et al. (1989) introduced a multireaction chemical nonequilibrium model into the mobile-immobile model. Selim and Ma (1995) modified the mobile-immobile model and found that the modified mobile-immobile approach offered improved predictions from that of the original model. This modified mobile-immobile model was further tested by Ma and Selim (1997).

Another type of physical nonequilibrium model is the two-flow domain approach, where flow distribution in the soil media was grouped into two mobile phases. One domain phase has greater flow velocity than the other (Skopp et al., 1981). Ma and Selim (1995) applied this concept to explain the bimodal breakthrough curve of tritium in several soils. Other examples of the two-flow domain models are Jarvis et al. (1991a,b), Chen and Wagenet (1992), and Gerke and van Genuchten (1993). Further classification of soil pores results in various physical nonequilibrium models, such as capillary bundle model (Lindstrom and Boersma, 1971), multiple flow domain model (Wilson et al., 1992; Hutson and Wagenet, 1995), and continuous flow distribution model (Skopp and Gardner, 1992). However, these physical nonequilibrium models have not been coupled with chemical nonequilibrium approaches.

MOBILE-IMMOBILE PHYSICAL NONEQUILIBRIUM MODELS (MIM)

This class of models was proposed in an effort to represent flow heterogeneity during solute transport in soils. A commonly used physical nonequilibrium model is the mobile-immobile (MIM) approach, where soil water was conceptualized as two fractions, one is mobile and the other is immobile. Retention mechanism of solutes from both mobile and immobile water was assumed to be identical. Solute transfer between the two water fractions was described by an empirical first-order diffusion with α as the mass transfer coefficient. Assuming a representative soil volume V (cm^3) with total soil mass P (g), we have a continuity equation for one-dimensional solute transport:

$$\frac{\Delta(M/V)}{\Delta t} = -\frac{\Delta J}{\Delta z} \tag{4.1}$$

where J is the solute flux ($\mu g\ cm^{-2}\ h^{-1}$) in the z direction, M is total solute mass in volume V (μg), and t is time (h). J can be expressed as:

$$J = -D_m \frac{\partial C_m}{\partial z} + q\, C_m \tag{4.2}$$

where q is water flux in the mobile region ($cm\ h^{-1}$), C_m is solute concentration in the mobile water ($\mu g/mL$), and D_m is the dispersion coefficient in the mobile water ($cm^2\ h^{-1}$). The total amount of solute (M) can be written as:

$$M = M_m + M_{im} + N_m + N_{im} \tag{4.3}$$

where M_m and M_{im} are amounts of solutes adsorbed from the mobile and immobile water (μg), respectively. N_m and N_{im} are the amounts of solutes present in the mobile and immobile water (μg), respectively. Assuming constant V, q, and D_m, the continuity equation can be rewritten as:

$$\frac{1}{V}\left[\frac{\Delta M_m}{\Delta t} + \frac{\Delta M_{im}}{\Delta t} + \frac{\Delta N_m}{\Delta t} + \frac{\Delta N_{im}}{\Delta t}\right] = -q\frac{\Delta C_m}{\Delta z} + \frac{\Delta}{\Delta z}\left[D_m \frac{\partial C_m}{\partial z}\right] \tag{4.4}$$

Let W_m and W_{im} be the water volume associated with the mobile and immobile phase (cm^3), respectively, we thus have:

$$N_m = W_m C_m; \quad \text{and} \quad N_{im} = W_{im} C_{im} \tag{4.5}$$

thus,

$$\frac{1}{V}\left[\frac{\Delta N_m}{\Delta t} + \frac{\Delta N_{im}}{\Delta t}\right] = \theta_m \frac{\Delta C_m}{\Delta t} + \theta_{im} \frac{\Delta C_{im}}{\Delta t} \tag{4.6}$$

where θ_m and θ_{im} are the mobile and immobile water contents ($cm^3\ cm^{-3}$) with $\theta = \theta_m + \theta_{im}$. When $\Delta \rightarrow 0$, the generalized convection-dispersion equation (CDE) is derived:

$$\frac{1}{V}\left[\frac{\partial M_m}{\partial t} + \frac{\partial M_{im}}{\partial t}\right] + \theta_m \frac{\partial C_m}{\partial t} + \theta_{im} \frac{\partial C_{im}}{\partial t} = -q \frac{\partial C_m}{\partial z} + D_m \frac{\partial^2 C_m}{\partial z^2} \tag{4.7}$$

This equation (4.7) can be used in two ways. One way is to divide the total soil mass (P) into P_m, soil mass in direct contact with the mobile water, and P_{im}, soil mass in direct contact with the immobile water. Therefore, a sorbed concentration S'_m based on P_m and S'_{im} based on P_{im} can be defined as,

$$M_m = S'_m P_m; \quad \text{and} \quad M_{im} = S'_{im} P_{im} \tag{4.8}$$

therefore,

$$f\rho \frac{\partial S'_m}{\partial t} + (1-f)\rho \frac{\partial S'_{im}}{\partial t} + \theta_m \frac{\partial C_m}{\partial t} + \theta_{im} \frac{\partial C_{im}}{\partial t} = -q \frac{\partial C_m}{\partial z} + D_m \frac{\partial^2 C_m}{\partial z^2} \tag{4.9}$$

where $f = P_m/P$ and $\rho = P/V$. Equation 4.9 was used by van Genuchten and Wierenga (1976) where a total mass balance can be expressed as,

$$\text{Total Mass} = \theta_m C_m + \theta_{im} C_{im} + \rho [f S'_m + (1-f) S'_{im}] \tag{4.10}$$

Another way of applying Eq. 4.7 is that of Selim and Ma (1995), where S_m and S_{im} are defined based on total soil mass (P) in the soil volume V, such that,

$$M_m = S_m P; \quad \text{and} \quad M_{im} = S_{im} P \tag{4.11}$$

Therefore, we have a modified CDE for the mobile-immobile model:

$$\rho \frac{\partial S_m}{\partial t} + \rho \frac{\partial S_{im}}{\partial t} + \theta_m \frac{\partial C_m}{\partial t} + \theta_{im} \frac{\partial C_{im}}{\partial t} = -q \frac{\partial C_m}{\partial z} + D_m \frac{\partial^2 C_m}{\partial z^2} \tag{4.12}$$

This modified CDE (Eq. 4.12) or that of Eq. 4.7 are used along with the following first-order mass transfer equation:

$$\theta_{im} \frac{\partial C_{im}}{\partial t} + \rho \frac{\partial S_{im}}{\partial t} = \alpha (C_m - C_{im}) \tag{4.13}$$

where α is referred to as a mass transfer coefficient (h^{-1}).

ESTIMATION OF θ_m AND α

Application of the MIM model requires two additional parameters over the traditional convective-dispersive equation (CDE). That is, F, fraction of mobile water (θ_m/θ_{im}), and α, the mass transfer coefficient between the mobile and immobile water phases. These two additional model parameters are difficult to estimate for most applications. It is desirable to obtain such parameters based on experimental measurements. Although this MIM is conceptual in nature, no experimental methodologies were proposed to estimate model parameters. Consequently, the MIM approach is often used to describe experimental results (model calibration) and model parameters were thus optimized.

Several examples are available in the literature on various parameter estimation efforts associated with the MIM approach. A common way of differentiating θ_m and θ_{im} is through curve-fitting (Li and Ghodrati, 1994; van Genuchten and Wierenga, 1977). De Smedt and Wierenga (1984) found θ_m=0.853θ is unique for unsaturated glass beads with diameters in the neighborhood of 100 µm. However, the definition of macropores and micropores is vague and somewhat arbitrary (Beven and Germann, 1982; Smettem and Kirkby, 1990). An experimental method to estimate the two porosities is to measure soil water content at some arbitrary water tension (ψ). Smettem and Kirkby (1990) used water content at ψ of 14 cm as the matching point between the inter-aggregate (macro-) and the intra-aggregate (micro-) porosity by examining the ψ-θ soil moisture characteristic curve. Jarvis et al. (1991b) estimated macroporosity from specific yield under water tension of 100 cm. Other water tensions used to differentiate macropores from micropores are 3 cm (Luxmoore, 1981), 10 cm (Wilson et al., 1992), 20 cm (Selim et al., 1987), and 80 cm (Nkedi-Kizza et al., 1982). A list of water tensions used by different authors was provided by Chen and Wagenet (1992). The equivalent diameters at these water tensions range from 10 to 10000 µm based on capillary flow. Other methods used to estimate visible macropores from fauna activity include dye tracing (Trojan and Linden, 1992; Booltink and Bouma, 1991), dental plaster casting (Wang et al., 1994), resin impregnation (Singh et al., 1991), macropore tracing on clear plastic sheet (Ela et al., 1992; Logsdon et al., 1990), X-ray computed tomography or CT-scan (Warner et al., 1989), and in situ photography of soil profile (Edwards et al., 1988). The macroporosity estimated using the above methods often ignores the continuity of macropore as well as their distribution with soil depth, and may not be suitable for transport modeling (Chen et al., 1993; Munyankusi et al., 1994; McCoy et al., 1994).

Another experimental measurement of θ_m is based on the following mass balance equation:

$$\theta C = \theta_m C_m + \theta_{im} C_{im} \tag{4.14}$$

When α is small enough to assume C_{im} = 0 and C_m = C_o (input concentration) after some elapsed (infiltration) time, an approximate equation is obtained,

$$\theta_m = \theta \frac{C}{C_m} = \theta \frac{C}{C_o} \tag{4.15}$$

Applications of this method can be found in Clothier et al. (1992) and Jaynes et al. (1995). By assuming that the tracer concentration in the mobile water phase (C_m) equals the input concentration (C_o), Jaynes et al. (1995) derived the following formula from Eqs. 4.13 and 4.14,

$$\ln\left(1 - \frac{C}{C_o}\right) = -\frac{\alpha}{\theta_{im}}\,t + \ln\left(\frac{\theta_{im}}{\theta}\right) \tag{4.16}$$

α and θ_{im} can be estimated by plotting ln(1-C/C_o) versus elapsed time. However, the assumption of C_m=C_o associated with this method is questionable and may not be correct as long as α>0. Slightly different from the approach of Jaynes et al. (1995) and Clothier et al. (1992), Goltz and Roberts (1988) estimated the fraction of mobile water as the ratio of velocity calculated from hydraulic conductivity to the velocity measured from tracer experiment.

Chen et al. (1993) estimated functional macropores by monitoring water content in the soil during water infiltration. They assumed that drainage from a saturated soil took place in three stages: initial drainage from macropores only, middle drainage from both macropore and micropores, and final drainage from micropores only. The demarcation lines between the three stages were based on some calculated characteristic time. The initial drainage was used to estimate macroporosity and macro-hydraulic conductivity. Another alternative way to study macropore continuity is to measure air permeability at certain water tension (Roseberg and McCoy, 1990). Nevertheless, the determination of θ_m is arbitrary. In fact, the concepts of macropore and micropore are relative and change with experimental conditions, such as flow velocity (van Genuchten and Wierenga, 1977; Nkedi-Kizza et al., 1983; Watson and Luxmoore, 1986, Li and Ghodrati, 1994). Nkedi-Kizza et al. (1983) found a decrease in *F* with flow velocity. On the other hand, van Genuchten and Wierenga (1977) found an increase in *F* with flow velocity. Li and Ghodrati (1994) observed no consistent relationship between *F* and flow velocity in a NO_3 study. This inconsistency may be explained by differing in soil type, flow velocity range, and applicability of the mobile-immobile approach.

Another important parameter is α. van Genuchten and Dalton (1986) and van Genuchten (1985) derived an estimation of α based on soil geometry:

$$\alpha_c = \frac{n\,(1-F)\,\theta\,D_e}{a_e^2} \tag{4.17}$$

where n is a geometry factor, a_e is an average effective diffusion length, D_e is the effective diffusion coefficient. Eq. 4.17 has been widely used to estimate α in modeling solute transport in porous media (Selim and Amacher, 1988; Goltz and Roberts, 1988). Another formula for α for spherical aggregates was derived by Rao et al. (1980a,b):

$$\alpha_t = \frac{D_e\,(1-F)\theta}{a^2}\,\alpha^* \tag{4.18}$$

α^* is a time-dependent variable. The α values for cubic aggregates were also derived by Rao et al. (1982) using an equivalent spherical radius such as a=0.6203l where l is the length of the side of a cubic. Such an approach was utilized by Selim and Ma (1995) to describe atrazine transport in an aggregated clay soil.

For a two-flow domain physical nonequilibrium model where the soil may be characterized by two flow regions having velocities (V_A, V_B) and water contents (θ_A, θ_B), Skopp et al. (1981) derived an α value from Duguid and Lee (1977) for steady-state flow condition:

$$\alpha_S = \frac{2\ K_B\ \theta_A\ (V_A - V_B)^2}{g\ a\ a_p\ \pi} \tag{4.19}$$

where g is the acceleration due to gravity (cm h^{-2}), a is the aggregate size (cm), a_p is the inter-aggregate pore size (cm), and K_B is the conductivity of the soil matrix (cm h^{-1}). This equation assumed that mass transfer is a function of pressure gradient between the two flow domains. Applying these two methods to the data of Anderson and Bouma (1977), Skopp et al. (1981) found that α estimated from the Duguid and Lee's model (Eq. 4.19) was two orders of magnitude smaller than that from Eq. 4.17.

Estimated α based on the aggregate geometry does not always provide adequate prediction of breakthrough curves (BTCs). Moreover, α is equally affected by experimental conditions. Additional adjustment of α to individual experimental conditions is usually made by fitting the mobile-immobile model to experimental data. Fitted α values increases with flow velocity (De Smedt and Wierenga, 1984; De Smedt et al., 1986; van Genuchten and Wierenga, 1977; Kookana et al., 1993; Li and Ghodrati, 1994). This velocity dependence is perhaps reasonable since turbulent mixing may occur at high flow velocities. In a laboratory study with glass beads under both saturated and unsaturated conditions, De Smedt and Wierenga (1984) found that α (day^{-1}) is linearly related to the mobile phase velocity v_m ($=q/\theta_m$, cm/day) as $\alpha=0.042v_m+2.2$. They also obtained $\alpha=0.02\ v_m$ in a later study with sand under unsaturated water conditions (De Smedt et al., 1986). When convective water transfer between flow domains significantly contributes to solute transfer, large α values may be expected (Gerke and van Genuchten, 1993; Jarvis et al., 1991a). Steenhuis et al. (1990) assumed that solute transfer between flow domains is mainly through convection transfer of water. Thus, when convective mass transfer exists, the empirical first-order diffusion transfer equation (Eq. 4.13) may not be sufficient. In addition to flow velocity, pore connectiveness also affects α values. Greater pore convectivity results in smaller α's (Skopp and Gardner, 1992).

GENERALIZED CHEMICAL RETENTION MODELS

This class of models is based on the assumption that adsorption affinities are different for the various constituents of the soil, and are represented by a system of consecutive and concurrent reactions. Figure 4.1 shows a schematic diagram of a general chemical nonequilibrium model, where C is solute concentration in soil solution (µg/mL), and S_e is the amount of solute retained by the soil matrix (µg/g soil) and is in equilibrium with C. The sorbed phases S_1 and S_2 are in direct contact with C and are governed by two types of kinetic reactions (µg/g soil), S_3 is a consecutive adsorption component from S_2 (µg/g soil), and S_{irr} refers to the irreversible adsorption sites (µg/g soil). Associated parameters are as follows: K_d is the equilibrium constant; k_1 and k_2 are the forward and backward reaction rate coefficients associated with S_1; k_3 and k_4 are the respective coefficients for S_2; k_5 and k_6 are the coefficients for S_3. In addition, k_{irr} is the rate coefficient for the irreversible reaction.

Nonlinear Multireaction Models (MRM)

In the nonlinear multireaction model, adsorption sites are assumed to be unlimited on the soil matrix. Thus, the reaction rates are only functions of solute concentration in soil solution. The chemical reactions governing the various retention mechanisms are (Ma and Selim, 1994b),

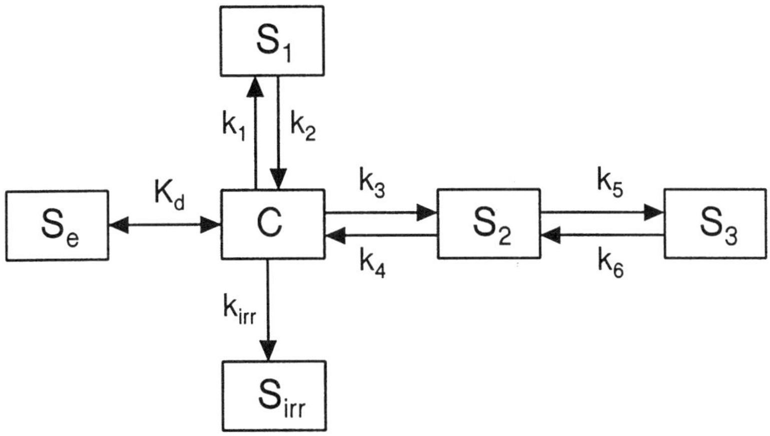

Figure 4.1. A schematic diagram of the proposed chemical nonequilibrium concept.

$$S_e = K_d \left(\frac{\theta}{\rho} \right) C^n \tag{4.20}$$

$$\frac{\partial S_1}{\partial t} = k_1 \left(\frac{\theta}{\rho} \right) C^n - k_2 \, S_1 \tag{4.21}$$

$$\frac{\partial S_2}{\partial t} = k_3 \left(\frac{\theta}{\rho} \right) C^n - k_4 \, S_2 - k_5 S_2 + k_6 S_3 \tag{4.22}$$

$$\frac{\partial S_3}{\partial t} = k_5 \, S_2 - k_6 \, S_3 \tag{4.23}$$

$$\frac{\partial S_{irr}}{\partial t} = k_{irr} \, C \tag{4.24}$$

where n is the reaction order (dimensionless), θ is the volumetric water content (cm^3 cm^{-3}) and ρ is the soil bulk density ($g\ cm^{-3}$). The term K_d refers to the equilibrium distribution coefficient ($cm^3\ g^{-1}$) and k_1 to k_6 and k_{irr} have units of h^{-1}.

The above equations formulate a generalized model with several possible processes. Depending on the purpose of a specific study and/or the data available, selected processes are chosen. For example, Selim and Ma (1995) developed an atrazine retention and transport model based on experimental evidences, which includes only K_d, k_3, k_4, and k_5. Selim et al. (1998) included only K_d, k_1 and k_2 in their metolachlor studies. Recently, Ma and Selim (1997) conducted a comprehensive comparison between models including various combinations of reaction processes. Table 4.1 provides a list of the model formulations and their description of a set of batch results. Each formulation was simultaneously fitted to a set of kinetic batch adsorption results from different initial concentrations (Ma and Selim, 1994a,b). Figure 4.2 shows an example of the fitted results for an initial atrazine concentration of 29.45 μg/mL. Equivalent model description was obtained from F3, F6, F7, F8, and F9. The one-site model (F1) provided the worst atrazine description. Although F2, F4, and F5 well predicted concentration decrease with reaction time, the predicted trend appears more linear than that measured. Table 4.2 lists the total sum of squared errors for predicting eight soil column breakthrough results (Ma and Selim, 1997) and Figure 4.3 shows examples of their prediction of atrazine BTCs. Generally, MRM is less sensitive to the number of model parameters and the predicted BTCs are indistinguishable.

Second-Order Models (SOM)

Another type of model formulation is the second-order model, which assumes that there are limited solute adsorption sites in the soil. Therefore, the reaction rates are dependent on solute concentration in solution as well as the amount of vacant (or available) adsorption sites on matrix surfaces. For clarity, similar symbols as MRM were used to identify similar mechanisms in the mathematical formulations although the units may be different as illustrated below. Denoting ϕ as the amount of sites available for solute adsorption, the associated retention mechanisms were (Ma and Selim, 1994a,b):

$$S_e = K_d \, \theta \, C \phi \tag{4.25}$$

$$\frac{\partial S_1}{\partial t} = k_1 \, \theta \, C \phi - k_2 \, S_1 \tag{4.26}$$

$$\frac{\partial S_2}{\partial t} = k_3 \, \theta \, C \phi - k_4 \, S_2 - k_5 \, S_2 + k_6 \, S_3 \tag{4.27}$$

$$\frac{\partial S_3}{\partial t} = k_5 \, S_2 - k_6 \, S_3 \tag{4.28}$$

$$\frac{\partial S_{irr}}{\partial t} = k_{irr} \, C \tag{4.29}$$

Here ϕ is related to the sorption capacity (S_{max}) by:

$$S_{max} = \phi + S_e + S_1 + S_2 \tag{4.30}$$

where ϕ and S_{max} are the unoccupied and total sorption sites on soil surfaces, respectively (μg solute per g soil). In addition, S_{max} was considered as an intrinsic soil property and is time-invariant. The unit for K_d is $cm^3 \, \mu g^{-1}$, k_1 and k_3 have a derived unit of $cm^3 \, \mu g^{-1} \, h^{-1}$; k_2, k_4, k_5, k_6, and k_{irr} are assigned with units of h^{-1}.

Table 4.3 lists the nine model formulations and model description for a set of batch kinetic adsorption results as used for MRM and Figure 4.4 shows an example of fitted results for initial atrazine concentration of 29.45 μg/mL. Again, F3, F6, F7, F8, and F9 offered the best description of atrazine adsorption kinetics. In Figure 4.5 are examples of predicted atrazine BTCs. As shown in Figure 4.5B, the differences among model formulations are more obvious when flow interruption was introduced. Compared to

Table 4.1. Goodness-of-Fit, Model Parameter Estimations, and 95% Confidence Intervals of MRM for Describing Atrazine Batch Adsorption Kinetics in a Sharkey Soil.[a]

Formulation	r^2	rmse	k_1	k_2	k_3	k_4	k_5	k_6	K_d	k_{irr}
			h^{-1}						(dimensionless)	h^{-1}
F1	0.980	0.951	0.43448±0.12866	0.42068±0.13236	—[c]	—	—	—	—	—
F2	0.997	0.351	—	—	—	—	—	—	0.76139±0.03084	0.00069±0.00007
F3	0.999	0.216	0.00290±0.00049	0.00389±0.00101	—	—	—	—	0.70865±0.02265	—
F4	0.998	0.282	0.53970±0.07277	0.67723±0.10191	—	—	—	—	—	0.00064±0.00006
F5	0.998	0.305	—	—	0.54002±0.07806	0.67256±0.10885	0.00150±0.00020	—	—	—
F6	0.999	0.189	0.00333±0.00135	0.01166±0.00756	—	—	—	—	0.69594±0.02403	0.00040±0.00013
F7	0.999	0.190	0.59922±0.07860	0.81326±0.12293	0.00252±0.00044	0.00332±0.00094	—	—	—	—
F8	0.999	0.190	—	—	0.60283±0.07693	0.81193±0.11974	0.00340±0.00068	0.00335±0.00095	—	—
F9	0.999	0.212	—	—	0.00394±0.00165	0.00855±0.00694	0.00181±0.00165	—	0.69727±0.02705	—

[a] After Ma and Selim, 1997.
[b] r^2 = coefficient of determination; rmse = root mean square error; values after ± are 95% confidence intervals.
[c] Not included in the formulation.

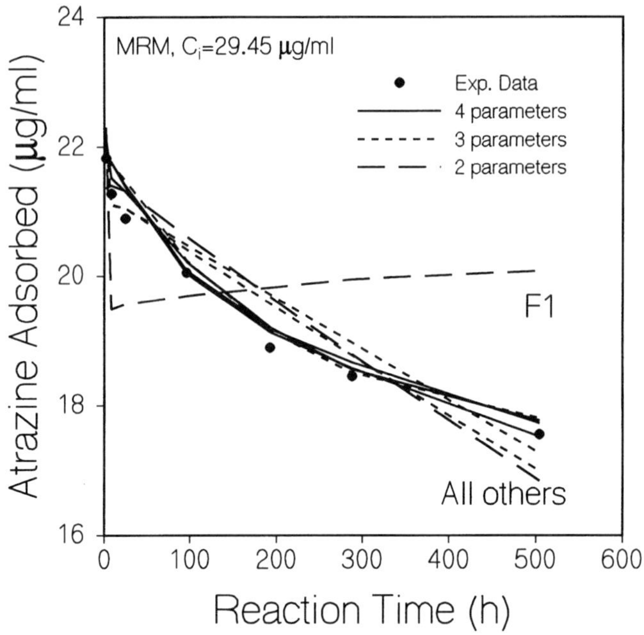

Figure 4.2. Atrazine concentration in soil solution versus reaction time in batch experiments for initial concentration of 29.45 μg/mL. Model was fitted across all the initial concentrations (2.95, 5.91, 11.94, 17.73, 23.48, and 29.45 μg/mL) simultaneously with MRM. Simulated results for other concentrations were not shown (after Ma and Selim, 1997).

MRM, SOM is more sensitive to the number of model parameters and generally provided better atrazine BTC predictions (Table 4.2).

COUPLING OF CHEMICAL EQUILIBRIUM MODELS TO MIM

In modeling reactive solutes, physical nonequilibrium models are commonly coupled with chemical retention mechanisms. Chemical retention is described by a series of kinetic reactions as described in earlier sections. Each reaction is featured by several rate coefficients or equilibrium constants. However, because of the uncertainty of the chemical characteristics of the two soil regions, it is difficult to justify different rate coefficients associated with each region. A commonly used approach is to assume the same rate coefficients to both dynamic and stagnant soil regions (Selim and Amacher, 1988; van Genuchten and Wierenga, 1976; Selim et al., 1987; Gaston and Selim, 1990; Selim and Ma, 1995). This assumption is reasonable if the two regions only differ in aggregate size (or pore size). The simplest chemical retention mechanism are the equilibrium isotherms. Based on the MRM approach (Eq. 4.20), we have

$$S_m = K_d\left(\frac{\theta_m}{\rho}\right)C_m{}^n \tag{4.31}$$

$$S_{im} = K_d\left(\frac{\theta_{im}}{\rho}\right)C_{im}{}^n \tag{4.32}$$

Table 4.2. Total Sum of Squared Errors (Over All Eight Soil Columns) of Predicted Atrazine Breakthrough Curves (BTCs).[a]

Formulation		Models					
		Multireaction Models (MRM)			Second-Order Model (SOM)		
MRM-α_t		MRM	MRM-MIM-α_c	MRM-MIM-α_t	SOM	SOM-MIM-α_c	SOM-MIM-α_t
F1	k_1,k_2	35.90	36.07	33.53	9.33	3.23	4.99
F2	K_d,k_{irr}	41.42	41.11	38.67	10.27	5.42	5.00
F3	k_1,k_2,K_d	41.03	41.20	38.45	4.17	4.81	1.87
F4	k_1,k_2,k_{irr}	39.34	39.61	37.07	7.67	4.15	3.83
F5	k_3,k_4,k_5	39.15	39.52	36.90	4.41	3.67	2.09
F6	k_1,k_2,k_{irr},K_d	40.75	40.99	38.19	5.16	4.59	2.26
F7	k_1,k_2,k_3,k_4	39.56	40.09	37.35	2.35	4.16	1.67
F8	k_3,k_4,k_5,k_6	39.57	40.09	37.35	3.81	4.05	1.83
F9	k_3,k_4,k_5,K_d	40.69	40.99	38.15	3.97	4.56	1.78

[a] After Ma and Selim 1997.

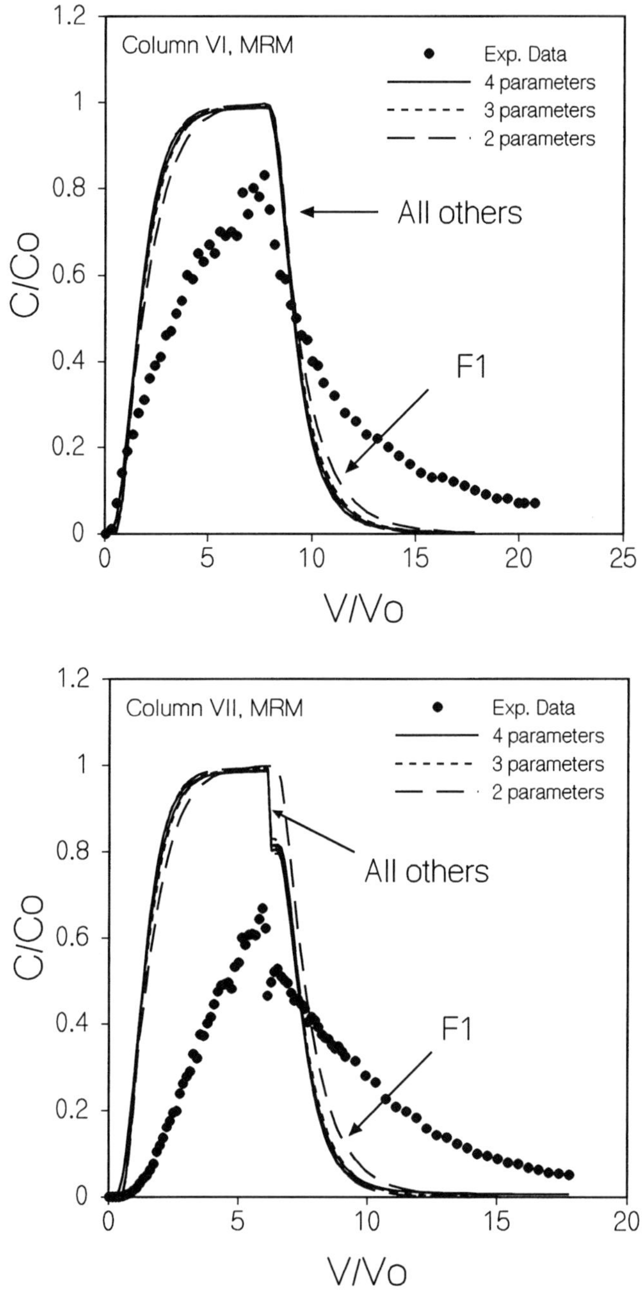

Figure 4.3. Example comparison of MRM for predicting atrazine BTCs in a Sharkey soil. The Freundlich (n=0.7635) was obtained from batch isotherm, and kinetic reaction coefficients were derived from Figure 4.2 and are listed in Table 4.1. (A) 10 cm column packed with 4–6 mm soil aggregates, no flow interruption. (B) 15 cm column packed with 2–4 mm soil aggregates, 16 d flow interruption after pulse input (after Ma and Selim, 1997).

Table 4.3. Goodness-of-Fit, Model Parameter Estimations, and 95% Confidence Intervals of SOM for Describing Atrazine Batch Adsorption Kinetics in a Sharkey Soil.[a,b]

Formulation	r^2	rmse	k_1 $(\mu g/mL)^{-1}$ h^{-1}	k_2 h^{-1}	k_3 $(\mu g/mL)^{-1}$ h^{-1}	k_4 h^{-1}	k_5 h^{-1}	k_6 h^{-1}	K_d $(\mu g/mL)^{-1}$	k_{irr} h^{-1}
F1	0.984	0.826	0.00828±0.00231	0.32759±0.09885	—	—	—	—	—	—
F2	0.998	0.311	—	—	—	—	—	—	0.01779±0.00081	0.00063±0.00007
F3	0.999	0.185	0.00007±0.00001	0.00347±0.00092	—	—	—	—	0.01638±0.00057	—
F4	0.999	0.246	0.01019±0.00132	0.54368±0.08017	—	—	—	—	—	0.00058±0.00006
F5	0.999	0.252	—	—	0.01015±0.00133	0.53694±0.08018	0.00129±0.00016	—	—	—
F6	0.999	0.155	0.00008±0.00003	0.00965±0.00521	—	—	—	—	0.01611±0.00054	0.00033±0.00012
F7	0.999	0.159	0.01134±0.00137	0.66550±0.09407	0.00007±0.00001	0.00302±0.00082	—	—	—	—
F8	0.999	0.125	—	—	0.01133±0.00108	0.65586±0.07352	0.00302±0.00045	0.00354±0.00073	—	—
F9	0.999	0.167	—	—	0.00011±0.00004	0.00947±0.00643	0.00163±0.00099	—	0.01604±0.00062	—

a After Ma and Selim 1997.

b r^2 = coefficient of determination; rmse = root mean square error; values after ± are 95% confidence intervals.

c Not included in the formulation.

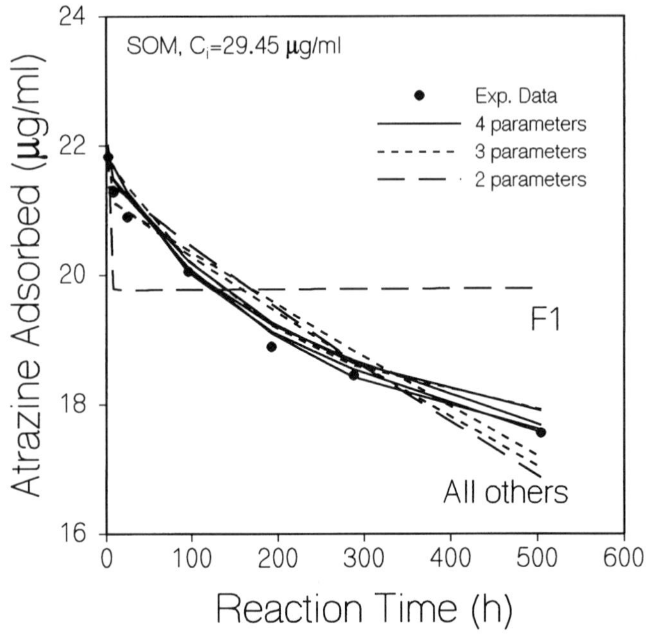

Figure 4.4. Atrazine concentration in soil solution versus reaction time in batch experiments for initial concentration of 29.45 µg/mL. Model was fitted across all the initial concentrations (2.95, 5.91, 11.94, 17.73, 23.48, and 29.45 µg/mL) simultaneously with SOM. Simulated results for other concentrations were not shown (after Ma and Selim, 1997).

Taking the derivatives of S_m and S_{im}, we have:

$$\frac{\partial S_m}{\partial t} = K_d \, n \, C_m^{\,n-1} \left(\frac{\theta_m}{\rho} \right) \frac{\partial C_m}{\partial t} \tag{4.33}$$

$$\frac{\partial S_{im}}{\partial t} = K_d \, n \, C_{im}^{\,n-1} \left(\frac{\theta_{im}}{\rho} \right) \frac{\partial C_{im}}{\partial t} \tag{4.34}$$

Similarly, Based on the SOM (Eq. 4.25), we have,

$$S_m = \frac{K_d \theta_m C_m S_{max}}{1 + K_d \theta_m C_m + K_d \theta_{im} C_{im}} \tag{4.35}$$

$$S_{im} = \frac{K_d \theta_{im} C_{im} S_{max}}{1 + K_d \theta_m C_m + K_d \theta_{im} C_{im}} \tag{4.36}$$

The above equations can be substituted into Eqs. 4.12 and 4.13 and are commonly solved using numerical schemes. Figures 4.6 and 4.7 are examples of the equilibrium models in describing atrazine transport breakthrough curves (BTCs). The reaction order n in the MRM model and S_{max} in the SOM model were derived from batch adsorption isotherms of Ma and Selim (1994b). The prediction curves are based on K_d values

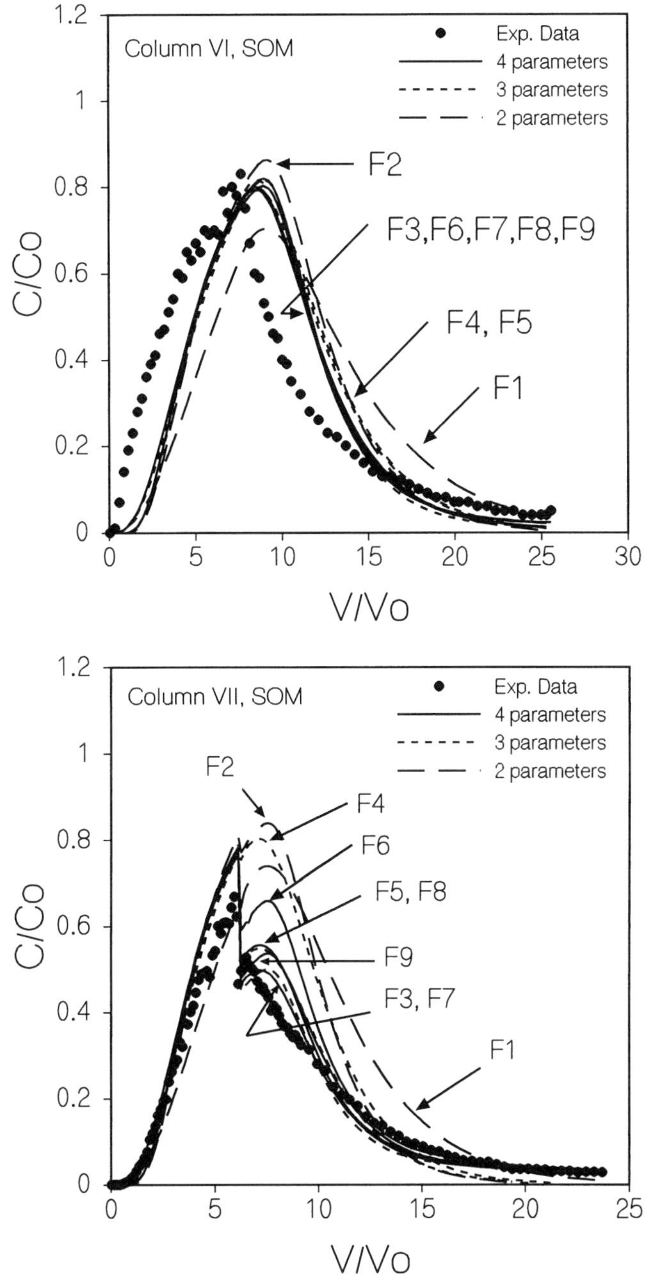

Figure 4.5. Example comparison of SOM for predicting atrazine BTCs in a Sharkey soil. The adsorption maximum (Smax=184.62 μg/g soil) was obtained from batch isotherm, and kinetic reaction coefficients were derived from Figure 4.4 and are listed in Table 4.3. (A) 10 cm column packed with 4–6 mm soil aggregates, no flow interruption. (B) 15 cm column packed with 2–4 mm soil aggregates, 16 d flow interruption after pulse input (after Ma and Selim, 1997).

from batch kinetic studies. When there is no flow interruption, the equilibrium model can describe the BTCs adequately (Figure 4.6). However, both MRM and SOM are less capable of describing the BTCs with flow interruptions (Figure 4.7). Generally, better model predictions were obtained with SOM than with the MRM.

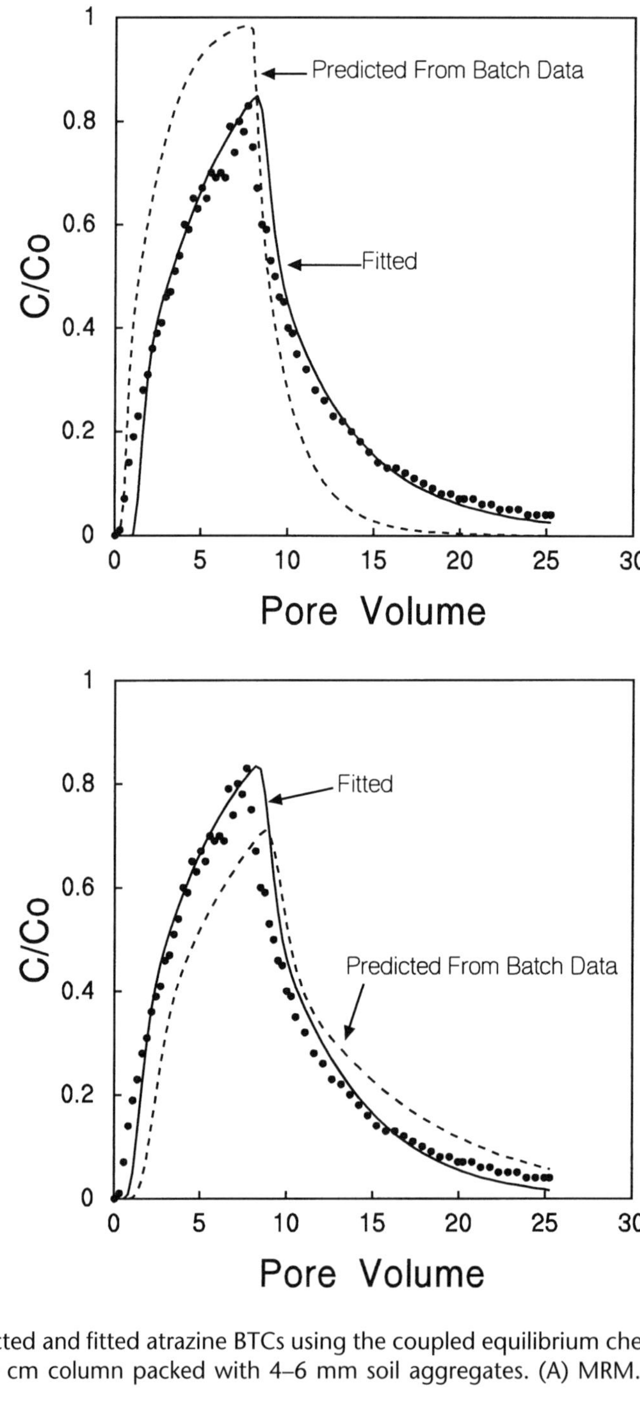

Figure 4.6. Predicted and fitted atrazine BTCs using the coupled equilibrium chemical reaction-MIM model for the 10 cm column packed with 4–6 mm soil aggregates. (A) MRM. (B) SOM.

COUPLING OF CHEMICAL NONEQUILIBRIUM TO THE MIM

Kinetic One-Site Adsorption Models

Since equilibrium models seldom provide adequate predictions of solute in soils, kinetic reactions are often formulated to describe solute retention during movement in soils (Selim and Amacher, 1997). The retention mechanisms associated with the

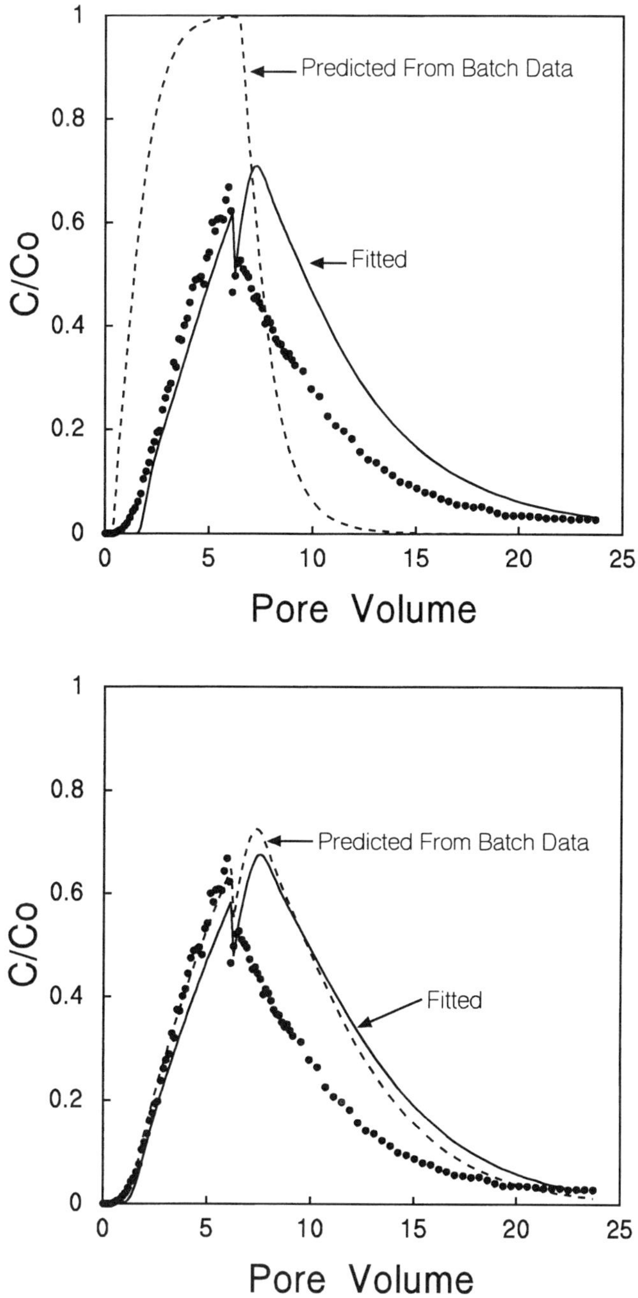

Figure 4.7. Predicted and fitted atrazine BTCs using the coupled equilibrium chemical reaction-MIM model for the 15 cm column packed with 2–4 mm soil aggregates and 16 d flow interruption. (A) MRM. (B) SOM.

mobile and immobile water phases were assumed to have similar rate coefficients. A one-site adsorption model can be derived based on Eq. 4.21.

$$\frac{\partial S_m}{\partial t} = k_1 \left(\frac{\theta_m}{\rho}\right) C_m{}^n - k_2 \; S_m \tag{4.37}$$

$$\frac{\partial S_{im}}{\partial t} = k_1 \left(\frac{\theta_{im}}{\rho}\right) C_{im}{}^{n} - k_2 S_{im} \tag{4.38}$$

Similarly, based on second-order formulation (SOM), the governing kinetic reaction for each soil region may be written as,

$$\frac{\partial S_m}{\partial t} = k_1 \theta_m C_m \phi - k_2 S_m \tag{4.39}$$

$$\frac{\partial S_{im}}{\partial t} = k_1 \theta_{im} C_{im} \phi - k_2 S_{im} \tag{4.40}$$

Subject to the constraint that a maximum adsorption capacity S_{max} is related to the amount of vacant and filled (or occupied) sites such that

$$S_{max} = \phi + S_m + S_{im} \tag{4.41}$$

Figure 4.8 provides an example of the one-site adsorption kinetic model. Compared to Figures 4.6 and 4.7, the two-parameter model provided improved description of atrazine transport, especially on the tailing side of the BTCs.

Multiple-Site Adsorption Models

Since the soil is chemically heterogeneous, multisite adsorption models have shown distinct improvement in the description of solute transport in soils over one-site models (Ma and Selim, 1994a,b; Selim et al., 1992). Applying the mobile-immobile concept to the MRM model, and defining all the adsorbed solute concentration (S) in terms of total soil mass in the system, we have,

for the mobile water phase,

$$(S_e)_m = K_d \left(\frac{\theta_m}{\rho}\right) C_m{}^{n} \tag{4.42}$$

$$\frac{\partial (S_1)_m}{\partial t} = k_1 \left(\frac{\theta_m}{\rho}\right) C_m{}^{n} - k_2 (S_1)_m \tag{4.43}$$

$$\frac{\partial (S_2)_m}{\partial t} = k_3 \left(\frac{\theta_m}{\rho}\right) C_m{}^{n} - k_4 (S_2)_m - k_5 (S_2)_m + k_6 (S_3)_m \tag{4.44}$$

$$\frac{\partial (S_3)_m}{\partial t} = k_5 (S_2)_m - k_6 (S_3)_m \tag{4.45}$$

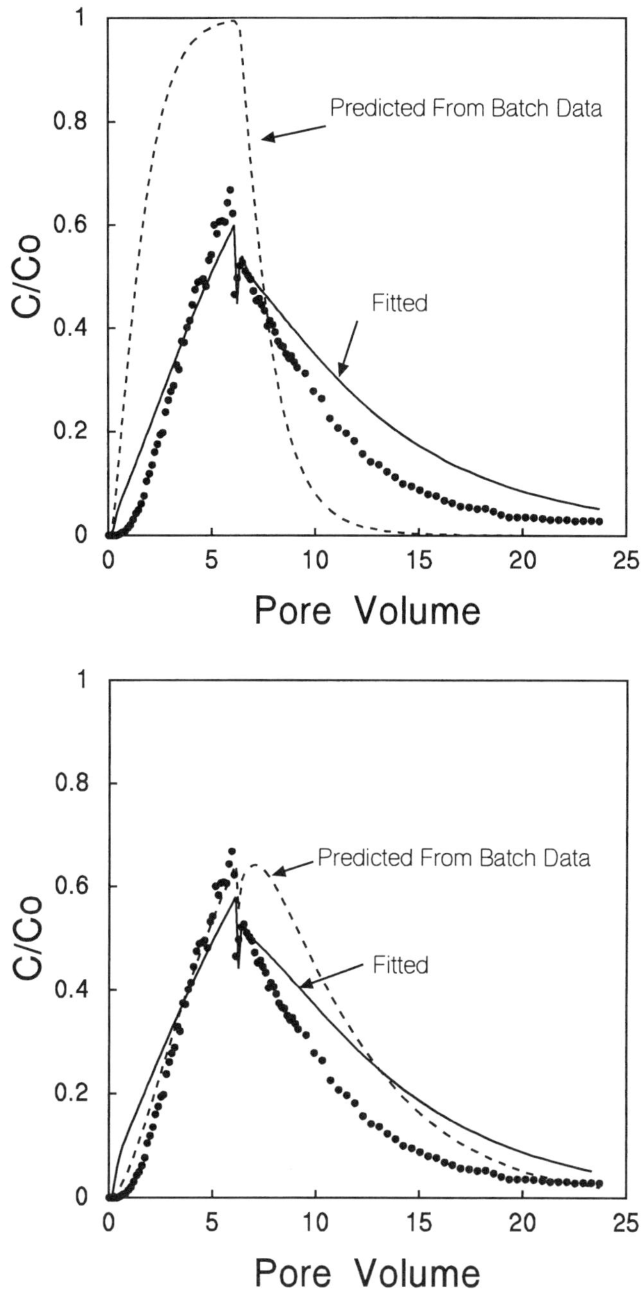

Figure 4.8. Predicted and fitted atrazine BTCs using the coupled one-adsorption kinetic chemical reaction-MIM model for the 15 cm column packed with 2–4 mm soil aggregates and 16 d flow interruption. (A) MRM. (B) SOM.

$$\frac{\partial (S_{irr})_m}{\partial t} = k_{irr} \ C_m \tag{4.46}$$

and for the immobile water phase:

$$(S_e)_{im} = K_d \left(\frac{\theta_{im}}{\rho}\right) C_{im}^{\ n} \tag{4.47}$$

$$\frac{\partial (S_1)_{im}}{\partial t} = k_1 \left(\frac{\theta_{im}}{\rho}\right) C_{im}^{\ n} - k_2 (S_1)_{im} \tag{4.48}$$

$$\frac{\partial (S_2)_{im}}{\partial t} = k_3 \left(\frac{\theta_{im}}{\rho}\right) C_{im}^{\ n} - k_4 (S_2)_{im} - k_5 (S_2)_{im} + k_6 (S_3)_{im} \tag{4.49}$$

$$\frac{\partial (S_3)_{im}}{\partial t} = k_5 (S_2)_{im} - k_6 (S_3)_{im} \tag{4.50}$$

$$\frac{\partial (S_{irr})_{im}}{\partial t} = k_{irr} C_{im} \tag{4.51}$$

The total amount of solute adsorbed from the mobile water (S_m) is:

$$S_m = (S_e)_m + (S_1)_m + (S_2)_m + (S_3)_m + (S_{irr})_m \tag{4.52}$$

and that from the immobile water (S_{im}) is:

$$S_{im} = (S_e)_{im} + (S_1)_{im} + (S_2)_{im} + (S_3)_{im} + (S_{irr})_{im} \tag{4.53}$$

Therefore, the total amount of solute adsorbed from both mobile and immobile water is

$$S = S_m + S_{im} \tag{4.54}$$

Table 4.2 presents the goodness-of-prediction of several MRM models for eight soil column BTCs in terms of total sum of squared errors (Ma and Selim, 1997) and Figure 4.9 shows examples of model prediction for F9. Generally, the coupling of MRM with MIM did not improve atrazine prediction. In addition, MRM-MIM is not sensitive to α values.

Extending the mobile-immobile concept to the second-order model (SOM), we have for the mobile phase,

$$(S_e)_m = K_d \, \theta_m \, C_m \, \phi \tag{4.55}$$

$$\frac{\partial (S_1)_m}{\partial t} = k_1 \, \theta_m \, C_m \, \phi - k_2 (S_1)_m \tag{4.56}$$

$$\frac{\partial (S_2)_m}{\partial t} = k_3 \, \theta_m \, C_m \, \phi - k_4 (S_2)_m - k_5 (S_2)_m + k_6 (S_3)_m \tag{4.57}$$

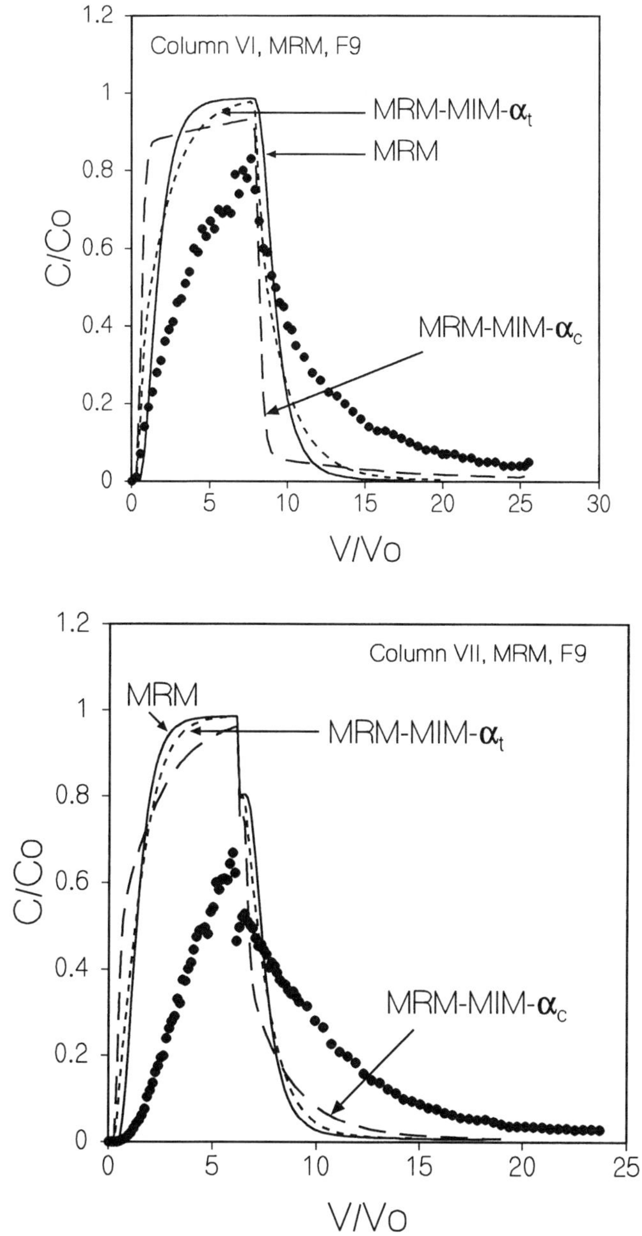

Figure 4.9. Example comparison of MRM with coupled MRM-MIM for predicting atrazine BTCs in a Sharkey soil. The Freundlich (n=0.7635) was obtained from batch isotherm, and kinetic reaction coefficients were derived from Figure 4.2 and are listed in Table 4.1. (A) 10 cm column packed with 4–6 mm soil aggregates, no flow interruption. (B) 15 cm column packed with 2–4 mm soil aggregates, 16 d flow interruption after pulse input (after Ma and Selim, 1997).

$$\frac{\partial (S_3)_m}{\partial t} = k_5 \, (S_2)_m - k_6 \, (S_3)_m \tag{4.58}$$

$$\frac{\partial (S_{irr})_m}{\partial t} = k_{irr} \; C_m \tag{4.59}$$

For the immobile water phase:

$$(S_e)_{im} = K_d\ \theta_{im}\ C_{im}\ \phi \tag{4.60}$$

$$\frac{\partial (S_1)_{im}}{\partial t} = k_1\ \theta_{im}\ C_{im}\ \phi - k_2\ (S_1)_{im} \tag{4.61}$$

$$\frac{\partial (S_2)_{im}}{\partial t} = k_3\ \theta_{im}\ C_{im}\ \phi - k_4\ (S_2)_{im} - k_5\ (S_2)_{im} + k_6\ (S_3)_{im} \tag{4.62}$$

$$\frac{\partial (S_3)_{im}}{\partial t} = k_5\ (S_2)_{im} - k_6\ (S_3)_{im} \tag{4.63}$$

$$\frac{\partial (S_{irr})_{im}}{\partial t} = k_{irr}\ C_{im} \tag{4.64}$$

Here ϕ is related to the sorption capacity (S_{max}) by:

$$S_{max} = \phi + S_m + S_{im} \tag{4.65}$$

Table 4.2 also lists the goodness-of-prediction of several SOM models for eight BTCs in terms of the root mean squared errors along with MRM model predictions. Figure 4.10 shows examples of predicted atrazine BTCs. The introduction of MIM into SOM considerably improved prediction of atrazine BTCs. Moreover, the two different α estimations resulted in significant differences in atrazine BTC predictions. A time-dependent α (Eq. 4.18) offered better BTC prediction than a constant α value (Eq. 4.17).

OTHER FORMULATIONS OF MIM, MRM, AND SOM

MIM

Over the years, various formulations of solute transport and retention models have been introduced in the literature. As mentioned above, van Genuchten and Wierenga (1976) introduced a parameter *f* in their MIM formulation (Eq. 4.9). This type of mobile-immobile model is the most commonly used as physical nonequilibrium approach in soil science (Brusseau et al., 1989; Selim et al., 1987). The difficulty with this MIM formulation is how to estimate the *f* parameters. There is no experimental estimation of *f*. The most commonly used assumption is $f=F$ (the fraction of mobile water) (Selim and Amacher, 1988; Brusseau et al., 1989; Gaston and Selim, 1990). Selim and Ma (1995) compared the MIM model of van Genuchten and Wierenga (1976) with previously described MIM (without *f*) and found that the MIM without *f* provided better BTC predictions (Figure 4.11). Rao et al. (1979) applied the MIM of van Genuchten and Wierenga (1976) along with the Freundlich adsorption isotherm and found that the model is not capable of predicting 2,4-D Amine BTCs in three soils.

MRM

So far, we discussed MRM, SOM, and MIM along the line of Eqs. 4.12–4.13 and 4.20–4.30. Since the models are conceptual in nature, other variations are possible

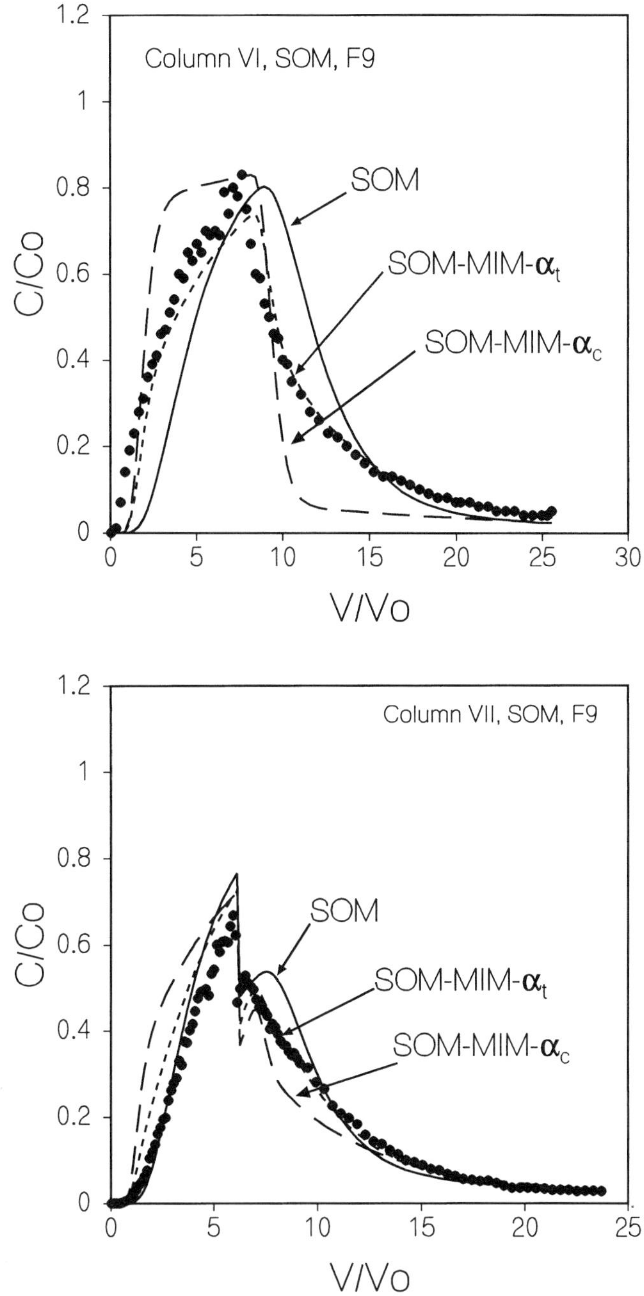

Figure 4.10. Example comparison of SOM with coupled SOM-MIM for predicting atrazine BTCs in a Sharkey soil. The adsorption maximum (Smax=184.62 μg/g soil) was obtained from batch isotherm, and kinetic reaction coefficients were derived from Figure 4.2 and are listed in Table 4.1. (A) 10 cm column packed with 4–6 mm soil aggregates, no flow interruption. (B) 15 cm column packed with 2–4 mm soil aggregates, 16 d flow interruption after pulse input (after Ma and Selim, 1997).

and have been proposed in the literature. In the MRM model of Selim et al. (1989), a Freundlich equation was directly used such as $S_e=K_dC^n$, rather than Eq. 4.20. Thus, S_e is not affected by θ and ρ. Brusseau et al. (1989) coupled a consecutive equilibrium

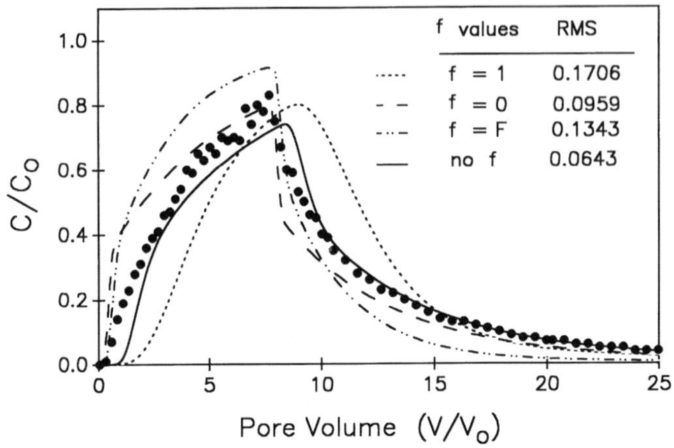

Figure 4.11. Measured (solid circles) and predicted atrazine BTC with 4–6 mm aggregated Sharkey soil, column length of 10 cm, velocity of 1.76 cm/h, and initial concentration of 11.33 μg/mL. Atrazine predictions were based on mobile-immobile approaches and the model parameters were from batch experiments. *F* is the fraction of mobile water (after Selim and Ma, 1995).

and first-order kinetic chemical nonequilibrium model to MIM of van Genuchten and Wierenga (1976), such as:

$$\frac{\partial(S_1)_m}{\partial t} = F_m K_m \frac{\partial C_m}{\partial t} \tag{4.66}$$

$$\frac{\partial(S_1)_{im}}{\partial t} = F_{im} K_{im} \frac{\partial C_{im}}{\partial t} \tag{4.67}$$

$$\frac{\partial(S_2)_m}{\partial t} = k_{m1}(S_1)_m - k_{m2}(S_2)_m = k_{m2}[(1-F_m) K_m C_m - (S_2)_m] \tag{4.68}$$

$$\frac{\partial(S_2)_{im}}{\partial t} = k_{im1}(S_1)_{im} - k_{im2}(S_2)_{im} = k_{im2}[(1-F_{im}) K_{im} C_{im} - (S_2)_{im}] \tag{4.69}$$

where F_m and F_{im} are the fractions of equilibrium sites in the mobile and immobile regions, respectively. K_m and K_{im} are the respective equilibrium constants (cm^3/g). k_{m1}, k_{m2}, k_{im1}, and k_{im2} are first-order sorption coefficients (h^{-1}). They found that such a multiprocess nonequilibrium model provided better descriptions to several data sets in the literature.

SOM

Selim and Amacher (1988) introduced an f_s coefficient to partition adsorption site between S_1 and S_2, such that,

$$\frac{\partial S_1}{\partial t} = k_1 \theta C \phi_1 - k_2 S_1 \tag{4.70}$$

$$\frac{\partial S_2}{\partial t} = k_3\,\theta\,C\,\phi_2 - k_4\,S_1 \tag{4.71}$$

where,

$$\phi_1 = f_s\,S_{max} - S_1$$
$$\phi_2 = (1 - f_s)\,S_{max} - S_2 \tag{4.72}$$

The f_s coefficient is usually derived from batch adsorption isotherm using the two-site Langmuir equation (Selim and Amacher, 1988; Selim et al., 1998). Figure 4.12 shows an example of predicted Cr BTCs with parameters obtained from batch experiment. Model prediction highly depends on initial concentrations of batch experiments. To eliminate such initial concentration effects, Selim et al. (1998) fitted all adsorption kinetics at several initial metolachlor concentrations simultaneously. Figure 4.13 shows fitted metolachlor concentrations in soil solution with reaction time for both SOM formulations (with and without f_s) with slightly better goodness-of-fit from SOM without f_s.

To apply the second-order model approach to MIM of van Genuchten and Wierenga (1976), Selim and Amacher (1988) used a one-adsorption kinetic model such as:

$$\frac{\partial S_m}{\partial t} = k_1\,\theta_m\,C_m\,\phi_m - k_2\,S_m \tag{4.73}$$

$$\frac{\partial S_{im}}{\partial t} = k_1\,\theta_{im}\,C_{im}\,\phi_{im} - k_2\,S_{im} \tag{4.74}$$

where

$$\phi_m = f\,S_{max} - S_m$$
$$\phi_{im} = (1 - f)\,S_{max} - S_{im} \tag{4.75}$$

Selim and Amacher (1988) found that the coupled SOM-MIM did not improve Cr BTC prediction (Figure 4.14). Selim et al. (1998) applied the second-order two adsorption site model to MIM of Selim and Ma (1995). The adsorption sites were assumed to be available to solutes in both mobile and immobile waters; however, they are divided between adsorption site-1 and site-2 according to Eq. 4.72. The chemical retention equations are:

$$\frac{\partial (S_1)_m}{\partial t} = k_1\,\theta_m\,C_m\,\phi_1 - k_2\,(S_1)_m \tag{4.76}$$

$$\frac{\partial (S_2)_m}{\partial t} = k_3\,\theta_m\,C_m\,\phi_2 - k_4\,(S_2)_m \tag{4.77}$$

$$\frac{\partial (S_1)_{im}}{\partial t} = k_1\,\theta_{im}\,C_{im}\,\phi_1 - k_2\,(S_1)_{im} \tag{4.78}$$

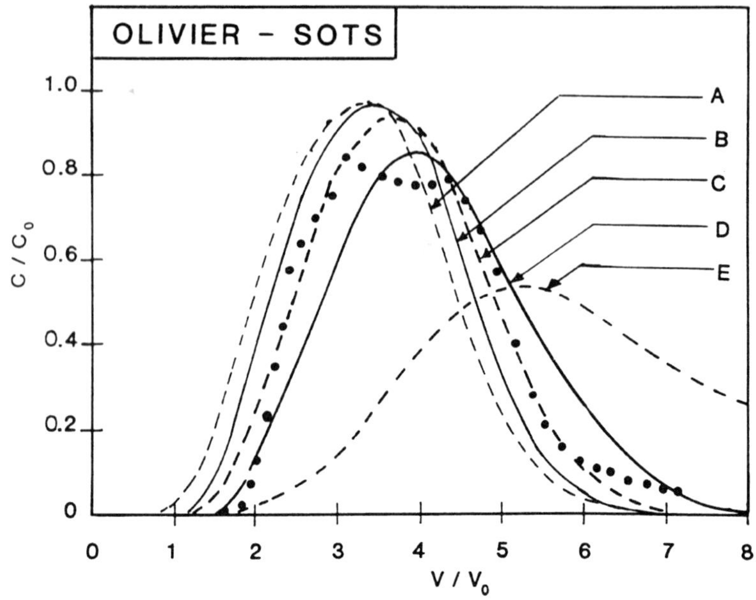

Figure 4.12. Effluent concentration distributions for Cr(VI) in Windsor soil. Curves A, B, C, D, and E are predictions using the second-order model with batch rate coefficients for C_o of 100, 25, 5, 2, and 1 μg/mL, respectively (after Selim and Amacher, 1988, with permission).

$$\frac{\partial (S_2)_{im}}{\partial t} = k_3 \, \theta_{im} \, C_{im} \, \phi_2 - k_4 \, (S_2)_{im} \tag{4.79}$$

where

$$\phi_1 = f_s \, S_{\max} - (S_1)_m - (S_1)_{im}$$

$$\phi_2 = (1 - f_s) \, S_{\max} - (S_2)_m - (S_2)_{im} \tag{4.80}$$

Figure 4.15 shows an example of metolachlor BTC predicted using parameters derived from Figure 4.13. Better predictions were obtained when no f_s was introduced. In addition, parameters derived from both adsorption and desorption kinetics offered better metolachlor BTC prediction than parameters from adsorption kinetics alone. Such a result is expected since both adsorption and desorption were taking place during column miscible experiments.

SUMMARY

The models described in this chapter are of the deterministic type where both chemical and physical nonequilibrium mechanisms are considered. Other types of chemical and physical nonequilibrium models include competitive ion exchange models (Selim et al., 1987; Mansell et al., 1988; Gaston and Selim, 1990; Selim et al., 1992), transfer function models (Utermann et al., 1990; Grochulska and Kladivko, 1994), two-flow domain models (Skopp et al., 1981; Jarvis et al., 1991a,b; Chen and Wagenet, 1992; Gerke and van Genuchten, 1993), and multiple flow domain models (Wilson et al., 1992; Hutson

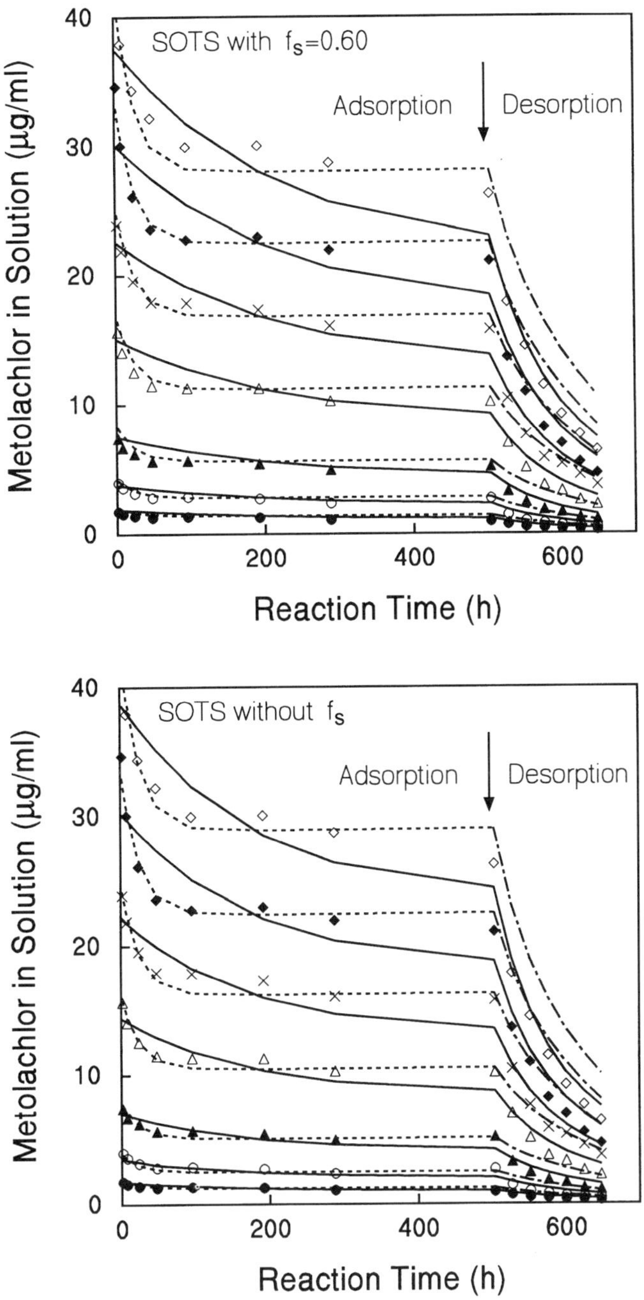

Figure 4.13. Metolachlor concentrations in soil solution versus reaction time. Symbols are measured concentrations for initial concentrations of 5, 10, 20, 40, 60, 80, and 100 µg/mL from the bottom to top; solid lines are fitted with the SOTS models using both adsorption and desorption data; dash lines are fitted with SOTS models using only adsorption results; dash-dot lines are predicted metolachlor concentrations for each desorption step using parameters from adsorption kinetics. (A) SOM with f_s=0.60, (B) SOM without f_s (after Selim et al., 1998).

and Wagenet, 1995; Gwo et al., 1996). Several of these approaches are discussed in detail in other chapters. Other models in the literature include two/three-dimensional trans-

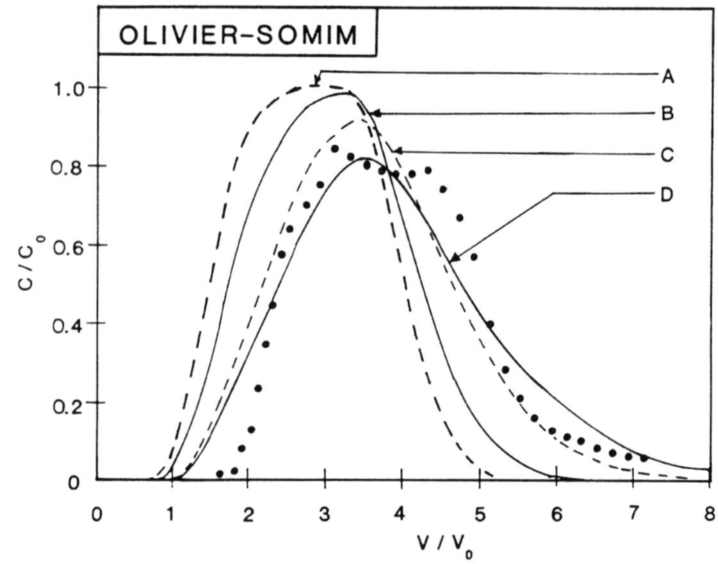

Figure 4.14. Effluent concentration distributions for Cr(VI) in Windsor soil. Curves A, B, C, and D are predictions using the second-order mobile-immobile model with batch rate coefficients for C_o of 25, 5, 2, and 1 μg/mL, respectively (after Selim and Amacher, 1988, with permission).

port (Goltz and Roberts, 1986, 1988), unsaturated water flow (Noborio et al., 1996; Ellsworth et al., 1996; Nassar and Horton, 1992), spatial variability (Ellsworth and Boast, 1996; Logsdon and Jaynes, 1996; van Wesenbeeck and Kachanoski, 1991; Starr, 1990), multilayer solute transport (van Wesenbeeck and Kachanoski, 1994; Leij and Dane, 1991; Leij et al., 1991; Selim et al., 1977), and large-scale/management modules (Toride and Leij, 1996a,b; Ahuja et al., 1993; Jury et al., 1990).

A major obstacle for most modelers is how to parameterize a model. Many conceptual parameters associated with MRM, SOM, and MIM are not easily identifiable (Ma and Selim, 1996). The examples presented in the chapter are solely on solute concentration in soil solution with less emphasis on the amount of solute adsorbed (e.g., S_1, S_2, S_3, S_{irr}, etc.). Gaston and Locke (1994) made an effort to quantify adsorbed alachlor in different phases using organic solvent extractions, and found that a consecutive retention model performed well when concentrations in soil solution as well as in the solid phases were used in model parameterization. However, as discussed earlier, it is difficult to discriminate among reaction mechanisms based on batch adsorption kinetics and BTCs (Selim and Amacher, 1997). Flow interruption during miscible experiments provided further evidences in revealing reaction mechanisms and/or physical nonequilibrium.

REFERENCES

Ahuja, L.R., D.G. DeCoursey, B.B. Barnes, and K.W. Rojas. Characteristics of macropore transport studied with the ARS root zone water quality model. *Trans. ASAE.* 36:369–380, 1993.

Amacher, M.C., H.M. Selim, and I.K. Iskandar. Kinetics of chromium (VI) and cadmium retention in soils: A nonlinear multireaction model. *Soil Sci. Soc. Am. J.* 52:398–408, 1988.

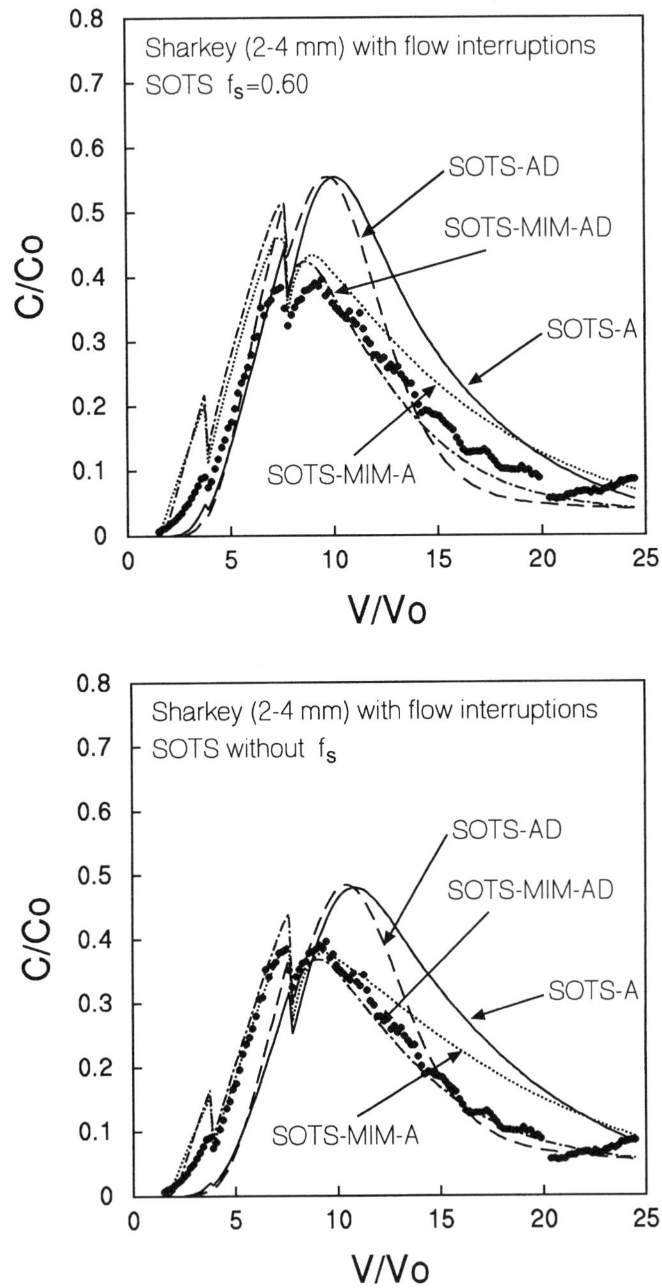

Figure 4.15. Predicted metolachlor breakthrough curves (BTCs) for a 15 cm column packed with 2–4 mm aggregates using SOM and SOM-MIM with parameters from adsorption (-A) and adsorption-desorption (-AD) batch results. Two four-day flow interruptions were introduced. One was after four pore volumes of metolachlor input and the other is after leaching with two pore volumes of metolachlor-free solution. (A) SOM with f_s=0.60, (B) SOM without f_s (after Selim et al., 1998).

Anderson, J.L. and J. Bouma. Water movement through pedal soils: 1. Saturated flow. *Soil Sci. Soc. Am. J.* 41:413–418, 1977.

Beven, K. and P. Germann. Macropores and water flow in soils. *Water Resour. Res.* 18:1311–1325, 1982.

Booltink, H.W.G. and J. Bouma. Physical and morphological characterization of bypass flow in a well-structured clay soil. *Soil Sci. Soc. Am. J.* 55:1249–1254, 1991.

Brusseau, M.L., R.E. Jessup, and P.S.C. Rao. Modeling the transport of solutes influenced by multiprocess nonequilibrium. *Water Resour. Res.* 25:1971–1988. 1989.

Chen, C. and R.J. Wagenet. Simulation of water and chemicals in macropore soils. I. Representation of the equivalent macropore influence and its effect on soil water flow. *J. Hydrol.* 130:105–126, 1992.

Chen, C., D.M. Thomas, R.E. Green, and R.J. Wagenet. Two-domain estimation of hydraulic properties in macropore soils. *Soil Sci. Soc. Am. J.* 57:680–686, 1993.

Clothier, B.E., M.B. Kirkham, and J.E. Mclean. In situ measurement of the effective transport volume for solute moving through soil. *Soil Sci. Soc. Am. J.* 56:733–736, 1992.

Coats, K.H. and B.D. Smith. Dead-end pore volume and dispersion in porous media. *Soc. Pet. Eng. J.* 4:73–84, 1964.

Deans, H.H. A mathematical model for dispersion in the direction of flow in porous media. *Soc. Pet. Eng. J.* 3:49–52, 1963.

De Smedt, F. and P.J. Wierenga. Solute transfer through columns of glass beads. *Water Resour. Res.* 20:225–232, 1984.

De Smedt, F., F. Wauters, and J. Sevilla. Study of tracer movement through unsaturated sand. *Geoderma.* 38:223–236, 1986.

Duguid, J.O. and P.C.Y. Lee. Flow in fractured porous media. *Water Resour. Res.* 13:558–566, 1977.

Edwards, W.M., L.D. Norton, and C.E. Redmond. Characterizing macropores that affect infiltration into no-tilled soil. *Soil Sci. Soc. Am. J.* 52:483–487, 1988.

Ela, S.D., S.C. Gupta, and W.J. Rawls. Macropore and surface seal interactions affecting water infiltration into soil. *Soil Sci. Soc. Am. J.* 56:714–721, 1992.

Ellsworth, T.R. and C.W. Boast. Spatial structure of solute transport variability in an unsaturated field soil. *Soil Sci. Soc. Am. J.* 60:1355–1367, 1996.

Ellsworth, T.R., P.J. Shouse, T.H. Skaggs, J.A. Jobes, and J. Fargerlund. Solute transport in unsaturated soil: Experimental design, parameter estimation, and model discrimination. *Soil Sci. Soc. Am. J.* 60:397–407, 1996.

Gaston, L.A. and H.M. Selim. Transport of exchangeable cations in an aggregated clay soil. *Soil Sci. Soc. Am. J.* 54:31–38, 1990.

Gaston, L.A. and M.A. Locke. Predicting alachlor mobility using batch sorption kinetic data. *Soil Sci.* 158:345–354, 1994.

Gerke, H.H. and M.Th. van Genuchten. A dual-porosity model for simulating the preferential movement of water and solutes in structured porous media. *Water Resour. Res.* 29:305–319, 1993.

Goltz, M.N. and P.V. Roberts. Three-dimensional simulations for solute transport in an infinite medium with mobile-immobile zone. *Water Resour. Res.* 22:1139–1148, 1986.

Goltz, M.N. and P.V. Roberts. Simulations of physical nonequilibrium solute transport models: Application to a large scale field experiment. *J. Contamin. Hydrol.* 3:37–63, 1988.

Gottschlich, C.F. Axial Dispersion in a packed bed. *A.I.Ch.E. J.* 9:88–92, 1963.

Grochulska, J. and E.J. Kladivko. A two-region model of preferential flow of chemicals using a transfer function approach. *J. Environ. Qual.* 23:498–507, 1994.

Gwo, J.P., P.M. Jardine, G.V. Wilson, and G.T. Yeh. Using a multiregion model to study the effects of advective and diffusive mass transfer on local physical nonequilibrium and solute mobility in a structured soil. *Water Resour. Res.* 32:561–570, 1996.

Hutson, J.L. and R.J. Wagenet. A multiregion model describing water flow and solute transport in heterogenous soils. *Soil Sci. Soc. Am. J.* 59:743–751, 1995.

Jarvis, N.J., P.-E. Jansson, P.E. Dik, and I. Messing. Modeling water and solute transport in macroporous soil. II. Model description and sensitivity analysis. *J. Soil Sci.* 42:59–70, 1991a.

Jarvis, N.J., L. Bergstrom, and P.E. Dik. Modeling water and solute transport in macroporous soil. II. Chloride breakthrough under non-steady flow. *J. Soil Sci.* 42:71–81, 1991b.

Jaynes, D.B., S.D. Logsdon, and R. Horton. Field method for measuring mobile/immobile water content and solute transfer rate coefficient. *Soil Sci. Soc. Am. J.* 59:352–356, 1995.

Jury, W.A., J.S. Dyson, and G.L. Butters. Transfer function model of field-scale solute transport under transient water flow. *Soil Sci. Soc. Am. J.* 54:327–332, 1990.

Kookana, R.S., R.D. Schuller, and L.A.G. Aylmore. Simulation of simazine transport through soil columns using time-dependent sorption data measured under flow conditions. *J. Contamin. Hydrol.* 14:93–115, 1993.

Leij, F.J. and J.H. Dane. Solute transport in a two-layer medium investigated with time moments. *Soil Sci. Soc. Am. J.* 55:1529–1535, 1991.

Leij, F.J., J.H. Dane, and M.Th. Van Genuchten. Mathematical analysis of one-dimensional solute transport in a layered soil profile. *Soil Sci. Soc. Am. J.* 55:944–953, 1991.

Li, Y. and M. Ghodrati. Preferential transport of nitrate through soil columns containing root channels. *Soil Sci. Soc. Am. J.* 58:653–659, 1994.

Lindstrom, F.T. and L. Boersma. A theory on the mass transport of previously distributed chemicals in a water saturated sorbing porous medium. *Soil Sci.* 111:192–199, 1971.

Logsdon, S.D. and D.B. Jaynes. Spatial variability of hydraulic conductivity in a cultivated field at different times. *Soil Sci. Soc. Am. J.* 60:703–709, 1996.

Logsdon, S.D., R.R. Allmaras, L. Wu, J.B. Swan, and G.W. Randall. Macroporosity and its relation to saturated hydraulic conductivity under different tillage practices. *Soil Sci. Soc. Am. J.* 54:1096–1101, 1990.

Luxmoore, R.J. Micro-, meso-, and macroporosity of soil. *Soil Sci. Soc. Am. J.* 45:671–672, 1981.

Ma, L. and H.M. Selim. Evaluation of nonequilibrium models for predicting atrazine transport in soils. *Soil Sci. Soc. Am. J.* 61:1299–1307, 1997.

Ma, L. and H.M. Selim. Atrazine retention and transport in soils. *Rev. Environ. Contamin. Toxicol.* 145:129–173, 1996.

Ma, L. and H.M. Selim. Transport of a nonreactive solute in soils: A two-flow domain approach. *Soil Sci.* 159:224–234, 1995.

Ma, L. and H.M. Selim. Predicting atrazine adsorption-desorption in soils: A modified second-order kinetic model. *Water Resour. Res.* 30:447–456, 1994a.

Ma, L. and H.M. Selim. Predicting the transport of atrazine in soils: Second-order and multireaction approaches. *Water Resour. Res.* 30:3489–3498, 1994b.

Mansell, R.S., S.A. Bloom, H.M. Selim, and R.D. Rhue. Simulated transport of multiple cations in soil using variable selectivity coefficients. *Soil Sci. Soc. Am. J.* 52:1533–1540, 1988.

McCoy, E.L., C.W. Boast, R.C. Stehouwer, and E.J. Kladivko. Macropore hydraulics: Taking a sledgehammer to classical theory. In *Soil Processes and Water Quality*. Adv. Soil Sci., Lal, R. and B.A. Stewart (Eds.), 1994, pp. 303–348.

Munyankusi, E., S.C. Gupta, J.F. Moncrief, and E.C. Berry. Earthworm macropores and preferential transport in a long-term manure applied typic hapludalf. *J. Environ. Qual.* 23:773–784, 1994.

Nkedi-Kizza, P., P.S.C. Rao, R.E. Jessup, and J.M. Davidson. Ion exchange and diffusive mass transfer during miscible displacement through and aggregate Oxisol. *Soil Sci. Soc. Am. J.* 46:471–476, 1982.

Nkedi-Kizza, P., J.W. Biggar, M.Th. van Genuchten, P.J. Wierenga, H.M. Selim, J.M. Davidson, and D.R. Nielson. Modeling tritium and chloride 36 transport through an aggregated Oxisol. *Water Resour. Res.* 19:691–700, 1983.

Nassar, I.N. and R. Horton. Simultaneous transfer of heat, water, and solute in porous media: I. Theoretical development. *Soil Sci. Soc. Am. J.* 56:1350–1356, 1992.

Noborio, K., K.J. McInnes, and J.L. Heilman. Two-dimensional model for water, heat, and solute transport in Furrow-irrigated soil: I. Theory. *Soil Sci. Soc. Am. J.* 60:1001–1009, 1996.

Rao, P.S.C., J.M. Davidson, R.E. Jessup, and H.M. Selim. Evaluation of conceptual models for describing nonequilibrium adsorption-desorption of pesticide during steady-state flow in soils. *Soil Sci. Soc. Am. J.* 43:22–28, 1979.

Rao, P.S.C., R.E. Jessup, D.E. Ralston, J.M. Davidson, and D.P. Kilcrease. Experimental and mathematical description of nonadsorbed solute transfer by diffusion in spherical aggregate. *Soil Sci. Soc. Am. J.* 44:684–688, 1980a.

Rao, P.S.C., D.E. Ralston, R.E. Jessup, and J.M. Davidson. Solute transport in aggregated porous media: Theoretical and experimental evaluation. *Soil Sci. Soc. Am. J.* 44:1139–1146, 1980b.

Rao, P.S.C., R.E. Jessup, and T.M. Addiscott. Experimental and theoretical aspects of solute diffusion in spherical and nonspherical aggregates. *Soil Sci.* 133:342–349, 1982.

Roseberg, R.J. and E.L. McCoy. Measurement of soil macropore air permeability. *Soil Sci. Soc. Am. J.* 54:969–974, 1990.

Selim, H.M. and M.C. Amacher. *Reactivity and Transport of Heavy Metals in Soils*. CRC/Lewis, Boca Raton, FL, 1997.

Selim, H.M. and M.C. Amacher. A second-order kinetic approach for modeling solute retention and transport in soils. *Water Resour. Res.* 24:2061–2075, 1988.

Selim, H.M. and L. Ma. Transport of reactive solutes in soils: A modified two-region approach. *Soil Sci. Soc. Am. J.* 59:75–82, 1995.

Selim, H.M., L. Ma, and H. Zhu. Transport of reactive solutes in soils: Evaluation of second-order two-site approaches. *Water Resour. Res.* (Submitted), 1998.

Selim, H.M., B. Buchter, C. Hinz, and L. Ma. Modeling the transport and retention of cadmium in soils: multireaction and multicomponent approaches. *Soil Sci. Soc. Am. J.* 56:1004–1015, 1992.

Selim, H.M., M.C. Amacher, and I.K. Iskandar. Modeling the transport of chromium (VI) in soil columns. *Soil Sci. Soc. Am. J.* 53:996–1004, 1989.

Selim, H.M., R. Schulin, and H. Fluhler. Transport and ion exchange of calcium and magnesium in an aggregated soil. *Soil Sci. Soc. Am. J.* 51:876–884, 1987.

Selim, H.M., J.M. Davidson, and P.S.C. Rao. Transport of reactive solutes through multi-layered soils. *Soil Sci. Soc. Am. J.* 41:3–10, 1977.

Selim, H.M., J.M. Davidson, and R.S. Mansell. Evaluation of a two-site adsorption-desorption model for describing solute transport in soils. Summer Computer Simulation Conference. 12–24, July, 1976. Washington, DC, 1976, pp. 444–448.

Singh, P., R.S. Kanwar, and M.L. Thompson. Macropore characterization for two tillage systems using resin impregnation technique. *Soil Sci. Soc. Am. J.* 55:1674–1679, 1991.

Skopp, J. and W.R. Gardner. Miscible displacement: An interacting flow region model. *Soil Sci. Soc. Am. J.* 56:1680–1686, 1992.

Skopp, J., W.R. Gardner, and E.J. Tyler. Solute movement in structured soils: Two-region model with small interaction. *Soil Sci. Soc. Am. J.* 45:837–842, 1981.

Skopp, J. and A.W. Warrick. A two-phase model for the miscible displacement of reactive solutes in soils. *Soil Sci. Soc. Am. J.* 38:545–550, 1974.

Smettem, K.R.J. and C. Kirkby. Measuring the hydraulic properties of a stable aggregated soil. *J. Hydrol.* 117:1–13, 1990.

Starr, J.L. Spatial and temporal variation of ponded infiltration. *Soil Sci. Soc. Am. J.* 54:629–636, 1990.

Steenhuis, T.S., J.-Y. Parlange, and M.S. Andreini. A numerical model for preferential solute movement in structured soils. *Geoderma.* 46:193–208, 1990.

Toride, N. and F.J. Leij. Convective-dispersive stream tube model for field-scale solute transport. I. Moment analysis. *Soil Sci. Soc. Am. J.* 60:342–352, 1996a.

Toride, N. and F.J. Leij. Convective-dispersive stream tube model for field-scale solute transport. II. Examples and Calibration. *Soil Sci. Soc. Am. J.* 60:352–361, 1996b.

Trojan, M.D. and D.R. Linden. Microrelief and rainfall effects on water and solute movement in earthworm burrows. *Soil Sci. Soc. Am. J.* 56:727–733, 1992.

Utermann, J., E.J. Kladivko, and W.A. Jury. Evaluating pesticide migration in tile-drained soil with transfer function model. *J. Environ. Qual.* 19:707–714, 1990.

van Genuchten, M.Th. A general approach for modeling solute transport in structured soils. Proc. 17th Int. Congress. IAH, Hydrogeology of Rocks of Low Permeability. Jan. 7–12, 1985, Tucson, AZ. *Mem. Int. Assoc. Hydrogeol.* 17:512–526, 1985.

van Genuchten, M.Th. and P.J. Wierenga. Mass transfer studies in sorbing porous media. II. Experimental evaluation with Tritium (^3H_2O). *Soil Sci. Soc. Am. J.* 41:272–277, 1977.

van Genuchten, M.Th. and P.J. Wierenga. Mass transfer studies in sorbing porous media: I. Analytical solutions. *Soil Sci. Soc. Am. J.* 40:473–480, 1976.

van Genuchten, M.Th., J.M. Davidson, and P.J. Wierenga. An evaluation of kinetic and equilibrium equations for the prediction of pesticide movement through porous media. *Soil Sci. Soc. Am. Proc.* 38:29–35, 1974.

van Genuchten, M.Th. and F.N. Dalton. Models for simulating salt movement in aggregated field soils. *Geoderma.* 38:165–183, 1986.

Van Wesenbeeck, I.J. and R.G. Kachanoski. Spatial scale dependence of in situ solute transport. *Soil Sci. Soc. Am. J.* 55:3–7, 1991.

Van Wesenbeeck, I.J. and R.G. Kachanoski. Effect of variable horizon thickness on solute transport. *Soil Sci. Soc. Am. J.* 58:1307–1316, 1994.

Wang, D., J.M. Norman, B. Lowery, and K. McSweeney. Nondestructive determination of hydrogeometrical characteristics of soil macropores. *Soil Sci. Soc. Am. J.* 58:294–303, 1994.

Warner, G.S., J.L. Nieber, I.D. Moore, and R.A. Geise. Characterizing macropores in soils by computed tomography. *Soil Sci. Soc. Am. J.* 53:653–660, 1989.

Watson, K.W. and R.J. Luxmoore. Estimating macroporosity in a forest watershed by use of a tension infiltrometer. *Soil Sci. Soc. Am. J.* 50:578–582, 1986.

Wilson, G.V., P.M. Jardine, and J.P. Gwo. Modeling the hydraulic properties of a multiregion soil. *Soil Sci. Soc. Am. J.* 56:1731–1737, 1992.

CHAPTER FIVE

Analytical Solutions for Nonequilibrium Transport Models

F.J. Leij and N. Toride

INTRODUCTION

Natural soils typically possess a complicated pore structure and are macroscopically heterogeneous, especially near the soil surface. The soil may be aggregated with small inter-particle pores inside aggregates and large intra-particle pore spaces (drying cracks, interpedal voids). There may be an abundance of macropores created by decomposed roots or as borings by animals such as earthworms and gophers. Continuous macropores will act as conduits for rapid movement of water and dissolved substances in the soil. On the other hand, there are micropores with little flow. As a consequence, solute movement occurs mainly by advection in the macropores and largely by diffusion in the micropores. Solutes will move relatively rapid in the larger pores, but they will be retarded by the nonequilibrium exchange between larger and smaller pores because there is no appreciable flow in the latter. Under most circumstances, the solute breakthrough curve for such a heterogeneous soil will have an early, high peak and a long tail. The conventional advection-dispersion equation, i.e., for a single continuum, cannot adequately describe solute movement in macroscopically heterogeneous media. An alternative transport equation needs to be formulated.

Nonequilibrium processes in soils are difficult to quantify because a detailed knowledge of the flow geometry and the distribution and composition of the solid phase is usually lacking. Furthermore, nonequilibrium flow and transport should be conceptualized in a way that is convenient for mathematical modeling. The bi-continuum or dual-porosity model has been very popular for nonequilibrium transport because of its straightforward formulation with a limited number of parameters. This chapter concerns analytical methods for solving the bi-continuum problem. Such methods can only be applied to linear transport problems, i.e., the dependent variable only appears with an exponent of 0 or 1. We impose the additional restriction that the coefficients of the dependent variable and its derivatives are constant and do not depend on the independent variables. It may therefore seem that the application of analytical methods is greatly limited. However, there are a number of important cases where these assumptions may be considered valid for practical purposes. One example is the description of breakthrough curves from homogeneous laboratory soil columns, where ideal conditions may be maintained. Another example is the study of solute movement over large temporal and spatial scales where, for lack of knowledge of the actual transport properties, one may as well select constant parameters.

A large part of the chapter is devoted to the derivation of closed-form expressions for one-dimensional transport with first-order exchange between the two continua. Nonequilibrium features in the solute distribution are attributed to differences in flow regions (physical nonequilibrium). The same transport equation may be used for the case where nonequilibrium is caused by two different adsorption mechanisms (chemical nonequilibrium). The derivation of the solution to the general boundary, initial, and production value problems is reviewed with particular attention to mathematical procedures specific for nonequilibrium transport. Many of the mathematical procedures for nonequilibrium models were established by physical chemists, petroleum and chemical engineers (Klinkenberg, 1948; Hiester and Vermeulen, 1952; Coats and Smith, 1964) prior to the use of such models for solute transport in soils.

If the soil consists of aggregates with a well-defined geometry, the mathematical model for nonequilibrium transport may be refined. A distinction is made between advective and dispersive transport in the mobile inter-aggregate zone, along the mean flow direction, and diffusive transport according to Fick's second law in the immobile intra-aggregate region, perpendicular to the aggregate surface. The mathematical solution for the problem tends to be more complicated than for the first-order approximation. The problem for well-defined media can be made equivalent to the problem assuming first-order transfer, by defining the immobile concentration as the average of the intra-aggregate concentration. Furthermore we define effective dispersion coefficients for equilibrium problems that are equivalent to nonequilibrium problems involving well-defined aggregate shapes or the first-order assumption.

Finally, we give a brief overview of using time moments and transfer functions to quantify the solute distribution for nonequilibrium transport. The mathematical procedure for obtaining moments or transfer functions is simpler than the derivation of a closed-form solution (Valocchi, 1985; Jury and Roth, 1990). The approach is well-suited to handle more complex transport models provided that they are linear or can be linearized. Moment analysis can be applied to gain insight in the importance of model parameters and the geometry of the solid on solute movement. We may formulate expressions for the residence time associated with a particular process in the transport model (Sardin et al., 1991). Temporal and spatial moments are also convenient tools for analyzing experimental solute distributions versus time and space, respectively.

FIRST-ORDER TRANSFER

Model Formulation

Nonequilibrium can be of a "physical" nature, i.e., the liquid phase consists of a mobile and an immobile region (van Genuchten and Wierenga, 1976), or a "chemical" nature, i.e., the sorbed phase contains sites with "instantaneous" and "kinetic" sorption (Selim et al., 1976). The governing equations for these two types of nonequilibrium are mathematically identical (Nkedi-Kizza et al., 1984); their analytical solution has been widely published (Lassey, 1988; van Genuchten and Wagenet, 1989; Toride et al., 1993). Transfer between the "equilibrium" and "nonequilibrium" continua is described as a first-order process for mathematical and physical reasons. The pore-size geometry and the flow and adsorption properties of most porous media are usually not known in sufficient detail to warrant the use of more refined transfer models. The first-order approach may include models that simultaneously consider physical and chemical nonequilibrium processes (Brusseau et al., 1989) or those that discern more than two continua. We will not discuss the analytical modeling of such extensions because no new concepts are involved.

Solute movement is studied analytically by making the usual restrictive assumption that the soil is isotropic and that there is one-dimensional steady flow—both the water velocity and the dispersion coefficients are constant with respect to time and space. Solutes may be subject to linear retardation, i.e., the equilibrium sorption between solution and sorbed phases is described with a linear isotherm; and zero- and first-order production or degradation. The first-order production is assumed to depend on position. The extension to time-dependent production is trivial and has been reported by Lindstrom and Boersma (1989).

Two-Region Model

The physical nonequilibrium model assumes that: (i) the liquid region can be partitioned into a mobile (or flowing) and immobile (or stagnant) region, (ii) solute movement occurs by advection and dispersion in the mobile region, and (iii) solute exchange between the two regions occurs by first-order diffusion (Coats and Smith, 1964; van Genuchten and Wierenga, 1976). The governing equations are

$$\begin{aligned}(\theta_m + f\rho k)\frac{\partial c_m}{\partial t} &= \theta_m D_m \frac{\partial^2 c_m}{\partial x^2} - \theta_m v_m \frac{\partial c_m}{\partial x} - \alpha(c_m - c_{im}) \\ &\quad -(\theta_m \mu_{l,m} + f\rho k \mu_{s,m} - c_m + \theta_m \lambda_{l,m}(x,t) + f\rho\lambda_{s,m}(x,t)\end{aligned} \tag{5.1}$$

$$\begin{aligned}[\theta_{im} + (1-f)\rho k]\frac{\partial c_{im}}{\partial t} &= \alpha(c_m - c_{im}) - [\theta_{im}\mu_{l,im} + (1-f)\rho k \mu_{s,im}]c_{im} \\ &\quad + \theta_{im}\lambda_{l,im}(x,t) + (1-f)\rho\lambda_{s,im}(x,t)\end{aligned} \tag{5.2}$$

where c is the volume-averaged solute concentration in the liquid phase (ML^{-3}); t is time (T); x is position (L); D is a dispersion coefficient (L^2T^{-1}); θ is the volumetric water content (L^3L^{-3}); v is the pore-water velocity (LT^{-1}); f represents the fraction of sorption sites in equilibrium with the fluid of the mobile region; ρ is the bulk density (ML^{-3}); k is a distribution coefficient for linear adsorption (ML^{-3}), α is a first-order mass transfer coefficient (T^{-1}); μ_l and μ_s are first-order decay coefficients for degradation in the liquid and sorbed phases, respectively (T^{-1}); λ_l ($ML^{-3}T^{-1}$) and λ_s ($MM^{-1}T^{-1}$) are position dependent zero-order rate coefficients for solute production in the liquid and sorbed phases, respectively; while the subscripts m and im refer to mobile and immobile liquid regions, and ℓ and s to the liquid and sorbed phases.

The (total) volumetric water content is given by $\theta = \theta_m + \theta_{im}$. The pore-water velocity with respect to the cross-sectional area of the entire liquid region (v) and the velocity in the mobile region (v_m) obeys the relationship $\theta v = \theta_m v_m$. Furthermore, it is customary to define $D = \theta_m D_m/\theta$ (Parker and van Genuchten, 1984).

Two-Site Model

In this case, nonequilibrium is attributed to differences in exchange sites, with kinetic or equilibrium sorption. The "chemical" nonequilibrium model may be written as

$$\left[1+\frac{f\rho k}{\theta}\right]\frac{\partial c}{\partial t}=D\frac{\partial^2 c}{\partial x^2}-v\frac{\partial c}{\partial x}-\frac{\alpha\rho}{\theta}[(1-f)kc-s_k]$$
$$-\mu_l c-\frac{f\rho k}{\theta}\mu_{s,e}c+\lambda_l(x)+\frac{f\rho}{\theta}\lambda_{s,e}(x) \tag{5.3}$$

$$\frac{\partial s_k}{\partial t}=\alpha[(1-f)kc-s_k]-\mu_{s,k}s_k+(1-f)\lambda_{s,k}(x) \tag{5.4}$$

where s_k is the concentration in the sorbed phase for "type-2" sites (MM^{-1}); f is the fraction of "type-1" exchange sites, α is a first-order kinetic rate coefficient (T^{-1}); while the subscripts e and k refer to equilibrium and kinetic exchange sites, and ℓ and s to the liquid and sorbed phases.

Dimensionless Model

The two-site and the two-region models can be cast in the same dimensionless form with the parameters listed in Table 5.1. The following dimensionless equations can be obtained, which we consider for a semi-infinite solution domain:

$$\beta R\frac{\partial C_1}{\partial t}=\frac{1}{P}\frac{\partial^2 C_1}{\partial X^2}-\frac{\partial C_1}{\partial X}+\omega(C_2-C_1)-\mu_1 C_1+\lambda_1(X) \qquad (0<X<\infty, T>0) \tag{5.5}$$

$$(1-\beta)R\frac{\partial C_2}{\partial T}=\omega(C_1-C_2)-\mu_2 C_2+\lambda_2(X) \tag{5.6}$$

where β is a partition coefficient; R is a retardation factor; C_1 is the dimensionless solution concentration of the mobile region (two-region model) or the entire liquid phase (two-site model) and C_2 is the solution concentration of the immobile region or the adsorbed concentration for type-2 sites; T is time; X is distance; P is the Peclet number; ω is a mass transfer coefficient; μ and λ are dimensionless first-order decay and zero-order production terms; and the subscripts 1 and 2 refer to phases with equilibrium and nonequilibrium processes, respectively. The coefficient ω, which quantifies the rate of nonequilibrium exchange to the bulk flow rate, is actually a Damköhler number (Boucher and Alves, 1959). The magnitude of several dimensionless parameters depends not only on the dimensional model parameters but also on the somewhat arbitrary constants c_0 and L.

In addition to selecting the proper transport equation, it is important to formulate appropriate mathematical conditions for the solution of the above problem. Because advective-dispersive transport occurs exclusively in phase 1, the boundary conditions are only cast in terms of C_1, nonequilibrium processes do not affect the specification of the mathematical conditions. The problem will be solved for a third-type inlet condition (a prescribed solute flux) and an infinite outlet condition. The resulting solution may be accurate enough for problems involving different conditions, such as a first-type inlet condition (a prescribed solute concentration) or a finite outlet (van Genuchten and Parker, 1984). Exact solutions for such conditions may be obtained by replacing the equilibrium part of the solution with an expression derived for the different conditions as published by, among others, van

Table 5.1. Dimensionless Parameters for the Two-Region and Two-Site Transport Models.[a]

Parameter	Two-Site Model	Two-Region Model
T	$\frac{vt}{L}$	$\frac{vt}{L}$
X	$\frac{x}{L}$	$\frac{x}{L}$
P	$\frac{v_m L}{D_m}$	$\frac{vL}{D}$
β	$\frac{\theta + f\rho k}{\theta + \rho k}$	$\frac{\theta_m + f\rho k}{\theta + \rho k}$
R	$1+\frac{\rho k}{\theta}$	$1+\frac{\rho k}{\theta}$
ω	$\frac{\alpha(1-\beta)RL}{v}$	$\frac{\alpha L}{\theta v}$
C_1	$\frac{c}{c_o}$	$\frac{c_m}{c_o}$
C_2	$\frac{s_k}{(1-f)kc_o}$	$\frac{c_{im}}{c_o}$
μ_1	$\frac{L}{\theta v}[\theta\mu_t + f\rho k\mu_{s,e}]$	$\frac{L}{\theta v}[\theta_m\mu_{t,m} + f\rho k\mu_{s,m}]$
μ_2	$\frac{L}{\theta v}[(1-f)\rho k\mu_{s,k}]$	$\frac{L}{\theta v}[\theta_{im}\mu_{t,im} + (1-f)\rho k\mu_{s,im}]$
λ_1	$\frac{L}{\theta v c_o}[\theta\lambda_t + f\rho\lambda_{s,e}]$	$\frac{L}{\theta v c_o}[\theta_m\lambda_{t,m} + f\rho\lambda_{s,m}]$
λ_2	$\frac{L}{\theta v c_o}[(1-f\rho\lambda_{s,k}]$	$\frac{L}{\theta v c_o}[\theta_{im}\lambda_{t,im} + (1-f)\rho\lambda_{s,im}]$

[a] c_o is a characteristic concentration and L is a characteristic length.

Genuchten and Alves (1982). The general initial and boundary conditions for which the above equations will be solved are given by

$$C_1(X,0) = C_2(X,0) = F(X) \tag{5.7}$$

$$\left(C_1 - \frac{1}{P}\frac{\partial C_1}{\partial X}\right)\bigg|_{X=0^+} = G(T) \tag{5.8}$$

$$\frac{\partial C_1}{\partial X}(\infty, T) = 0 \tag{5.9}$$

where F and G are arbitrary functions which will be specified later, along with λ, to illustrate pertinent transport problems.

The linearity of the problem makes it convenient to distinguish a boundary (BVP), initial (IVP), and production (PVP) value problem. These can be solved separately and added to obtain the solution to the overall problem. In the BVP all conditions are homogeneous except for the input function (i.e., λ_1 and λ_2 in (5.5) and (5.6) and $F(X)$ in (5.7) are equal to zero). Similarly, the IVP is solved by having a zero input, $G(T)$, and no solute production, while the PVP is based on a zero input and initial distribution. This superposition principle can also be applied, if necessary, to solve an individual BVP, IVP, or PVP by breaking up complicated expressions for G, F, or λ_1 and λ_2, respectively.

Concentrations obtained from the effluent of finite soil columns or from solution samplers are usually viewed as representing C_1. On the other hand, soil coring or other nondestructive techniques such as TDR or electrical conductivity will yield results that are best described with a total resident concentration. Such a concentration is defined as mass of solute per unit volume of soil, with weighed contributions from both C_1 and C_2:

$$C_T = \beta R C_1 + (1-\beta) R C_2 \tag{5.10}$$

Flux-averaged or flowing concentrations need to be used if the solute concentration is obtained as the ratio of the solute and water fluxes; for example, for effluent curves of finite soil columns. The flux-averaged concentration is obtained from the resident concentration for phase 1 according to the well-known transformation (Kreft and Zuber, 1978):

$$C^f = C^r - \frac{1}{P}\frac{\partial C^r}{\partial X} \tag{5.11}$$

where the superscripts r and f denote the resident and flux modes.

Solution Procedure

The solution of (5.5) and (5.6), subject to (5.7) through (5.9), will be obtained in the standard manner for semi-infinite domains by applying a Laplace transform with respect to X and T (Lindstrom and Narasimhan, 1973; Toride et al., 1993). First, we apply the following Laplace transform with respect to time

$$\mathcal{L}_t[C(X,T)] = \overline{C}(X,s) = \int_0^\infty C(X,T)\exp(-sT)dT \tag{5.12}$$

to (5.5), (5.6), (5.8), and (5.9). After substituting initial condition (5.7), we obtain

$$\beta R(s\overline{C}_1 - F) = \frac{1}{P}\frac{d^2\overline{C}_1}{dX^2} - \frac{d\overline{C}_1}{dX} + \omega(\overline{C}_2 - \overline{C}_1) - \mu_1\overline{C}_1 + \frac{\lambda_1}{s} \tag{5.13}$$

$$(1-\beta)R(s\overline{C}_2 - f) = \omega(\overline{C}_1 - \overline{C}_2) - \mu_2\overline{C}_2 + \frac{\lambda_2}{s} \tag{5.14}$$

subject to

$$\left(\overline{C}_1 - \frac{1}{P}\frac{d\overline{C}_1}{dX}\right)\Bigg|_{X=0^+} = \overline{G}(s) \tag{5.15}$$

$$\frac{d\overline{C}_1}{dX}(\infty, s) = 0 \tag{5.16}$$

The Laplace transform of C_2 is

$$\overline{C}_2 = \frac{1}{s+b}\left[a_1\overline{C}_1 + F + \frac{\lambda_2}{(1-\beta)Rs}\right] \tag{5.17}$$

where

$$a_1 = \frac{\omega}{(1-\beta)R} \qquad b = \frac{\omega+\mu_2}{(1-\beta)R} \tag{5.18a,b}$$

We can invert this expression with the convolution theorem (Spiegel, 1965), to formulate C_2 in the regular time domain in terms of C_1:

$$\begin{aligned} C_2(X,T) &= \int_0^T a_1 \exp[-b(T-\tau)]C_1(X,\tau)d\tau \\ &+ F(X)\exp(-bT) + \frac{\lambda_2(X)}{\omega+\mu_2}[1-\exp(-bT)] \end{aligned} \tag{5.19}$$

Equation 5.13 is cast in terms of one dependent variable by substituting 5.17 for $\overline{C}_2$. This yields

$$\frac{1}{P}\frac{d^2\overline{C}_1}{dX^2} - \frac{d\overline{C}_1}{dX} + \left(\frac{a_1\omega}{s+b} - \omega - \mu_1 - \beta Rs\right)\overline{C}_1 + \left(\frac{\omega}{s+b} + \beta R\right)F + \frac{\lambda_1}{s} + \frac{a_1\lambda_2}{(s+b)s} = 0 \tag{5.20}$$

We opt to solve the ordinary differential equation with the following Laplace transform

$$\mathcal{L}_X[\overline{C}(X,s)] = \tilde{\overline{C}}(r,s) = \int_0^\infty \overline{C}(X,s)\exp(-rX)dX \tag{5.21}$$

After applying this transform to Eq. 5.20 and using inlet condition (Eq. 5.15), we can obtain the following algebraic expression for the (double) transformed concentration

$$\tilde{\overline{C}}_1(r,s) = \left\{r\overline{C}_1(0,s) - P\left[\overline{G}(s) + \left(\frac{\omega}{s+b} + \beta R\right)\tilde{F}(r) + \frac{\tilde{\lambda}_1(r)}{s} + \frac{a_1\tilde{\lambda}_2(r)}{s+b}\right]\right\} / (r-r_1)(r-r_2) \tag{5.22}$$

with

$$r_{1,2} = \frac{P}{2} \pm \sqrt{\frac{P^2}{4} - P\left(\frac{a_1\omega}{s+b} - \omega - \mu_1 - \beta R s\right)} \tag{5.23}$$

The inversion procedure is most conveniently carried out by first considering the transform with respect to X. The right-hand side of Eq. 5.20 is inverted term by term using a table of Laplace transforms (Spiegel, 1965) to obtain

$$\begin{aligned}\overline{C}_1(X,s) = \frac{1}{r_2 - r_1}\Big\{&\left[r_2 \exp(r_2 X) - r_1 \exp(r_1 X)\right]\overline{C}_1(0,s) - \left[\exp(r_2 X) - \exp(r_1 X)\right]P\overline{G}(s)\\ &- P\int_0^X \left[\exp[r_2(X-\xi)] - \exp[r_1(X-\xi)]\right]\left[\left(\frac{\omega}{s+b} + \beta R\right)F(\xi) + \frac{\lambda_1(\xi)}{s} + \frac{a_1\lambda_2(\xi)}{(s+b)s}\right]d\xi\Big\}\end{aligned} \tag{5.24}$$

Outlet condition (Eq. 5.16) is used for evaluating $\overline{C}_1(0,s)$ by differentiating Eq. 5.24 according to Leibniz' rule. Substitution of this result in Eq. 5.24 allows us to obtain the general solution in the transformed time domain, which may be written as the following sum of solutions to the BVP (first line), IVP (second and third line), and PVP (fourth and fifth line):

$$\begin{aligned}C_1(X,s) = {}& \frac{P\overline{G}(s)}{r_1}\exp(r_2 X)\\ &+ \frac{P}{r_2 - r_1}\left(\frac{\omega}{s+b} + \beta R\right)\Big\{\frac{r_2}{r_1}\exp(PX)\int_0^\infty \exp[-r_1(X+\xi)]F(\xi)d\xi\\ &- \int_X^\infty \exp[r_1(X-\xi)]F(\xi)d\xi - \int_0^X \exp[r_2(X-\xi)]F(\xi)d\xi\Big\}\\ &+ \frac{P}{r_2 - r_1}\Big\{\frac{r_2}{r_1}\exp(PX)\int_0^\infty \exp[-r_1(X+\xi)]\left[\frac{\lambda_1(\xi)}{s} + \frac{a_1\lambda_2(\xi)}{(s+b)s}\right]d\xi\\ &- \int_X^\infty \exp[r_1(X-\xi)]\left[\frac{\lambda_1(\xi)}{s} + \frac{a_1\lambda_2(\xi)}{(s+b)s}\right]d\xi - \int_0^X \exp[r_2(X-\xi)]\left[\frac{\lambda_1(\xi)}{s} + \frac{a_1\lambda_2(\xi)}{(s+b)s}\right]d\xi\Big\}\end{aligned} \tag{5.25}$$

We rewrite Eq. 5.23 as (Villermaux and van Swaaij, 1969; De Smedt and Wierenga, 1979; Toride et al., 1993):

$$r_{1,2} = \frac{P}{2} \pm a_0\sqrt{s + b + \frac{a_2}{s+b} - d} \tag{5.26}$$

with

$$a_0 = \sqrt{\beta RP},\ a_2 = \frac{\omega}{\beta R} a_1 = \frac{\omega^2}{(1-\beta)\beta R^2},\ d = \frac{1}{\beta R}\left(\frac{P}{4} + \omega + \mu_1\right) - b \qquad (5.27a,b,c)$$

For the inversion with respect to time, we make use of the property that the inverse of the iterated Laplace transform of a function is equal to the generalized convolution integral of the function (Walker, 1987; Sneddon, 1995). The Laplace transform and its inverse may be applied sequentially to a particular variable. Consider a sequence of two, where we write t for t_1 in at least one place in the expression for the concentration, and replace t by t_2 in all remaining places. The corresponding transformation variable s is split up into s_1 and s_2. The sequential inversion of a concentration that is transformed in this manner may be written as

$$h(t_1,t_2) = \mathcal{L}_{t_1}^{-1}\left\{\mathcal{L}_{t_2}^{-1}\left[\overline{C}(s_1,s_2)\right]\right\} \qquad (5.28)$$

The concentration in the regular time domain is equal to the convolution of the function:

$$C(x,t) = \int_0^t h(\tau, t-\tau)d\tau \qquad (5.29)$$

An example of this approach for nonequilibrium transport is given on page 159 of Jury and Roth (1990). The following identities may be established with the generalized convolution theorem (Lindstrom and Narasimhan, 1973):

$$\mathcal{L}_T\left\{\exp(-bT)\int_0^T I_0\left[2\sqrt{a_2(T-\tau)\tau}\right]h(\tau)d\tau\right\} = \frac{1}{s+b}\overline{h}\left(s+b-\frac{a_2}{s+b}\right) = \frac{\overline{h}(s^*)}{s+b} \qquad (5.30)$$

$$\mathcal{L}_T\left\{\exp(-bT)\frac{\partial}{\partial T}\int_0^T I_0\left[2\sqrt{a_2(T-\tau)\tau}\right]h(\tau)d\tau\right\} = \overline{h}\left(s+b-\frac{a_2}{s+b}\right) = \overline{h}(s^*) \qquad (5.31)$$

where I_0 is the modified zero-order Bessel function while $h(\tau)$ and $\overline{h}(s^*)$ denote regular and transformed functions with s^* equal to $s+b-a_2/(s+b)$. We will continue by deriving general solutions for the three problems.

Boundary Value Problem

The general solution for the BVP according to Eq. 5.25 can now be written as

$$\overline{C}_1(X,s) = \frac{P\overline{G}(s)}{a_0}\exp\left(\frac{PX}{2}\right)\overline{h}_1(X,s^*+d) \qquad (5.32)$$

where the Laplace domain function, $\overline{h}_1(X,s^*+d)$, is given by

$$h_1(X,s^* + d) = \frac{\exp\left[-a_0 X\sqrt{s^* + d}\right]}{\sqrt{s^* + d} + P/2a_0} \tag{5.33}$$

With the help of Eq. 5.31 we can obtain

$$\begin{aligned} C_1(X,T) &= \frac{P}{a_0}\exp\left(\frac{PX}{2}\right)\int_0^T G(T-\tau)\exp(-b\tau)\frac{\partial}{\partial\tau}\left\{\int_0^\tau I_0\left[2\sqrt{a_2(\tau-\eta)\eta}\right]h_1(X,\eta)d\eta\right\} \\ &= \frac{P}{a_0}\exp\left(\frac{PX}{2}\right)\int_0^T G(T-\tau)\exp(-b\tau)\left\{h_1(X,\tau) + \int_0^\tau\sqrt{\frac{a_2\eta}{\tau-\eta}}I_1\left[2\sqrt{a_2(\tau-\eta)\eta}\right]h_1(X,\eta)d\eta\right\} \end{aligned} \tag{5.34}$$

where I_1 denotes the modified first-order Bessel function. The auxiliary function $h_1(X,T)$ is defined as

$$h_1(X,T) = \exp(-dT)\left[\frac{1}{\sqrt{\pi T}}\exp\left(-\frac{a_0^2 X^2}{4T}\right) - \frac{P}{2a_0}\exp\left(\frac{P^2 T}{4a_0^2} + \frac{PX}{2}\right)\text{erfc}\left(\frac{a_0^2 X + PT}{2a_0\sqrt{T}}\right)\right] \tag{5.35}$$

The expressions for C_1 and, using Eq. 5.19, C_2 as the result of an arbitrary input function, $G(T)$, are hence:

$$\begin{aligned} C_1(X,T) = \int_0^T G(T-\tau)&\left[\exp\left(-\frac{\omega\tau}{\beta R}\right)G_1(X,\tau)\right. \\ &\left. + \int_0^\tau \frac{\omega}{R}\sqrt{\frac{\eta}{(1-\beta)\beta(\tau-\eta)}}H_1(\eta;\tau)G_1(X,\eta)d\eta\right] \end{aligned} \tag{5.36}$$

$$C_2(X,\tau) = \frac{\omega}{(1-\beta)R}\int_0^T \exp\left[-\frac{\omega+\mu_2}{(1-\beta)R}(T-\tau)\right]C_1(X,\tau)d\tau \tag{5.37}$$

where the auxiliary function $G_1(X,\tau)$ and $H_1(\eta;\tau)$ are given by

$$G_1(X,\tau) = \exp\left(-\frac{\mu_1\tau}{\beta R}\right)\left[\sqrt{\frac{P}{\pi\beta R\tau}}\exp\left(-\frac{(\beta RX-\tau)^2}{4\beta R\tau/P}\right) - \frac{P}{2\beta R}\exp(PX)erfc\left(\frac{\beta RX+\tau}{\sqrt{4\beta R\tau/P}}\right)\right] \tag{5.38}$$

$$H_1(\eta,\tau) = \exp\left[-\frac{\omega\eta}{\beta R} - \frac{\omega+\mu_2}{(1-\beta)R}(\tau-\eta)\right]I_1\left[\frac{2\omega}{R}\sqrt{\frac{(\tau-\eta)\eta}{(1-\beta)\beta}}\right] \tag{5.39}$$

The solution for a flux-averaged concentration can be readily obtained from this solution by applying the transformation given by Eq. 5.11 to $G_1(X,\tau)$.

Initial Value Problem

The solution for the IVP is slightly longer than for the BVP, but the inversion to the regular time domain follows the same pattern. The second and third line of Eq. 5.25 can be rewritten as:

$$\overline{C}_1(X,s) = \frac{P}{2a_0}\left(\frac{\omega}{s+b} + \beta R\right)$$
$$\left\{\int_0^\infty \left[\overline{h}_1(X+\xi, s^* + d) - \frac{P}{2a_0}\overline{h}_2(X+\xi, s^* + d\right]\exp\left[\frac{P}{2}(X-\xi)\right]F(\xi)d\xi \right.$$
$$+ \int_X^\infty \overline{h}_3(\xi - X, s^* + d)\exp\left[\frac{P}{2}(X-\xi)\right]F(\xi)d\xi \tag{5.40}$$
$$\left. + \int_0^X \overline{h}_3(X-\xi, s^* + d)\exp\left[\frac{P}{2}(X-\xi\right]F(\xi)d\xi\right\}$$

where the additional transformed auxiliary functions are given by

$$\overline{h}_2(X, s^* + d) = \frac{\exp\left(-a_0 X\sqrt{s^* + d}\right)}{\left(\sqrt{s^* + d} + P/2a_0\right)\sqrt{s^* + d}} \tag{5.41}$$

$$\overline{h}_3(X, s^* + d) = \frac{\exp\left(-a_0 X\sqrt{s^* + d}\right)}{\sqrt{s^* + d}} \tag{5.42}$$

The inversion is conducted with the aid of Eqs. 5.30 and 5.31. After the same type of differentiation as for the BVP and some algebraic manipulations, we can obtain the following solutions for a general initial distribution $F(X)$:

$$C_1(X,T) = \int_0^\infty F(\xi)\left\{\exp\left(-\frac{\omega T}{\beta R}\right)G_2(X,T,\xi)\right.$$
$$\left. + \int_0^T \left[\frac{\omega}{\beta R}H_0(\tau;T) + \sqrt{\frac{a_2\tau}{T-\tau}}H_1(\tau;T)\right]G_2(X,\tau,\xi)d\tau\right\}d\xi \tag{5.43}$$

$$C_2(X,T) = F(X)\exp\left(-\frac{\omega+\mu_2}{(1-\beta)R}T\right)$$
$$+\frac{\omega}{(1-\beta)R}\int_0^\infty\int_0^T F(\xi)\left[H_0(\tau;T) + \frac{\omega}{\beta R}\sqrt{\frac{T-\tau}{a_2\tau}}H_1(\tau;T)\right]G_2(X,\tau,\xi)d\tau d\xi \tag{5.44}$$

The inverse transforms of Eqs. 5.41 and 5.42 are as follows

$$h_2(X,T) = \exp\left[-dT + \frac{P^2T}{4a_0^2} + \frac{PX}{2}\right]\mathrm{erfc}\left[\frac{a_0^2X + PT}{2a_0\sqrt{T}}\right] \tag{5.45}$$

$$h_3(X,T) = \frac{1}{\sqrt{\pi T}}\exp\left[-dT - \frac{a_0^2X^2}{4T}\right] \tag{5.46}$$

The expression for $G_2(X,\tau,\xi)$ is

$$\begin{aligned} G_2(X,T,\xi) = {} & \exp\left(-\frac{\mu_1 T}{\beta R}\right)\left\{\sqrt{\frac{\beta RP}{4\pi\tau}}\exp\left(PX - \frac{[\beta R(\xi + X) + \tau]^2}{4\beta R\tau/P}\right)\right. \\ & \left. -\frac{P}{2}\exp(PX)\mathrm{erfc}\left[\frac{\beta R(\xi + X) + \tau}{\sqrt{4\beta R\tau/P}}\right] + \sqrt{\frac{\beta RP}{4\pi\tau}}\exp\left(-\frac{[\beta R(\xi - X) + \tau]^2}{4\beta R\tau/P}\right)\right\} \end{aligned} \tag{5.47}$$

while $H_0(\tau;T)$ is defined by

$$H_0(\tau;T) = \exp\left[-\frac{\omega\tau}{\beta R} - \frac{\omega + \mu_2}{(1-\beta)R}(T-\tau)\right]I_0\left[2\sqrt{a_2(T-\tau)\tau}\right] \tag{5.48}$$

Production Value Problem

The Laplace transform of the solution for the PVP is very similar to that for the IVP as can be seen from Eq. 5.25. The solution procedure for the PVP is therefore identical to that for the IVP and we will proceed by immediately stating the concentrations for depth dependent production terms:

$$\begin{aligned} C_1(X,T) = {} & \frac{1}{\beta R}\int_0^\infty\int_0^T\left\{\lambda_1(\xi)\exp\left(-\frac{\omega\tau}{\beta R}\right)G_2(X,\tau,\xi)\right. \\ & \left. + \int_0^\tau\left[a_1H_0(\eta;\tau)\lambda_2(\xi) + \sqrt{\frac{a_2\eta}{\tau-\eta}}H_1(\eta;\tau)\lambda_1(\xi)\right]G_2(X,\eta,\xi)d\eta\right\}d\tau d\xi \end{aligned} \tag{5.49}$$

$$C_2(X,T) = \frac{\lambda_2}{\omega + \mu_2}[1 - \exp(-bT)] + \frac{\omega}{(1-\beta)R}\int_0^T\exp[-b(T-\tau)]C_1(X,\tau)d\tau \tag{5.50}$$

The solutions of the PVP may also be written in terms of Goldstein's *J*-function. Because this function is often employed in analytical solutions for bi-continuum nonequilibrium transport, we will review this alternative solution procedure. The integration with respect to τ in Eq. 5.49 is carried out according to Fubini's theorem (Lindstrom and Stone, 1974; Taylor, 1985):

$$\int_0^T \int_0^\tau f(\eta,\tau)d\eta d\tau = \int_0^T \int_\eta^T f(\eta,\tau)d\tau d\eta \tag{5.51}$$

to rewrite Eq. 5.49 as

$$C_1(X,T) = \frac{1}{\beta R}\int_0^\infty \left\{ \int_0^T \exp\left(-\frac{\omega\tau}{\beta R}\right)\lambda_1(\xi)G_2(X,\tau,\xi)d\tau + \int_0^T \int_\eta^T \exp\left(-p' - \frac{\omega\mu_2\eta}{(\omega+\mu_2)\beta R} - q'\right) \right.$$
$$\left. \times\left[\lambda_1(\xi)\frac{\omega+\mu_2}{(1-\beta)R}\sqrt{\frac{p'}{p'}}I_1\left[2\sqrt{p'q'}\right] + a_2\lambda_2(\xi)I_0\left[2\sqrt{p'q'}\right]\right]G_2(X,\eta,\xi)d\tau d\eta \right\} d\xi \tag{5.52}$$

where

$$p' = \frac{\omega^2}{(\omega+\mu_2)\beta R}\eta \qquad q' = \frac{\omega+\mu_2}{(1-\beta)R}(\tau-\eta) \tag{5.53a,b}$$

The second integral with respect to τ is rewritten with the help of the *J*-function (Goldstein, 1953), which is defined as:

$$J(p,q) = 1 - \exp(-q)\int_0^p \exp(-w)I_0\left[2\sqrt{qw}\right]dw \tag{5.54}$$

with

$$p = \frac{\omega^2}{(\omega+\mu_2)\beta R}\tau \qquad q = \frac{\omega+\mu^2}{(1-\beta)R}(T-\tau) \tag{5.55a,b}$$

Utilizing the properties of the *J*-function (van Genuchten, 1981), we can establish that

$$\int_\eta^T \frac{\omega+\mu_2}{(1-\beta)R}\exp(-p'-q')\sqrt{\frac{p'}{q'}}I_2(2\sqrt{p'q'})d\tau = J(p',q'') - \exp(-p') \tag{5.56}$$

$$\int_\eta^T \exp(-p'-q')I_0\left(2\sqrt{p'q'}\right)d\tau = \frac{\omega+\mu_2}{(1-\beta)R}[1-J(q'',p')] \tag{5.57}$$

where

$$q'' = \frac{\omega+\mu_2}{(1-\beta)R}(T-\eta) \tag{5.58}$$

After some elementary algebra, the solution for C_1 becomes

$$C_1(X,T) = \int_0^{\infty}\int_0^{T} \exp\left(-\frac{\omega\mu_2\tau}{(\omega+\mu_2)\beta R}\right)\left(\lambda_1(\xi)J(p,q) + \frac{\omega\lambda_2(\xi)}{\omega+\mu_2}[1-J(p,q)]\right)\frac{G_2(X,\tau,\xi)}{\beta R}\,d\tau d\xi \qquad (5.59)$$

The expression for C_2 can be derived from this solution. Substitution of Eq. 5.59 into Eq. 5.50 yields a term with triple integrals. The double integration with respect to τ and ξ can be simplified with Fubini's theorem and integration by parts. This involves differentiation of the *J*-function and integration of the modified Bessel functions (Abramowitz and Stegun, 1970). The following results may be obtained:

$$C_2(X,T) = \frac{\lambda_2}{\omega+\mu_2}\left[1-\exp\left(-\frac{\omega+\mu_2}{(1-\beta)R}T\right)\right] + \frac{\omega}{\beta R(\omega+\mu_2)}\int_0^{\infty}\int_0^{T}\exp\left(-\frac{\omega\mu_2\tau}{(\omega+\mu_2)\beta R}\right)$$
$$\times\left[[1-J(q,p)]\lambda_1(\xi) + \frac{\omega\lambda_2(\xi)}{\omega+\mu_2}\left(1-J(q,p)+\sqrt{\frac{q'}{p'}}I_1\left(2\sqrt{pq}\right)\right)\right]G_2(X,\tau,\xi)\,d\tau d\xi \qquad (5.60)$$

Specific Solutions

The previous general solutions can be further evaluated for simple boundary, initial, or production functions. Specific analytical solutions will be derived for a Heaviside-type distribution for *F*, *G*, and λ. Solutions for a Dirac and exponential distribution were presented, among others, by Toride et al. (1993) while Toride et al. (1995) treat estimation of nonequilibrium parameters from observed concentrations.

Boundary Value Problem

As an example, consider the input of step input according to

$$G(T) = \begin{cases} 0 & T<0 \\ G_0 & T>0 \end{cases} \qquad (5.61)$$

Substitution of this condition into Eq. 5.36 and application of Fubini's theorem and properties of the Goldstein function yield the following relatively simple solution

$$C_1(X,T) = G_o\int_0^{T}\exp\left(-\frac{\omega\mu_2\tau}{(\omega+\mu_2)\beta R}\right)J(p,q)G_1(X,\tau)\,d\tau \qquad (5.62)$$

To derive the solution for C_2, we substitute the solution for C_1 into Eq. 5.37 and change the order of integration. Furthermore, we use the identity:

$$\int_{\eta}^{T}\exp\left[-\frac{\omega+\mu_2}{(1-\beta)R}(T-\tau)\right]J(p',q')\,d\tau = \frac{(1-\beta)R}{\omega+\mu_2}[1-J(q,p)] \qquad (5.63)$$

to derive the following solution

$$C_2(X,T) = \frac{\omega G_o}{\omega + \mu_2} \int_0^T \exp\left[-\frac{\omega \mu_2 \tau}{(\omega + \mu_2)\beta R}\right] [1 - J(q, p)] G_1(X, \tau) d\tau \tag{5.64}$$

The solution for a step input ($G[T]=G_o$ if $0<T<T_o$ and otherwise zero) can be directly written down with the above expressions according to the superposition principle:

$$C(X,T) = C(X,T) - C(X,T - T_o) \tag{5.65}$$

Consider the solution of a boundary value problem, for which we want to illustrate the effect of the transfer rate, α, and the immobile water content, θ_m, on the shape of the breakthrough curve. Figure 5.1 shows the flux-averaged concentration versus time at x=50 cm resulting from the application of a 2-day solute pulse at x = 0. Relevant transport parameters are as follows: v = 20 cm/d, D = 10 cm²/d, R = 5, and θ = 0.5. There is no solute production or decay. The behavior of C_1 and C_2 is shown for four combinations of α and θ_m corresponding to ω = 0.2, 1, 5, or ∞ and β = 0.25, 0.5, 0.75, and 0.99, respectively. Note that we assumed $f=\theta_m/\theta$. The case where θ_m = 0.495 and α approaches infinity corresponds to equilibrium transport; the breakthrough curve is fairly symmetric with the highest concentration occurring after 13.5 days, which follows directly from the center of mass of the solute pulse and the retarded movement through the soil, while the C_1 and C_2 profiles are similar. With decreasing α and θ_m, there is earlier breakthrough with more nonsymmetric spreading. Furthermore, there is an increasing difference between the C_1 and C_2 profiles if the degree of nonequilibrium increases. The solute will move faster through the mobile region, at a speed of v/β, but once solute ends up in the more prevalent immobile region it will reside there for longer times due to the decreased exchange rate.

Figure 5.2 deals with the same problem but we are now interested in the value of f. This parameter symbolizes the fraction of sorption sites that is in direct contact with the mobile region in case of reactive transport. It is plausible that a majority of the sorption sites are only accessible from the stagnant region since this region tends to be closer to the solid. The assumption that $f = \theta_m/\theta$ may hence be an overestimation. All transport parameters are the same as for Figure 5.1 while we use ρk = 2 to obtain R=5 for θ = 0.5. The two different concentrations are again plotted, but now for five values of f using α = 0.2 d^{-1}, θ_m = 0.25, and ω = 1. The accompanying value for β is shown as well in Figure 5.2. The breakthrough curves demonstrate that an increase in f results in slower solute movement due to the enhanced opportunity for sorption. The shapes of the breakthrough curve are similar since the nonequilibrium does not pertain to sorption; the variation in f is manifested as a change in R. The implication is that solute retardation may be less than expected for transport of a reactive solute in a fractured medium.

Initial Value Problem

The *Heaviside* initial profile is given by

$$F(X) = F_o[U(X - X_1) - U(X - X_2)] = \begin{cases} F_o & X_1 < X < X \\ 0 & \text{otherwise} \end{cases} \tag{5.66}$$

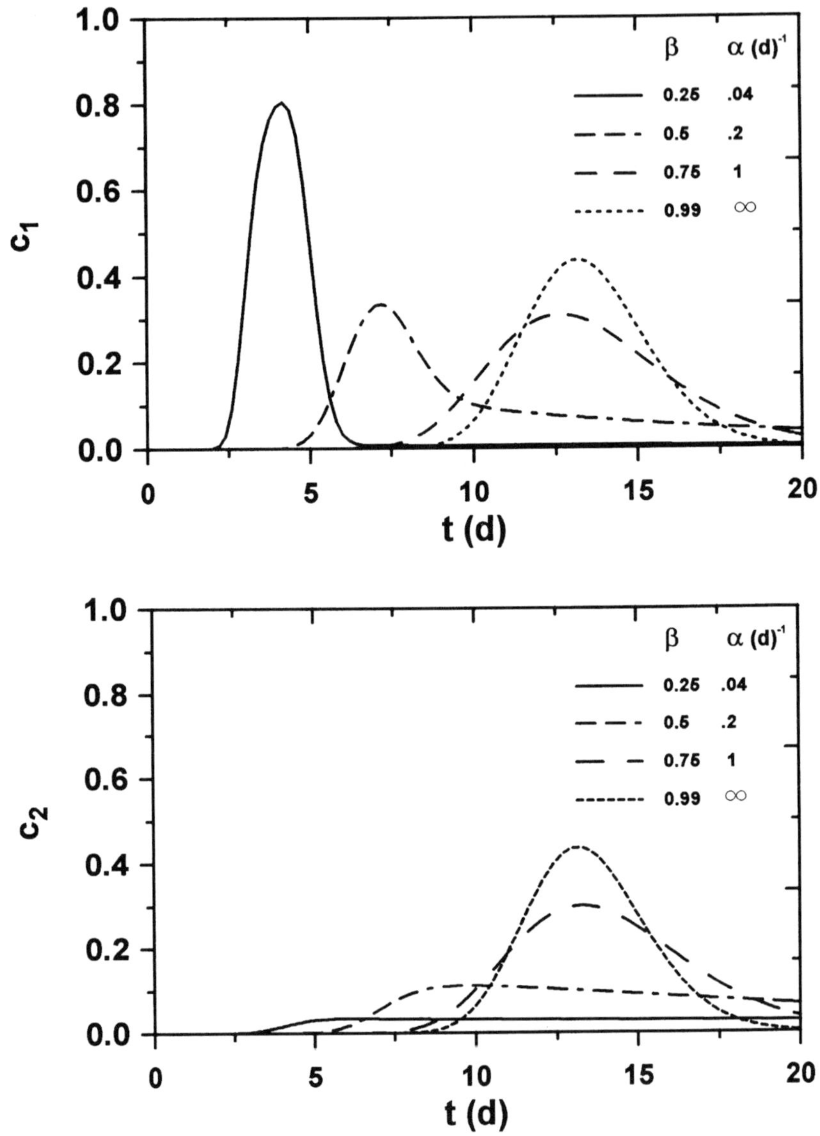

Figure 5.1. Concentrations for phases 1 and 2 versus time at x=50 cm for a 2-d pulse input calculated with the first-order nonequilibrium model using four different combinations of the mobile water content, $\beta=\theta_m/\theta$, and the transfer rate, α.

where $U(X)$ denotes the unit-step or Heaviside function. The following solution is obtained by substituting this condition into Eq. 5.43 and subsequent integration:

$$
\begin{aligned}
C_1(X,T) &= F_o \exp\left(-\frac{\omega T}{\beta T}\right)[G_3(X,T;X_1) - G_3(X,T;X_2)] \\
&+ F_o \int_0^T \left[\frac{\omega}{\beta R} H_0(\tau;T) + \sqrt{\frac{a_2 \tau}{T-\tau}} H_1(\tau;T)\right][G_3(X,\tau;X_1) - G_3(X,\tau;X_2)]d\tau
\end{aligned}
\tag{5.67}
$$

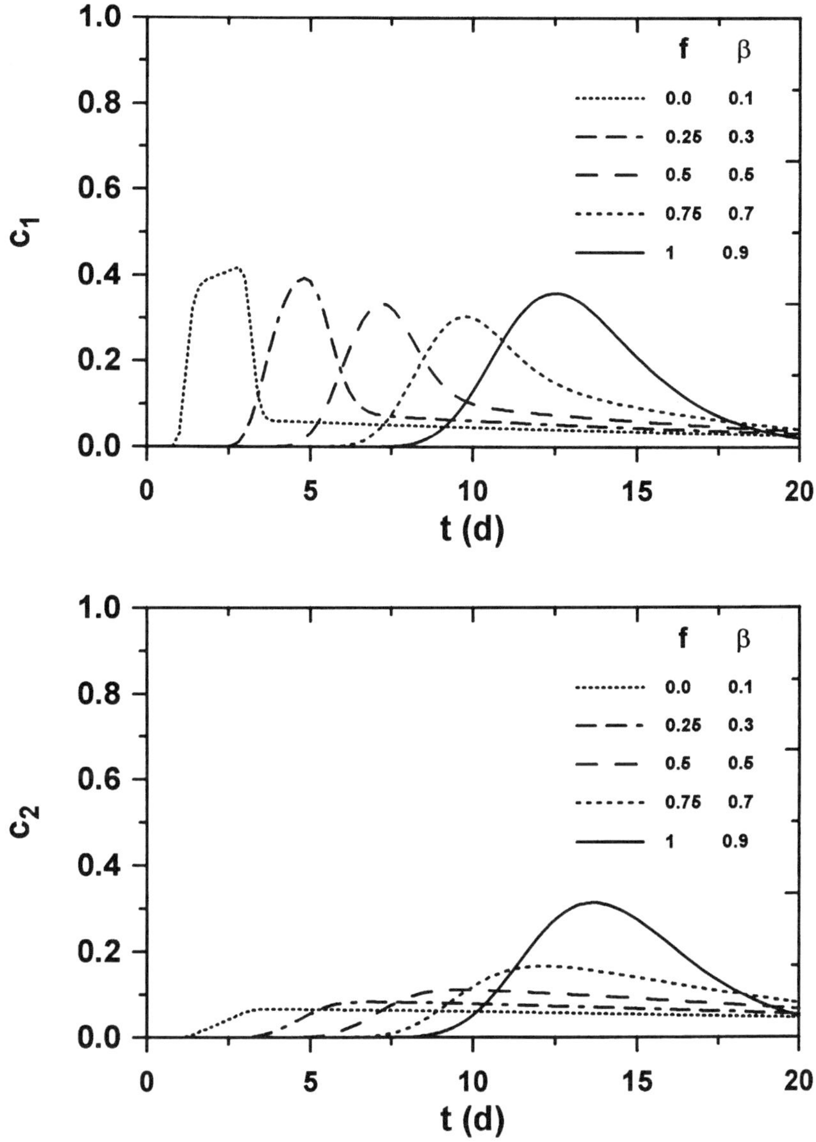

Figure 5.2. Concentrations for phases 1 and 2 versus time at x=50 cm for a 2-d pulse input calculated with the first-order nonequilibrium model using five different fractions of sorption sites, f, that equilibrate with the mobile region.

where

$$G_3(X,T,X) = \exp\left(-\frac{\mu_1\tau}{\beta R}\right)\left\{1-\frac{1}{2}\operatorname{erfc}\left(\frac{\beta R(X-X_i)-\tau}{\sqrt{4\beta R\tau/P}}\right)+\frac{1}{2}\left(1+P(X+X_i)+\frac{P\tau}{\beta R}\right)\right.$$
$$\left.\times\exp(PX)\operatorname{erfc}\left[\frac{\beta R(X_i+X)+\tau}{\sqrt{4\beta R\tau/P}}\right]-\sqrt{\frac{P\tau}{\pi\beta R}}\exp\left(PX-\frac{[\beta R(X+X_i)+\tau]^2}{4\beta R\tau/P}\right)\right\} \tag{5.68}$$

Similarly, the expression for C_2 is written as

$$C_2(X,T) = F_o[U(X-X_1) - U(X-X_2)]\exp\left(-\frac{\omega-\mu_2}{(1-\beta)R}T\right)$$
$$+\frac{\omega F_o}{(1-\beta)R}\int_0^T\left[H_0(\tau;T)+\frac{\omega}{\beta R}\sqrt{\frac{T-\tau}{a_2\tau}}H_1(\tau;T)\right][G_3(X,\tau;X_1)-G_3(X,\tau;X_2)]d\tau \tag{5.69}$$

The previous one-dimensional solutions can be readily adapted to describe three-dimensional transport because there are no further nonequilibrium processes due to the additional coordinates (Leij et al., 1993). The solution may be written as

$$C_1(X,Y,Z,T) = F_o\exp\left(-\frac{\omega T}{\beta R}\right)[G_3(X,T;X_1)-G_3(X,T;X_2)\Gamma(Y,Z,T)$$
$$+F_o\int_0^T\left[\frac{\omega}{\beta R}H_0(\tau;T)+\sqrt{\frac{a_2\tau}{T-\tau}}H_1(\tau;T)\right][G_3(X,\tau;X_1)-G_3(X,\tau;X_2)]\Gamma(Y,Z,\tau)d\tau \tag{5.70}$$

$$C_2(X,T) = F_o[U(X-X_1)-U(X-X_2)][U(Y-Y_1)-U(Y-Y_2)]$$
$$\times[U(Z-Z_1)-U(Z-Z_2)]\exp\left(-\frac{\omega+\mu_2}{(1-\beta)R}T\right)$$
$$+\frac{\omega F_o}{(1-\beta)R}\int_0^T\left[H_0(\tau;T)+\frac{\omega}{\beta R}\sqrt{\frac{T-\tau}{a_2\tau}}H_1(\tau;T)\right] \tag{5.71}$$
$$\times[G_3(X,\tau;X_1)-G_3(X,\tau;X_2)]\Gamma(Y,Z,\tau)d\tau$$

for a solute that is initially present in the region $X_1<X<X_2$, $Y_1<Y<Y_2$, and $Z_1<Z<Z_2$. The transversal auxiliary function is given by

$$\Gamma(Y,Z,T) = \frac{\beta R\sqrt{P_Y P_Z}}{4\pi T}\exp\left(-\frac{\beta R(P_Y Y^2 + P_Z Z^2)}{4T}\right) \tag{5.72}$$

where P_Y and P_Z are Peclet numbers for the two transverse coordinates. Consider the initial value problem sketched in the first item of Figure 5.3 where the solute is originally located in the regions $5<x<15$ (F=1) and $25<x<35$ (F=0.5) for $15<y<25$ and $-\infty<z<\infty$. The problem is in fact two-dimensional. There is uniform flow in the x-direction with v=50 cm/d, while R=1, and the dispersion coefficient for the x-direction, D_x, is 20 cm²/d and those for the y and z directions are D_y=D_z=5 cm²/d. The nonequilibrium parameters are β=0.5 and ω=1, which corresponds to α=0.25 d^{-1} and θ_m=0.25 if θ=0.5 and L=100. The remainder of Figure 5.3 consists of the plots for C_1, C_2, and C_T at t=0.5 d. The first concentration, C_1, follows a predictable pattern, the initial peaks are moved downstream as result of advection; the initially distinct peaks are blurred as the result of dispersion. Because of nonequilibrium exchange, solute remains in the mobile region across a large part of the x-direction. The C_2 profile is still quite similar to the

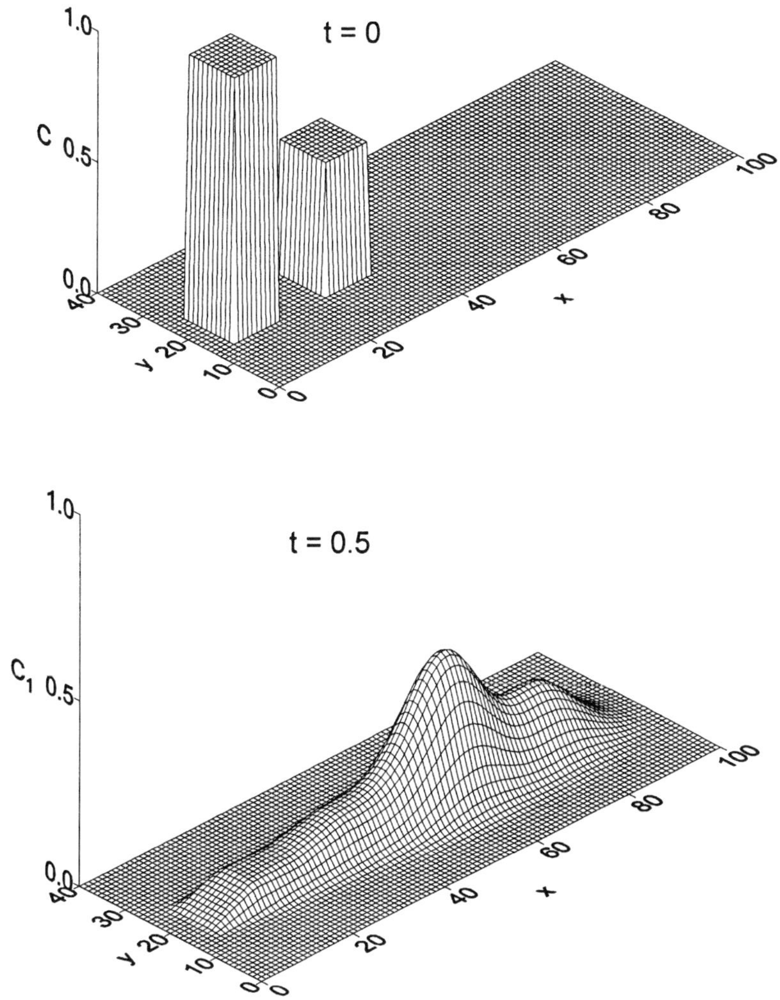

Figure 5.3. Initial concentration and C_1 at t=0.5 d as a function of longitudinal (x) and transversal (y) direction for a nonreactive solute according to the first-order nonequilibrium model with β=0.5 and ω=1.

initial distribution, the solute arrives through movement in the mobile region and subsequent transfer to the immobile region. Finally, the total concentration, C_T consists of weighted contributions from the equilibrium and nonequilibrium concentrations. It is apparent that the complete displacement of solutes from the soil profile will be more difficult to achieve as a result of nonequilibrium phenomena.

Production Value Problem

The *Heaviside* production profile is written as

$$\lambda_k(X) = \lambda_{k0}[U(X - X_1) - U(X - X_2)] = \begin{cases} \lambda_{k0} & X_1 < X < X_2 \\ 0 & \text{otherwise} \end{cases} \quad (k = 1,2) \tag{5.73}$$

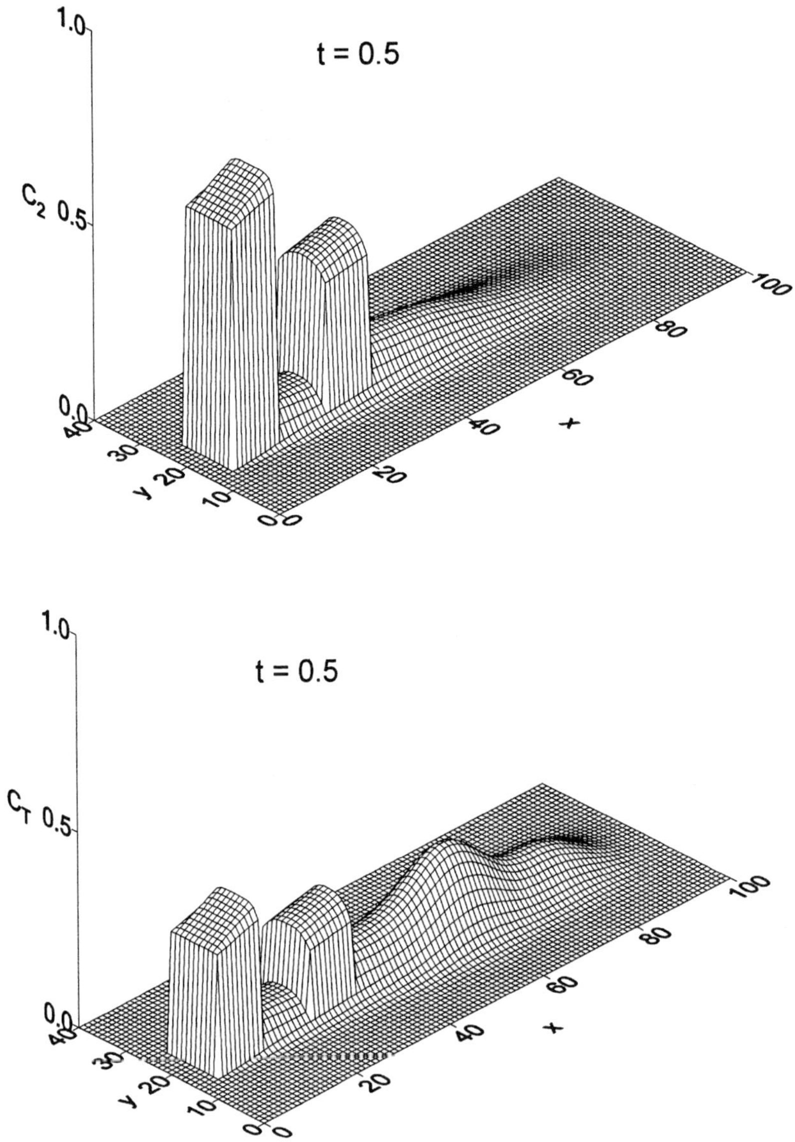

Figure 5.3. (Continued) C_2 and C_T at t=0.5 d as a function of longitudinal (x) and transversal (y) direction for a nonreactive solute according to the first-order nonequilibrium model with β=0.5 and ω=1.

where k is equal to 1 for the equilibrium phase and 2 for the nonequilibrium profile. The corresponding concentrations are given by:

$$C_1(X,T) = \frac{1}{\beta R}\int_0^T \exp\left(-\frac{\omega\mu_2\tau}{(\omega+\mu_2)\beta R}\right)\left(\lambda_{10}J(p,q) + \frac{\omega\lambda_{20}}{\omega+\mu_2}[1-J(q,p)]\right) \times [G_3(X,\tau;X_1) - G_3(X,\tau;X_2)]d\tau \tag{5.74}$$

$$C_2(X,T) = \frac{\lambda_2}{\omega+\mu_2}[U(X-X_1)-U(X-X_2)]\left[1-\exp\left(\frac{\omega+\mu_2}{(1-\beta)R}T\right)\right]$$
$$+\frac{\omega}{\beta R(\omega+\mu_2)}\int_0^T [G_3(X,\tau,X_1)-G_3(X,\tau,X_2)]\exp\left(-\frac{\omega\mu_2\tau}{(\omega+\mu_2)\beta R}\right)$$
$$\times\left[[1-J(q,p)]\lambda_{10}+\frac{\omega\lambda_{20}}{\omega+\mu_2}\left(1-J(q,p)+\sqrt{\frac{q}{p}}I_1\left(2\sqrt{pq}\right)\right)\right]d\tau \tag{5.75}$$

WELL-DEFINED AGGREGATES

Model Formulation

For well-defined structured or aggregated porous media—i.e., media for which the size and geometry of all aggregates are known and can be easily defined—we can model the diffusion process inside the aggregate for the physical nonequilibrium model. An analytical solution may be available for transport in media composed of aggregates with a simple shape (spheres, slabs, cylinders). This allows a more detailed description of the concentration inside the aggregate. The dependency of the transfer coefficient α or ω on the diffusion coefficient and the aggregate geometry can hence be elucidated (van Genuchten and Dalton, 1986). The derivation and formulation of the analytical solutions is somewhat more complicated than for the first-order model, but is generally accomplished by taking the Laplace transform with respect to time, solution of the ensuing ordinary differential equation with Bessel functions, and inversion to the regular time domain with complex variable theory. In the following we will only state the mathematical problem for various aggregates and provide references for further details.

For transport in the mobile region we can generalize the previous expression for first-order transfer according to Eq. 5.1. If zero- and first-order rate processes are excluded, the formulation in dimensional parameters becomes:

$$\theta_m R_m \frac{\partial c_m}{\partial t} + \theta_{im} R_{im} \frac{\partial c_{im}}{\partial t} = \theta_m D_m \frac{\partial^2 c_m}{\partial x^2} - \theta_m v_m \frac{\partial c_m}{\partial x} \tag{5.76}$$

where $R_m = 1 + \rho_m k_m/\theta_m$ and $R_{im} = 1 + \rho_{im} k_{im}/\theta_{im}$ are the retardation factors for the mobile and immobile regions exclusively (Parker and Valocchi, 1986). This notation obviates the use of f as in Eqs. 5.1 and 5.2. It is assumed that transverse concentration gradients in the mobile region can be neglected and that we can use a one-dimensional transport equation; the approach can be readily extended to higher dimensional problems (Goltz and Roberts, 1986). The same equation for the mobile region can be used for the different types of aggregates listed below.

Spherical Aggregates

Solute movement in the immobile phase of spherical aggregates is given by Fick's law, given in dimensional form as

$$R_{im}\frac{\partial c_a}{\partial t} = \frac{D_a}{r^2}\frac{\partial}{\partial r}\left(r^2\frac{\partial c_a}{\partial r}\right) \qquad (0 \le r \le a_s) \tag{5.77}$$

where r is the radial coordinate in a spherical system, a_s is the radius of the spherical aggregate, c_a is the concentration inside the aggregate, and D_a is an effective diffusion coefficient for the aggregate. The two boundary conditions for the diffusion equation are based on continuity at the aggregate surface and symmetry inside the aggregate. The concentration is prescribed at the interface of the macropore and aggregate while there is no flux at the center of the aggregate:

$$c_m(x,t) = c_a(x, r = a_s, t) \qquad \frac{\partial c_a}{\partial r}(x, r = 0, t) = 0 \tag{5.78a,b}$$

Note that the concentration in the aggregates depends on three independent variables (x,r,t), whereas the concentration in the macropore depends on only two variables (x,t). In dimensionless form the equation for intra-aggregate transport becomes

$$\frac{\partial C_a}{\partial T} = \frac{\gamma_s}{\xi^2}\frac{\partial}{\partial \xi}\left(\xi^2 \frac{\partial C_a}{\partial \xi}\right) \qquad (0 \le \xi \le 1) \tag{5.79}$$

with $T=vt/L$, $C=c/c_o$, $\xi=r/a_s$, and $\gamma_s=LD_a/a_s^2vR_{im}$. An analytical solution for the mobile concentration for this problem was presented by van Genuchten (1985) for a step input and an infinite outlet condition. Rasmuson and Neretnieks (1980) provided a simpler solution for the breakthrough curve in medium with a zero-gradient at the outlet. An immobile concentration equivalent to that in the first-order approximation is obtained by averaging the aggregate concentration according to

$$c_{im} = \frac{3}{a_s^3}\int_0^{a_s} r^2 c_a(x,r,t)dr \qquad \text{or} \qquad C_{im} = 3\int_0^1 \xi^2 C_a(X,\xi,T)d\xi \tag{5.80a,b}$$

Rectangular Aggregates

In case of a rectangular aggregate/void geometry where the solid consists of slabs with half-width a_r, transport inside the aggregate is governed by the customary diffusion equation for Cartesian coordinates:

$$R_{im}\frac{\partial c_a}{\partial t} = D_a \frac{\partial^2 c_a}{\partial y^2} \qquad (0 \le y \le a_r) \tag{5.81}$$

where y is a transverse coordinate. The boundary conditions are

$$c_m(x,t) = c_a(x, y = a_r, t) \qquad \frac{\partial c_a}{\partial y}(x, y = a_r, t) = 0 \tag{5.82a,b}$$

The dimensionless equation is given by

$$\frac{\partial C_a}{\partial T} = \gamma_r \frac{\partial^2 C_a}{\partial Y^2} \qquad (0 \le Y \le 1) \tag{5.83}$$

where $Y=y/a_r$ and $\gamma_r=LD_a/a_r^2vR_{im}$. Solutions for this problem were provided by Sudicky and Frind (1982) and van Genuchten (1985). The problem can again be made entirely one-dimensional by defining the following immobile concentration:

$$c_{im}=\frac{1}{a_r}\int_0^{a_r}c_a(x,y,t)dy \qquad \text{or} \qquad C_{im}=\gamma_r\int_0^1 C_a(X,Y,T)dY \tag{5.84a,b}$$

Solid Cylindrical Aggregates

In case of flow and transport around solid but porous cylindrical aggregates, the diffusion inside the aggregates is described by the following problem:

$$R_{im}\frac{\partial c_a}{\partial t}=\frac{D_a}{r}\frac{\partial}{\partial r}\left(r\frac{\partial c_a}{\partial r}\right) \qquad (0\le r\le a_c) \tag{5.85}$$

subject to the familiar conditions

$$c_m(x,t)=c_a(x,r=a_c,t) \qquad \frac{\partial c_a}{\partial r}(x,r=0,t)=0 \tag{5.86a,b}$$

where r is now a radial coordinate in a cylindrical system and a_c is the radius of the cylindrical solid. The dimensionless equation for this case is

$$\frac{\partial C_a}{\partial T}=\frac{\gamma_c}{\xi}\frac{\partial}{\partial \xi}\left(\xi\frac{\partial C_a}{\partial \xi}\right) \qquad (0\le \xi\le 1) \tag{5.87}$$

with $\gamma_c=LD_a/a_c^2vR_{im}$ and $\xi=r/a_c$. The analytical solution for this problem can also be found in van Genuchten (1985). The immobile concentration is defined as

$$c_{im}=\frac{2}{a_c^2}\int_0^{a_c}rc_a(x,r,t)dr \qquad \text{or} \qquad C_{im}=2\int_0^1\xi C_a(X,\xi,T)d\xi \tag{5.88a,b}$$

Hollow Cylindrical Aggregates

An important problem is the flow from a macropore surrounded by the soil matrix. This scenario has been investigated by van Genuchten et al. (1984) by considering a cylindrical macropore of radius a_p surrounded by a cylindrical soil mantle of radius a_o. The mathematical formulation is very similar to the previous case except that there is somewhat of a role reversal with a cylindrical mobile region and an annular geometry for the aggregate:

$$R_{im}\frac{\partial c_a}{\partial t}=\frac{D_a}{r}\frac{\partial}{\partial r}\left(r\frac{\partial c_a}{\partial r}\right) \qquad (a_p\le r\le a_o) \tag{5.89}$$

If the concentration is continuous at the interface of the macropore and the aggregate and if there is no diffusion across the outer surface of the aggregate, the boundary conditions become

$$c_m(x,t) = c_a(x, r = a_p, t) \qquad \frac{\partial c_a}{\partial r}(x, r = a_o, t) = 0 \tag{5.90a,b}$$

The dimensionless equation for solute transport from a cylindrical pore is

$$\frac{\partial C_a}{\partial T} = \frac{\gamma_p}{\xi}\frac{\partial}{\partial \xi}\left(\xi \frac{\partial C_a}{\partial \xi}\right) \qquad (1 \le \xi \le \xi_o) \tag{5.91}$$

with $\gamma_p = LD_a/a_p^2 vR_{im}$, $\xi = r/a_p$, and $\xi_o = a_o/a_p$. The fairly complicated analytical solution for this problem is given by van Genuchten et al. (1984). The dimensional and dimensionless immobile concentration is given as

$$c_{im} = \frac{2}{a_o^2 - a_p^2}\int_{a_p}^{a_o} r c_a(x,r,t)dr \qquad \text{or} \qquad C_{im} = \frac{2}{\xi_o^2 - 1}\int_1^{\xi_o} \xi C_a(X,\xi,T)d\xi \tag{5.92a,b}$$

Equivalency Approaches

Shape Factors

Aggregated media often consist of particles with different shapes for which there is no closed-form solution for the solute concentration. Rather than approximating the results by applying the first-order transfer model, it may be possible to derive shape factors that can be substituted in the analytical solution for a well-defined medium to obtain approximate results for other types of aggregates. In the following, we will illustrate the use of shape factors to extend analytical models to porous media consisting of uniform aggregates with an arbitrary geometry. Note that it is not necessary for the aggregates to be uniformly distributed; Rasmuson (1985) investigated the case of an arbitrary aggregate shape for different size groups.

A soil consisting of arbitrary aggregates is conceptualized as a medium of spheres so that the diffusion characteristics of the actual and equivalent soil are equivalent. Other than spherical aggregates may be used for this purpose. The equivalence is conveniently obtained by expanding the hyperbolic functions in the Laplace domain solution for the averaged aggregate concentration, C_{im}. Consider the example of van Genuchten and Dalton (1986) for a rectangular aggregate. The dimensional and dimensionless relationship between the mobile and immobile transformed concentrations for a sphere may be given by

$$\overline{c}_{im} = \left(1 - \frac{sL}{15\gamma_s v}\right)\overline{c}_m \qquad \text{or} \qquad \overline{C}_{im} \approx \left(1 - \frac{s}{15\gamma_s}\right)\overline{C}_m \tag{5.93a,b}$$

and for a rectangular aggregate by

$$\bar{c}_{im} = \left(1 - \frac{sL}{3\gamma_r v}\right)\bar{c}_m \qquad \text{or} \qquad \overline{C}_{im} = \left(1 - \frac{s}{3\gamma_r}\right)\overline{C}_m \tag{5.94a,b}$$

For a given mobile concentration—the transport equation in the mobile region is independent of the aggregate type—the immobile concentrations are equivalent provided that

$$\gamma_s = \gamma_r / 5 \qquad \text{or} \qquad a_s = 2.24 a_r \tag{5.95a,b}$$

where 2.24 is the shape factor, $f_{r,s}$, to express transport in a rectangular aggregate with a spherical model (or vice versa). The breakthrough curve for a medium of rectangular aggregates with a thickness of 0.446 cm will be similar to a medium of 1-cm diameter spheres.

The shape factor approach involves several approximations, most noticeably the neglect of the time dependency of the shape factor since it was derived in the Laplace domain (note that the parameter α in the first-order model is also independent of time). During transient conditions, the shape factor will change somewhat along with the immobile concentration. Van Genuchten (1985) obtained shape factors by matching average concentrations in aggregates in the absence of flow. This amounts to pairing different diffusion problems—of which solutions exist for a wide range of aggregates (Carslaw and Jaeger, 1959). In this manner van Genuchten found that $f_{r,s}$=2.54 while $f_{c,s}$=1.37 for the conversion from cylinders to spheres. Figure 5.4 shows calculated breakthrough curves at the exit, Ce(T), for transport in soil containing rectangular (Figure 5.4a) and solid cylindrical (Figure 5.4b) aggregates using the exact and equivalent model with γ_r=6.4γ_s and γ_c=2.06γ_s, respectively. The agreement between the exact and approximate curves seems reasonable, considering the approximations that are involved. A better result is obtained for cylinders, which have a surface/volume ratio more similar to spheres than rectangles do. By equating the solutions for the simple mobile-immobile and aggregate diffusion models, approximate expressions can also be obtained for the transfer coefficient α in the first-order nonequilibrium model (van Genuchten, 1985; Parker and Valocchi, 1986; van Genuchten and Dalton, 1986).

Equilibrium Transport Model

The conventional advection-dispersion equation (ADE) for equilibrium transport may be used to describe nonequilibrium transport by defining an equivalent dispersion coefficient. This equivalent of effective dispersion coefficient, D_e, is obtained by modifying the standard dispersion coefficient. D. Bolt (1982) outlined how the effects of mechanical dispersion, molecular diffusion, and nonequilibrium spreading may be lumped together into one overall dispersion coefficient (Passioura, 1971; Davidson et al., 1983). This approach has been popular in the past for predicting breakthrough curves because the mathematical simplifications that are achieved by considering a single continuum for transport.

We will illustrate the procedure by considering a sphere, following van Genuchten and Dalton (1986). For an initially solute-free soil, the Laplace transform of the ADE given by Eq. 5.76 is

$$\theta_m R_m s\bar{c}_m + \theta_{im} R_{im} s\bar{c}_{im} = \theta_m D_m \frac{d^2\bar{c}_m}{dx^2} - \theta_m v_m \frac{d\bar{c}_m}{dx} \tag{5.96}$$

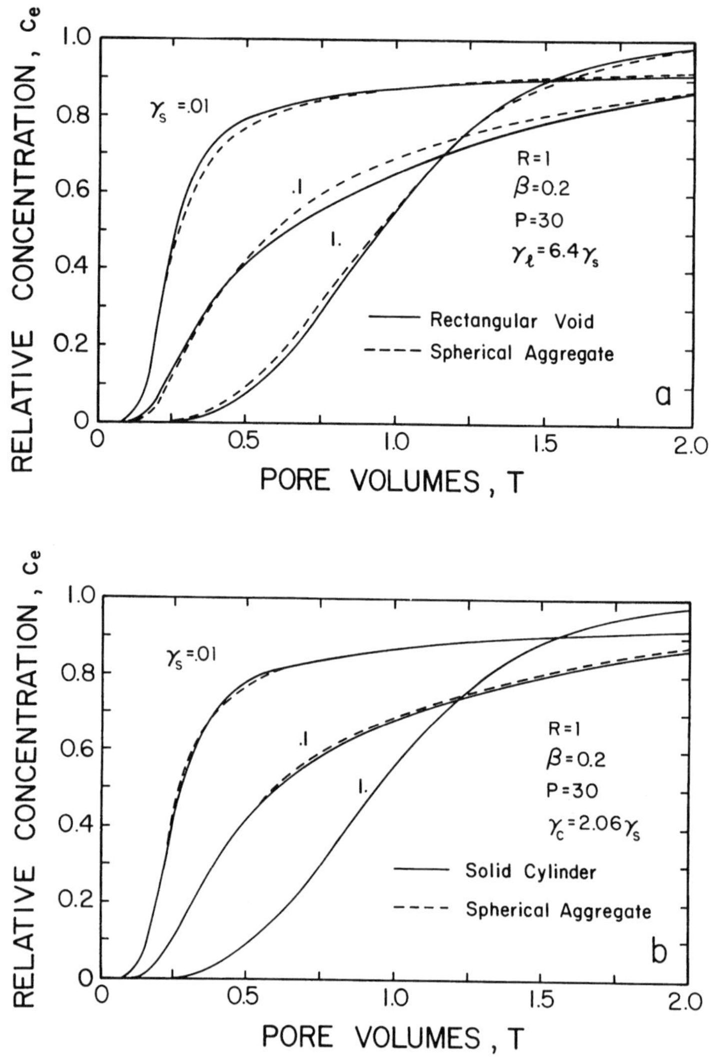

Figure 5.4. Breakthrough curves calculated with the equivalent spherical model and the exact solution for (a) rectangular voids/aggregates and (b) cylindrical aggregates (after van Genuchten, 1985).

If we substitute the expression for the transformed immobile concentration according to (Eq. 5.93a) we can rewrite this in terms of one dependent variable, c_m. After inversion to the real-time domain we obtain:

$$R\frac{\partial c_m}{\partial t} = D\frac{\partial^2 c_m}{\partial x^2} - v\frac{\partial c_m}{\partial x} + \frac{(1-\beta)RL}{15\gamma_s v}\frac{\partial^2 c_m}{\partial t^2} \tag{5.97}$$

where $\beta=\theta_{im}R_{im}/\theta_m R_m$ and $R=\theta_m R_m/\theta+\theta_{im}R_{im}/\theta$. The second time derivative may be approximated by applying the continuity operator twice, i.e.,

$$R^2\frac{\partial^2 c_m}{\partial t^2} = v^2\frac{\partial^2 c_m}{\partial x^2} \tag{5.98}$$

This approximation allows nonequilibrium transport to be described by the following linear equilibrium equation:

$$R\frac{\partial c_m}{\partial t} = D_e\frac{\partial^2 c_m}{\partial x^2} - v\frac{\partial c_m}{\partial x} \tag{5.99}$$

where the subscript for the concentration is usually dropped, while the dispersion coefficient is defined by

$$D_e = D + \frac{(1-\theta_m/\theta)a_s^2 v^2 R_{im}^2}{15 D_a R^2} \tag{5.100}$$

The effective dispersion coefficient increases with aggregate size, flow rate, and the immobile sorption capacity. On the other hand, increased diffusion inside the aggregate reduces the macroscopic dispersion. Effective dispersion coefficients may also be obtained through moment analysis. Parker and Valocchi (1986) obtained the same result for D_e; and they estimated the error associated with this approach using moment analysis. It should be noted again that the time-dependency has been eliminated.

The same procedure may be followed to obtain effective dispersion coefficients for media containing different shapes of aggregates. In the following, we state the effective dispersion coefficient according to van Genuchten and Dalton (1986). For rectangular aggregates:

$$D_e = D + \frac{(1-\theta_m/\theta)a_r^2 v^2 R_{im}^2}{3 D_a R^2} \tag{5.101}$$

for solid cylinders:

$$D_e = D + \frac{(1-\theta_m/\theta)a_c^2 v^2 R_{im}^2}{8 D_a R^2} \tag{5.102}$$

for cylindrical macropores:

$$D_e = D + \frac{[2\ln(a_o/a_p)-1](1-\theta_m/\theta)a_o^2 v^2 R_{im}^2}{4 D_a R^2} \tag{5.103}$$

and for spherical aggregates:

$$D_e = \frac{\theta_m D_m}{\theta} + \frac{(1-\theta_m/\theta)a^2 v^2 R_{im}^2}{15 D_a R^2} \tag{5.104}$$

The first-order nonequilibrium model may also be described by a linear equilibrium model by specifying the following effective dispersion coefficient:

$$D_e = D + \frac{(1-\theta_m/\theta)^2 v^2 R_{im}^2 \theta}{\alpha R^2} \tag{5.105}$$

FURTHER APPLICATION OF THE LAPLACE TRANSFORM

Time Moments

Moments are frequently used to characterize statistical distributions such as those of solute particles (concentrations) versus time or position. Numerical values for time moments are readily obtained through integration of the breakthrough curve. Algebraic expressions for time moments may be used to elucidate transient solute transport and to analyze breakthrough curves.

The nth (time) moment of the breakthrough curve at $x=L$ for a solute application at $x=0$ is defined as

$$m_n(L) = \int_0^{\infty} t^n c(L,t)dt \qquad (n = 0,1,2,...) \tag{5.106}$$

Numerical moments are obtained by integrating experimental concentrations versus time whereas algebraic moments may be obtained by integrating the product of the appropriate power of time and the analytical solution. The zeroth moment is proportional to the total solute mass, the first moment quantifies the mean displacement, the second moment is indicative of the variance (dispersion), whereas the third moment quantifies the asymmetry or skewness of the breakthrough curve. Normalized moments, μ_n, are obtained as follows:

$$\mu_n = m_n / m_0 \tag{5.107}$$

The mean breakthrough time is given by μ_1. Central moments are defined with respect to the mean according to

$$\mu'_n(L) = \frac{1}{m_0}\int_0^{\infty} (t-\mu_1)^n C(L,t)dt \tag{5.108}$$

The variance of a breakthrough curve, which can be used to assess solute dispersion, is given by the second central moment, μ'_2. The degree of asymmetry of the breakthrough curve is indicated by its skewness, $\mu'_3/(\mu'_2)^{3/2}$.

Values for transport parameters may be obtained by equating experimental (numerical) with theoretical (algebraic) moments (Leij and Dane, 1992; Jacobsen et al., 1992). This procedure is not reliable if experimental moments of higher order (e.g., $n>2$) are needed, since even small deviations between experimental and modeled concentrations at larger times will greatly bias such moments (Kreft and Zuber, 1978; Neretnieks, 1983).

Theoretical moments are normally obtained by evaluating the integrals using properties of the Laplace transform (Spiegel, 1965). The following equality can be established:

$$m_n(x) = (-1)^n \lim_{s\to 0} \frac{d^n}{ds^n} \bar{c}(x,s) \tag{5.109}$$

where, as before, $\bar{c}(x,s)$ is the concentration in the Laplace domain, and s is the (complex) transformation variable. Expressions for moments can hence be obtained by differentiating the solution in the Laplace domain and letting the Laplace variable go to zero (Aris, 1958; van der Laan, 1958). The advantage of moment analysis is largely due to the fact that concentrations in the Laplace domain tend to be easier to obtain and to manipulate, especially with mathematical software, than those in the regular time domain. An example of the derivation is given on page 194 of Jury and Roth (1990).

Consider again the first-order model for bi-continuum nonequilibrium transport

$$\theta_m R_m \frac{\partial c_m}{\partial t} + \theta_{im} R_{im} \frac{\partial c_{im}}{\partial t} = \theta_m D_m \frac{\partial^2 c_m}{\partial x^2} - \theta_m v_m \frac{\partial c_m}{\partial x} \tag{5.110}$$

$$\theta_{im} R_{im} \frac{\partial c_{im}}{\partial t} = \alpha (c_m - c_{im}) \tag{5.111}$$

to be applied for instantaneous solute application according to

$$c(x,0) = 0 \qquad 0 \le x < \infty \tag{5.112}$$

$$c(0,t) = \frac{m_o}{v} \delta(t) \tag{5.113}$$

$$\frac{\partial c}{\partial x}(\infty, t) = 0 \tag{5.114}$$

where m_o is the solute mass that is applied per unit area of soil solution and $\delta(t)$ is a Dirac delta function describing instantaneous solute application at t=0. The first-type inlet condition is used to describe flux-averaged concentrations obtained from effluent samples (Parker and van Genuchten, 1984). The mean breakthrough time, which is also the residence time in this case, is obtained by substituting the algebraic solution for the mobile concentration in Eq. 5.109. After differentiating and taking the limit it can be established that

$$\mu_1 = \frac{Rx}{v} \tag{5.115}$$

It appears that the nonequilibrium conditions do not affect the mean travel time. The variance, μ'_2, of the predicted breakthrough curve may be obtained in a similar manner

$$\mu'_2 = \frac{2\theta_m D_m R^2 x}{\theta v^3} + \frac{2\theta(1-\theta_m/\theta)^2 R^2 x}{\alpha v} \tag{5.116}$$

This expression clearly demonstrates how nonequilibrium phenomena, as quantified by α and β, will enhance solute spreading. An expression for the effective dispersion coefficient is derived by equating variances for the equilibrium and nonequilibrium transport model. Moments may also be obtained for a higher order or for more complex transport problems (Valocchi, 1985).

Transfer Functions

Concepts

We are interested in using transfer functions (TFs) to predict solute transport based upon linear mechanistic nonequilibrium models. This *a priori* approach contrasts the use of implicit TFs that are calibrated with experimental data where no conceptualization of solute transport is required.

Conventional numerical or analytical solutions of the transport equation provide the solute concentration for an arbitrary time and position. However, this is more than required if we are merely interested in the concentration as a function of time for a given depth (the converse case of spatial dependency for a given time will not be considered here). The temporal domain is discretized to obtain a residence time distribution (RTD) for the solute—the corresponding TF is used to describe the breakthrough curve. It is assumed that (i) water flow is steady and macroscopically deterministic, (ii) the soil system consists of one inflow and outflow region, (iii) there are no solute sources or sinks in the soil, and (iv) the inlet and outlet concentrations are of the flux-averaged type (the concentration is equal to the ratio of a specified or measured solute flux and the water flux).

The solute response $y(L,t)$ as result of an input $y(0,t)$ is of particular interest. The input function can be a solute pulse applied to a soil column or soil surface, whereas the output function is the solute breakthrough curve at the column outlet or any particular depth in the soil profile. The Laplace transform of the signal is given by

$$\bar{y}(x,s) = \int_0^\infty y(x,t)\exp(-st)dt \tag{5.117}$$

and the corresponding TF is defined as

$$G(L,s) = \frac{\bar{y}(L,s)}{\bar{y}(0,s)} \tag{5.118}$$

It can be shown that the expression for the TF does not depend on the shape of the input and the resulting output signals. The function characterizes the overall transport behavior of the soil between the input and output locations. Expressions for the TF are conveniently obtained for a wide variety of problems, as mentioned earlier, because concentrations in the Laplace domain are relatively easy to derive.

Before examining ways to obtain results in the real-time domain, let us briefly establish the concept of the residence time distribution (RTD). Water flows at a constant flow rate, $q=\theta v$, through the inlet and outlet surfaces of an arbitrary soil volume while solute starts to be applied at the inlet when t=0. The residence time of a water molecule, t_r, is the difference between time of output and input (i.e., the time needed to traverse the soil volume). The distribution of t_r is of particular interest, and it is usually quantified with tracer studies. The RTD can be viewed as a probability density function (pdf) denoting the likelihood that a tracer particle will reside in the soil within a certain time period. The amount of tracer with residence time t_r exiting the soil during a small time period, dt_r, is given by $qy(L,t_r)$ dt_r. A continuous pdf is obtained by normalizing the tracer flux with respect to the total amount of applied tracer:

$$f(L,t_r) = \frac{qy(L,t_r)}{\int_0^\infty qy(L,t_r)dt_r} \qquad \text{and} \qquad \int_0^\infty f(L,t_r)dt_r = 1 \tag{5.119}$$

A Dirac type input is most convenient for determining $f(L,t_r)$ because no particles with different residence times will exit simultaneously; the RTD is equal to the travel time probability density function (Jury, 1982).

Consider instantaneous application of an amount m_o of perfect tracer per unit cross-sectional area of a soil column. The amount of tracer exiting the soil per unit time is given by $qy(L,t)=qy(L,t_r)$ and the RTD may be defined according to

$$f(L,t_r) = \frac{qy(L,t)}{\int_0^\infty qy(L,t)dt} = \frac{q}{m_o} y(L,t) \tag{5.120}$$

The RTD may hence directly be determined from the output signal for instantaneous solute input. The transfer function, $G(L,s)$, for a Dirac delta input, $y(0,t)=m_o\delta(t)/q$, may be obtained according to Eq. 5.118:

$$G(L,s) = \frac{\bar{y}(L,s)}{\frac{m_o}{q}\int_0^\infty \delta(t)\exp(-st)dt} = \bar{f}(L,s) \tag{5.121}$$

We have the desirable property that the transfer function is the Laplace transform of the RTD, regardless of the input function.

The concentration in the real-time domain for an arbitrary input can be readily obtained by taking the convolution of input signal and RTD (Jury and Sposito, 1985):

$$y(L,t) = \mathcal{L}^{-1}[G(L,s)\bar{y}(0,s)] = \int_0^t f(L,\tau)y(0,t-\tau)d\tau \tag{5.122}$$

This expression can be quite useful when the RTD is first determined from experimental input and output concentrations at a certain position to predict later output concentrations for other positions (Jury et al., 1982). If the RTD is not known explicitly, solute transport can be quantified by numerically inverting the TF or by using moment analysis.

Time moments of the RTD are given by

$$\mu_n(x) = \int_0^\infty t_r^n f(x,t_r)dt_r \qquad (n = 0,1,2,\ldots) \tag{5.123}$$

Note that these moments are already normalized. The mean residence time is given by

$$< t_r(x) > = \int_0^{\infty} t_r(x) f(x, t_r) dt_r \tag{5.124}$$

This value is the ensemble average of all possible (residence) times. The variance and the coefficient of variation (reduced variance) are given by

$$\sigma^2(x) = \int_0^{\infty} (t_r(x) - < t_r(x) >)^2 f(x, t_r) dt_r, \quad \sigma'^2(x) = \sigma^2(x) / < t_r(x) >^2 \tag{5.125a,b}$$

Similar to the concentrations of a solute breakthrough curve, we can use the Laplace transform of the RTD, $G(x,s)$, to derive moments of the RTD:

$$\mu_n(x) = (-1)^n \lim_{s \to 0} \frac{d^n}{ds^n} G(x,s) \tag{5.126}$$

It can be shown that $G(x,0)$ corresponds to the ratio of the output to the input signal for steady state conditions.

With the definition of the TF (Eq. 5.118), the mean residence time and variance associated with solute transport in a soil system with length L can be obtained from the mean and variance of the input and output signals:

$$\begin{aligned} < t_r >_{\text{soil}} &= < t_r(L) > - < t_r(0) > \\ \sigma^2_{\text{soil}} &= \sigma^2(L) - \sigma^2(0) \end{aligned} \tag{5.127}$$

This principle can be used to model more complicated transport processes by distinguishing subprocesses with separate transfer functions, $G_i(x,s)$ (Sardin et al., 1991).

Nonequilibrium Transport

Transfer functions allow us to quantify the effect of nonequilibrium or other transport phenomena with characteristic times. As an example, we select the case of nonequilibrium transport of a nonreactive solute (R=1) as described by Eqs. 5.110 through 5.114. After taking the Laplace transform of Eq. 5.111, we can write

$$\bar{c}_{im} = \frac{\bar{c}_m}{1 + s t_M} \qquad \left(t_M = \frac{\theta_{im}}{\alpha} \right) \tag{5.128}$$

where t_M is a characteristic mass transfer time for solute movement between the mobile and immobile regions. The corresponding TF is obtained from the ratio of the "output" and "input" concentrations in the Laplace domain:

$$M(s) = \frac{\theta_{im} \bar{c}_{im}}{\theta_m \bar{c}_m} = \frac{K_{im}}{1 + s t_M} \qquad (K_{im} = \theta_{im} / \theta_m) \tag{5.129}$$

where K_{im} can be viewed as a distribution coefficient. If we transform Eq. 5.110 for the mobile region and substitute the above results we may write

$$D_m \frac{d^2\bar{c}_m}{dx^2} = v_m \frac{d\bar{c}_m}{dx} + s[1+M(s)]\bar{c}_m \tag{5.130}$$

The inlet and outlet boundary conditions and the corresponding unit Dirac input and output signals are

$$\bar{c}_m(0^+,s) - \frac{D}{v}\frac{d\bar{c}_m}{dx}(0^+,s) = \bar{y}(0^-,s) \quad , \quad \bar{y}(0^-,s) = 1 \tag{5.131a,b}$$

$$\frac{d\bar{c}_m}{dx}(L^-,s) = 0 \quad , \quad \bar{y}(L^+,s) = \bar{c}_m(L^-,s) \tag{5.132a,b}$$

After solving this system for $\bar{c}_m(x,s)$, we may obtain the following TF from the ratio of the outflow and inflow concentrations:

$$G(L,s) = \frac{4r\exp[P(1-r)/2]}{(1+r)^2 - (1-r)^2\exp(-Pr)} \quad , \quad r = \sqrt{1 + \frac{4t_m s}{P}[1+M(s)]} \tag{5.133a,b}$$

where $P=v_mL/D_m=vL/D$ is the Peclet number, and $t_m=L/v_m$ is the characteristic advection time in the mobile phase.

The mean residence time and variance are obtained by differentiating this TF following Eq. 5.126:

$$< t_r(L) > = t_m(1+K_{im}) \tag{5.134}$$

$$\sigma'^2(L) = \frac{\sigma^2(L)}{< t_r(L) >^2} = \frac{2}{P} - \frac{2}{P^2}[1-\exp(-P)] + \frac{2K_{im}}{1+K_{im}}\frac{t_M}{< t_r >} \approx \frac{2}{t_m}(t_D + t'_M) \tag{5.135}$$

with t_M' as a modified mass transfer time. The mean residence time depends on the characteristic convection time and the volume fractions of the immobile and mobile liquid phases. Equation 5.134 also illustrates the well-known finding that this time does not depend on mass transfer kinetics and dispersion. The variance of the RTD consists of contributions from hydrodynamic dispersion (first two terms of the right-hand side of Eq. 5.135, the second one may be neglected if P>5) and mass transfer kinetics (last term). An increase in the mass transfer time, t_M, leads to additional spreading.

Figure 5.5 illustrates the influence of t_M'/t_m on the RTD as a function of dimensionless time (t/t_m) for a Dirac delta input predicted for P=20. All curves have the same mean residence time of $<t_r(L)>=3t_m$ with $K_{im}=2$ ($\theta_{im}=2\theta_m$). The curves were obtained from $G(L,s)$ using a fast Fourier transform for inversion to the real-time domain (Press et al., 1986). For small t_M'/t_m, where hydrodynamic dispersion is important compared to nonequilibrium transfer, the breakthrough curves are fairly symmetrical. If nonequilibrium becomes more important, the curves are increasingly nonsigmoidal with solute appearing at the outlet over a longer period of time t/t_m.

Transfer functions are convenient to quantify linear transport in media consisting of well-defined aggregates because they are developed from the Laplace transformed concentrations. Such transformed concentrations are considerably easier to obtain than regular concentrations if intra-aggregate diffusion is included in the model. Valocchi (1985) provided a TF for spheres while Sardin et al. (1991) presented the TF and an equivalent shape factor for rectangular, spherical, and cylindrical aggregates. The superposition principle facilitates the assembly of an overall TF from small-scale TFs to characterize transport in an arbitrary number of pore domains (Valocchi, 1990).

SUMMARY

In this chapter we have reviewed classical analytical methods for modeling nonequilibrium transport in soils according to the advection-dispersion equation. The analytical results may offer valuable insight in nonequilibrium transport of solute in natural soils even if such results are strictly speaking only valid for idealized media where transport may be described by a linear model. We considered the widely studied bi-continuum transport problem where the liquid phase of the porous medium consists of a mobile and an immobile region (physical nonequilibrium). The analytical methods to solve this problem can be readily applied to solve the mathematically equivalent case of chemical nonequilibrium and they may also be used to tackle more complex nonequilibrium models.

A detailed overview was given of the solution of the general bi-continuum model for first-order approximation of solute transfer between mobile and immobile regions. The solution was obtained by taking the Laplace transform with respect to time and position. Particular attention was paid to steps that are pertinent for the nonequilibrium case such as the use of Goldstein's *J*-function and the generalized convolution theorem. The problem was split up into a boundary, initial, and production value problem. A specific solution was provided for each problem in case of a Heaviside distribution for the input, initial, and production function, respectively. Examples were given on the effect of α, which governs solute transfer between immobile and mobile regions, and f, the fraction of sorption sites in contact with the mobile region, on the breakthrough curves for C_1 and C_2. A decrease in α, signifying increased nonequilibrium, resulted in a solute breakthrough curve with a relatively early and high peak and substantial tailing. Lower values for f caused early breakthrough due to the reduced opportunity of sorption by the solid from the mobile region.

For well-defined aggregates we briefly reviewed further refinements of the nonequilibrium model by combining the advection-dispersion equation for inter-aggregate transport in the mobile region and Fick's law for intra-aggregate transport. The derivation of the solution for such problems is somewhat lengthier than for the first-order approximation; several references are provided for such solutions. We should mention that it is also convenient to study transport in fractured media by analytical methods in the time domain (Laplace transform) and by numerical methods in the spatial domain (finite difference or finite element methods). Such an approach has not been reviewed here, but further details may be found in Sudicky (1990) and Gallo et al. (1996). Furthermore, we reviewed equivalency approaches, also based on Laplace analysis, to broaden and simplify the use of analytical techniques for transport in structured media. Shape factors derived from stagnant diffusion problems allow the use of solutions derived for simple aggregates to other types of aggregates. We further discussed using the equation for equilibrium transport to describe nonequilibrium transport by modifying the dispersion coefficient.

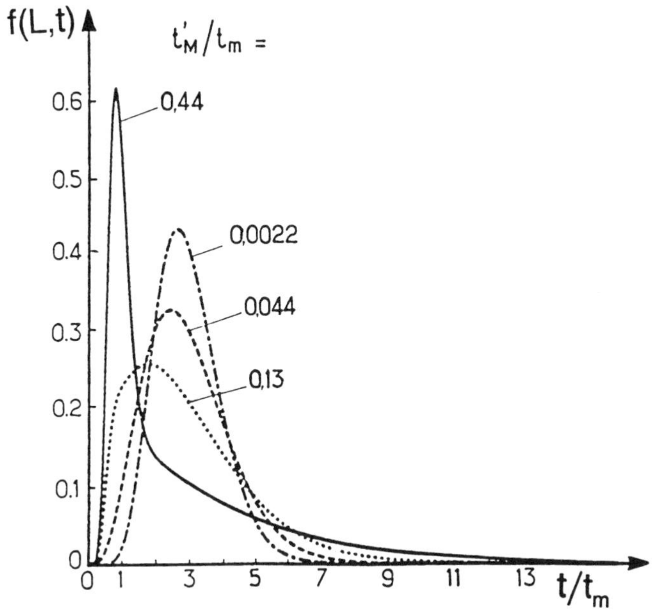

Figure 5.5. Breakthrough curves resulting from a Dirac input for P=20, K_{im}=2, and four different ratios of the characteristic mass transfer and advection times, t_M'/t_m (after Sardin et al., 1991).

Finally, we introduced the use of moment analysis and transfer functions as methods for studying nonequilibrium transport. In contrast to closed-form expressions that allow solute concentrations to be quantified at arbitrary times and positions, these methods provide important algebraic or numerical information regarding the behavior of the solute in the soil. The methods were also based on Laplace analysis, but the mathematics is simplified because there is no need for inversion to the real-time domain, which is typically the most difficult step. Time moments are useful to analyze experimental solute breakthrough curves and to quantify the effect of model parameters on solute movement. Transfer functions, and the concept of the residence time distribution on which it is based, are convenient for estimating the timescale for different transport processes. Both moments and transfer functions are promising tools for nonequilibrium solute transport described by relatively complex deterministic or stochastic models.

LIST OF SYMBOLS

a	radius (Eq. 5.77)
a_0,a_1,a_2	constants (Eq. 5.27)
b	constant (Eq. 5.18b)
c	solute concentration of the liquid phase [ML^{-3}] (Eq. 5.1)
c_o	characteristic concentration [ML^{-3}] (Table 5.1)
C	dimensionless solute concentration (Table 5.1)
D	dispersion coefficient [L^2T^{-1}] (Eq. 5.1)
f	fraction of exchange sites equilibrating with the mobile liquid phase for the two-region model, or to be at equilibrium for the two-site model (Table 5.1)

$f(x,t_r)$	probability density function (pdf) of t_r (i.e., RTD) (Eq. 5.123)
$F(X)$	initial concentration for the IVP (Eq. 5.7)
$G(L,s)$	transfer function (Eq. 5.118)
$G(T)$	input concentration for the BVP (Eq. 5.8)
G_i	auxiliary equilibrium part of final solution of general or specific solutions (i = 1,2,3) (Eqs. 5.38, 5.47, and 5.68)
h_i	auxiliary equilibrium functions for derivation of analytical solutions (i = 1,2,3) (Eq. 5.35, 5.45, and 5.46)
$H_i(\tau;T)$	auxiliary expressions for nonequilibrium transport (i = 0,1) (Eqs. 5.48 and 5.39)
I_0, I_1	modified Bessel function of orders zero and one (Eqs. 5.31 and 5.34)
J	Goldstein's J-function (Eq. 5.54)
k	distribution coefficient [$M^{-1}L^3$] (Eq. 5.1)
K	distribution coefficient (Eq. 5.129)
L	characteristic length (Table 5.1)
m_n	nth-order time moment [$ML^{-3}T^{n+1}$] (Eq. 5.106)
p	variable in Goldstein's J-function (Eq. 5.55a)
P	Peclet number (Table 5.1)
q	variable in Goldstein's J-function (Eq. 5.55b), water flow rate [LT^{-1}] (Eq. 5.119)
R	retardation factor (Table 5.1)
r	Laplace variable [–] (Eq. 5.21), radial coordinate [L] (Eq. 5.77)
s	Laplace variable [T^{-1}] or [–] (Eq. 5.12)
s_k	concentration due to kinetic sorption [MM^{-1}] (Eq. 5.3)
s^*	modified Laplace variable (Eq. 5.30)
t	time [T] (Eq. 5.1)
T	dimensionless time (Table 5.1)
U	Heaviside unit-step function (Eq. 5.66)
v	average pore-water velocity [LT^{-1}] (Eq. 5.1)
x	(longitudinal) distance [L] (Eq. 5.1)
X	dimensionless distance (Table 5.1)
y	(transverse) distance [L] (Eq. 5.81)
Y	dimensionless transverse distance (Eq. 5.83)

GREEK

α	first-order rate coefficient [T^{-1}] (Eq. 5.1)
β	dimensionless variable for partitioning in nonequilibrium transport models (Table 5.1)
γ	shape factor (Eq. 5.95b)
Γ	auxiliary transverse function (Eq. 5.72)
δ	Dirac delta function [T^{-1}] (Eq. 5.113)
θ	volumetric water content [L^3L^{-3}] (Eq. 5.1)
λ	zero-order production term [$ML^{-3}T^{-1}$] (Eq. 5.1)
μ	first-order degradation term [T^{-1}] (Eq. 5.1), normalized moment (Eq. 5.107)
ξ	dimensionless radial distance (Eq. 5.79)
ρ	bulk density [ML^{-3}] (Eq. 5.1)
σ^2	variance (Eq. 5.125a)
ω	dimensionless mass transfer coefficient (Table 5.1)

SUBSCRIPTS

a	aggregate (Eq. 5.77)
c	(solid) cylinder (Eq. 5.85)
e	equilibrium (Eq. 5.3), effective (Eq. 5.99)
im	immobile phase (Eq. 5.1)
k	kinetic (Eq. 5.3)
ℓ	liquid phase (Eq. 5.1)
m	mobile phase (Eq. 5.1)
M	mass transfer (Eq. 5.128)
n	order of moment (Eq. 5.109)
o	reference (Table 5.1)
p	(hollow) cylinder (Eq. 5.89)
r	residence (Eq. 5.119)
s	adsorbed phase (Eq. 5.1), sphere (Eq. 5.77)
T	total (resident) (Eq. 5.10)
1	phase 1 (Table 5.1)
2	phase 2 (Table 5.1)

SUPERSCRIPTS

f	flux mode (Eq. 5.11)
r	resident mode (Eq. 5.11)

SPECIAL SYMBOLS

$\bar{}$	Laplace transform with respect to time (Eq. 5.12)
$\tilde{}$	Laplace transform with respect to distance (Eq. 5.21)
$\mathcal{L}$	Laplace transform (Eq. 5.12)

ACRONYMS

ADE	advection-dispersion equation
BVP	boundary value problem
IVP	initial value problem
pdf	probability density function
PVP	production value problem
RTD	residence time distribution
TDR	time-domain reflectometry
TF	transfer function

REFERENCES

Abramowitz, M. and I.A. Stegun. *Handbook of Mathematical Functions*. Dover, New York, 1970.

Aris, R. On the dispersion of linear kinematic waves. *Proc. R. Soc. London A* 245, 268–277, 1958.

Bolt, G.H. Movement of solutes in soils: Principles of adsorption/exchange chromatography. In *Soil Chemistry. B. Physico-Chemical Models*, Bolt, G.H. (Ed.), Elsevier, Amsterdam, 1892.

Boucher, D.F. and G.E. Alves. Dimensionless numbers. *Chem. Eng. Progr.*, 55:55–64, 1959.

Brusseau, M.L., R.E. Jessup, and P.S.C. Rao. Modeling the transport of solutes influenced by multi-process nonequilibrium. *Water Resour. Res.*, 25:1971–1988, 1989.

Carslaw, H.S. and J.C. Jaeger. *Conduction of Heat in Solids*. Oxford University Press, New York, 1959.

Coats, K.H. and B.D. Smith. Dead-end pore volume and dispersion in porous media. *Soc. Petrol. Eng. J.*, 4:73–84, 1964.

Davidson, J.M., P.S.C. Rao, and P. Nkedi-Kizza. Physical processes influencing water and solute transport in soils. In *Chemical Mobility and Reactivity in Soil Systems*. Soil Sci. Soc. Amer., Spec. Publ. 11. Madison, WI, 1983.

De Smedt, F. and P.J. Wierenga. A generalized solution for solute flow in soils with mobile and immobile water. *Water Resour. Res.*, 15:1137–1141, 1979.

Gallo, C., C. Paniconi, and G. Gambolati. Comparison of solution approaches for the two-domain model of nonequilibrium transport in porous media. *Adv. Water Resour.*, 19:241–253, 1996.

Goldstein, S. On the mathematics of exchange processes in fixed columns. I. Mathematical solutions and asymptotic expansions. *Proc. Roy. Soc. London A*, 219:151–185, 1953.

Goltz, M.N. and P.V. Roberts. Three-dimensional solutions for solute transport in an infinite medium with mobile and immobile zones. *Water Resour. Res.*, 22:1139–1148, 1986.

Hiester, N.K. and T. Vermeulen. Saturation performance of ion-exchange columns and adsorption columns. *Chem. Eng. Prog.*, 48:505–516, 1952.

Jacobsen, O.H., F.J. Leij, and M.Th. van Genuchten. Parameter determination for chloride and tritium transport in undisturbed lysimeters during steady flow. *Nordic Hydrology*, 23:89–104, 1992.

Jury, W.A. Simulation of solute transport using a transfer function model. *Water Resour. Res.* 18:363–368, 1982.

Jury, W.A. and K. Roth. *Transfer Functions and Solute Movement Through Soil: Theory and Applications*. Birkhäuser Verlag, Basel, 1990.

Jury, W.A. and G. Sposito. Field calibration and validation of solute transport models for the unsaturated zone. *Soil Sci. Soc. Amer. J.*, 49:1331–1341, 1985.

Jury, W.A., L.H. Stolzy, and P.J. Shouse. A Field test of the transfer function model for predicting solute transport. *Water Resour. Res.*, 18:369–375, 1982.

Klinkenberg, A. Numerical evaluation of equations describing transient heat and mass transfer in packed solids. *Ind. Eng. Chem.*, 40:1992–1994, 1948.

Kreft, A. and A. Zuber. On the physical meaning of the dispersion equation and its solutions for different initial and boundary conditions. *Chem. Eng. Sci.*, 33:1471–1480, 1978.

Lassey, K.R. Unidimensional solute transport incorporating equilibrium and rate-limited isotherms with first-order loss. 1. Model conceptualizations and analytic solutions. *Water Resour. Res.*, 3:343–350, 1988.

Leij, F.J. and J.H. Dane. Moment method applied to solute transport with binary and ternary exchange. *Soil Sci. Soc. Am. J.*, 56:667–674, 1992.

Leij, F.J., N. Toride, and M.Th. van Genuchten. Analytical solutions for non-equilibrium solute transport in three-dimensional porous media. *J. Hydrol.*, 151:193–228, 1993.

Lindstrom, F.T. and L. Boersma. Analytical solutions for convective-dispersive transport in confined aquifers with different initial and boundary conditions. *Water Resour. Res.*, 25:241–255, 1989.

Lindstrom, F.T. and M.N.L. Narasimhan. Mathematical theory of a kinetic model for dispersion of previously distributed chemicals in a sorbing porous medium. *SIAM J. Appl. Math.*, 24:496–510, 1973.

Lindstrom, F.T. and W.M. Stone. On the start up or initial phase of linear mass transport of chemicals in a water saturated sorbing porous medium. *SIAM J. Appl. Math.*, 26:578–591, 1974.

Neretnieks, I. A note on fracture flow dispersion mechanisms in the ground. *Water Resour. Res.,* 19:364–370, 1983.

Nkedi-Kizza, P., J.W. Biggar, H.M. Selim, M.Th. van Genuchten, P.J. Wierenga, J.M. Davidson, and D.R. Nielsen. On the equivalence of two conceptual models for describing ion exchange during transport through an aggregated oxisol. *Water Resour. Res.,* 20:1123–1130, 1984.

Parker, J.C. and A.J. Valocchi. Constraints on the validity of equilibrium and first-order kinetic transport models in structured soils. *Water Resour. Res.,* 22:399–407, 1986.

Parker, J.C. and M.Th. van Genuchten. Flux-averaged and volume-averaged concentrations in continuum approaches to solute transport. *Water Resour. Res.,* 20:866–872, 1984.

Passioura, J.B. Hydrodynamic Dispersion in Aggregated Media. 1. Theory. *Soil Sci.,* 111:339–344, 1971.

Press, W.H., B.P. Flannery, S.A. Teukolsky, and W.T. Vetterling. *Numerical Recipes.* Cambridge Univ. Press, New York, 1986.

Rasmuson, A. The effect of particles of variable size, shape and properties on the dynamics of fixed beds. *Chem. Eng. Sci.,* 40:621–629, 1985.

Rasmuson, A. and I. Neretnieks. Exact solution of a model for diffusion in particles and longitudinal dispersion in packed beds. *AIChE J.* 26:686–690, 1980.

Sardin, M., D. Schweich, F.J. Leij, and M.Th. van Genuchten. Modeling the nonequilibrium transport of linearly interacting solutes in porous media: A review. *Water Resour. Res.,* 27:2287–2307, 1991.

Selim, H.M., J.M. Davidson, and R.S. Mansell. Evaluation of a two-site adsorption-desorption model for describing solute transport in soils. In *Proc. Summer Computer Simulation Conf., Washington, DC,* 1976.

Sneddon, I.H. *Fourier Transforms.* Dover, New York, 1995.

Spiegel, M.R. *Theory and Problems of Laplace Transforms.* Schaum's Outline Ser., McGraw-Hill, New York, 1965.

Sudicky, E.A. and E.O. Frind. Contaminant transport in fractured porous media: Analytical solution for a system of parallel fractures. *Water Resour. Res.,* 18:1634–1642, 1982.

Sudicky, E.A. The Laplace transform Galerkin technique for efficient time-continuous solution of solute transport in double-porosity media. *Geoderma,* 46:209–232, 1990.

Taylor, A. *General Theory of Functions and Integration.* Dover, New York, 1985.

Toride, N., F.J. Leij, and M.Th. van Genuchten. A comprehensive set of analytical solutions for nonequilibrium solute transport with first-order decay and zero-order production. *Water Resour. Res.,* 29:2167–2182, 1993.

Toride, N., F.J. Leij, and M.Th. van Genuchten. The CXTFIT Code for Estimating Transport Parameters from Laboratory or Field Tracer Experiments. *Res. Rep. No. 137,* U.S. Salinity Lab., USDA, ARS, Riverside, CA, 1995.

Valocchi, A.J. Validity of the local equilibrium assumption for modeling sorbing solute transport through homogeneous soils. *Water Resour. Res.,* 21:808–820, 1985.

Valocchi, A.J. Use of temporal moment analysis to study reactive solute transport in aggregated porous media. *Geoderma,* 46:233–247, 1990.

van der Laan, E.T. Notes on the diffusion-type model for the longitudinal mixing in flow. *Chem. Eng. Sci.,* 7:187–191, 1958.

van Genuchten, M.Th. Non-Equilibrium Transport Parameters from Miscible Displacement Experiments. *Res. Rep. No. 119,* U.S. Salinity Lab., USDA, ARS, Riverside, CA, 1981.

van Genuchten, M.Th. A general approach for modeling solute transport in structured soils. In *Hydrology of Rocks of Low Permeability. Proc. 17th Int. Congr. Int. Assoc. Hydrogeol.* 17:513–526, 1985.

van Genuchten, M.Th. and W.J. Alves. Analytical solution of the one-dimensional convective-dispersive solute transport equation. *U.S.D.A. Technical Bulletin* 1661, 1982.

van Genuchten, M.Th. and F.N. Dalton. Models for simulating salt movement in aggregated field soils. *Geoderma,* 38:165–183, 1986.

van Genuchten, M.Th. and J.C. Parker. Boundary conditions for displacement experiments through short laboratory soil columns. *Soil Sci. Soc. Am. J.* 48:703–708, 1984.

van Genuchten, M.Th., D.H. Tang, and R. Guennelon. Some exact solutions for solute transport through soils containing large cylindrical macropores. *Water Resour. Res.,* 20:335–346, 1984.

van Genuchten, M.Th. and R.J. Wagenet. Two-site/two-region models for pesticide transport and degradation: theoretical development and analytical solutions. *Soil Sci. Soc. Am. J.,* 53:1303–1310, 1989.

van Genuchten, M.Th. and P.J. Wierenga. Mass transfer studies in sorbing porous media. I. Analytical solutions. *Soil Sci. Soc. Am. J.,* 40:473–481, 1976.

Villermaux, J. and W.P.M. van Swaaij. Modèle représentatif de la distribution des temps de séjour dans un réacteur semi-infini à dispersion axiale avec zones stagnantes. Application à l'écoulement ruisselant dans des colonnes d'anneaux Raschig. *Chem. Eng. Sci.* 24:1097–1111, 1969.

Walker, G.R. Solution to a class of coupled linear partial differential equations. *IMA J. Appl. Math.,* 38:35–48, 1987.

Chapter Six

Use of Fractals to Describe Soil Structure

H.W.G. Booltink, J. Bouma, and P. Droogers

INTRODUCTION

When studying soils in detail they never act completely as isotropic, homogeneous porous media, which implies that the formal requirements for applying Richards' equation often are not met. Rather than make assumptions about soil and boundary conditions, we can also try to characterize the soil fabric with morphological methods and use such characterizations to derive mechanistic models that explain and predict water and solute movement. Specifically, soil pores can be described in terms of type, shape, size and continuity, allowing a distinction of different flow domains in the soil. Aggregates are often homogeneous internally, so even though classic flow theory should not be applied to the entire heterogeneous soil, it can be applied to the aggregates which will be wetted through surrounding macropores. Flow patterns through these macropores are a function of many factors (e.g., Bouma, 1981,1990).

Two approaches can be followed when characterizing bypass flow, using morphological techniques: (i) direct measurements can be made in large cores (e.g., Bouma et al., 1981). The role of soil morphology consists of defining representative measurement volumes or measurement techniques (e.g., Bouma et al., 1982; Ehlers, 1975), and: (ii) parameters can be derived to be used in mechanistic flow models which consider soils to be composed of different interacting flow domains. Procedures differ in their degree of complexity (e.g., Bouma and de Laat, 1981; Hoogmoed and Bouma, 1980; Booltink et al., 1993). In the second section, measurement techniques for characterizing morphological features related to bypass flow on laboratory scale will be discussed. Staining patterns reflecting soil morphology will be analyzed in terms of fractal dimensions and volumes of active macropores. In a subsequent section, the obtained physical and morphological characteristics will be implemented in simulation models. These simulation models are applied on laboratory scale, and by means of a Monte Carlo simulation procedure, extended to a field site. Soil structure plays an important role in soil quality and sustainability of agricultural systems. The role of soil structure will be subsequently illustrated with examples on the effect on availability and accessibility of water for plants and the impact of soil structure on nitrate leaching from a field soil. The last section demonstrates the use of simulation models to explore the effects of different soil structure types with the aim to optimize soil structure and define guidelines for soil management. Future research needs are finally discussed.

QUANTITATIVE CHARACTERIZATION OF SOIL STRUCTURE

Within soil survey, soil structure is usually described in qualitative terms such as: "weak sub-angular blocky" (e.g., Soil Survey Staff, 1975). Although these terms describe soil structure conscientiously, they can hardly be used for calculations or model simulation purposes. A more quantitative approach was followed by Jongerius et al. (1972), Murphy et al. (1977), and Bullock and Murphy (1980). In these studies macropore spaces were visualized or measured and expressed in terms of pore size distributions. Bouma et al. (1977) and Ringrose-Voase and Bullock (1984) developed a technique for measuring macropores in undisturbed soil, using methylene blue staining patterns. They stratified macropores according to Brewer (1964) in categories of channels, vughs, and planar voids and discussed functional physical properties of different types of macropores. This, however, still in descriptive terms. Moran et al. (1989) and McBratney and Moran (1990) used fluorescent dye tracers in combination with digital binary image production to interpret soil macropore structure.

Several studies focused on combined application of soil physical characteristics and morphological features. Bullock and Thomasson (1979) used image analysis of thin sections to calculate macroporosity. Total macroporosity, however, is not a very relevant property in relation to water flow processes such as bypass flow since very few macropores, that contribute only a little to total macroporosity, dominate the transport of water and solutes (Bouma et al., 1977).

Spaans et al. (1990) measured changes of hydraulic conductivities in a Costa Rican soil before and after the clearing of tropical rain forest. In that study, differences in hydraulic conductivity were illustrated with micromorphometric analyses of thin sections. The effects of tillage on soil morphology and porosity were investigated by Shipitalo and Protz (1987). These studies, however, have in common that soil structure was used to illustrate physical properties rather than linking morphology to soil physical properties in quantitative terms.

Bouma (1981) suggested to functionally characterizing bypass flow by means of dye tracers in combination with macromorphometric techniques. This approach was followed by Kooistra et al. (1987) who combined soil, morphological measurements with computer simulations on moisture deficits. In that study, the effects of horizontal planar voids on upward unsaturated flow of water were stressed. Van Stiphout et al. (1987) demonstrated, in a field study, the occurrence of bypass flow and infiltration of water from discontinuous macropores. This subsurface infiltration process is generally referred to as "internal catchment." In structured soils the use of morphometric data in combination with soil physical characteristics is particularly relevant. Fractal theory seems to be a promising tool in characterizing macropore systems and soil aggregates, since size and shape of different combinations of functional macropores can be quantified and coupled to physical characteristics. Bartoli et al. (1991) use fractal dimensions to characterize soil structure. Perfect (1996) demonstrated a fractal model describing different soil aggregates. Crawford et al. (1993) used fractal theory to explain diffusion processes in heterogeneous soils. Hatano et al. (1992) and Booltink et al. (1993) linked fractal dimensions of stained macropores to measured bypass flow characteristics and incorporated this approach in a simulation model. Advanced three-dimensional scanning techniques can be applied. Chen et al. (1992) and Schaafsma et al. (1992) used Magnetic Resonance Imaging (MRI), to measure flow phenomena in soils. Computed Tomography (CT) has been used more widely. Hopmans et al. (1992) used CT to characterize soil water distribution in one-step outflow experiments for measuring soil retentivity and conductivity. Joschko et al. (1991) quantified earthworm burrow systems in three-dimensions. Warner et al. (1989) characterized macropores in soils using CT-techniques. Heijs et al. (1995) used CT images before and after a bypass flow experiment. Using three-dimensional image analysis tools, redistribution of water after the experiment could be quantified.

Experimental Setup

The experimental laboratory setup is shown in Figure 6.1. Soil samples were placed on a funnel connected to an outflow collector equipped with a pressure transducer. To prevent erosion of the sample, a disk, perforated with approximately 75 2.0-mm holes, was placed at the bottom. Rain showers were applied on top of the soil samples by using a small rain simulator, consisting of an adjustable tube pump and a needle irrigator.

Two tensiometers were installed, one at the top and one at the bottom of the sample. These tensiometers provide information on: (i) the initial pressure head in the soil; (ii) the time when water starts to infiltrate in macropores (top tensiometer) and; (iii) the hydraulic conditions at the bottom of the soil column. Installation of the tensiometer cups should be carried out at a slightly upward angle to prevent water flow from macropores into the cavity in which the cups are placed. Tensiometers should be sufficiently large to represent hydraulic soil conditions adequately, and were 5.0 cm long with a diameter of 5.0 mm.

All measurements were controlled by a computer equipped with a 14- bit 16-channel analog-digital converter. The measurement interval is flexible with a minimum of 5 sec. Output was shown graphically on the screen and stored in a file.

To measure breakthrough curves and/or adsorption of chemical species, the basic setup can be extended easily with, for example, a specific ion electrode or a spectrophotometer. Soil samples used in the studies described were taken at the Kandelaar experimental farm in Eastern Flevoland in the Netherlands. The soil was classified as a mixed mesic Hydric Fluvaquent (Soil Survey Staff, 1975).

Analysis of Staining Patterns

Fractal Analysis

The two-dimensional fractal dimension of staining patterns, Ds2, can be derived by fractionalizing a horizontal, stained, cross section into squares (pixels) of side r, and then counting the number of stained pixels N2(r). By repeating this process with increasing r values, a range of N2(r) values is obtained. Double logarithmic plots of N2(r) against r show a straight line (Hatano et al., 1992) from the slope of which the fractal dimension Ds2 is estimated. This model is written in a general form as:

$$\log N(r) = -D_S \log r + c \tag{6.1}$$

where c is constant, N is the number of stained pixels (2-D) or voxels (3-D), r is the pixel (or voxel) width, and D_s is the fractal dimension.

Measurement of the three-dimensional fractal dimension (D_s3) requires three-dimensional image data, which requires vertical sampling, on horizontal cross sections, at an interval with adequate resolution. Pixel size was 1 mm in the case of the measurement using a digitizer. It is impossible to create such thin horizontal cross sections. Therefore, D_s3 was calculated by interpolating two-dimensional image data measured on cross sections. The value of D_s3 can be obtained from the double logarithmic plot of side width r and the number of voxels stained $N_3(r)$. If $N_3(r)$ is multiplied by the volume of a cube of side r, an approximation to the total volume of stained parts is obtained. When r = 1, the size of the smallest pixel, this approximates the total volume of stained macropores.

Morphological Methods

Before starting an experiment the soil surface was exposed carefully by chipping away soil aggregates in order to create an undisturbed, natural infiltration surface. Water was applied on top of the sample contained 0.01 *M* methylene blue using the setup as described below and depicted in Figure 6.1.

After the experiment, the sample was peeled off in layers of 2 cm thickness. The methylene blue patterns were (in our experiments) photographed and later analyzed by an image analyzer (Nexus 6400). The areas and perimeters of all individual stains were calculated directly by the Nexus. The bitmaps produced by the image analyzer were used for further analyses. The 2-D fractal dimensions of the staining patterns were determined as described above. Three-dimensional fractal dimensions (D_s3) and the total volume of stained patterns (Vs) were calculated by vertical integration of the 2-D data, measured on cross sections (Hatano and Booltink, 1992).

The criteria for the ratio between area and perimeter ($Ar/Pe^2 < 0.015$), as described by Bouma et al. (1977), were used to stratify staining patterns in terms of vertical cracks and horizontal pedfaces of structure elements. Area of the cracks was calculated following:

$$A = \sum_{j=1}^{m} \sum_{i=1}^{n} \Delta Z * Pe \tag{6.2}$$

in which A is the total sample area of the vertical cracks, j and i are respectively the layer number and stain number, Z segment thickness, and Pe the perimeter of the stain. As an example, relating fractal dimensions to staining patterns, images of staining patterns of methylene blue in horizontal cross sections at depths of 45 mm and 110 mm in the cores of A, B, and D are presented in Figure 6.2. Although three images at a depth of 45 mm in the cores of A, B, and D have almost the same stained area, the staining pattern in core A is longer and more slender, the staining pattern in core B is more extensive, and the staining pattern in core D is more crowded than the others. D_s2 values of 1.18 for A, 1.35 for B, and 1.30 for D reflect well these staining characteristics. Three images at a depth of 110 mm have smaller stained areas but show similar staining patterns; a long and slender pattern for core A, a fragmentary pattern for core B, and an extensive pattern for core D. Their D_s2 values are 1.09, 1.02, and 1.67, respectively.

Relating Soil Structure to Hydraulic Characteristics

Case Study I: Morphological Characteristics to Predict Amounts of Bypass Flow

Using an experimental setup similar to the one showed Figure 6.1, five cylinders (200 mm in diameter and 200 mm long) were brought to identical initial soil physical conditions (Booltink and Bouma, 1991). Initial moisture content of the five samples varied between 0.42 to 0.48 $m^3\ m^{-3}$. Each of the five samples was subjected to a rain shower of 10 mm, at an intensity of 13 mm hr^{-1}. After termination of the experiment

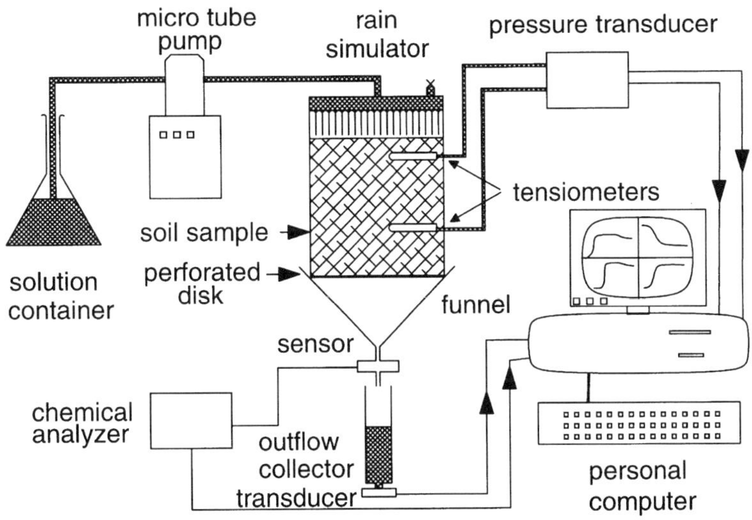

Figure 6.1. Schematic setup of the computer-controlled device for measuring bypass flow.

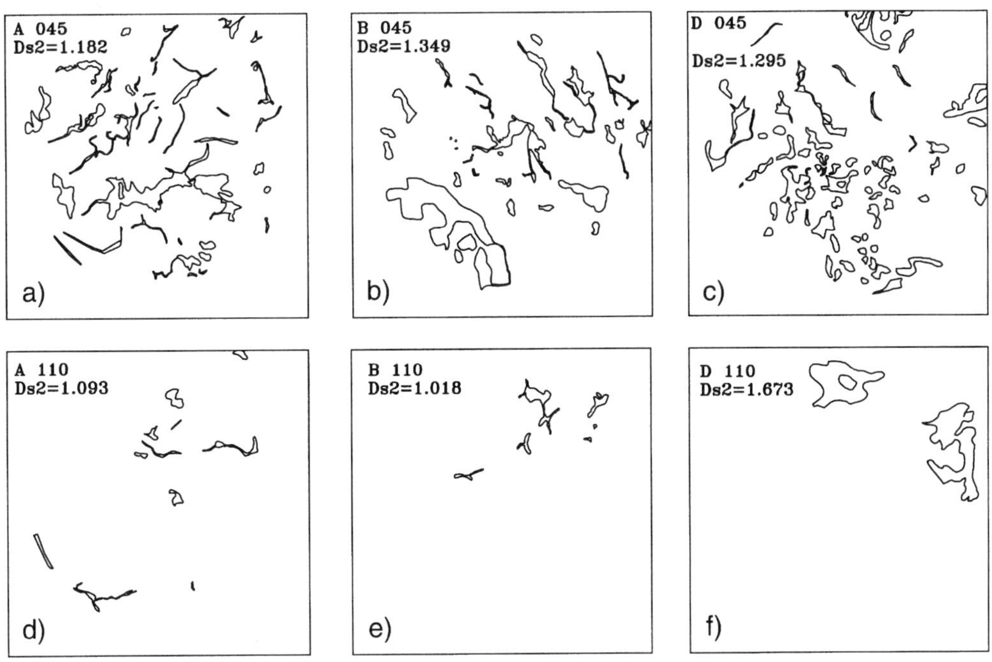

Figure 6.2. Some typical examples of staining patterns in horizontal cross-sections in soil cores at depths of 45 mm (a-c) and 110 mm (d-f), respectively.

the sample was sectioned in layers at depths: 15, 25, 45, 60, 90, 110, 125, and 160 mm from the top of the sample; for every layer a cross section of the methylene blue staining patterns was analyzed using software. A geographical information system and morphological characteristics and fractal dimensions were calculated.

Table 6.1 presents measured amounts of outflow from the cores. Outflow from all cores started after a significant time lag (T_{lag}) from the start of irrigation. Outflow rates

Table 6.1. Summary of Soil Physical and Soil Morphological Results.[a]

Sample	O_m (mm)	T_{lag} (s)	D_s3_u-1 (-)	D_s3_l -1 (-)	Vs_u (-)	Vs_l (-)
A	4.02	1920	2.29	2.08	0.087	0.007
B	5.01	1560	2.36	2.20	0.083	0.003
C	6.07	1260	2.39	2.28	0.090	0.021
D	4.34	1860	2.38	2.58	0.098	0.037
E	6.54	1140	2.43	2.04	0.085	0.004

[a] Om represents the total amount of bypass flow at the end of the experiment; T_{lag} the time lag before initial breakthrough of the bypass front at the bottom of the soil core; D_s3-1 represents the depth weighted 3D-fractal dimension for, respectively, the upper ($_u$) and lower part ($_l$) of the soil sore; Vs depicts the fraction of methylene blue stained parts for, respectively, the upper ($_u$) and lower part ($_l$) of the soil sore.

for each core, during irrigation of 2400 s, were almost equal to the applied irrigation rate of 13 mm hr^{-1}. This implies bypass flow through vertically continuous macropores; absorption processes, surface infiltration, and lateral absorption in these macropores have been minimal. As soon as irrigation ceased, outflow reduced remarkably and stopped after a few minutes. Furthermore, the total amount of outflow (O_m) that ranged from 4.0 to 6.5 mm can be represented as a function of time lag (T_{lag}):

$$O_m = -01818 T_{\log} + 9.85 \tag{6.3}$$

with a correlation coefficient of 0.996. This indicates that the time lag for outflow was induced by internal catchment (storage of water in vertically noncontinuous macropores) and surface storage.

On the other hand, morphological data on staining patterns can be regarded as a result of both bypass flow and internal catchment. The volume fraction of staining (Vs) is, therefore, important in explaining bypass flow and internal catchment. Vs was obtained from the number of the smallest unit cubes [$N_3(1)$ in Eq. 6.1]. Values of Vs for the upper and lower half of the core as well as for the whole core are summarized in Table 6.1.

Table 6.2 shows the results of some trials to explain the total amount of outflow (O_m). Although there are no direct physical relationships between O_m and Vs, Vs and D_s3–1, they can explain the amount of outflow. Equation 6.3, with a correlation coefficient of 0.894, was the best predictor for O_m when using only information from the upper part of the cores. The shape of this equation indicates that smaller values of Vs and greater values of D_s3–1 increased the total amount of outflow, indicating that a quasi 3-D structure is able to describe the total amount of outflow.

The amount of outflow cannot be explained by using only morphological data for the lower half of the core. However, data for both the upper and lower halves of the core predicted outflow with great accuracy (Equation 6 from Table 6.2.) The shape of this equation indicates that smaller values of Vs and greater values of D_s3–1 increased the total amount of outflow, indicating that a quasi 3-D structure is able to describe the total amount of outflow. The amount of outflow cannot be explained by using only morphological data for the lower half of the core. However, data for both the upper and lower halves of the core predicted outflow with great accuracy (Equation 6 from Table 6.2.)

Table 6.2. Examples to Explain the Amount of Outflow (O_m, Table 6.1) Using the Fractal Dimension (D_s3-1) and the Volumetric Fraction of Stained Parts (Vs).[a]

Nr.	Equation	Correlation Coefficient
1.	$Om = -62.1\ Vs_u + 10.7$	0.340
2.	$Om = -19.8\ Vs_l + 5.5$	0.262
3.	$Om = -187.1\ (Vs^{Ds3-1})_u + 12.0$	0.894*[b]
4.	$Om = 16.2\ (Vs^{Ds3-1})_l + 5.1$	0.033
5.	$Om = -4.6\ Vs_u + 1.7\ Vs_l + 43.7$	0.516
6.	$Om = -230.6\ (Vs^{Ds3-1})_u + 232.4\ (Vs^{Ds3-1})_l + 12.6$	0.995**[b]

[a] Subscripts of $_u$ and $_l$ represent the upper and lower half of core.
[b] Subscripts of * and ** represent that the equation is significant at 5% and 1% levels, respectively.

The shape of this equation indicates that the contribution of staining patterns in the lower half of the cores to bypass flow differs from that in the upper half of core; that is, greater values of Vs and smaller values of D_s3–1 in the lower half of the column increase the total amount of outflow. A physical morphological explanation for this phenomenon is that stains in the upper half of the cores develop around relatively small peds and are mostly not continuous. In the lower half, where the number of stains have been sharply decreased, only vertical continuous macropores, which dominate bypass flow, remain. This tendency was also found by Hatano et al. (1992).

Case Study II: Morphological Characteristics to Predict Outflow Curves

The incorporation of measured soil geometrical characteristics, such as fractal dimensions, to predict bypass breakthrough and outflow curves and thus allow them to be used in simulation models can be facilitated by the use of pedotransfer functions which relate easily obtainable soil characteristics to more complex ones (Bouma and Van Lanen, 1986). Analysis of the data obtained in this experiment augmented by data from previous experiments (Booltink and Bouma, 1991), suggested that the time lag for initial breakthrough at the bottom of the soil column, $T_{(d)}$, could be estimated with a linear regression equation (Eq. 6.4) containing the parameters Vs_l (the volume stained at the lower part of soil column) and D_s3–1 (the depth weighted fractal dimension of the lower part of the soil samples).

Thus:

$$\frac{1}{T_{(d)}} = a * \left[\frac{Vs_l}{100}\right](D_s 3-1) + b \tag{6.4}$$

Parameters a and b are empirical constants.

High values of Vs_l indicate either a large number of small macropores, or a few big water-conducting macropores, will lead to a reduction of $T_{(d)}$. The fractal dimension, on the other hand, gives information on the geometry of water-conducting macropores. A macropore system with a D_s3 value of 2 consists of mainly vertically oriented cracks; a value of 3, on the other hand, indicates that horizontal cracks dominate the system (Hatano and Booltink, 1992).

The value of $T_{(d)}$, computed with Eq. 6.4 can be used to calculate the propagation of the waterfront in macropores.

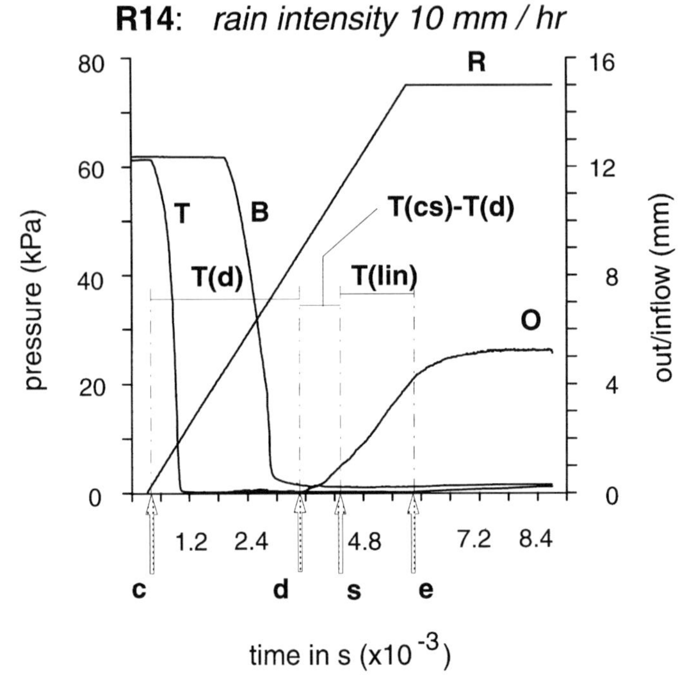

Figure 6.3. Results of a typical bypass flow. Arrows indicate characteristic points in time: **c** refers to start of water entering the macropores ($T_{(c)}$), **d** to the point of initial breakthrough ($T_{(d)}$), **s** to the starting point of the linear outflow ($T_{(cs)}$), and **e** to the end point of linear outflow ($T_{(ce)}$). **B** and **T** are referring to the bottom and top tensiometer, respectively. **R** refers to the cumulative rain applied, and **O** the cumulative outflow.

Figure 6.3 shows the experimental results of typical samples with rain intensities of 10 mm hr^{-1}. For all experiments, irrigation was started after monitoring initial conditions for 300 s and was continued until 15 mm of water was applied. The top tensiometers all react within a few minutes of irrigation starting, indicating that water has entered the macropores. After the experiment, this was confirmed by the presence of methylene blue stains on the cups. These top tensiometers provide, therefore, reliable information on the time $T_{(c)}$, expressed in seconds after starting the rain shower, when water enters the macropores. The time intervals (T) were determined by scanning the computer-measured output file (measuring interval 5 sec). The measured cumulative outflow curves all show identical behavior. Beyond $T_{(d)}$, when outflow starts, an exponential progression between the times $T_{(d)}$ and $T_{(cs)}$ (start of linear outflow) can be seen. Between $T_{(cs)}$ and $T_{(ce)}$ (end of linear outflow) there is a nearly linear outflow pattern, and after $T_{(ce)}$ an exponential fadeout pattern is shown. T-values are expressed in seconds after starting the rain shower. The five elements constituting bypass flow will now be discussed separately.

Surface Storage

When rain intensity exceeds the infiltration capacity of the soil matrix, a water layer starts to build up on the soil surface. The thickness of this water layer, surface storage, is highly variable due to the roughness of the soil surface and therefore cannot be measured directly. A minimum and a maximum surface storage is distinguished.

The minimum value indicates when the very first macropore starts to conduct water. Infiltration until the start of macropore flow (indicated by a pressure increase of the tensiometer in the top of the sample) was calculated by using the two-parameter infiltration equation as developed by Stroosnijder (1976). Minimum surface storage can now be calculated by subtracting the infiltrated amount of water from the applied amount (Table 6.3). By following this procedure for all cores, an average surface storage of 0.74 mm with a standard deviation of 0.30 mm water was calculated.

The minimum surface storage is an important boundary condition in the entire bypass flow process, as it determines the time of incipient bypass flow and the amount of water which infiltrates at the soil surface after cessation of bypass flow. Since not all the macropores present start to conduct water after $T_{(c)}$, we also defined a maximum surface storage; i.e., the level of ponding at which all macropores, accessible for bypass flow, conduct water. No direct measurements for this parameter were carried out during the experiment. An estimate can be made by determining the amount of water leaving the soil column after the rain shower has stopped. This water represents the amount of water still present on the soil surface and in the macropores. For these experiments, maximum surface storage averaged 2.67 mm, with a standard deviation of 0.62 mm. This value agreed fairly well with visual observations during the experiments.

Initial Breakthrough

The time of initial breakthrough [$T_{(d)}$], at the bottom of the soil cylinder, is related strongly to the applied rain intensity. However, the rather large variation between $T_{(d)}$ values within one rain intensity; e.g., sample R08 and sample R11 (Table 6.3) indicates that geometry of the water conducting macropores is an important variable in the flow process.

By applying a regression analysis, parameters a and b in the pedotransfer function (Eq. 6.4) were calculated to be 3.46506 and 0.02849, respectively. The correlation coefficient r was 0.812, which is significant at 1% (n=20). Although this pedotransfer function is strictly empirical it has a clear physical meaning. An increase in the volume of stained macropores (Vs_l) shortens the time lag. A high fractal dimension (D_s3-1), significant for a soil system dominated by horizontal pedfaces, on the contrary, extends the time lag.

The regression analysis was also performed for the upper part of the soil column. No significant relation could be found there. This makes clear that the lower parts of the samples restrict bypass flow.

Exponential Start of Outflow

At low rain intensities the exponential start of outflow (ESO), between $T_{(d)}$ and $T_{(cs)}$ is clearly present; e.g., sample R14 in Figure 6.3 and Table 6.3. At high rain intensities; e.g., sample R15, ESO is hardly visible. Bouma and Dekker (1978) found that water flowing into macropores, during bypass flow, formed small bands on the vertical walls of structure elements. The number and length of the bands were strongly related to the amount and intensity of rain applied. The development of new bands, and the creation of new initial infiltration areas, are important at low intensities. At high intensities, due to the fast propagation of the macropore wetting front, the available capacity for developing new bands is used relatively fast. This latter capacity is defined as the rain intensity dependent, maximum area of the structure elements in contact with bypass flow water.

Table 6.3. Physical Results.[a]

Sample	Minimum Surf. Stor (mm)	Maximum Surf. Stor (mm)	T(d) (s)	T(cs)-T(d) (s)	ESO (mm)	Or/Ir (-)	Total Outflow (mm)
			10 mm hr-1				
R13	1.17	2.37	22.98	22.62	2.17	0.67	5.1
R14	0.39	1.69	31.98	8.82	1.24	0.74	5.2
			20 mm hr-1				
R03	0.89	2.79	15.12	5.88	0.82	0.69	5.0
R04	0.39	2.09	12.30	3.90	0.67	0.78	6.2
R06	0.56	1.46	15.72	8.88	0.51	0.39	2.2
R07	1.00	3.40	19.80	1.56	0.46	0.70	5.2
			30 mm hr-1				
R08	0.84	3.14	5.88	6.12	1.00	0.56	4.7
R09	0.34	2.34	7.98	2.82	0.55	0.71	6.6
R10	0.67	3.17	4.08	10.02	1.33	0.84	6.9
R11	1.17	2.37	10.50	3.30	0.80	0.56	4.4
R12	1.25	2.35	7.62	7.98	0.86	0.38	3.1
			40 mm hr-1				
R15	0.55	3.05	8.40	2.16	0.46	0.71	5.7
R16	0.45	3.45	7.08	0.72	0.69	0.77	8.5
R17	0.67	3.27	3.96	4.32	0.63	0.66	7.1

[a] T(d) refers to the time of initial breakthrough, T(cs)-T(d) represents the duration of the exponential start of outflow (ESO), and Or/Ir is the outflow rate expressed as a fraction of the inflow rate. T values are expressed in seconds after starting the rain shower.

ESO is also influenced by the initial pressure head in the sample. A low initial pressure head reduces absorption along the macropore walls, which leads to a short ESO period; e.g., sample R04 in Table 6.3.

Internal catchment, defined as the accumulation of water in vertically discontinuous macropores (Van Stiphout et al., 1987), leads to high ESO values, especially in the early stage of the flow process, when dead-end macropores are first filled. Applying the morphological analysis Eq. 6.2, a distinction could be made between vertically oriented and horizontally oriented macropores. A sample whose structure is dominated with vertically oriented macropores will enhance bypass flow, while horizontally oriented macropores will increase internal catchment.

Linear Outflow

In the time span $T_{(cs)}$ and $T_{(ce)}$, outflow is linear ($T_{(lin)}$ in Figure 6.3) for all samples (i.e., the time derivative for outflow is constant) as has been observed consistently in earlier measurements (e.g., Bouma et al., 1981; Van Stiphout et al., 1987). A quasi steady-state condition is thus obtained between absorption processes and applied rain intensity. In Table 6.3 outflow rate during steady state flow is expressed as a fraction of the inflow rate (Or/Ir).

Relevant parameters which determine the absorption rate (rain intensity minus the outflow rate) have been examined using multiple regression techniques (Statistical Package for Social Studies, SPSS, 1991). The total explained variance when correlating the absorption rate with the initial pressure head and the soil morphological characteristics such as Vs_u, Vs_l D_s3_u, and D_s3_l the parameters, was 0.688 (n=15). This analysis showed that geometry of the macropore system (Vs_l^{Ds3l}), defined as the volume stained (Vs) raised to the power of the fractal dimension (D_s3), is even more important in explaining the absorption rate than the initial pressure head of the soil. The low correlation of stained volume and geometry in the upper parts of the soil cylinders indicates once more that bypass flow is governed by the most restrictive layer.

Total Outflow

Cumulative total outflow at the end of the experiment is presented in Table 6.3. Average measured outflow over the 15 samples was 5.6 mm (standard deviation 1.6 mm) which is 37% of the applied amount. Although rain intensity has an important effect on several individual processes, there was no significant relationship between the applied rain intensity and the total amount of outflow. At low rain intensities the time of initial breakthrough is much larger compared to high intensities, but this effect is compensated by the duration of the rain shower. The initial pressure head seems to have some impact, but not significant statistically.

MODELING FLOW PROCESSES USING SOIL STRUCTURE

The occurrence of large cracks and channels in soils (macropores) has a major impact on water and solute movement. Observation of rapid vertical movement of free water along macropores in unsaturated soils was already reported in drainage studies at Rothamsted (UK) by Lawes et al., 1982. Since water flow within macropores can follow different flow modes such as film-flow (Germann, 1987), it cannot be described accurately by Richards' equation, which is based on Darcian flow theory. These theories were developed for ideal homogeneous isotropic flow systems (e.g., Taylor, 1953;

Klute, 1973), as pointed out by White (1985). A clear definition of a macropore is difficult to give. Beven and Germann (1982) summarized literature and concluded that macropore sizes varied in a range between 30 to 3000 mm The effects of macropore systems on water flow and solute transport are highly variable and cannot be characterized by pore sizes only, as concluded by Beven and Germann (1982). A functional characterization is, therefore, required (Bouma, 1981). Germann and Beven (1982) quantified macroporosity from conductivity measurements at lower suctions. This type of data, however, does not provide any information on functioning of macropores in terms of conducting and absorbing water. To describe preferential flow-paths dye tracers are useful tools. Anderson and Bouma (1973) and Bouma and Dekker (1978) used methylene blue to characterize contact areas between bypass flow water and macropore walls.

Although dye tracers provide good possibilities for characterizing flow patterns in structured soils, so far only little effort has been made to use this kind of information as input for simulation models describing bypass flow and related processes. Edwards et al. (1979) used a two-dimensional model which simulates the flow patterns of water infiltrating into the surface of a soil column. Effects of macropores were simulated by means of a cylindrical hole from which water moved away radially and vertically. A system of uniformly distributed channels with different widths and depths was used by Beven and Clark (1986) to simulate infiltration into a homogeneous soil matrix containing macropores. Chen and Wagenet (1992) simulated water and solute transport by combining Darcian flow for transport in the soil with the Hagen-Poiseuille and Chezy-Manning equations for water transport in macropores. Hatano and Sakuma (1991) designed a combined capacity-bypass model to simulate transport of water and solutes in an aggregated soil. In their approach, soil structure was simulated by plate-like elements which reflected contact areas and exchange capacities of water and solutes. The interchange between the mobile and immobile phase occurred in an additional mixing phase.

Two-region models describing convective dispersive transport in a mobile and immobile phase were used by van Genuchten and Wieringa (1976), DeSmedt et al. (1986), and Vanclooster et al. (1994). The use of this type of model in structured soils is, however limited, since nonideal flow modes, such as film-flow, along macropore wall, cannot be described accurately by means of convective dispersive flow. Especially in moderate climate, with relatively low rain intensities, these nonideal flow modes are common.

Germann and Beven (1985) and Germann (1990) simulated bypass flow using kinematic wave theory in combination with a sink term for water absorption. In their approach the soil matrix and macropores are considered as one domain through which water is transported as a kinematic pulse.

All above models have in common that real soil structure, defined as the physical constitution of a soil material, as expressed by the size, shape, and arrangements of the elementary particles and voids (Booltink et al., 1993), is not considered. Bronswijk (1988) calculated macropore sizes by means of swelling and shrinkage characteristics measured on structure elements in the laboratory. In his approach, excess of water on the soil surface during a rain event was immediately transported to the groundwater with interaction with the surrounding soil. Jarvis and Leeds-Harrison (1987, 1990) developed a model which calculates water balances in two domains, a soil matrix and macropore domain. Soil structure was represented by means of cube-shaped soil aggregates and transport of water through the macropores was calculated using an empirical tortuosity factor. Interaction between soil aggregates and macropore water is a function of aggregate sorptivity. Hutson and Wagenet (1995) developed a model in which a series of adjacent columns, each with specific hydraulic characteristics, was used to simulate bypass flow and solute transport. Lateral interaction between the columns allowed the simulation of chemical exchange processes.

Hoogmoed and Bouma (1980) developed a model based on measured soil morphological characteristics, describing surface storage and lateral absorption of water along macropore walls and drainage. Although this model is based on morphological observations it does not include the effects of tortuous water transport through macropores. The model is only capable of calculating bypass flow in small columns with strict boundary conditions.

Detailed modeling of bypass flow was described by Heijs and Lowe (1997). They used the lattice Boltzmann equation to calculate permeability of a clay soil. Data on soil structure was obtained from CT-imaginary (Heijs et al., 1995). Di Pietro (1993) used lattice-gas techniques to model transport of water through a macropore. This technique is described elsewhere in this book.

Case Study III: Modeling on Laboratory Scale

Model Description

The quasi three-dimensional simulation model of Hoogmoed and Bouma (1980), was applied to simulate the bypass flow experiments as described in the previous section. The model considers the following processes:

A. Vertical infiltration of water in the upper soil surface and water flow between underlying layers, simulated with the Darcy equation.

$$z = 0 \quad -K\delta H/\delta z = R \quad 0 < T < T_{(p)} \tag{6.5}$$

$$0 < H < w \quad T_{(p)} \leq T \leq T_{(c)} \tag{6.6}$$

$$z = L \quad -K\delta H/\delta z = \text{constant} \tag{6.7}$$

where z = depth (m) in a soil column with length L, $T_{(p)}$ = time (s) when ponding starts, $T_{(c)}$ = time when water enters the macropores, w = surface storage (m), R = rain intensity (m s^{-1}), and H = hydraulic head (m water).

B. Flow of water in macropores when a threshold value for surface storage is reached.

C. Lateral absorption of water into vertical crack walls (Van der Ploeg and Benecke, 1974).

$$T > T_{(c+lag)} \quad 0 < x \leq x_t \quad \frac{\delta}{\delta x}\left[D\frac{\delta\theta}{\delta x}\right] = \frac{\delta\theta}{\delta t} \tag{6.8}$$

With the following boundary conditions:

$$T > T_{(c+lag)} \quad x = 0 \quad \frac{\delta\theta}{\delta x} = 0 \tag{6.9}$$

$$T > T_{(c+lag)} \quad x = x_t \quad \frac{\delta\theta}{\delta x} = 0 \tag{6.10}$$

Where x = horizontal distance (m) from the infiltration surface, q = volumetric moisture content, x_t= penetration depth of the wetting front at time t, lag is the time lag between $T_{(c)}$ and the arrival of the wetting front at depth z, and D is the diffusivity.

D. Water not absorbed at the bottom of the soil column was calculated as drainage. An extended model description was provided by Hoogmoed and Bouma (1980), Van der Ploeg and Benecke (1974), and Booltink and Bouma (1993). To improve the description of the bypass flow processes, some model modifications were made:

1. At the start of a bypass flow event not all the vertical macropores are instantly available for conducting water due to surface roughness. The accessibility of macropores at the soil surface was proportionally divided between a minimum and a maximum threshold value for surface storage, as derived from visual observations.
2. Vertical water absorption on the horizontal pedfaces of structure elements inside the soil is largely dominated by gravity forces. To simulate the latter process a module which calculates a Darcy subsurface infiltration similar to the infiltration of water into the upper soil surface was added.
3. The model of Hoogmoed and Bouma did not consider the effects of tortuous water transport in terms of absorption and breakthrough. They assumed cracks to be vertically continuous and straight. Water running into such cracks exits very rapidly. In reality, pathways are tortuous. The pedotransfer function (Eq. 6.4) for estimating the time lag for initial breakthrough was incorporated in the model and used to calculate the rate of propagation of the water front in macropores.

Simulation Results

Figure 6.4 presents simulation results using the model in combination with the pedotransfer function for predicting time lag for initial breakthrough (Eq. 6.4). Simulated and measured outflow have been plotted as a scatter diagram.

To test the quality of the model: (i) the average measured outflow was compared with average simulated outflow by means of a *t*-test. Simulation results did not differ significantly from the measured data with a probability of 95%. (ii) The regression coefficient (r=0.66) deviates from the 1:1 line (r=1.0). However, r was significant at a 1% probability level, indicating that the relation is adequately described. Outflow is the most important term in the mass balance (6.3 mm average with a standard deviation of 1.49 mm). Due to the relatively dry initial condition and the water remaining on top of the sample, surface infiltration is high compared with previous experiments (5.4 mm average with a standard deviation of 1.15 mm) (Booltink and Bouma, 1991). Absorption into horizontal pedfaces seems hardly relevant (0.7 mm avg, 0.64 mm std). For sample R06 where the area of horizontal pedfaces dominates the system, this becomes, however, an important sink term.

Measured time lags in Table 6.4, for the samples R03, R08, R14, and R16, were compared with time lags obtained from the pedotransfer function (Eq. 6.4). To check the pedotransfer function independently, the parameters a and b were calculated excluding data of the samples R03, R08, R14, and R16; time lags calculated, agreed fairly well with measured time lags. The apparently large deviation for sample R14 is related to the duration of the rain shower and, therefore, of the same order of magnitude as the time lag of sample R16. Since time lag has no influence on surface infiltration, this term was disregarded. Although differences in time lag can be considerable (e.g., R14) the effect on the simulated sink terms is rather small, indicating that the model is not

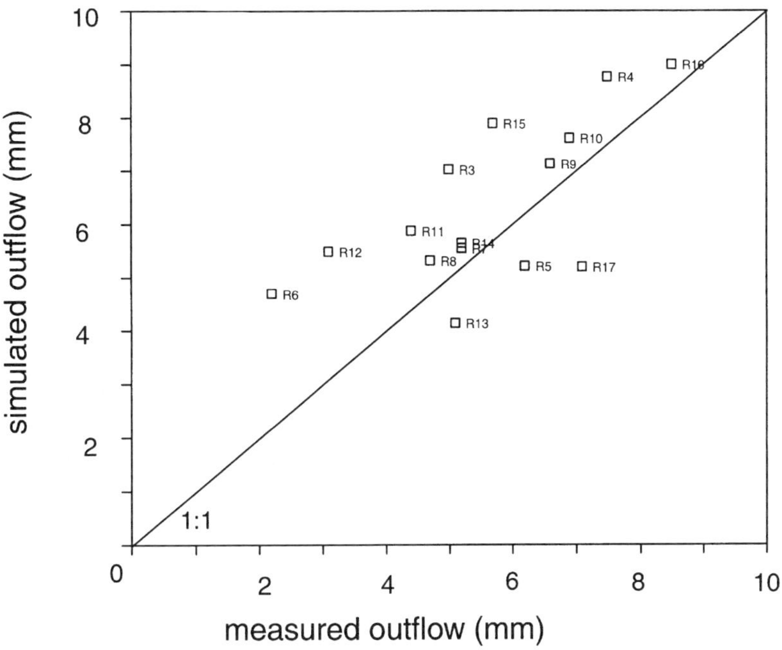

Figure 6.4. Scatter diagram for measured versus simulated outflow by using the model of Hoogmoed and Bouma (1980) extended with the pedotransfer function (Eq. 6.4) to predict propagation of water through the macropore domain.

Table 6.4. Sensitivity of the Extended Simulation Model of Hoogmoed and Bouma (1980) for Measured and Predicted Values of the Time Lag to Initial Breakthrough at the Bottom of the Soil Cores.

Sample		Time Lag (s)	Drainage (mm)	Absorption Vertical Cracks (mm)	Horizontal Pedfaces (mm)
R03	measured	1512	7.1	1.8	0.5
	predicted	1488	7.0	1.8	0.5
R08	measured	588	5.3	3.7	0.7
	predicted	594	5.3	3.7	0.7
R14	measured	3198	5.8	1.5	0.4
	predicted	2550	5.6	1.7	0.4
R16	measured	7.08	0.88	0.19	0.5
	predicted	8.58	0.90	0.17	0.4

very sensitive to small variations of this characteristic, especially in combination with relatively low rain intensities.

Case Study IV: Modeling on Field Scale

Model Description

The numerical simulation model LEACHM (Leaching Estimation and Chemistry Model) was used as the base model (Wagenet and Hutson, 1989). LEACHM considers different processes in a variable soil profile, with or without plant growth. Processes included are: transient fluxes of water, fluxes and transformation of nitrogen, pesticides and salts; evapotranspiration and rainfall. In this study, the water flow sub model LEACHW was used (Wagenet et al., 1989; Wagenet and Hutson 1989).

In LEACHW, water flow is calculated using a finite-difference solution of the Richards' equation:

$$\frac{\delta\theta}{\delta t} = \frac{\delta}{\delta z}\left[K(\theta)\frac{\delta H}{\delta z}\right] - U(z,t) \tag{6.11}$$

Where θ is the volumetric water content ($m^3\ m^{-3}$), t is time (day), K is the hydraulic conductivity ($m\ day^{-1}$), H is hydraulic head (m) composed of matrix potential y and profile depth z, U is a sink term representing water uptake by plants (day^{-1}).

The original Campbell K-θ-ψ equations to describe soil hydraulic properties (Campbell, 1974; Hutson and Cass, 1987), were replaced by van Genuchtens' closed form equations (van Genuchten, 1980).

Since LEACHW originally does not consider bypass flow, the model was extended with the following modifications:

A. For every rain or irrigation event (30 minutes interval), the amount of water that could not infiltrate through the soil surface during the time period (amount divided by intensity of the rain event) was determined. This surplus is stored on the soil surface and when a certain threshold value for surface storage (*MinSS*) is exceeded, water starts to flow into the macropores. Not all macropores present are equally accessible, due to small differences in micro relief (Booltink et al., 1993). A maximum surface storage (*MaxSS*) was therefore defined. Above this level, water flows directly into the macropores. Between *MinSS* and *MaxSS* excess water is divided proportionally between surface storage and bypass flow. Water remaining on the surface continues to infiltrate after the rain has stopped.

B. In the macropore domain, water transport is simulated using a tipping-bucket approach. Propagation of the water front in the macropores (Bp_{prop} [mm day^{-1}]), i.e., tipping-bucket switching times, is calculated using a regression equation (Eq. 6.12) based on rain intensities (R_{inten}) and measured morphological properties (Booltink et al., 1993)

$$Bp_{prop} = a_{trv} * R_{inten} + d_{trv} * G_{fac} + b_{trv} \tag{6.12}$$

Parameters a_{trv}, b_{trv}, and d_{trv} are empirical constants. The dimensionless geometry factor (G_{fac}) is defined as:

$$\frac{1}{G_{fac}} = \left[\frac{Vs_l}{100}\right](D_s 3_l - 1) \tag{6.13}$$

where Vs_l is the volume of methylene blue stains (% vol.) in the limiting soil layer and D_s3_l-1 is the depth-weighed fractal dimension of the stains. High values of Vs_l, which are an indication of either a large number of small pores or few big water-conducting macropores, will lead to a reduction of Bp_{prop}. The fractal dimension, on the other hand, gives information on the geometry of water-conducting macropores. A macropore system with a D_s3_l-1 value of 1 consists mainly of vertically oriented cracks; a value of 2, on the other hand, indicates that horizontal cracks dominate the system.

C. When water flows into macropores it is absorbed into macropore walls. Lateral absorption of water from vertically continuous macropores is based on absorption described by a diffusivity equation (Eq. 6.14). Vertical absorption of water on horizontal pedfaces of structural elements is dominated by gravity forces, described by the Darcy equation:

$$Qp_{hor} = -D(\theta)\frac{\Delta\theta}{\Delta\left[\frac{X_{struc}}{2}\right]} \tag{6.14}$$

$$Qp_{ver} = -K(\psi)\left[\frac{\Delta\psi}{\Delta\left[\frac{X_{struc}}{2}\right]}+1\right] \tag{6.15}$$

where Qp_{hor} and Qp_{ver} represent the potential horizontal and vertical absorption fluxes (m day^{-1}), respectively, D is soil water diffusivity in m^2 day^{-1}, and X_{struc} is the structure element diameter (m). *Dy* and *Dq* were calculated as a geometrical mean between saturation (macropore wall) and the actual value, respectively, *of y* and *q* in the adjacent soil matrix as simulated by the Richards' equation. The potential absorption fluxes were reduced for the area of soil in contact with bypass water as indicated by the occurrence of stains. Contact areas for horizontal and vertical absorption were determinated by using dye tracers in laboratory experiments. This procedure was described and is based on stratification of methylene blue stained macropores into sets of horizontal and vertical oriented macropores using the ratio between area and perimeter ($Ar/Pe^2 < 0.015$) (Bouma et al., 1977).

The reduced, actual absorption fluxes were added as an additional sink term to Richards' equation. The surplus of water, not absorbed during the residence time in a given compartment, is added to the next compartment and finally to the groundwater.

D. Originally LEACHW did not provide a special bottom boundary condition which can simulate dynamic groundwater levels. Simulation of such a groundwater level was realized by combining a freely draining profile with an impermeable soil layer at the bottom of the profile. Deep drainage toward ditches (*Qd*) was simulated using a third-order polynomial:

$$Qd(h) = Cd_0 + Cd_1 h + Cd_2 h^2 + Cd_3 h^3 \tag{6.16}$$

where h is the pressure potential in the bottom compartment and the constants Cd_0, Cd_1, Cd_2, and Cd_3 are fitting parameters. A similar procedure was followed to simulate discharge through tile drains (*Qtd*). Parameter h in Eq. 6.16, in that case, represents the groundwater level above a specified drain depth and constants are indicated as: Ctd_0, Ctd_1, Ctd_2, and Ctd_3, respectively.

Sensitivity Analysis

The modified LEACHW-model was tested using data from the Kandelaar experimental farm. The objective of this test was to: (i) obtain information on the performance of the model and, (ii) calibrate the model on measured time series of groundwater levels. For this purpose, a Monte Carlo analysis (MCA) was used. In this

MCA, the uncertainty of model inputs is characterized in terms of distributions with or without correlations among the various parameters.

From exploratory simulation runs the maximum compartment thickness was determined to be 0.02 m. The total profile depth was set at 1.5 m and consists, therefore, of 75 compartments. During the exploratory runs, using the soil layers as determined in the soil survey, bypass flow occurred only in a limited number of cases. An additional top soil layer of 0.02 m, which reflected surface resistance, caused by surface smearing during ploughing and small loose structure elements, was, therefore, added.

Model simulation outputs were evaluated in a combined uncertainty and sensitivity analysis. In the sensitivity analysis, the variation of the various parameters was tested against simulated water balance terms by means of a linear regression model:

$$y(k) = \beta_0 + \beta_1 x_1(k) + \ldots \beta_p x_p(k) + e(k) \quad k = 1,\ldots,n \tag{6.17}$$

where b_0, b_1,...,b_p denote the ordinary regression coefficients, obtained by minimizing the sum of squares error (e(k)). The values y(k) and $x_1(k)$,...,$x_p(k)$ represent the model output and input parameters, respectively, in the k-th model simulation. *n* denotes the total number of simulation runs.

The sensitivity analysis was carried out on the simulated mass balance terms for the winter season 1989–1990. Since not all parameter combinations show linear relations with the various mass balance terms, ordinary regression coefficients were replaced by their rankings, where the smallest value gets ranking 1, the next one 2, etc. This technique can be used to linearize nearly nonlinear relations (Iman and Conover, 1982). The saturated hydraulic conductivity of top soil dominated mostly in Eq. 6.17. Parameters regulating tortuous water transport in macropores, a_{trv}, b_{trv}, and d_{trv} from Eq. 6.12, also have significant impact on cumulative absorption into macropores walls.

Leachate at the bottom of the soil profile is influenced mainly by the regression parameters of the deep drainage characteristic (Cd_0, Cd_1, Cd_2, and Cd_3). Most significant model parameters were selected for further model calibration. Selection was based on their importance in the sensitivity analysis as well as on their contribution to a process.

Model Calibration

Although model input parameters and their distributions reflect individual variability, not all combinations of these input parameters necessarily result in realistic simulation results. Further calibration is therefore required. In this study a set-theoretic approach, developed by Jansen (1993) and Jansen et al. (1993b) was used. This ROtated RAndom SCan (RORASC) procedure, based on previous work of Keesman and Van Straaten (1988, 1989) and Keesman (1989, 1990) uses an iterative Monte Carlo search procedure to reduce the parameter space by updating the currently available set of acceptable parameters. In this way, one gradually and efficiently zooms in on the acceptable parameter set.

The RORASC procedure consists of three basic steps (Jansen, 1993):

1. *Initialization.* Before starting the iterative procedure, an initial set of parameters has to be determined, which serves as a starting point for the subsequent simulations. In this study the initial set of 150 model runs, used for the sensitivity and uncertainty analysis, was used for the first calibration step.
2. *Generation* of new candidate samples. Simulated groundwater levels for the winter season 1989/1990 were compared to measured time series for the same period. By

visual inspection of the results, the simulation runs were divided in a population of acceptable and nonacceptable results. The generation of a new set of parameters is achieved by transforming or rotating the original parameter space to focus on the current set of acceptable parameters. This transformation is based on the decomposition of the covariance matrix of the accepted parameters, as described in detail by Jansen (1993) and Jansen et al. (1993). Subsequently a random-scan on the basis of uniform random sampling in this transformed space is performed which serves as a new set of candidate samples.

3. *Simulation and acceptance.* Model simulations are performed for the newly generated candidate samples and results are again divided into a population of acceptable and nonacceptable results. The file containing the total number of acceptable results is updated and the transformation and random-scan procedures are repeated iteratively until simulation results are satisfactory.

To obtain a reliable covariance matrix when using LHS in combination with RORASC, a number of samples $N > 10p$ is recommended, where p is the number of parameters to be sampled. Since the number of selected parameters was 8 totally, 80 simulation runs were executed in every RORASC cycle. Parameters not varied were fixed at values obtained from the best simulation run in the first Monte Carlo session.

Calibration was performed on measured cumulative drain discharges from the Kandelaar experimental farm in the Netherlands during the winter season 1989–1990 and the measured time series of groundwater levels in the same period. In the RORASC optimization procedure only three iterative steps had to be performed. The third step did not lead to significantly better simulation results, indicating that the parameters had reached their limits. It also indicates that the initial parameter space was close to the calibrated one.

Ill-defined parameters, i.e., parameters which cannot be measured accurately, and which were for that reason represented by a uniform distribution, slowly concentrate around their most likely values. The uniform distribution becomes log normal.

In Figure 6.5, measured groundwater levels are presented with the 99% confidence intervals of the simulation results. The effect of the RORASC procedure is demonstrated by the smaller band between the upper and lower 1% confidence intervals which is especially obvious from days 60 to 110. When interpreting these data, one has to consider the difference in sensitivity for the groundwater above and below the depth of the tile-drains of 0.82 m. Whereas a simulated deviation of a few centimeters below the tile-drain level will have little effect on the water balance, if the groundwater level is above tile-drain depth an equivalent deviation will have a large effect on the drain discharge because of the high drain discharge characteristic.

Model Validation

Using the input parameters as determined in the third RORASC step, a Monte Carlo simulation of 80 model runs was used to validate the model.

In Figure 6.6, measured groundwater levels in the winter season 1991–1992 are compared to the simulation results. Except for the bypass flow event around day 65, all events were well simulated by the model. The model overestimates the groundwater level from day 45 to 60, although this does not lead to large mass balance errors, as discussed in the model calibration section.

The generally small difference between the 95 and 99% confidence intervals in Figure 6.6 indicates that the calibrated parameter space was within the physical limits and, therefore, does not lead to large simulation errors or outliers. Only around

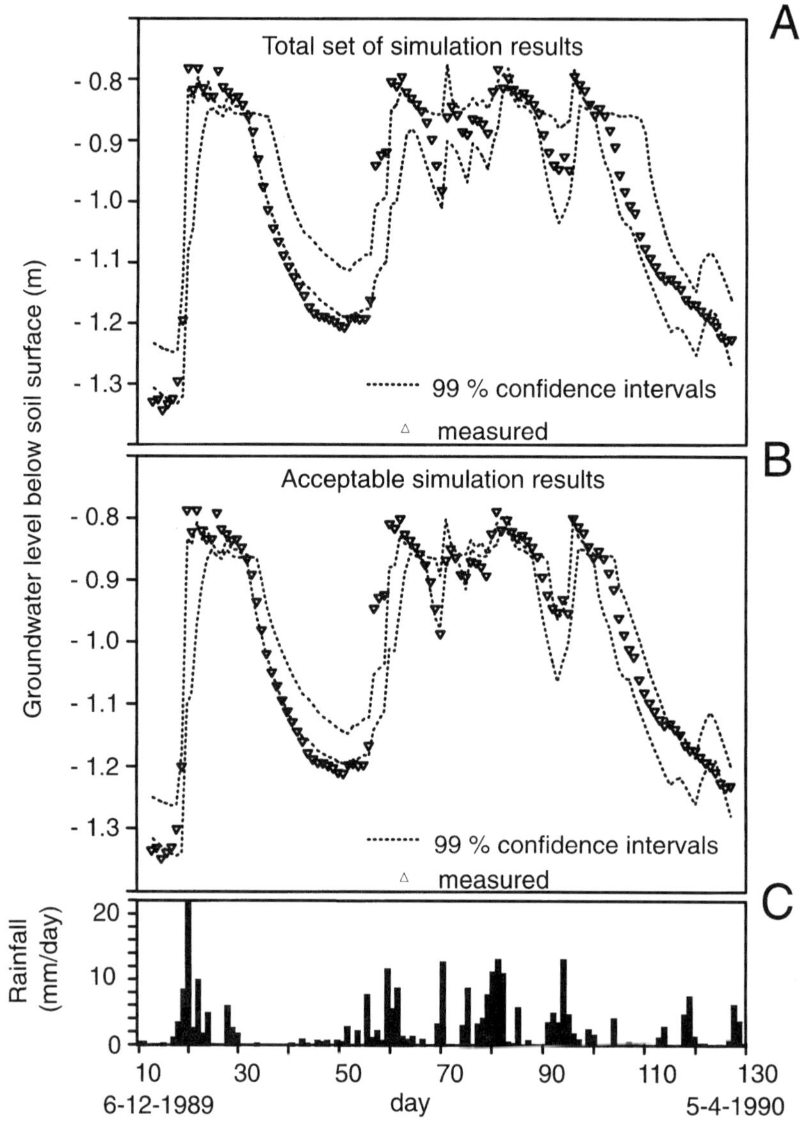

Figure 6.5. Comparison of measured groundwater levels below soil surface for the winter season 1989–1990 with upper and lower 99% confidence intervals. The total of all simulation results are presented in **a**; the acceptable simulation results, only, are shown in **b**; **c** represents the daily precipitation during the simulation period.

day 95 can a distinct difference between the 95 and 99% levels be seen. The average simulated cumulative drain discharge at the end of the period of 79.9 mm was close to the measured cumulative drain discharge of 86.9 mm. Cumulative surface infiltration is the most important term in the water balance. Bypass flow is also substantial and was calculated to be 48.1 mm. Absorption along macropores in this winter period was relatively small at 8.7 mm. The mass balance error was small; for most simulations the value was 0.6 mm with only a few outliers.

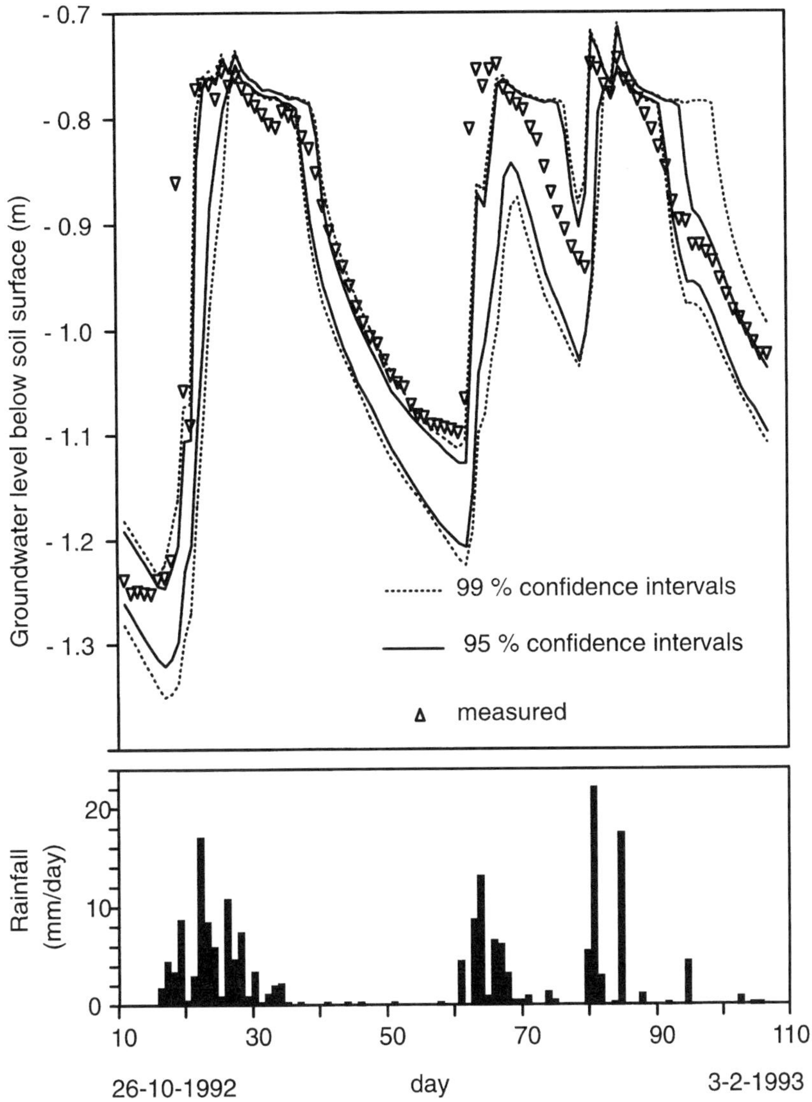

Figure 6.6. Validation results of simulated groundwater levels below soil surface for the winter season 1991–1992, **a**. The daily precipitation during the simulation period is represented in **b**.

IMPACT OF NONEQUILIBRIUM FLOW ON SOIL AND WATER QUALITY

Case Study V: Effects of Soil Structure on Root Development and the Extraction of Water

Soils that are identical from a genetic and taxonomic point of view can form considerably different soil structure types as a result of applied management (Kooistra et al., 1985; Droogers et al., 1996). Soil structure has an important impact on soil quality and sustainability. A comparative study between different management types and soil structures within one soil type can therefore be used to deduce sustainable management systems. The amount of water and dissolved solutes which is "available" for root water uptake is often defined as the amount of soil water between field capacity

Table 6.5. Basic Soil Properties for the Three Distinguished Soil Structure Types for the Top Soil.[a]

	Bulk Density Mg m^{-3}		Organic Matter %		Porosity m^3 m^{-3}	
	avg	std	avg	std	avg	std
Bio	1.47	0.065	3.3	0.59	0.42	0.015
Conv	1.68	0.061	1.7	0.05	0.36	0.021
Perm	1.38	0.109	5.0	0.46	0.46	0.023

[a] *Bio* is biodynamic temporary grassland, *Conv* is conventional temporary grassland and *Perm* is permanent grassland.

and wilting point. Bouma (1990) makes a distinction between "available" and "accessible" water for root uptake and introduced a methodology to take this accessibility into account during modeling activities. Bouma and Van Lanen (1989) presented an exploratory case study which clearly shows that accessibility can have a strong impact on crop growth and water dynamics of a soil.

Accessibility of soil water depends, on the one hand, on root density and root distribution, and on the other hand, on the ability of the soil to supply roots with water at a certain rate. Rooting patterns are mainly influenced by the penetration resistance, and thus by the density and aggregation of the soil, and can directly be observed. Availability is mainly a function of the soil hydraulic characteristics.

In this study, that is fully described by Droogers et al. (1997), three soil structure types resulting from different management practices in one type of soil were compared in terms of their capacity to allow crop transpiration (Table 6.5).

Experimental Layout

Three sites were selected, each under a different management system but all within the same soil type, a loamy mixed mesic Typic Fluvaquent (Soil Survey Staff, 1975), in the southwestern part of the Netherlands. First site, abbreviated as *Bio*, has been managed for 70 years according to biodynamic principles (Reganold, 1995), which implies that no chemical fertilizer or chemical crop protection was utilized, but that soil quality was maintained by a favorable crop rotation and by use of organic fertilizer. However, soil tillage did not deviate from conventional systems and consisted of plowing to a 0.30 m depth. The second site, denoted as *Conv*, belonged to an experimental farm with a conventional management system for the region, including a crop rotation of potatoes, grassland, sugarbeet, and grains and use of agrochemicals. On both sites, a field with temporary grassland, applied as part of crop rotation, was used for this study. The *Bio* field was used as temporary grassland for two years; the *Conv* field for three years. Finally, a site was selected which had been permanently under grassland since 1947 (*Perm*).

The 15 undisturbed samples (five per structure type) for the transpiration experiment were saturated and some additional nutrients were added to ensure growth without nutrient limitations. In every sample seven barley seedlings were planted and the top and bottom of the samples were covered with plastic to prevent soil evaporation. Samples were placed in a greenhouse and were weighed at two-day intervals, thus providing information on transpiration. No additional water was added.

After the plants failed to take up any more water due to drought stress, three randomly selected samples from each structure type were sliced horizontally at 0.10 m and 0.15 m depth. Rooting patterns were recorded in the middle of these sections and

Table 6.6. Total measured transpiration and initial moisture content for the three structure types and number of roots and root density observed on 6 horizontal sections per structure type.

	Initial Moisture Content ($m^3\ m^{-3}$)	Measured Transpiration ($m^3\ m^{-3}$)	Number of Roots (–)		Root Density (m^{-2})
			avg	std	
Bio	0.342	0.210	74	15.3	5139
Conv	0.340	0.225	76	14.8	5278
Perm	0.357	0.251	90	9.9	6250

were quantified by determining the relation between a series of hypothetical extraction zones and the remaining *"dead volume"* (De Willigen and Van Noordwijk, 1987). The hypothetical extraction zone was defined as a circle around each root and the "dead volume" as the soil fraction which was not located within the extraction zone of any root. This characteristic relation between the hypothetical extraction zone and the dead volume will be denoted *as "rooting pattern characteristic,"* and is a function of the number of roots and their distribution. Van Noordwijk et al. (1993) showed that this method leads to a frequency distribution of "nearest neighbor distances" (Diggle, 1983).

Accessibility of Water as a Result of Soil Structure and Rooting Patterns

The total amount of measured transpiration was significantly different for the three structure types and was highest for *Perm* followed by, respectively, *Conv* and *Bio* (Table 6.6). To illustrate the effect of soil structure on rooting patterns and, therefore, accessibility of water for plants, two different examples are selected from 18 observed rooting patterns. First, the observed rooting patterns from the grid of 12×12 cm with grid sizes of 0.5 cm are shown Figure 6.7. *Bio6-10* could be regarded as an example of a heterogeneously distributed rooting pattern, while roots in *Perm6-10* were more homogeneously distributed. Secondly, the distance from each grid cell to the nearest root was measured, providing plots (Figure 6.8). Clearly, *Bio6-10* had more regions which were at greater distances from roots than *Perm6-10*. Droogers et al. (1997) calculated a distribution function of distances to the nearest root. This distribution function showed that 50% of the soil volume was located more than 1 cm from the nearest root for *Bio6-10*, while this was only 20% for *Perm6-10*.

The two samples presented were not necessarily representative, but were selected to clarify the applied method. Results from all the samples showed that the number of roots for *Perm* was highest, followed by *Bio* and *Conv* (Table 6.6) Differences were not significant at a 5% level but at 10% level, *Perm* differed from *Bio* and *Conv*. When explaining differences in transpiration among treatments, two factors have to be distinguished: the number and distribution of roots and the ability of the soil to supply water to these roots. The first factor, the rooting pattern, was favorable for *Perm*, which had a fine and loose soil structure, resulting in a high number of roots and a homogeneous distribution. The *Bio* and *Conv* field had fewer roots and a more clustered distribution pattern than *Perm*, although distribution for *Bio* was slightly more homogeneous than for *Conv*. The second factor, the ability of the soil to supply roots with water was obviously better for *Conv* than for *Bio* and even for *Perm*. Differences can be explained by comparing hydraulic conductivity curves. In the wet range conductivity was lowest for *Conv*, but in the dry range conductivity was highest, resulting in the highest fluxes when conditions were relatively dry. Combination of the two factors showed that

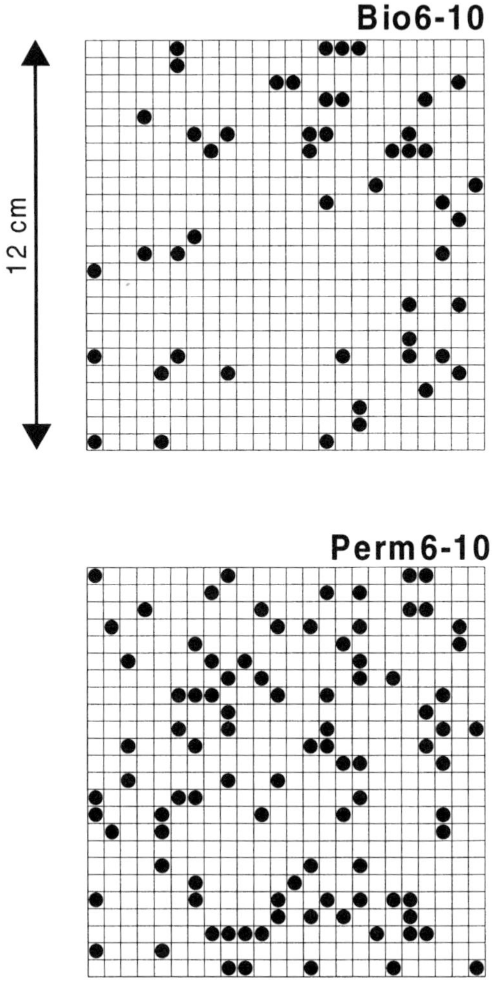

Figure 6.7. Rooting patterns of two different cross sections from a *Bio* and a *Perm* section at 10 cm depth. For each grid-cell of 0.5 x 0.5 cm, a dot indicates the presence of a root.

accessibility was almost equal for *Perm* and *Conv*, but significantly lower for *Bio*. For *Perm*, favorable rooting properties resulted in this higher accessibility, while the ability of the soil to supply roots with water was the determining factor for *Conv*.

Case Study VI: Field Effects of Soil Structure on Bypass and Nitrate Leaching

To obtain insight in the process of nitrate leaching in structured clay soils, both water outflow and soil mineral nitrogen profiles have to be determined to allow interpretations of nitrate contents in drainage water. Agricultural management also has a large impact on nitrate leaching as indicated by, for example, Steenvoorden (1989). Time and amount of application of organic and inorganic fertilizers strongly influence the nitrate concentration in groundwater. The application of catchcrops to reduce nitrate leaching was studied by Finke (1993). In this case study (Booltink, 1995), long-term monitoring of bypass flow and soil mineral nitrogen contents was used to: (i), investigate bypass flow and related nitrate leaching in conjunction with different fer-

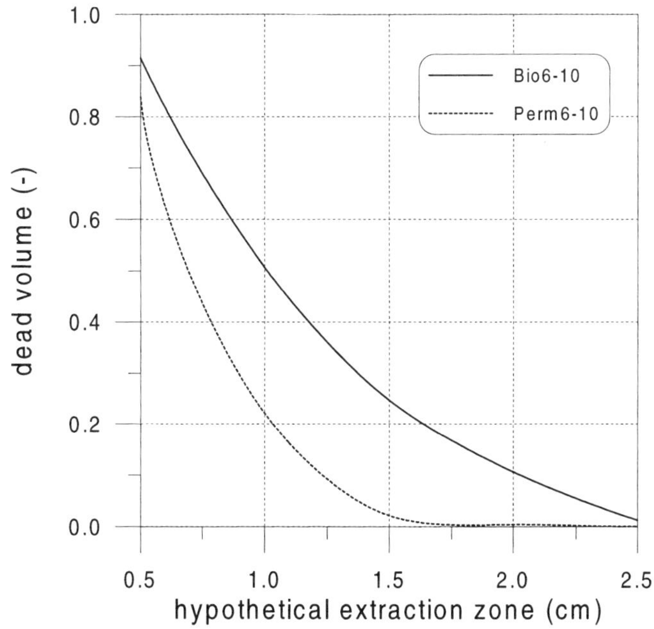

Figure 6.8. Measured distances to the nearest root for each grid-cell based on the rooting patterns as shown in Figure 6.7.

tilizing and land management treatments, and (ii) define procedures for farmers to reduce nitrate leaching and maximize fertilizer use efficiency.

Monitoring Site

The study area was located at the experimental farm The Kandelaar in Eastern Flevoland, the Netherlands. The soil, reclaimed in 1958 and first used for agricultural purposes in 1967, was characterized as a mixed, mesic Hydric Fluvaquent (Soil Survey Staff, 1975).

Field experiments with iodide tracers (Van Ommen et al., 1988), in combination with methylene blue laboratory experiments (Booltink and Bouma, 1991; Booltink et al., 1993) showed that, as a result of annual tillage practices, the number of continuous macropores rapidly decreased at the bottom of the Ap horizon. Macropores present below that depth originated from irreversible shrinkage during reclamation, and were continuous over the profile until the sandy subsoil at approximately 1 m depth.

A CR10-datalogger (Campbell, 1974) was used for continuous monitoring of drain discharge on one tile drain in the center of the study area. Drain depth varied from approximately 0.90 m at the beginning to 1.10 m at the end of the drain. Catchment area of the drain was approximately 7200 m^2, discharge resolution of the drain outflow apparatus was, therefore, 0.03 mm water head. Water samples were taken every 0.06 mm of discharge, using a sequential water sampler (ISCO). Eight samples were mixed in a sample bottle by the sampler. In this way, flow was proportionally sampled representing 0.48 mm of discharge per sample bottle. Drain water samples were analyzed for NO^{3-}, NH^{4+}, and Cl^- in the laboratory. Rain amount and groundwater level was continuously 0.20 m high), conditions for nitrate leaching in the subsoil can be compared to those in the winter period.

Table 6.7. Total Drain Outflow Related to the Total Precipitation During Periods of Bypass Flow.[a]

Day Number	GW_{init} (m-surf.)	Precipitation (mm)	Drain Outflow (mm)	Drain/Rain (-)
348–362	1.16	50.6	32.2	0.636
380–396	1.13	29.8	19.8	0.664
411–418	0.78	17.2	14.5	0.843
497–498	1.18	4.6	1.4	0.312
670–672	1.23	22.1	6.5	0.296
784–792	1.03	14.5	12.2	0.842
905–912	1.12	92.4	41.7	0.452
1041–1052	1.25	85.5	33.6	0.393
1082–1092	1.22	31.3	22.3	0.713
1100–1106	1.02	42.0	32.5	0.773

[a] Initial groundwater level before the bypass flow event is indicated as GW_{init}, Drain/Rain represents the ratio between drain discharge and the amount of precipitation during the period considered (1 January 1989 is day nr. 1).

Bypass Flow Measurements

Bypass flow is illustrated by the nearly immediate reaction of groundwater to high precipitation intensities (e.g., as presented earlier in Figure 6.5 and Figure 6.6). Instant drain discharge occurs when groundwater is at a depth less than 0.80 m below soil surface. During the measuring period of 2.5 years a total of 38 bypass events were registered. Time lags for initial breakthrough of bypass flow to the groundwater was fast, within 1800 sec, if rain intensity is high or, since the nonstructured sand horizon reduces the breakthrough peak, the groundwater level is within 1.00 m depth.

Table 6.7 presents data on drain discharge during clearly distinguished periods with bypass flow events. The discharge/rain ratio varies between 0.30 to 0.84. This difference is mainly caused by the storage of bypass flow water in the sand horizon, as indicated by the initial groundwater levels just before the bypass flow event.

Bypass flow is also demonstrated by measured tensiometer profiles. In Figure 6.9 fortnightly tensiometer data at 0.50 m depth are combined with measured rain intensities. A nearly immediate response on rain events can be seen for, e.g., plot 56 and 115. Plot 16 and 179, on the contrary, show a strongly delayed reaction to rain events. For plot 179 it even takes two months before equilibrium is reached. Cups with a fast reaction can therefore be regarded as being close to water-conducting macropores (Booltink and Bouma, 1991). Slowly reacting cups, on the contrary, are located inside structure elements where only little water displacement occurs due to diffusion processes. These techniques are, therefore, unsuitable for predictive purposes such as nitrate leaching, or for validation of solute transport models.

Nitrate Leaching as a Result of Bypass Flow

In Figure 6.10 drainage discharge peaks, over a period of 5 days in February 1990 (day 411 to 416) is taken as an example to illustrate solute fluxes during that period. The tile drains discharged 14.5 mm (84%) of the 17.2 mm precipitation in the period of 5 days. Even though the profile was relatively wet before this bypass flow period, water storage capacity, measured by tensiometers, was approximately 0.10 m. In a nonmacroporous soil this rain event would, therefore, not have led to drain discharge. In Figure 6.10 drain discharge rate is presented in combination with chloride and

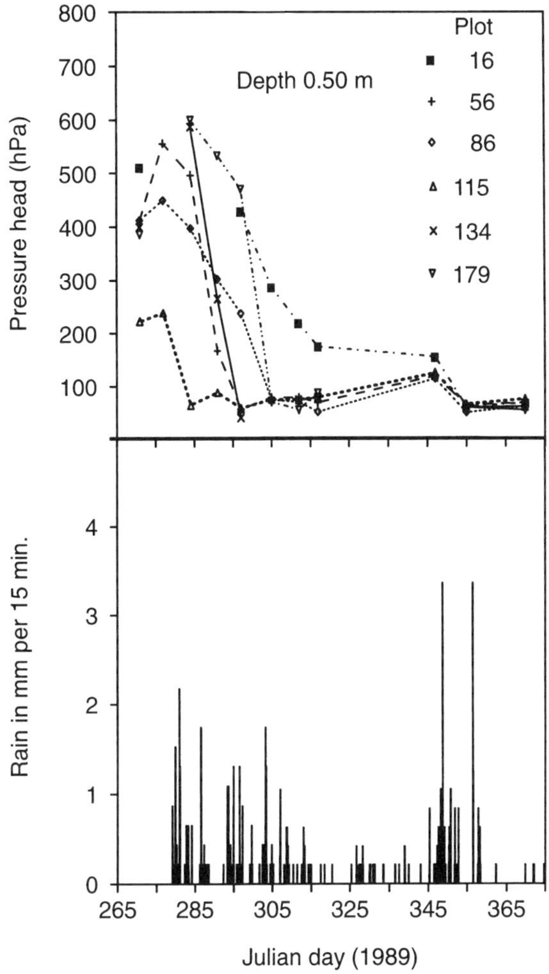

Figure 6.9. Measured rain intensities with corresponding tensiometer reactions at 0.50 m depth. Period 5 October to 31 December 1991 (1 January 1989 is day nr. 1).

nitrate concentration. Due to dilution of groundwater with initially high chloride concentration, by bypass flow water with low chloride concentration, the chloride concentration in drain discharge decreases proportionally with the discharge rate. Nitrate, on the contrary, shows the opposite trend because of the dissolution of easily accessible nitrate in the top layer of the soil. Due to the relatively small structure elements in the plow layer, there is a good contact between bypass water and soil material, which allows dissolution of nitrate as was previously suggested by Addiscot and Cox (1976). Increasing nitrate concentration with discharge rates was also reported by MacDuff and White (1984), Martin and White (1982), and White et al. (1983). However, the origin of this nitrate was not determined.

Although there is a considerable amount of mineral nitrogen in the subsoil, this nitrogen has no effect on the leaching process, as is illustrated by a bypass event in June 1991 (Figure 6.11). During this event, nitrate concentrations in drain water remained low. Nitrate in the topsoil, either applied by spring fertilizing or mineralized from the soil organic matter, had all been consumed by an active growing winter wheat crop. Since rooting depth at that time had not yet reached the subsoil (plants

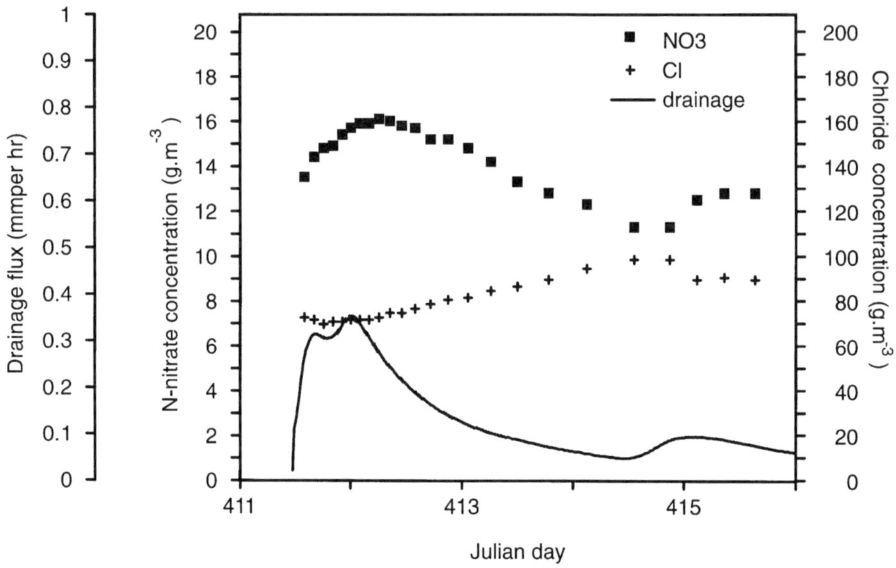

Figure 6.10. Drain discharge and concentrations of nitrate and chloride in the drainage water in February 1990 (1 January 1989 is day nr. 1).

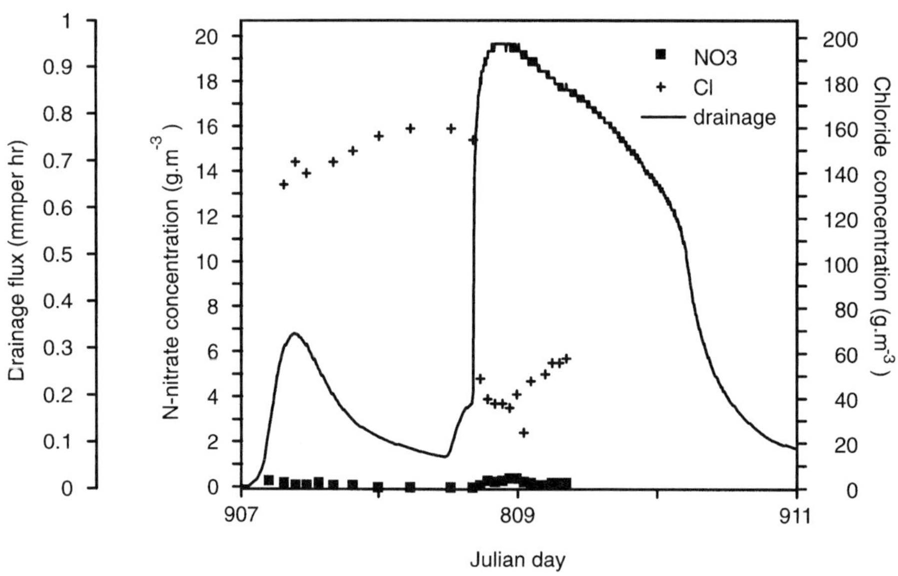

Figure 6.11. Drain discharge and concentrations of nitrate and chloride in the drainage water in June 1991 (1 January 1989 is day nr. 1).

were 0.20 m high), conditions for nitrate leaching in the subsoil can be compared to those in the winter period.

Case Study VII: Simulation Modeling to Explore Field Effects of Soil Structure

Effects of soil tillage on soil structure have been studied in many research projects. Usually, soil physical measurements are made after applying a certain type of tillage, and soil-structure data thus obtained are interpreted using threshold values of these data derived for plant growth or other soil uses. Development of dynamic simulation models has allowed expressions of time series of data rather than single values. For instance, rather than define the static air content at a defined pressure head of –10 kPa as a threshold value for plant growth, simulation allows continuous expressions for the air content during the growing season, allowing a more dynamic characterization. Still, in both cases characterization follows experiments and different, often costly, treatments have to be realized before measurements and evaluations can be made.

This section will explore a new procedure in which simulation of water movement is used to define the optimal structure and physical characteristics of a clay soil, which can be defined as: "allowing adequate infiltration and retention of water, while bypass flow along continuous macropores and associated rapid downward transport of solutes is minimized." Since soil structure cannot be defined in terms of size and shape of structure elements, structure is expressed in terms of hydraulic characteristics, such as hydraulic conductivity and moisture retention and other parameters that have successfully been used to define water movement in clay soils. Once an optimal structure has been defined, development of management procedures and tillage practices can be focused on realizing these structures. Thus, rather than trying to characterize the effects of different types of tillage and management on soil structure using physical measurements and simulation, the proposed procedure uses the latter two procedures to define optimal soil structure providing a challenge to both soil tillage and soil management to realize these structures under field conditions.

Soil, management, and weather data from the previously mentioned site at the Kandelaar experimental farm were used in combination with the modified version of LEACHW (Wagenet et al., 1989; Wagenet and Hutson, 1992) which includes the simulation of bypass flow, internal catchment, and drain discharge. Model calibration and validation was briefly described in Section 0. A detailed description of the calibration and validation procedure is given by Booltink (1994).

To demonstrate the simulated effects of soil management on different land qualities, three types of scenarios were explored: (i) scenarios related to the size and shape of soil structure elements, (ii) scenarios related to the soil physical characteristics, and (iii) scenarios related to the effect of management types. In Table 6.8, the explored scenarios are briefly described. From the Monte Carlo calibration study the best model fit of all 310 executions was selected for the initial parameter settings. This model execution is referred to as the "*original*" simulation; parameter settings of this scenario were used as a base for all the other scenarios.

To decide whether or not simulated scenarios could be regarded as successful, the following criteria for judging the simulation results were used: No ponding was allowed on the soil surface for a period longer than 24 hours. Since bypass flow can lead to enhanced nitrate leaching, bypass flow should be as low as possible. Workability should be in the spring; i.e., the time when a farmer can till his field without negative effects on soil structure. To meet this criterion, pressure head at 0.05 m below the soil surface should be –10 kPa or less. To avoid reduction processes in the top soil and unfavorable rooting conditions, water content in the top 0.50 m of the soil profile was not allowed to exceed the level of 90% of complete saturation for a period longer than 5 days.

Structure Scenarios

As can be seen in Figures 6.5 and Figure 6.6, drain discharge only occurs if the groundwater level is higher than approximately 80 cm below the soil surface. Reduc-

Table 6.8. Simulation Scenario Overview.[a]

Scenario	Description
0	Base scenario according to the best parameter set from the calibration study
Structure scenarios	
Improvement of soil structure would mean that soil structure elements should become smaller (scenario 1) and less blocky which implies that contact between bypass flow water and soil matrix (macropore walls) will increase (scenario 2). Smaller structure elements will also imply that macropores will become more tortuous which will increase the travel time of bypass flow in macropores (scenario 4)	
1	Structure element size reduced with 50%
2	Doubling of the horizontal and vertical contact areas
3	Combination of scenario 1 and 2
4	Reduction of the travel time
5	Combination of scenario 3 and 4
Soil physical scenarios	
Optimal tillage will result in higher porosities (scenario 6) and a more permeable soil (scenario 7).	
6	Increased pore volume with 5 % for the plough layer
7	Increased conductivity from 10 to 15 cm day^{-1}
8	Combination of scenario 6 and 7
Management scenarios	
Soil management can also include other options such as the creation of a mulch layer on top of the soil surface to reduce the rainfall energy and act simultaneously as a buffer for water storage (scenario 9). In scenario 10, finally, the effect of mixing the heavily textured topsoil with sand to obtain a better structure and texture is simulated. This last scenario was added to illustrate the maximum effect one can expect from different management types, indicating that the space for improved management can be found between the "original" simulation and scenario 10.	
9	Additional 2-cm mulch layer on the top of the soil surface
10	Replacement of the plough layer with a more sandy layer

All perturbations in the model parameters are based on the original base scenario (scenario 0).

tion of the amount of drain discharge can, therefore, be obtained by reducing the bypass flow peaks which will lead to rapid increase of the groundwater level. In Table 6.9, most relevant simulated water balance terms are presented. Since the upper boundary of the simulations was not changed compared to the original, the total amount of water which infiltrated remained constant. Different amounts of water absorption in the macropores can, however, be seen. In scenario 1 (50% decreased structure element diameters), an increase of the amount of absorbed water is simulated. An increase of the amount of absorbed water is also simulated as a result of the increase of the contact between macropore walls and bypass flow water. Combining these effects (scenario 3) leads to an amplification of the amount of absorbed water. A reduced propagation velocity (scenario 4 and 5) in the macropores does not have a significant effect on the amount of absorbed water in the macropore walls. This is mainly caused by the effect that bypass flow is a process which operates on a time scale of hours, while the other processes have a more daily fluctuation. Increasing absorption automatically leads to a decrease of bypass flow, since all water entering the macropores is either absorbed or added to the groundwater (bypass). So bypass could be reduced from 84 mm in the original scenario to 36 mm in the combined scenarios. However, this reduced amount of bypass does not lead to a change in the simulated amount of drain discharge. When analyzing the water balance terms on a daily basis, it appeared that absorption of water

Table 6.9. Water Balance Terms of the Applied Scenarios in mm Over the Simulation Period (November 27 1989 until April 4 1990).[a]

Scenario	Infiltration	Absorption	Bypass	Drain	Leachate
0	160	24	84	126	83
Structure scenarios					
1	159	40	68	127	81
2	159	46	63	126	82
3	158	72	36	126	81
4	160	24	84	127	82
5	158	73	36	127	81
Soil physical scenarios					
6	159	72	37	126	82
7	216	29	23	115	82
8	229	20	19	110	85
Management scenarios					
9	259	1	19	110	88
10	270	0	0	0	148

[a] Infiltration refers to the cumulative amount of surface infiltration, Absorption to the water absorbed in the horizontal and vertical macropores, Bypass to the amount of water which was routed directly to the groundwater through macropores, Drain to the total amount of drain discharge and leachate to the leaching to deeper soil layers. All amounts are in mm.

was only important in the beginning of the simulation period when the soil was relatively dry and, therefore, had a relevant absorption gradient for horizontal as well as for vertical absorption. When the soil wetted up during the winter season, absorption became an unimportant process that hardly contributed to the water balance and could not reduce the drainage peaks. On the other hand, increased absorption leads to increased matrix flow, meaning that water now reaches the groundwater through the micropore system rather than the macropore system. With respect to leaching of agrichemicals this can make significant difference; bypass flow can lead to high concentrations of nitrate leaching from the top soil directly to the groundwater.

As a result of the high macroporosity and the nonswelling behavior of the soil, ponding did not occur in this soil type; the threshold of 24 hours of ponding was, therefore, never met. To ensure an adequate workability, the pressure head at 0.05 m depth should be –10 kPa or less. As a result of the increased absorption in the relative dry soil the pressure head at 0.05 m depth varies a little in the beginning (around day number 25). For the rest of the winter season differences among the structure scenarios (1, 2, 3, 4, and 5) are hardly different from the original scenario. In spring, the threshold is no longer exceeded on day number 110 (March 14, 1990) for all structure scenarios. As a result of the high macroporosity, the criteria to prevent anaerobiosis was never met; highest simulated average moisture content in the first 0.50 m of the profile was at some peaks 0.52 m^3 m^{-3}, whereas the defined threshold value was 0.55 m^3 m^{-3}.

Soil Physical Scenarios

Changing the soil physical properties leads to considerable changes in the hydraulic behavior of the soil. A higher pore volume increases the amount of absorbed water in macropores due to the higher water-holding capacity and higher absorption gradient (Table 6.9). More absorption automatically results in a lower amount of bypass flow. Compared to the original scenario it has no effect on the other water balance terms.

An increased saturated hydraulic conductivity of the top layer results in a lower ponding probability and, therefore, more infiltration, less absorption along macropore walls, and less bypass flow. Leaching is not affected by an increased saturated hydraulic conductivity and drain discharge is 11 mm less compared to the original scenario. The combined effect of an increased pore volume and an increased saturated conductivity (scenario 8) has a reinforcing effect. Infiltration increases to 229 mm and absorption and bypass flow are proportionally decreased. Compared to the previous scenarios, drain discharge is also decreased. This shows that the saturated hydraulic conductivity very strongly determines the partitioning of water on the upper boundary and can therefore be considered as a crucial characteristic in improving soil and water quality. The workability and critical moisture levels are not affected and are not significantly different from the original scenario.

Management Scenarios

The simulation of an additional 0.02 m mulch layer on top of the soil surface reduced macropore flow strongly (Table 6.9). Absorption was close to zero and bypass was reduced to 19 mm over the entire winter season. As result of that, surface infiltration, and therefore matrix flow, strongly increased. Drain discharge was 16 mm lower compared to the original scenario, which seems insignificant, but this reduction was mainly caused by lower discharge peaks that have, as illustrated before, a dominating role in nitrate concentrations during bypass flow events. Scenario 10 in which the entire topsoil had been replaced by a somewhat lighter texture shows a different behavior. The infiltration capacity was always sufficient to let all the rain water infiltrate. Bypass flow and macropore flow is, therefore, zero. All water flow is matrix flow. The 0.30 m tick light textured layer buffers infiltrated water and slowly passes it through to deeper soil layers. For that reason no drain discharge occurs and more water transported as leachate to deeper soil layers or ditches. Due to the buffering capacity of the thin mulch layer the pressure head at 0.05 m depth for scenario 9 shows higher fluctuations. Scenario 10, on the contrary, more rapidly transports water to deeper soil layers and becomes, therefore, less dynamic than the original. The date in spring when the critical boundary of –10 kPa is no longer exceeded has been delayed by 4 days.

Summarizing Scenario Results

The simulation scenarios in which only soil structure geometrical characteristics were varied show little effect on bypass flow and related processes. Soil physical changes as a result of different management can have a relevant impact on the behavior of heavily textured clay soils that are susceptible to bypass flow. The saturated hydraulic conductivity is an especially key parameter in controlling bypass flow. Management measures which influence that characteristic in combination with higher water storage capacity of the plow layer are most efficient. This scenario study further demonstrates the use of a simulation model as a predictive tool for soil management practices. Using models in this way allows us to more specifically define field experiments in which details that are not described in the model can be tested and the practical implementation of suggested measures can be investigated.

FUTURE RESEARCH NEEDS

- Quantifying soil structure by means of structure models is necessary to allow flexible and dynamic simulation of soil structure. Fractal dimensions could be used to quantify soil structures. However, a soil structure type can never be derived from a fractal dimension alone, which implies that fractal dimensions are not adequate to simulate dynamical behavior of soil structure. New statistical analysis techniques, such as wavelets, which represent an exact structure in terms of scale and position parameters, are promising tools and need to be tested in this type of application.
- The chemical interaction of a structured soil during bypass flow events is strongly dominated by the contact area and retention time of bypass flow water along macropore walls. As indicated in this and numerous other studies, this contact area is generally only a small percent of the total surface area of natural soil aggregates. Using bulk chemical characteristics such as cation exchange capacity (C.E.C.) will result in erroneous simulation results which have a low reality level.
- Dynamic simulation of soil structure is crucial if the long-term effect of soil management, such as tillage, crop rotations, or farm management types (e.g., organic versus conventional farming) have to be simulated.
- Simulation models capable of simulating the effects of soil structure should be used in a proactive way. These models can be used to explore the effects of management types. Using models in this predictive mode can be a valuable research tool.

REFERENCES

Addiscot, T.M. and D. Cox. Winter leaching of nitrate from autumn-applied calcium nitrate, ammonium sulphate, urea and sulphur-coated urea in bare soil. *J. Agric. Sci. Camb.*, 87:381–389, 1976.

Anderson, J.L. and J. Bouma. Relationships between saturated hydraulic conductivity and morphometric data of an argillic horizon. *Soil Sci. Soc. Am. Proc.*, 37:408–413, 1973.

Bartoli, F., R. Philippy, M. Dirisse, S. Niquet, and M. Dubuit. Structure and self-similarity in silty and sandy soils: the fractal approach. *J. Soil Sci.*, 42:167–185, 1991.

Beven, K. and P. Germann. Macropores and water flow in soils. *Water Resour. Res.*, 18:1311–1325, 1982.

Beven, K. and R.T. Clark. On the variation of infiltration into a homogeneous soil matrix containing a population of macropores. *Water Resour. Res.*, 18:1311–1325, 1986.

Booltink, H.W.G. Field-scale modeling of bypass flow in a heavily textured clay soil. *Hydrol.*, 163:65–84, 1994.

Booltink, H.W.G. Field monitoring of nitrate leaching and water flow in a structured clay soil. *Agric. Ecosystems Environ.*, 52:251–261, 1995.

Booltink, H.W.G. and J. Bouma. Physical and morphological characterization of bypass flow in a well-structured clay soil. *Soil Sci. Soc. Am. J.*, 55:1249–1254, 1991.

Booltink, H.W.G. and J. Bouma. A sensitivity analysis on processes affecting bypass flow. *Hydrol. Proc.*, 7:33–43, 1993.

Booltink, H.W.G., R. Hatano, and J. Bouma. Measurement and simulation of bypass flow in a structured clay soil: A physico-morphological approach. *J. Hydrol.*, 148:149–168, 1993.

Bouma, J. Soil morphology and preferential flow along macropores. *Agric. Water Manag.*, 3:235–250, 1981.

Bouma, J. Using morphometric expressions for macropores to improve soil physical analyses of field soils. *Geoderma*, 46:3–11, 1 990.

Bouma, J., A. Jongerius, O. Boersma, A. de Jager, and D. Schoonderbeek. The function of different types of macropores during saturated flow through four swelling soil horizons. *Soil Sci. Soc. Am. J.,* 41:945–950, 1977.

Bouma, J. and L.W. Dekker. A case study on infiltration into dry clay soil. I. Morphological observations. *Geoderma,* 20:27–40, 1978.

Bouma, J., L.W. Dekker, and C.J. Muilwijk. A field method for measuring short-circuiting in clay soils. *J. Hydrol.,* 52:347–354, 1981.

Bouma, J. and P.J.M. de Laat. Estimation of the moisture supply capacity of some swelling clay soils in the Netherlands. *J. Hydrol.,* 49:247–259, 1981.

Bouma, J., C.F.M. Belmans, and L.W. Dekker. Water infiltration and redistribution in a silt loam subsoil with vertical worm channels. *Soil Sci. Soc. Am. J.,* 46:917–921, 1982.

Bouma, J. and H.A.J. Van Lanen. Transfer function and threshold values: From soil characteristics to land qualities, In *Quantified Land Evaluation Procedures,* Beek, K.J., P.A. Burrough, and D.E. McCormack (Eds.), ITC, International Institute for Aerospace Survey and Earth Sciences, Publication No. 6. Enschede, 1986, pp. 106–110.

Brewer, R. *Fabric and Mineral Analysis of Soils.* John Wiley & Sons, New York, 1964.

Bronswijk, J.J.B. Modelling of waterbalance, cracking and subsidence of clay soils. *J. Hydrol.,* 97:199–212, 1988.

Bullock, P. and A.J. Thomasson. Rothamsted studies of soil structure II. Measurement and characterization of macroporosity by image analysis and comparison with data from water retention measurements. *J. Soil Sci.,* 30:391–413, 1979.

Bullock, P. and C.P. Murphy. Towards quantification of soil structure. *J. Microscopy,* 120:317–328, 1980.

Campbell, G. A simple method for determining unsaturated conductivity from moisture retention data. *Soil Sci.,* 117:311–314, 1974.

Chen, C. and R.J. Wagenet. Simulation of water and chemicals in macropore soils. Part I. Representation of the equivalent macropore influence and its effect on soil water flow. *J. Hydrol.,* 130:105–126, 1992.

Chen, S., K.-H. Kim, F.-F. Qin, and A.T. Watson. Quantitative NMR imaging of multiphase flow in porous media. *Magn. Reson. Imaging,* 10:815–826, 1992.

Crawford., J.W., K.R. Ritz, and I. Young. Quantification of fungal morphology, gaseous transport and microbial dynamics in soil. An integrated frame work utilizing fractal geometry. *Geoderma,* 56:157–172, 1993.

DeSmedt, F., F. Wauter, and J. Sevilla. Study of tracer movement through unsaturated sand. *Geoderma,* 38:223–236, 1986.

De Willigen P. and M. Van Noordwijk. Roots, Plant Production and Nutrient Efficiency. PhD thesis. Wageningen Agricultural University, Wageningen, 1987.

Diggle P.J. *Statistical Analysis of Spatial Point Patterns.* Academic Press, London, 1983.

Di Pietro, L. Transferts d'eau dans milieux a porosite bimodale: Modelisation par la methode de gaz sur reseaux. PhD thesis Universite Montpellier, 1993, pp. 251.

Droogers P., A. Fermont, and J. Bouma. Effects of ecological soil management on workability and trafficability of a loamy soil in the Netherlands. *Geoderma,* 73:131–145, 1996.

Droogers, P., F.B.W. van der Meer, and J. Bouma. Water accessibility to plant roots in different soil structures occurring in the same soil type. *Plant and Soil,* (in press), 1997.

Edwards, W.M., R.R. van der Ploeg, and W. Ehles. A numerical study of the effects of noncapillary-sized pores upon infiltration. *Soil Sci. Soc. Am. J.,* 43:851–856, 1979.

Ehlers, W. Observation on earthworm channels and infiltration on tilled and untilled loess soil. *Soil Sci.,* 119:242–249, 1975.

FAO 1993 FESLM: An international framework for evaluating sustainable land management. World Soil Resources Report 73. FAO, Rome, 1993.

Finke, P.A. Field scale variability of soil structure and its impact on crop growth and nitrate leaching in the analysis of fertilizer scenario's. *Geoderma,* (accepted for publication) 1993.

Germann, P.F. The three modes of water flow through a vertical pipe. *Soil Sci.,* 144,153–154, 1987.

Germann, P.F. Preferential flow and the generation of runoff. 1. Boundary flow theory. *Water Resour. Res.,* 26:3055–3063, 1990.

Germann, P.F. and K. Beven. Kinematic wave approximation to infiltration into soils with sorbing macropores. *Water Resour. Res.,* 21:990–996, 1985.

Hatano, R. and H.W.G. Booltink. Using fractal dimensions of stained flow in a clay soil to predict bypass. *J. Hydrol.,* 135:121–131, 1992.

Hatano, R. and T. Sakuma. A plate model for solute transport through aggregated soil columns. I. Theoretical descriptions. *Geoderma,* 50:13–23, 1991.

Hatano, R., N. Kawamura, J. Ikeda, and T. Sakuma. Evaluation of the effect of morphological features of flow paths on solute transport by using fractal dimensions of methylene blue staining pattern. *Geoderma,* 53:31–44, 1992.

Heijs, A.W.J., J. de Lange, J.F.Th. Schoute, and J. Bouma. Computed tomography as a tool for non-destructive analysis of flow patterns in macroporous clay soils. *Geoderma,* 64:183–196, 1995.

Heijs, A.W.J. and C.P. Lowe. Numerical evaluation of the permeability and the Kozeny constant for two types of porous media. *Phys. Let. Rev.,* (accepted for publication) 1977.

Hoogmoed, W.B. and J. Bouma. A simulation model for predicting infiltration into a cracked clay soil. *Soil Sci. Soc. Am. J.,* 44:458–461, 1980.

Hopmans, J.W., T. Vogel, and P.D. Koblik. X-ray tomography of soil water distribution in one-step outflow experiments. *Soil Sci. Soc. Am. J.,* 56:355–362, 1992.

Hutson, J.L. and A. Cass. A retentivity function for use in soil water simulation models. *J. Soil Sci.,* 38:487–498, 1987.

Hutson, J.L. and R.J. Wagenet. A multi region model describing water flow and solute transport in heterogeneous soils. *Soil Sci. Soc. Am. J.,* 59:743–751, 1995.

Iman, R.L. and W.J. Conover. A distribution free approach to inducing rank correlations among input variables. *Communications in Statistics,* Vol. B11, 1982, pp. 311–334.

Jansen, P.H.M. RORASC an SELACC: Software for Performing the Rotated-Random-Scanning Calibration Procedure. CWM memorandum, in preparation. RIVM, Bilthoven, The Netherlands, 1993.

Jansen, P.H.M., P.S.C. Heuberger, and R. Sanders. Rotated-Random Scanning: A Simple Method for Set Valued Model Calibration. RIVM internal report, in preparation, RIVM, Bilthoven, The Netherlands, 1993.

Jarvis, N.J. and P.B. Leeds-Harrison. Modelling water movement in drained clay soil. I. Description of the model, sample output and sensitivity analysis. *J. Soil Sci.,* 38:487–498, 1987.

Jarvis, N.J. and P.B. Leeds-Harrison. Field test of a water balance model of cracking clay soils. *J. Hydrol.,* 112:203–218, 1990.

Jongerius, A., D. Schoonderbeek, A. Jager, and S. Kowalinski. Elecro-optical soil porosity investigations by means of a Quantimet B equipment. *Geoderma,* 7:177–198, 1972.

Joschko, M., O. Graff, P.C. Müller, K. Kotzke, P. Lindner, D.P. Pretscher, and O. Larink. A non-destructive method for the morphological assessment of earth worm burrow systems in three dimensions by X-ray computed tomography. *Biol. Fert. Soils,* 11:88–92, 1991.

Keesman, K.J. A Set-Membership Approach to the Identification and Prediction of Ill-Defined Systems: Application to a Water Quality System. Ph.D. Thesis, Univ. of Twente, Enschede, The Netherlands, 1989.

Keesman, K.J. Membership-set estimation using random scanning and principal component analysis. *Mathematics and Computers in Simulation,* 32:535–543, 1990.

Keesman. K.J. and G. van Straaten. Embedding of random scanning and principal component analysis in set-theoretic approach to parameter estimation. Proc. 12th IMACS World Congress, Paris, 1988.

Keesman, K.J. and G. van Straaten. Identification and prediction propagation of uncertainty in models with bounded noise. Intern. *J. Control,* 49:2259–2269, 1989.

Klute, A. Soil water flow theory and its application in field situations. In *Field Soil Water Regime,* Bruce, R.R. et al., (Eds.), SSSA Spec. Publ. 5. SSSA, Madison, WI, 1973, pp. 9–31.

Kooistra, M.J., J. Bouma, O.H. Boersma, and A. Jager. Soil-structure differences and associated physical properties of some loamy typic fluvaquents in the Netherlands. *Geoderma,* 36:215-228, 1987.

Lawes, J.B., J.H. Gilbert, and R. Warington. *On the Amount and Composition of the Rain and Drainage Water Collected at Rothamsted,* Williams, Clowes and Sons Ltd., London, U.K., 1982.

MacDuff, J.H. and R.E. White. Components of the nitrogen cycle measured for cropped and grassland soil-plant systems. *Plant Soil,* 76:35–47, 1984.

Martin, R.P. and R.E. White. Automatic sampling of stream water during storm events in small remote catchments. *Earth. Surf. Proc. Landforms,* 7:53–61, 1982.

McBratney, A.B. and C.J. Moran. A rapid method for soil macropore structure. II. Stereo logical model, statistical analysis and interpretation. *Soil Sci. Soc. Am. J.,* 54:509–515, 1990.

Moran, C.J., A.B. McBratney, and J. Koppi. A rapid analysis method for soil pore structure. I. Specimen preparation and digital binary image production. *Soil Sci. Soc. Am. J.,* 53:921–928, 1989.

Murphy, C.P., P. Bullock, and R.H. Turner. The measurement and characterization of voids in soil thin sections by image analysis. I. Principles and techniques. *J. Soil Sci.,* 35:217–229, 1977.

Perfect E. The menger sponge as a conceptual model for soil structure. 88th annual meeting of the ASA-CSSA-SSSA, Indianapolis, IN, November 3–8, 1996.

Reganold, J.P. Soil quality and profitability of biodynamic and conventional systems: A review. *Am. J. Altern. Agric.,* 10:36–45, 1995.

Ringrose-Voase, A.J. and P. Bullock. The automatic recognition and measurement of soil pore types by image analysis and computer programs. *J. Soil Sci.,* 35:673–684, 1984.

Schaafsma, T.J., H. van As, W.D. Palstra, J.E.M. Snaar, and P.A. de Jager. Quantitative measurement and imaging of transport processes in plants and porous media by ^{1}H NMR. *Magn. Reson. Imaging.,* 10:827–836, 1992.

Shipitalo, M.J. and R. Protz. Comparison of morphology and porosity of a soil under conventional and zero tillage. *Can. J. Soil Sci.,* 67:445–456, 1987.

Soil Survey Staff 1975 Soil Taxonomy: A basic system of soil classification for making and interpreting soil surveys. USDA-SCS Agric. Handbook. 436. U.S. Gov. Print. Office, Washington, DC, 1975.

Spaans, E.J.A., J. Bouma, A.L..E. Lansu, and W.G. Wielemaker. Measuring soil hydraulic properties after clearing of tropical rain forest in a Costa Rican soil. *Trop. Agric.,* 67:61–65, 1990.

Steenvoorden, J.H.A.M. Agricultural practices to reduce nitrogen losses via leaching and surface runoff. In *Management Systems to Reduce Impact of Nitrates.* Elsevier Applied Science, Amsterdam, The Netherlands, 1989, pp. 72–80.

Stroosnijder, L. Cumulative infiltration and infiltration rate in homogeneous soil. *Agric. Res. Rep.,* 847:69–99, 1976.

Taylor, G.L. Dispersion of soluble matter in solvent flowing slowly through a tube. *Proc. Roy. Soc.* 219:186–203, 1953.

Vanclooster, M., P. Viaene, J. Diels, and K. Christiaens. WAVE, A, Mathematical Model for Simulating Water and Agrochemicals in the Soil and Vadose Environment. Reference and User's Manual. Institute for Land and Water Management, Leuven, 1994.

Van der Ploeg, R.R. and P. Benecke. Unsteady, unsaturated n-dimensional moisture flow in soil: A computer simulation program. *Soil Sci. Soc. Am. Proc.,* 38:881–885, 1974.

van Genuchten, M.Th. A closed-form equation for predicting the hydraulic conductivity of unsaturated soils. *Soil Sci. Soc. Am. J.,* 44:892–898, 1980.

van Genuchten, M.Th. and P.J. Wieringa. Mass transfer studies in sorbing porous media. I. Analytical solutions. *Soil Sci. Soc. Am. J.,* w0:473–480, 1976.

Van Noordwijk, M., G. Brouwer, and K. Harmanny. Concepts and methods for studying interactions of roots and soil structure. *Geoderma,* 56:351–375, 1993.

Van Ommen, H. C., L. W. Dekker, R. Dijksma, J. Hulshof, and W.H. van der Molen. A new technique for evaluating the presence of preferential flow paths in non-structured soils. *Soil Sci. Soc. Am. J.,* 52:1192–1194, 1988.

Van Stiphout, T.P.J., H.A.J. Van Lanen, O.H. Boerma, and J. Bouma. The effect of bypass flow and internal catchment of rain on the water regime in a clay loam grassland soil. *J. Hydrol.,* 95(1/2):1–11, 1987.

Wagenet, R.J. and J.L. Hutson. *L*eaching *E*stimation and *Ch*emistry *M*odel: A Process-Based Model of Water and Solute Movement, Transformations, Plant Uptake and Chemical Reactions in the Unsaturated Zone. Continuum Water Resources Institute. Cornell University, 1989.

Wagenet, R.J. and J.L. Hutson. LEACHM: A Processone. Centre of Environmental Research, Cornell University, Ithaca, NY, 1992.

Wagenet, R.J., J.L. Hutson, and J.W. Biggar. Simulating the fate of a volatile pesticide in unsaturated soil: a case study with DBCP. *J. Environ. Qual.,* 18:78–84, 1989.

Warner, G.S., J.L. Nieber, I.D. Moore, and R.A. Giese. Characterizing macropores in soil by computed tomography. *Soil Sci. Soc. Am. J.,* 53:653–660, 1989.

White, R.E. The influence of macropores on the transport of dissolved and suspended matter through soil. *Adv. Soil Sci.,* 3:95–121, 1985.

White, R.E., S.R. Wellings, and J.P. Bell. Seasonal variations in nitrate leaching in structured clay soils under mixed land use. *Agric. Cult. Water Manag.,* 7:391–410, 1983.

Wild, A. Nitrate leaching under bare fallow at a site in northern Nigeria. *J. Soil Sci.,* 23:315–324, 1972.

CHAPTER SEVEN

Modeling the Impact of Preferential Flow on Nonpoint Source Pollution

N. Jarvis

BACKGROUND

Preferential flow may significantly enhance the risk of leaching of surface-applied contaminants to surface and groundwater, since much of the buffering capacity of the biologically and chemically reactive topsoil may be bypassed. In this way, chemicals can quickly reach subsoil layers where degradation and sorption processes are generally less effective. The term *preferential flow* implies that, for various reasons, infiltrating water does not have sufficient time to equilibrate with slowly moving "resident" water in the bulk of the soil matrix. For example, in structured soils, macropores (shrinkage cracks, worm channels, root holes) may dominate the soil hydrology, particularly in fine-textured soils, where they operate as high-conductivity flow pathways bypassing the denser impermeable soil matrix (Beven and Germann, 1982). Preferential flow also occurs in unstructured sandy soils in the form of unstable wetting fronts or "fingering" (Hillel, 1987; Glass and Nicholl, 1996), caused by profile heterogeneities such as horizon interfaces or by water repellency (Hendrickx et al., 1993). Thus, physical nonequilibrium conditions can occur in virtually all types of soils (Flury et al., 1994) caused by heterogeneities at scales ranging from the single pore to the pedon.

An increasing number of simulation models have been developed which can account for physical nonequilibrium processes. Some of these models are now quite widely used and well established in the research sphere. Many of the concepts underlying the models are well accepted. One important task for the future is to encourage the transfer of these models to the realm of management applications. This would promote sound environmental decision-making and allow quantitative evaluation of alternative strategies to minimize environmental degradation.

This chapter reviews the development and application of models to predict the impact of physical nonequilibrium processes on agricultural nonpoint source pollution. Some concepts and principles concerning preferential flow and transport are first outlined, followed by a brief review of the various modeling approaches which have been adopted. Some illustrative simulations using the dual-porosity model MACRO (Jarvis, 1994) then form the basis for a general discussion of the likely consequences of nonequilibrium flow behavior for agricultural nonpoint source pollution, using nitrate and pesticide leaching, and the salt balance in a salinized soil profile as examples. Finally, some applications of physical nonequilibrium models to evaluate the impacts of various management practices (e.g., irrigation, tillage) on agricultural nonpoint source pollution are discussed.

GENERAL PHYSICAL PRINCIPLES

Classical theory of water flow (Richards' equation) and solute transport (convection-dispersion equation) is based on continuum physics, with the underlying assumption that unique values of soil water pressure and solute concentration can be defined for a representative elementary volume of soil. Physical nonequilibrium occurs when soil heterogeneities result in the generation of lateral differences either in water pressures or solute concentrations, or both. More specifically, preferential *flow* and/or *transport* result when rates of lateral equilibration of water pressures and/or solute concentrations, respectively, are very slow in relation to the vertical flux rates (Jury and Flühler, 1992; Flühler et al., 1996). From these definitions, it can be noted that preferential transport can occur without preferential flow; for example, in water-saturated soil characterized by a broad pore size distribution and thus a large range of pore water velocities.

From the foregoing, it should be apparent that preferential flow is a generic term encompassing a range of different processes with similar characteristics and consequences for solute leaching in soils, although the underlying physical mechanisms may be different. Figure 7.1 presents a broad, tentative, classification of the various preferential flow processes and accompanying underlying mechanisms which have been observed and/or postulated. For illustrative purposes, Figure 7.1 groups soils into homogeneous, scale heterogeneous, and geometrically dissimilar porous media, each group tending to exhibit characteristic preferential flow phenomena. However, it should be stressed that the boundaries between these groups may be diffuse and far from watertight, and that more than one preferential flow mechanism may occur in any given soil.

Finger flows can be initiated by small- and large-scale heterogeneities within the soil and/or by flow concentration at the surface either due to interception by vegetation and stem flow, or water repellency (Figure 7.1). When dry, many soils possess water repellent properties caused by the presence of hydrophobic organic materials and coatings on soil particles (Jamison, 1946; Bond, 1964; Dekker and Jungerius, 1990; Bisdom et al., 1993). Rain falling on nearly air-dry water repellent surface soil tends to accumulate in shallow depressions where, aided by the hydrostatic pressure, it eventually infiltrates as finger flows (Ritsema et al., 1993; Dekker and Ritsema, 1996). It is also well known that layer and horizon interfaces can generate fingers, particularly where a coarse-textured sand layer underlies a finer material (Hill and Parlange, 1972). The downward movement of a wetting front is temporarily interrupted at the interface, since the water pressure must increase to the "water-entry" pressure of the coarse sand (i.e., near saturation). Due to local heterogeneities, the water-entry pressure may be exceeded at one or several points, rather than uniformly along the interface. This leads to the development of fingers moving rapidly into the subsoil layer at a rate slightly less than the saturated conductivity of the sand, while the remainder of the soil remains dry. Once formed, the fingers can only persist if lateral dispersion due to capillary forces is relatively weak. In an illustrative simulation study, Nieber (1996) showed that hysteresis in the soil water characteristic curve of narrow-graded sands could sustain finger flow, if the water-entry pressure on the wetting curve was larger (i.e., closer to zero pressure) than the air-entry pressure on the draining curve. If this condition was not satisfied, then the simulated finger quickly dissipated due to lateral capillary

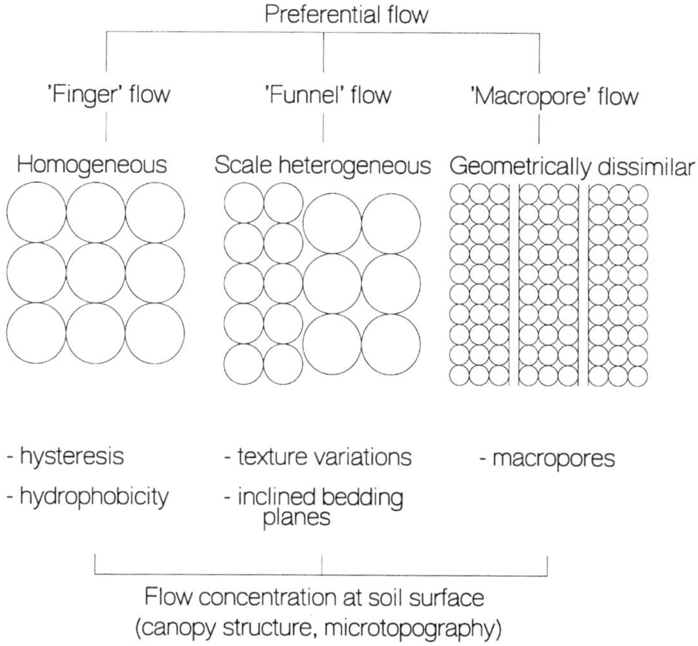

Figure 7.1. Preferential flow processes in soils: occurrence and mechanisms.

pressure gradients. Finally, it can be noted that lateral solute equilibration between wet fingers and surrounding dry soil may be severely restricted because of the small water content in the bulk soil, implying a negligible effective diffusion coefficient.

We should expect finger flows to occur predominantly, though not necessarily exclusively, in rather homogeneous narrow-graded sands. This is because such soils are characterized by a very steep portion of the soil water characteristic (large $d\theta/d\psi$) on the primary wetting curve close to saturation. This means that small differences in water pressure close to saturation can result in very large differences in water saturation, hydraulic conductivity and flux rates. Finger flow should be much less common in heterogeneous poorly-sorted materials with a broad pore-size distribution, since, in such soils, lateral dispersion cannot easily be prevented. In heterogeneous soils characterized by lenses and admixtures of various particle size fractions, preferential "funnel" flow as described by Kung (1990a,b) may occur (Figure 7.1). In this case, the flow pathways taken by the infiltrating water should depend on the relative values of unsaturated hydraulic conductivity of the component materials at the applied water flux. For example, at saturation, the coarsest sand fraction would comprise the preferred flow region, while at small flux rates under unsaturated conditions, the somewhat finer-textured materials may conduct all the water, since soil water pressures may not increase sufficiently to saturate the coarse sand.

Macropores are large, continuous, structural pores which constitute preferred flow pathways for infiltrating water in most soils. At the macroscopic scale of measurement, this is reflected in large increases in unsaturated hydraulic conductivity across a small soil water pressure head range close to saturation (Clothier and Smettem, 1990; Jarvis and Messing, 1995). At the pore scale, macropore flow is generated when the water pressure locally increases to near saturation at some point on the interface with the surrounding soil matrix, such that the water-entry pressure of the pore is exceeded. Macropore flow can be sustained if the vertical flux rates in the macropore are large in relation to the lateral infiltration losses into the matrix due to the prevailing capillary pressure gradient. This is most likely to be the case in clay soils with an impermeable matrix. These lateral losses can be further restricted by relatively impermeable interfaces between macropores and the bulk soil, including cutans on aggregate surfaces, and organic linings in biotic pores (e.g., Thoma et al., 1992). Thus, macropore flow can sometimes be significant even in lighter-textured soils of large matrix hydraulic conductivity.

The impact of macropore flow on solute transport depends strongly on the nature of the solute ion under consideration, particularly the size of the molecule as it affects diffusion rates between pore domains, its sorption characteristics, and the nature of any source/sink terms, including biological transformations, which affect the transport process (i.e., whether the solute is surface-applied and consumed in the soil, or whether it is produced within the soil). Later in this paper, the model described by Jarvis (1994) is used to illustrate the influence of some of these factors and to draw some general conclusions concerning the likely effects of macropore flow on leaching of various types of solutes.

EXISTING MODELS ACCOUNTING FOR PHYSICAL NONEQUILIBRIUM

In this section, some existing models accounting for preferential flow and transport are described. This review is not intended to be exhaustive, but rather to give some idea of the different types of models available, their advantages and limitations, and their potential as management tools.

Statistical models

Probabilistic particle-tracking or random walk methods attempt to mimic the effects of spatially variable preferential flow pathways in soil in a nonmechanistic way (e.g., Roth et al., 1990; Flury and Flühler, 1995). Such approaches may provide some interesting insights into solute transport processes occurring in heterogeneous soils, but are unlikely to prove useful for management purposes in the foreseeable future, mainly due to the difficulty in relating the model parameters to measurable (macroscopic) soil properties, but also because steady water flow is assumed.

Two- and Multidomain Models

These models take a macroscopic, continuum, approach by lumping individual preferential flow pathways in the soil into two or more pore domains within a one-dimensional numerical scheme. A large number of models of this type have been developed in recent years. They vary with regard to the number of flow domains considered (two or multidomain models), the degree of simplification and empiricism involved in process descriptions (e.g., functional vs. mechanistic models) and also the manner in which exchange between flow domains is represented. In the following sections, some of these existing models are described.

Functional Models

In recent years, a number of simple empirical (or functional) models have been developed which include some treatment of physical nonequilibrium conditions (e.g., Addiscott, 1977; Corwin et al., 1991; Hall, 1993). For example, Barraclough (1989a) described a simple three-region model based on a capacity ("tipping bucket") approach to describe water flow, the mobile-immobile water concept to characterize solute transport in two domains (see next section), and a simple instantaneous bypass routine to account for macropore flow. The model was apparently implemented only for nonreactive solute transport, although it was compared to nitrate leaching data during three winter seasons (Barraclough, 1989b).

It is sometimes not recognized that functional models may be rather limited in their ability to simulate preferential flow due to simplifications in the treatment of soil water flow. For example, these models usually assume that water outflow from each soil layer is zero until filled to field capacity. Thus, rapid and deep-penetrating bypass flow in dry, macroporous, soils cannot be simulated. Difficulties are also caused by the fact that the normal time step in these models is one day, whereas an appropriate timescale to characterize preferential flow and transport processes would be of the order of hours, or even minutes.

Dual-Porosity/Single Permeability Models

Analytical models have been developed based on the two-region (mobile-immobile water) convection-dispersion equation (CDE) for both nonreactive and sorbing and degrading solutes (van Genuchten and Wierenga, 1976; van Genuchten and Wagenet, 1989; Gamerdinger et al., 1990; Leij et al., 1993). These analytical solutions of the CDE require idealized initial and boundary conditions and assume steady state water flow. Thus, although such an approach may provide some theoretical insights into solute transport processes occurring in structured soils, they are not appropriate to field situations characterized by time-varying soil water content due to evaporation, root water uptake and intermittent inputs of rainfall. For this reason, their use as management tools is limited.

Numerical solutions of the mobile-immobile water concept have also been implemented. However, even when the mobile-immobile water CDE is solved numerically, it is often either coupled to Richards' equation, or steady water flow is assumed (Lafolie and Hayot, 1993). Such an approach can simulate preferential solute transport, but not preferential water flow. For transient conditions, one unresolved question is how best to define the fractional mobile water content in relation to the time-varying total water content. Zürmuhl and Durner (1996) discussed this problem and recommended defining a variable ratio between the mobile water content and the total water content based on the shape of the $K(\theta)$ function, such that the conductivity at the boundary between mobile and immobile water represented a constant fraction of the conductivity at the current water content. van Dam et al. (1990) presented a simple solution which may be appropriate to water repellent soils where a given fraction of the soil matrix remains unwetted during rain infiltration. In their model, the volume of soil containing mobile water is given as a constant fraction of the total soil volume, while water and solute storage in the immobile region is ignored.

Jarvis (1989) and Armstrong et al. (1995) described a model which is particularly suited to cracking clay soils. In this model (CRACK), vertical water flow and solute transport occur in a crack system which is defined, for each layer in the soil profile, by a fixed spacing and a porosity which varies due to swelling and shrinkage. Vertical transport is neglected in the soil aggregates, which act as sinks for water and sources/sinks for solute. In the case of solute, the aggregates are divided into numerical slices and lateral diffusion and exchange is modeled explicitly using Fick's law. In the latest version of the model, subroutines have been implemented to describe nitrate transformations and pesticide sorption and degradation (Armstrong et al., 1995).

Dual-Porosity/Dual Permeability Models

In these models, two pore domains are each characterized by a porosity, a water pressure (or water content) and solute concentration. In contrast to the mobile-immobile water approach, vertical water flow and solute transport are calculated in each domain, with mass exchange between domains treated as source/sink terms in the one-dimensional (vertical) model structure. This mass exchange between domains is calculated using approximate first-order equations, either based on an effective diffusion path length related to the macroscale soil geometry (aggregate size) or empirical mass transfer coefficients.

Gerke and van Genuchten (1993a) described a model of water flow and nonreactive solute transport based on the van Genuchten/Mualem model for the soil hydraulic functions (van Genuchten, 1980; Mualem, 1976), applied to a bimodal pore size distribution (Othmer et al., 1991; Durner, 1992). Richards' equation and the CDE are used to calculate water flow and solute transport, respectively, in both pore domains, while mass exchange of water and solutes is calculated using approximate first-order equations accounting for both convective and diffusive transfer (Gerke and van Genuchten, 1993b). Due to the geometry of the flow system, the rate of mass exchange between the domains is inversely proportional to the square of an effective diffusion pathlength (van Genuchten, 1985; Youngs and Leeds-Harrison, 1990).

Jarvis (1994) described a dual-porosity model (MACRO) in which Richards' equation and the convection-dispersion equation are used to model soil water flow and solute transport in soil micropores, while a simplified capacitance-type approach is used to calculate water and solute movement in macropores. This description of gravity-driven water flow in macropores can be considered as the numerical equivalent of the analytical kinematic wave model described by Germann and Beven (1986). Later in this paper, MACRO is used to illustrate some general aspects of the effects of physical nonequilibrium flow and transport on nonpoint source pollution. Therefore, this particular model is now described in more detail.

In MACRO, a simple "cut and join" or "two-line" method is used to define the hydraulic functions (Smettem and Kirkby, 1990), rather than the additive superimposition of two pore regions employed, for example, by Gerke and van Genuchten (1993a). The "boundary" between the two pore domains is conveniently defined by the air-entry pressure head in the Brooks and Corey (1964) equation, with an equivalent water content and hydraulic conductivity defining the saturated state of the micropores (see Figures 7.2 and 7.3). The boundary conditions at the soil surface are an important part of any nonequilibrium flow and transport model. In MACRO, a flux boundary condition is used to partition incoming rainfall between the two pore regions, depending on the infiltration capacity of the micropores. With respect to solute transport, the water infiltrating in macropores is characterized by a concentration c_{ma} calculated assuming complete mixing with solute stored in the solution phase of a shallow "mixing" depth z_{d}:

$$c^*_{ma} = \frac{\left(\left(\frac{z_d}{\Delta z}\right)Q_s\right) + (R\,c_r)}{R + \left(z_d\left(\theta + \left((1-f)\gamma\,k_d\right)\right)\right)} \tag{7.1}$$

where Q_s is the amount of solute stored in the top layer of the soil, Δz is the layer thickness, R is the amount of rainfall reaching the surface, c_r is the concentration in the rain, θ and γ_b are the water content and bulk density in the top layer, respectively, k_d is the sorption constant, and f is the fraction of sorption sites in the macropore region.

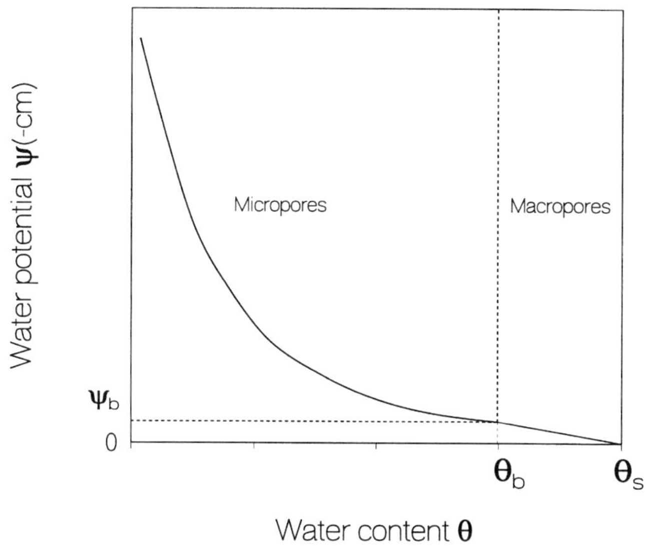

Figure 7.2. Soil water retention function in MACRO (ψ_b is the pressure head at the boundary between macro- and micropores, θ_b is the equivalent water content, θ_s is the saturated water content).

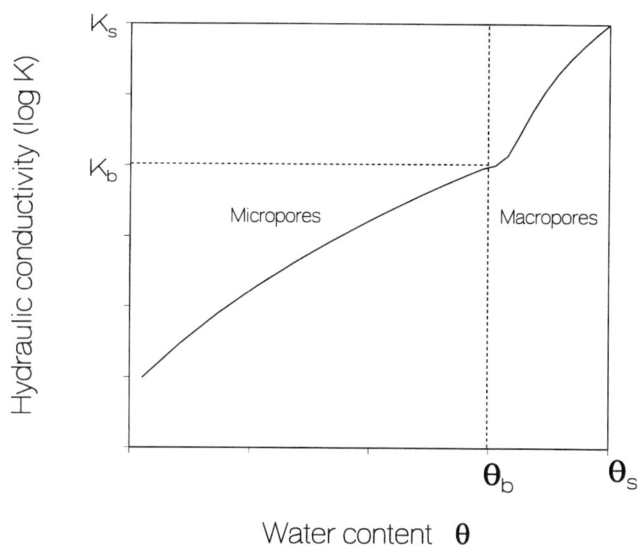

Figure 7.3. Hydraulic conductivity function in MACRO (K_b is the saturated hydraulic conductivity of the micropores, K_s is the total saturated hydraulic conductivity, other symbols as in Figure 7.2).

Water and solute exchange rates between the flow domains (M_w and M_s, respectively) are calculated as a function of an effective aggregate half-width (or diffusion pathlength) d, using approximate physically-based first-order expressions (Booltink et al., 1993; Gerke and van Genuchten, 1993b). For solute, both diffusive and convective mass transfer are considered:

$$M_w = \left(\frac{3\, D_w \gamma_w}{d^2}\right)(\theta_b - \theta_{mi}) \tag{7.2}$$

and

$$M_s = \left(\frac{3\, D_e\, \theta_{mi}}{d^2} \right) (c_{ma} - c_{mi}) + M_w\, c' \tag{7.3}$$

where the subscripts *ma* and *mi* refer to macropores and micropores, respectively, θ_b is the saturated water content of the micropores (Figures 7.2 and 7.3), γ_w is a scaling factor introduced to match the approximate and exact solutions to the diffusion problem (van Genuchten, 1985), c is the solute concentration where the prime notation indicates the direction of convective transfer, D_e is an effective diffusion coefficient, and D_w is an effective water diffusivity given by:

$$D_w = \left(\frac{D_{\theta_b} + D_{\theta_{mi}}}{2} \right) S_{ma} \tag{7.4}$$

where $D_{\theta b}$ and $D_{\theta mi}$ are the diffusivities at θ_b and θ, respectively, and S_{ma} is the degree of saturation in the macropores, which is introduced to account for the effects of incomplete wetted contact area between the two pore domains.

A full water balance is considered in MACRO, including precipitation (rain on an hourly or daily basis and/or irrigation), evapotranspiration and root water uptake, deep seepage and lateral fluxes to tile drains in saturated soil. Solute transport and transformation processes in the model include convective-dispersive transport, linear equilibrium sorption with sorption sites partitioned between the two pore domains, microbial degradation (four rate coefficients are specified for micro- and macropores and liquid and solid phases), plant uptake, and canopy interception/washoff.

In recent years, MACRO has been extensively tested against field and lysimeter measurements made in a range of contrasting soil types and for different solute species (Andreu et al., 1994; Saxena et al., 1994; Jarvis et al., 1994; Jarvis, 1995a; Gottesbüren et al., 1995; Jarvis et al., 1995a; Bergström, 1996; Andreu et al., 1996; Brown et al., 1997b). Valuable experience has been gained in terms of both the difficulties and possibilities of parameterizing MACRO for field applications, and some of these aspects are discussed in a later section.

Multiporosity/Multipermeability Models

Steenhuis et al. (1990) described a multidomain model of solute movement based on a piece-wise linear approximation to the $K(\theta)$ function and the assumption of a unit hydraulic gradient. Convective solute mixing between pore classes was calculated using the "mixing fraction" concept, whereby a fixed fraction of the water (and solute) in each pore class is mixed in a common "pool" before being redistributed. If this mixing fraction is set to zero, the multidomain model reduces to the stream-tube or capillary-bundle model, whereas if it is set to unity (i.e., complete lateral mixing), it becomes equivalent to the convection-dispersion equation.

Hutson and Wagenet (1995) described a multiregion model of water flow and solute transport (TRANSMIT) based on the existing single-domain model LEACHM (Wagenet and Hutson, 1987). TRANSMIT can be used to investigate the impacts of preferential flow occurring at the pore-scale (e.g., macropore flow) by simulating multiple pore domains, and also physical nonequilibrium induced by soil heterogeneity through modeling multiple, interacting, soil columns. A wide range of solutes can be dealt with, including pesticides and nitrate. Convective and diffusive mass exchange between all combinations of columns and/or pore domains is calculated using empirical expressions.

Gwo et al. (1995) describe a multiporosity model of water flow and solute transport (MURF/MURT) essentially based on the same overlapping pore continua concept developed by Gerke and van Genuchten (1993a). Again, Richards' equation and the CDE are used to model vertical transport in each pore domain, while water exchange is regulated by empirical exchange coefficients in simple first-order expressions. Solute exchange is driven by both convective transfer and first-order diffusive exchange.

Multiporosity models can, of course, default to the dual-porosity case when required, and there may be good reasons for doing so. As the number of pore classes in the model increases, the description of mass exchange processes necessarily becomes more uncertain and less mechanistic. This is because the nature and geometry of the interactions in the system are difficult to specify in a multidomain framework. For example, should mass exchange be calculated between all domains or, instead, sequentially from the largest pore class to the smallest? Also, should mass transfer coefficients be set equal for all interacting pairs of pore classes or be allowed to vary? In the latter case, the parameter estimation problem becomes very difficult. Thus, although multidomain models may provide valuable insights into the nature of solute transport in heterogeneous soils, and also allow much greater flexibility in matching the observed behavior of solute transport in soil compared to two-domain models (Gwo et al., 1995), they may require extensive calibration and cannot easily be used predictively for management purposes (Hutson and Wagenet, 1995).

Two-Dimensional Models

It is possible to model preferential flow and transport processes using single-domain models within two-dimensional finite element numerical schemes, when the characteristic spatial scale of the flow process is sufficiently large in relation to the size of the elementary volumes. Thus, it is feasible to numerically model finger flows and heterogeneity-induced preferential flow in this way (e.g., Nieber, 1996), since the scale of the process may be of the order of tens of centimeters, but probably not macropore flow which occurs at the pore scale.

Nieber and Misra (1995) describe a two-dimensional, dual-porosity, finite element model of water flow and nonreactive solute transport in tile-drained soil. The mass exchange terms between the domains were taken from Gerke and van Genuchten (1993a,b). A sensitivity analysis for the model was presented showing how chemical mass flux arriving at the drains was strongly regulated by the strength of interaction between the domains.

APPLICATIONS OF PHYSICAL NONEQUILIBRIUM MODELS

Model Suitability and Requirements for Management Applications

A model which is to be used for management applications within the field of nonpoint source pollution in agriculture should satisfy some requirements with respect to scope, degree of detail (and mechanism) and ease-of-use. These aspects are now briefly discussed.

Scope

A management model must deal with transient water flow, and a full dynamic water balance, including evapotranspiration and root water uptake. It should also allow predictions for many types of solute (i.e., nitrate, pesticides). Most models accounting for preferential flow and transport are restricted to either nonreactive tracers or only one solute which has real management interest.

The model should also include treatment of agricultural management practices in the model (e.g., tillage, irrigation, drainage). At present, few models explicitly describe such practices. Drainage is a particularly important feature. This is because the most significant impact of macropore flow on solute transport occurs in structured clay soils where the conductivity of the textural pore space is negligible. Poorly permeable clay soils are widely used for intensive agriculture where they are artificially drained to remove excess water. Largely due to macropore flow, drainage systems in such heavy clay soils constitute a significant route of entry of nonpoint source pollutants to surface water bodies.

Degree of Detail

The model should be as simple as possible to describe the problem of interest. In other words, the number of parameters should be minimized to make the problem of model parameterization tractable. In the case of multiporosity models, it may be possible to characterize three flow domains, although in most cases the problem of parameterization will be sufficiently difficult and uncertain in only two domains.

At the same time, the model should be as mechanistic as possible, with the degree of empiricism minimized. This is because almost all management applications will require some extrapolation in either time or space (i.e., so-called "what if" simulations). The model must be mechanistic if the user is to have some confidence in the accuracy of such model extrapolations.

Ease-of-Use

The efficient use of any model is enhanced by a user-interface which is easy to understand and learn. Management models which are used by industry and public authorities must be especially efficient in terms of demands on human resources (Russell et al., 1994). For example, the MACRO model (Jarvis, 1994) has a flexible, user-friendly interface with on-line help menus. Parameter estimation is made easier by default parameter values available within the model, while unreasonable parameter settings are flagged by warning and error messages. Outputs are selected by the user from a total of over 300 possible variables, and a graphics program is included for rapid analysis of the model outputs.

Parameterization, Calibration, and Validation

Calibration and Validation

The first step in the modeling process is to derive by calibration, values for those model parameters for which no measurements are available, either because of lack of time or money, or because the parameter is too difficult, or in some cases, impossible to measure. Traditionally, the second step is then to validate the model using an independent data set. There is continuing discussion and controversy as to whether solute transport models can be validated, in this classical sense. This discussion is even more relevant for models which deal with preferential flow and transport processes operating at the pore scale, since the nature of the preferential flow pathways cannot be known *a priori* (Flury et al., 1994) and are likely to be highly variable, both spatially and temporally. Questions therefore arise as to how these models can be used for management applications, and as to what might constitute a "validated" physical nonequilibrium flow model. The answer to this problem may lie in an alternative definition of a validated model as one which has been successfully applied and tested, with the minimum necessary calibration, at a number of sites representing a range of different environmental conditions, thereby establishing some confidence in the minds of the users that the model represents a sufficiently reasonable description of the complex and highly variable soil system. The "validated" model can then be used to answer real management problems, first by calibrating against available field data for the site of interest, and then by using the model predictively to evaluate the effects of alternative management strategies. It should again be noted here that if a model is to be used in this way, it must be mechanistic with the minimum amount of empiricism to give confidence in the accuracy of such predictions. Some examples of this type of management application are presented in a later section.

For some management applications, site-specific input data may not be available, so that model calibration may not possible. Here, physical nonequilibrium flow models must be used entirely predictively, with no possibility of calibration for "difficult" parameters. Estimation procedures or pedotransfer functions to derive parameter values from widely available soils data are then required. This is only likely to be possible for the simplest case of dual-porosity models, since the number of model parameters multiplies disproportionately as the number of domains increases (e.g., Gwo et al., 1995).

Parameter Estimation

Compared to the one-domain approach, several extra parameters are required in two-domain models such as MACRO. These are the diffusion pathlength regulating exchange between domains, the soil hydraulic properties defining the macropore region and, for reactive solutes, the fraction of sorption sites in each domain. Some tentative suggestions concerning appropriate measurement techniques and/or estimation procedures for these parameters now follow.

Parameters defining the hydraulic properties of the macropore region can be determined by curve-fitting the model functions to measurements of K made across an appropriate range of soil water pressure heads near saturation (say, zero to –100 cm H_2O). In the absence of such detailed data, the saturated micropore conductivity (i.e., K_b in Figure 7.3) can be estimated from disk permeameter or tension infiltrometer measurements (White et al., 1992; Jarvis and Messing, 1995) by making an arbitrary assumption concerning the pressure head defining the boundary between macropores and micropores (say, $\psi_b = -10$ cm H_2O). Sensitivity analyses show that model predictions are not especially sensitive to the actual value chosen for ψ_b, but are very sensitive to the value of K at ψ_b (= K_b). In the absence of any site measurements, the estimation procedure presented by Mayr et al. (1997) could be used to calculate K_b from soil water retention curve parameters.

It may be possible to estimate the diffusion pathlength from field observations of soil morphology, including the size, shape, and strength of development of the aggregate structure and the frequency or mean spacing of biotic macropores. In some reported applications of physical nonequilibrium models, the effective diffusion pathlengths required to match the field observations of water flow and/or solute transport are significantly larger than visual inspection of the soil morphology and structure would indicate (e.g., Saxena et al., 1994). As noted earlier, this may be due to cutans and organic linings at macropore/matrix interfaces, which restrict mass exchange processes between pore domains (e.g., Thoma et al., 1992).

One difficult-to-measure parameter for reactive solutes describes the partitioning of sorption sites between the pore domains. In the absence of experimental data, one simple approximation would be to distribute the sorption sites in proportion to the relative volumes of the pore domains. In many cases, macropore interfaces and linings should be more reactive than the surface area of the bulk soil (e.g., Mallawatantri et al., 1996), but this may be compensated for by the much smaller surface area per soil volume (Luxmoore et al., 1990). Another simple solution which may be appropriate for some purely predictive applications would be to set sorption in the fast-flow region to zero, giving "worst-case" estimates of nonpoint source pollution.

The Impact of Preferential Flow/Transport on Nonpoint Source Pollution

General Principles

In this section, the MACRO model (Jarvis, 1994) is used to illustrate some general principles concerning the impact of preferential flow and transport on nonpoint source pollution due to agricultural activities. A soil profile 1 m deep is simulated, with a zero pressure head at the bottom boundary (i.e., fixed water table). A common parameter set representing a clay soil (Table 7.1) is used throughout, together with a two-year series of daily weather data for southern England. A spring-sown crop is simulated in both years, with emergence on April 10th and harvest on August 18th. Evapotranspiration is calculated internally in the model using the generalized Penman-Monteith combination equation (Monteith, 1965).

Model simulations are performed for values of the effective aggregate half-width (or diffusion pathlength) d varying from 1 to 200 mm. A value of 1 mm implies complete equilibrium of both soil water pressures and solute concentrations, and this was confirmed by visual inspection of the model outputs. Increasing values of d represent an increasing degree of physical nonequilibrium related to the size and/or degree of soil structural development. The impact of preferential flow on the soil water balance and water recharge is first demonstrated. Simulation results are then presented for a nonreactive tracer, and also for some of the major types of nonpoint source pollutants in agriculture (salt balance in a saline soil profile, nitrate and pesticide leaching). Table 7.2 presents the boundary conditions and solute inputs for each of the simulation scenarios. With respect to the initial conditions, an initially solute-free soil was assumed for the nonreactive tracer and pesticide, while in the simulations of salt and nitrate leaching, the solute concentrations predicted in the soil at the end of a preliminary one-year simulation (with d set to 100 mm) were used as the initial condition (i.e., a "quasi-equilibrium" state).

Water Balance and Water Flow

As an example, Figure 7.4 shows the soil water contents predicted at 10–15 cm depth for three values of d. It demonstrates that soil water contents are generally reduced by physical nonequilibrium. This is particularly apparent following summer rainstorms, when the rewetting of the topsoil is rendered inefficient due to preferential flow. Indeed, physical nonequilibrium conditions result in a significantly drier soil

Table 7.1. Soil Hydraulic and Physical Properties Used in the Simulations.

Horizon	Parameter θ_s (m^3 m^{-3})	θ_b (m^3 m^{-3})	[a]θ_w (m^3 m^{-3})	[b]θ_r (m^3 m^{-3})	K_s (cm h^{-1})	K_b (cm h^{-1})	ψ_b (cm)	[c]λ	γ_b (g cm^{-3})
0–25 cm	0.50	0.45	0.25	0.0	10	0.01	12	0.1	1.35
25–100 cm	0.50	0.49	0.25	0.0	1	0.01	12	0.1	1.35

[a] Wilting point (assumed limit of water extraction by plant roots).
[b] Residual water content.
[c] Pore size distribution index in the Brooks-Corey (1964) equation.

Table 7.2. Boundary Conditions/Solute Applications for the Simulations.[a]

Boundary Condition	Simulation		
	Salt Balance	Nitrate Leaching	Nonreactive Tracer/Pesticide
Concentration at bottom boundary (M m^{-3})	100.0	0.3	0.0
Concentration in rain (M m^{-3})	1.0	1.0[b]	0.0
Solute application (M m^{-2})	None	10[c]	100[d]

[a] **Note that arbitrary mass units are used throughout: appropriate mass units could be kilograms for salt balance, grams for nitrate and the nonreactive tracer, and milligrams for pesticide.**
[b] Plus dry deposition of 0.001 M m^{-2} day^{-1}.
[c] On 31st March in both years.
[d] On 31st March in the first year only.

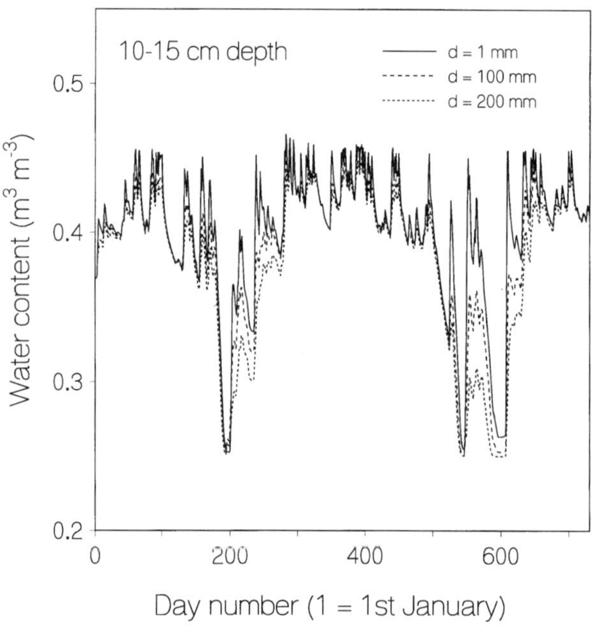

Figure 7.4. Soil water contents predicted at 10–15 cm depth as a function of the effective diffusion pathlength, *d*.

profile for most of the year (Figure 7.5) and an increased groundwater recharge (Figure 7.6). Compared to the equilibrium case, groundwater recharge was increased by ca. 9% with *d* set to 200 mm. Although not shown here, this small increase in percolation is matched by a reduction in evapotranspiration, because water storage in the profile must be in quasi-steady-state in the long term (Figure 7.5). Again, although not shown here, transpiration was unaffected by *d*, with reductions in soil evaporation accounting for the decrease in total evapotranspiration.

Nonreactive Tracer

Figure 7.7 shows the breakthrough curves at 1 m depth of the single application of nonreactive solute for values of *d* varying from 1 to 200 mm. With complete physical

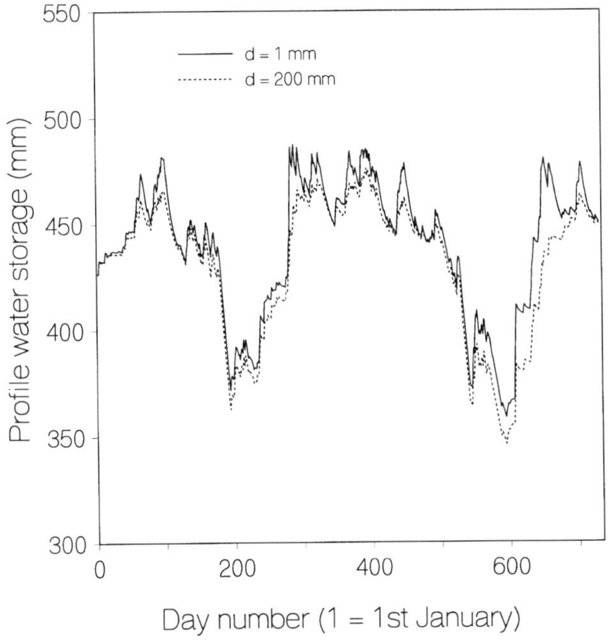

Figure 7.5. Predicted profile water storage (to 1 m depth) as a function of the effective diffusion pathlength, *d*.

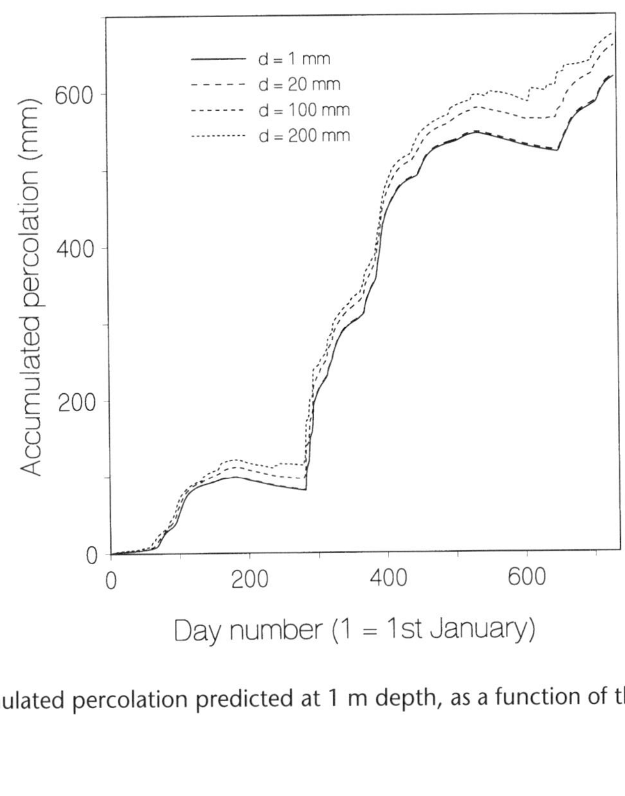

Figure 7.6. Accumulated percolation predicted at 1 m depth, as a function of the effective diffusion pathlength, *d*.

equilibrium (d = 1 mm), classical convective-dispersive behavior is noted, with breakthrough occurring first ca. 200 days after application, while more than 90% of the

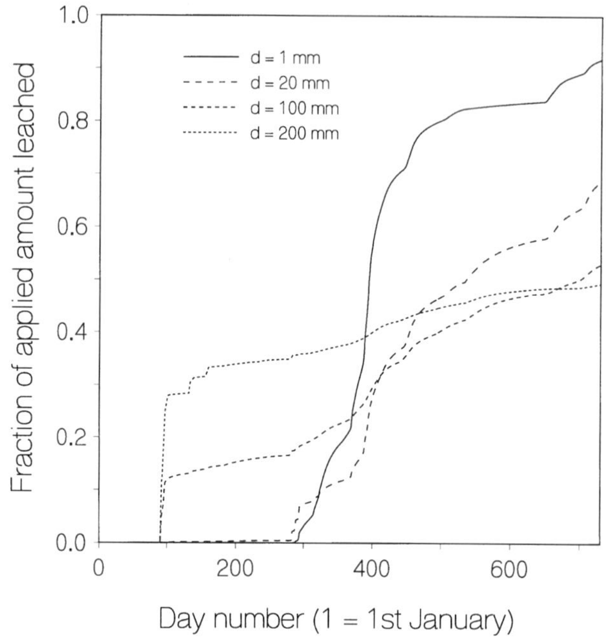

Figure 7.7. Predicted leaching at 1 m depth of the nonreactive tracer (application on day 90), as a function of the effective diffusion pathlength, d .

applied amount is leached by the end of the simulation period. At larger d values, the breakthrough curves illustrate the well-known features of nonequilibrium behavior, with early breakthrough and long tailing. At $d = 200$ mm, 30% of the solute reached 1 m depth only a few days after application. Figure 7.8 shows that this rapid breakthrough was due to ca. 30 mm of percolate generated by c. 47 mm of rain occurring in the first 10 days after application. At long times, physical nonequilibrium reduces leaching, since incoming water bypasses the solute stored in micropores. Thus, only ca. 50% of the applied amount is lost from the profile by the end of the simulation for d values of 100 and 200 mm. This is further illustrated in Figure 7.9 which compares, at one depth in the soil, solute concentrations predicted in micropores and macropores for $d = 100$ mm. At short times, solute concentrations are larger in macropores than in micropores. At long times, the reverse is the case.

The comparison between simulations with d values of 1 and 20 mm shows how preferential transport can affect solute leaching, even when soil water pressures are in equilibrium. Figure 7.6 shows that percolation from the profile was virtually identical at d values of 1 and 20 mm, whereas the amount of solute leached after day 300 was markedly smaller for $d = 20$ mm compared to $d = 1$ mm (Figure 7.7).

Nitrate Leaching

The effects of macropore flow on nitrate leaching are not well investigated. In uncropped and nonfertilized soils, leaching loads should be reduced, since nitrate is indigenous to the soil, being produced by mineralization of soil organic matter. However, under fertilized conditions, nitrate leaching loads may, on occasions, be considerably increased if heavy rain or irrigation immediately follows fertilizer application. The net long-term effects of preferential flow and transport on nitrate leaching are

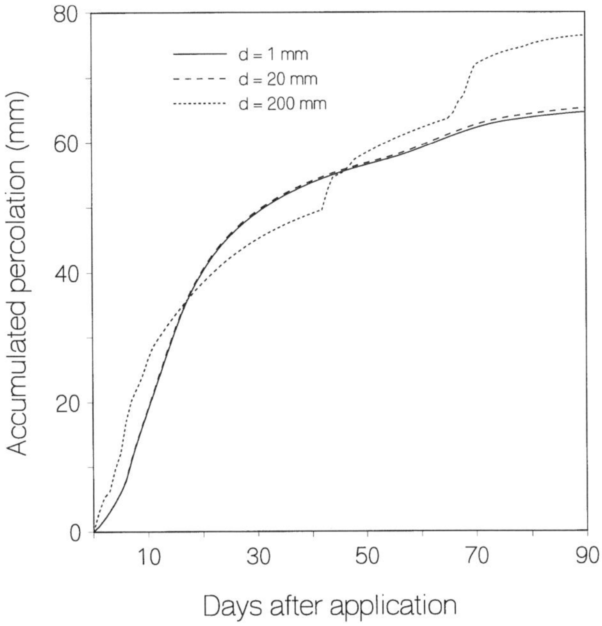

Figure 7.8. Accumulated percolation predicted at 1 m depth, as a function of the effective diffusion pathlength, *d*, for the first 90 days following solute application.

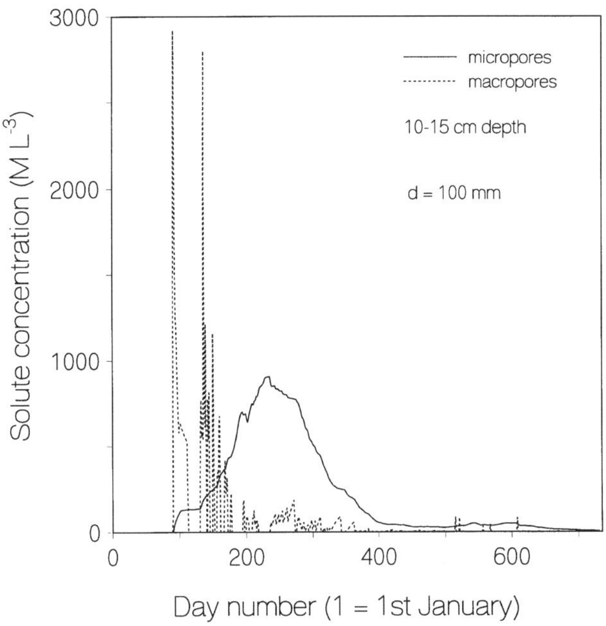

Figure 7.9. Predicted concentrations of a nonreactive tracer in micro- and macropores at 10–15 cm depth for *d* = 100 mm.

therefore difficult to predict from *a priori* knowledge. In an attempt to improve our understanding and to gain some insight into these processes, Larsson et al. (1996) recently coupled the MACRO model to the nitrogen turnover model SOILN (Johnsson

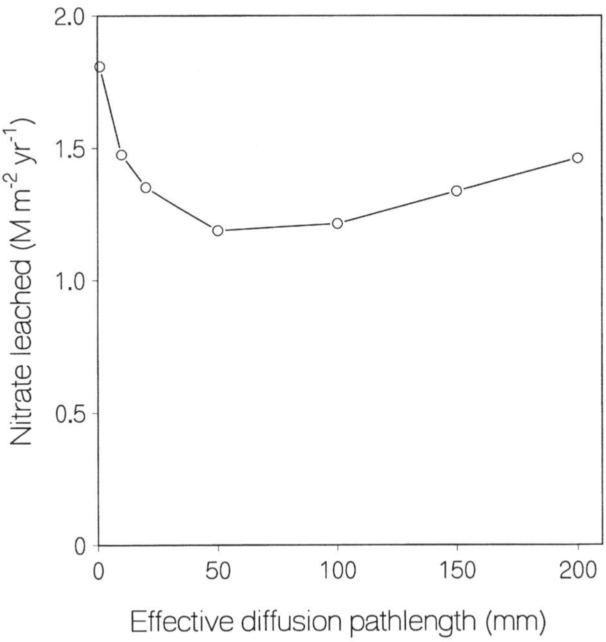

Figure 7.10. Predicted total annual nitrate leaching loss as a function of the effective diffusion pathlength, *d*.

et al., 1987). It is this combined modeling system which has been used in the example simulations presented here. Parameters in the SOILN model regulating nitrogen mineralization and immobilization, denitrification and crop uptake have been set to values considered appropriate to spring-sown cereals.

Figure 7.10 shows the annual leaching loss as a function of the effective diffusion pathlength. An interesting pattern emerges whereby leaching is reduced to a minimum value at d = 50 mm, and is ca. 52 and 23% larger, respectively, when complete physical equilibrium is assumed (d = 1 mm) and when d is set to 200 mm. Such a pattern is clearly due to the competing effects of physical nonequilibrium on leaching which occur when the solute under consideration is both surface-applied (fertilizer application once per year) and also indigenous to the soil (produced by mineralization of soil organic matter). Figure 7.11 shows that for large values of d, considerable losses of nitrate occur due to preferential flow immediately following fertilization on day 90 in both years (see Figure 7.8), while winter leaching losses are significantly reduced compared to the equilibrium case. Intermediate d values (in this example, 50 mm) represent the optimum state for minimizing nitrate leaching.

Pesticide Leaching

Figure 7.12 summarizes the simulation results for a pesticide with an assumed sorption k_{oc} value of 100 cm^3 g^{-1}. The organic carbon contents in the topsoil and subsoil were assumed to be 1.0 and 0.1%, giving sorption k_d constants of 1.0 and 0.1 cm^3 g^{-1}, respectively. In addition, degradation half-lives in the topsoil and subsoil were assumed to be 6.9 and 69 days, respectively. Figure 7.12 emphasizes the overwhelming significance of physical nonequilibrium flow for surface-applied, sorbing and degrading solutes, in that predicted leaching increases by nearly 2 orders of magnitude as d

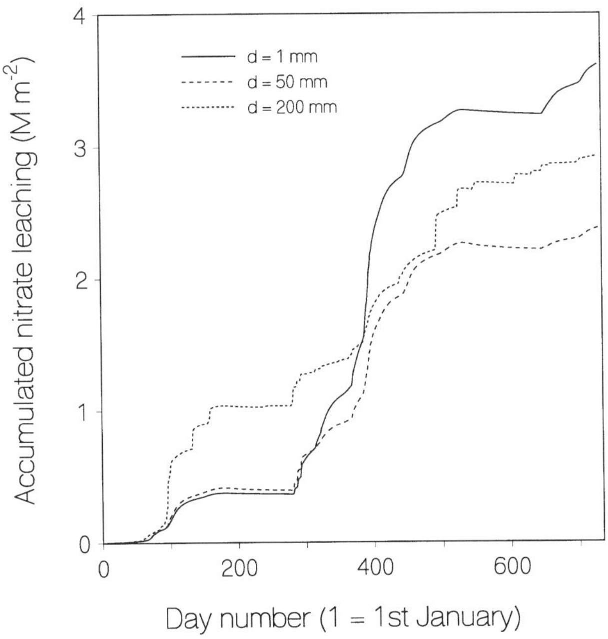

Figure 7.11. Accumulated nitrate leaching predicted as a function of the effective diffusion pathlength, *d*.

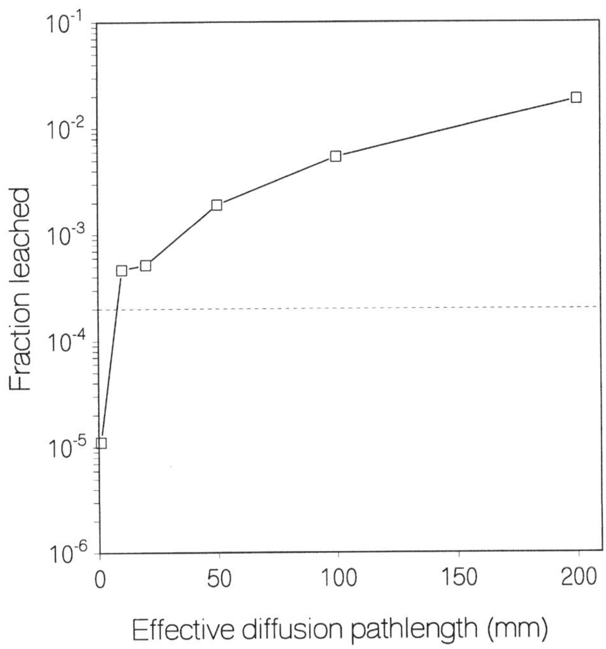

Figure 7.12. Total pesticide leaching loss predicted as a function of the effective diffusion pathlength, *d* (the dashed line indicates the leaching loss which is approximately equivalent to a mean concentration of 0.1 μg l^{-1} in the percolate).

increases from 1 mm to only 10 mm. For *d* values greater than 150 mm, leaching is ca. 3 orders of magnitude larger than for the case of complete physical equilibrium. Clearly, this has serious management implications, particularly as models are increasingly being used as decision-support tools in the pesticide registration process (Russell et al., 1994; Jarvis et al., 1995b). The hypothetical climate/soil scenario represented by Figure 7.12 would have resulted in approval of the test compound had the model disregarded physical nonequilibrium behavior. Accounting for physical nonequilibrium results in simulated leachate concentrations clearly exceeding 0.1 $\mu g\ l^{-1}$, which is the allowable concentration for a single pesticide in drinking water supplies in the EU legislative context.

In a more comprehensive simulation study, Jarvis (1995b) explored the implications for pesticide registration of using simulation models which account for physical nonequilibrium. Using the MACRO model, leachate concentrations at 1 m depth were simulated for 15 pesticides covering a wide range of properties, in soil profiles representing the main USDA textural classes. It was concluded that clay and silty clay soils constituted the "worst-case" for pesticide leaching due to preferential flow in macropores, followed by the freely-drained sandy soils, while loamy soils represented the "best-case." In the clay and silty clay soils, none of the fifteen compounds were classified as "low risk" (mean concentrations < 0.001 $\mu g\ l^{-1}$), while 12 of the 15 compounds were classified as "high risk" with mean concentrations exceeding 1 $\mu g\ l^{-1}$. Clearly, many pesticides in widespread use today would not be approved on the basis of worst-case scenarios that account for preferential flow. In this context, there is an urgent need for methods of coupling physical nonequilibrium flow models to pedological and geographical information in order to identify the likely magnitude and extent of preferential leaching of pesticides to groundwater and surface water bodies at a regional and national scale. Jarvis et al. (1996) described one such decision-support tool, or expert system, based on the MACRO model, in which the diffusion pathlength describing transfer between flow domains is related to observed soil horizon morphology, while a geohydrological classification scheme (Boorman et al., 1991) is used to assess likely routes of contamination (e.g., surface water vs. groundwater). For example, using this scheme, it is estimated that ca. 30% of the total land area of the United Kingdom (UK) is covered by soils which are strongly susceptible to bypass flow, while only 3% of these susceptible soils actually overlie important groundwater aquifers (J. Hollis, pers. comm.). Thus, in the UK, streams and rivers in lowland clay catchments are the ecosystems most at risk from rapid preferential pesticide movement.

Salt Balance and Salt Leaching

Figure 7.13 shows the general effects of physical nonequilibrium conditions on the salt balance in a saline soil profile. As expected, net leaching of salt from the profile is more effective assuming physical equilibrium (i.e., $d = 1$ mm), despite reduced percolation. This is due to the fact that encroachment of salt into the soil profile is exacerbated by nonequilibrium flow which allows penetration of fresh water in the macropores to the deeper subsoil layers of high salt content, leading to an enhanced upward diffusive transport of solute from the groundwater.

Using Nonequilibrium Models to Assess the Impact of Management Practices on Nonpoint Source Pollution

In this section, specific case studies are presented of the use of physical nonequilibrium models to evaluate the impacts of agricultural management practices on nonpoint source pollution.

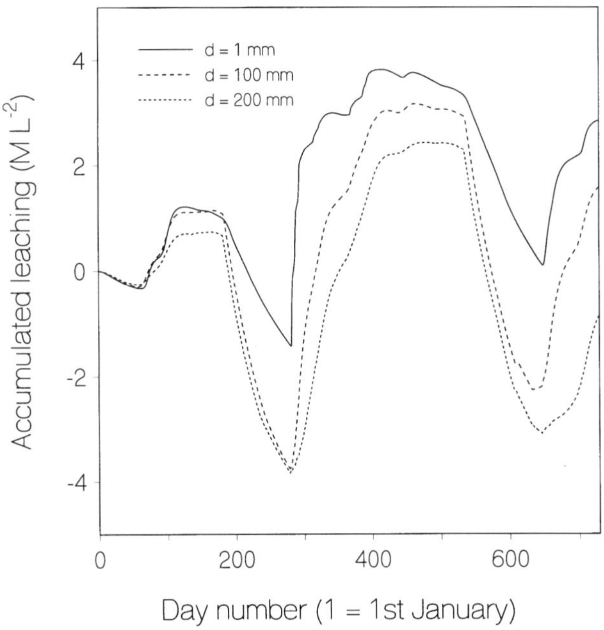

Figure 7.13. Accumulated salt leaching predicted as a function of the effective diffusion pathlength, *d*. The initial storage of salt in the profile was 3.78 ML^{-2}.

Salt Leaching in a Drained and Irrigated Saline Clay Soil

Using MACRO, Andreu et al. (1996) investigated the likely consequences of reduced irrigation inputs on the water and salt balance and cotton growth in a drained, saline cracking clay soil in a semiarid region of southern Spain. Enforced reductions in the supply of irrigation water may occur due to climate change, and also because of increased competition for water from industrial and domestic users. MACRO was first calibrated against detailed measurements of drainflows, soil water contents, water table depths, and chloride concentrations in the drainflow made in a two-week period during and immediately after an irrigation event in July 1989 (Andreu et al., 1994). The results clearly demonstrated the significant impact of physical nonequilibrium conditions on chloride leaching. The model was then successfully validated against field measurements of the soil water and chloride balance, water table depths, and drain outflows for the entire 1989 growing season (Andreu et al., 1996). Three-year scenario simulations were then performed assuming two different irrigation amounts (60 and 75% reductions from the 1989 amount) and two different frequencies (12 or 6 irrigations per growing season). The model predictions suggested that reduced irrigation may lead to up to 15% increases in the chloride content of the soil profile after 3 years and, despite overall reductions in water discharges, slight increases in chloride leaching via field drains (*ca.* 4 to 8%).

Soil Tillage and Nonpoint Source Pollution

Brown et al. (1997a,b) present the results of lysimeter experiments carried out to investigate the effects of tillage on leaching of bromide and the herbicide isoproturon in a Denchworth series heavy clay soil at Brimstone Farm (UK). Four replicate lysimeters were sampled from a field plot following sequential passes with a chisel disk, a moldboard plow, a Cambridge roll, and a power harrow penetrating to ca. 10 cm depth. Four additional lysimeters were taken from the same plot following two extra passes with the power harrow, this time to ca. 20 cm depth, in order to generate a deeper and finer tilth. This difference in seedbed condition was confirmed by subsequent measurements of the aggregate size distribution in the two treatments (Brown et al., 1997a). This information was used to parameterize the MACRO model (Jarvis, 1994), together with the pedotransfer functions described by Mayr et al. (1997). The pattern of water outflow was reasonably well predicted by the model without any calibration. The rather small differences in bromide leaching between tillage treatments were satisfactorily predicted by the model, but only when both the total soil porosity and the hydraulic conductivity of the micropores were increased for the finer-tilth lysimeters (Brown et al., 1997b). A threefold reduction in isoproturon leaching from the fine-tilth lysimeters was then well-simulated by the model, assuming that 1% of the sorption sites were located in macropores, without any further adjustments to parameter values. One interesting result of this study is that tillage effects on preferential flow and transport, and subsequent leaching, were shown to be much more significant for the chemically and biologically reactive solute isoproturon than for the inert tracer bromide, and that a simple two-region model was able to accurately reflect this different behavior.

Matthews et al. (1997) used the CRACK model (Armstrong et al., 1995) to assess the effects of soil tillage on nitrate leaching from the same Denchworth clay soil studied by Brown et al. (1997a,b). The model was shown to reproduce the pattern of water discharges and nitrate concentrations in mole drain outflow from a conventionally plowed plot at the site during a 15-day period in one winter season. A five-month winter period in another year was then simulated, assuming aggregate sizes of 10 and 20 cm to represent plowed and nonplowed (direct drilled) plots, respectively. The model predicted a 13% reduction in nitrate leaching in the nonplowed treatment, which matched the observed reduction of 20% reasonably well. A comparison of these studies performed on the same soil type highlights a potential difficulty in using tillage management to reduce nonpoint source pollution in structured soils, in that there may be conflicting requirements concerning the optimum tilth and tillage management to minimize leaching of pesticides and nitrate.

SUMMARY

Nonequilibrium conditions occur in field soils due to a variety of physical causes and must be considered the rule rather than the exception. Therefore, many models which can account for preferential flow and transport have been developed over the last 15 years. This is particularly the case for models dealing with macropore flow in structured soils. In contrast, development of models which can describe unstable "finger flow" in coarse sandy soils under field conditions is still at a very early stage, although some attempts have recently been made (e.g., Hendrickx and Yao, 1996). Much of the theoretical work on unstable flow has concentrated on understanding the processes as they occur in "ideal" porous media in the laboratory (Glass et al., 1989). It is still unclear to what extent this experience and knowledge can be transferred to the field situation, so that the conditions under which finger flow can be expected in the field cannot be predicted *a priori* and nor can its consequences (Glass and Nicholl, 1996). Thus, at least for the foreseeable future, preferential flow and transport in the matrix of coarser-textured field soils will probably be modeled using the model concept of mobile-immobile water, the parameters of which must be determined by calibration against measurements.

Although dual- and, to a lesser extent, multiporosity models are now well-established in the research arena, there are still only a few examples of management applications in the literature, and these are all very recent. Despite the extra difficulties of parameterizing these models, it is important that they be used for management purposes, simply because the consequences of preferential flow and transport for nonpoint source pollution are often dramatic, and because models that do not treat these processes are limited in their applicability. Hopefully, as we gain experience in parameterizing physical nonequilibrium models, they will be increasingly used to help decision-makers evaluate alternative management strategies to minimize nonpoint source pollution problems in agriculture. It should be stressed again here, that a reliable and usable model for field conditions must be able to deal with both preferential flow and transport. The example simulations presented in this paper clearly show that models which only deal with preferential transport may give misleading predictions when rapid bypass flow of water occurs in macropores in dry soil.

REFERENCES

Addiscott, T.M. A simple computer model for leaching in structured soils. *J. Soil Sci.*, 28:554–563, 1977.

Andreu, L., F. Moreno, N.J. Jarvis, and G. Vachaud. Application of the model MACRO to water movement and salt leaching in drained and irrigated marsh soils, Marismas, Spain. *Agric. Water Manage.*, 25:71–88, 1994.

Andreu, L., N.J. Jarvis, F. Moreno, and G. Vachaud. Simulating the impact of irrigation management on the water and salt balance in drained marsh soils (Marismas, Spain). *Soil Use Manage.*, 12:109–116, 1996.

Armstrong, A.C., A.M. Matthews, A.M. Portwood, and N.J. Jarvis. CRACK-NP. A model to predict the movement of water and solutes from cracking clay soils. *ADAS Report*, Land Research Centre, Gleadthorpe, Mansfield, Notts., UK, 1995.

Barraclough, D. A usable mechanistic model of nitrate leaching I. The model. *J. Soil Sci.*, 40:543–554, 1989a.

Barraclough, D. A usable mechanistic model of nitrate leaching II. Application. *J. Soil Sci.*, 40:555–562, 1989b.

Bergström, L. Model predictions and field measurements of Chlorsulfuron leaching under non-steady-state flow conditions. *Pest. Sci.*, 48:37–45, 1996.

Beven, K. and P. Germann. Macropores and water flow in soils. *Water Resour. Res.*, 18:1311–1325, 1982.

Bisdom, E.B.A., L.W. Dekker, and J.F.T. Schoute. Water repellency of sieve fractions from sandy soils and relationships with organic material and soil structure. *Geoderma*, 56:105–118, 1993.

Bond, R.D. The influence of the microflora on the physical properties of soils. II. Field studies on water repellent sands. *Austr. J. Soil Res.*, 2:123–131, 1964.

Booltink, H.W.G., R. Hatano, and J. Bouma. Measurement and simulation of bypass flow in a structured clay soil: A physico-morphological approach. *J. Hydrol.*, 148:149–168, 1993.

Boorman, D.B., J.M. Hollis, and A. Lilly. The production of the hydrology of soil types (HOST) data set. In *British Hydrological Society, 3rd National Hydrology Symposium*, Southampton, UK, 1991, pp. 6.7–6.13.

Brooks, R.H. and A.T. Corey. Hydraulic properties of porous media. *Hydrology Paper no. 3*, Colorado State Univ., Ft. Collins, CO, 1964.

Brown C.D., V. Marshall, A.D. Carter, A. Walker, D. Arnold, and R.L. Jones. Investigation into the effect of tillage on solute movement through a heavy clay soil. I. Lysimeter experiment. *Soil Use Manage.*, in press, 1997a.

Brown C.D., V. Marshall, A. Deas, A.D. Carter, D. Arnold, and R.L. Jones. Investigation into the effect of tillage on solute movement through a heavy clay soil. II. Interpretation using a radio-scanning technique, dye tracing and modeling. *Soil Use Manage.*, in press, 1997b.

Clothier, B.E. and K.R.J. Smettem. Combining laboratory and field measurements to define the hydraulic properties of soil. *Soil Sci. Soc. Am. J.*, 54:299–304, 1990.

Corwin, D.L., B.L. Waggoner, and J.D. Rhoades. A functional model of solute transport that accounts for bypass. *J. Environ. Qual.*, 20:647–658, 1991.

Dekker, L.W. and P.D. Jungerius. Water repellency in the dunes with special reference to the Netherlands. *Catena Suppl.*, 18:173–183, 1990.

Dekker, L.W. and C.J. Ritsema. Uneven moisture patterns in water repellent soils. *Geoderma*, 70:87–99, 1996.

Durner, W. Predicting the unsaturated hydraulic conductivity using multi-porosity water retention curves. In *Proc. of the International Workshop, Indirect Methods for estimating the hydraulic properties of unsaturated soils*, van Genuchten, M.T., F. Leij, and L. Lund (Eds.), Univ. California, Riverside, 1992, pp. 185–201.

Flühler H., W. Durner, and M. Flury. Lateral solute mixing processes—A key for understanding field-scale transport of water and solutes. *Geoderma*, 70:165–183, 1996.

Flury, M., H. Flühler, W.A. Jury, and J. Leuenberger. Susceptibility of soils to preferential flow of water. *Water Resour. Res.*, 30:1945–1954, 1994.

Flury, M. and H. Flühler. Modeling solute leaching in soils by diffusion-limited aggregation: Basic concepts and application to conservative solutes. *Water Resour. Res.*, 31:2443–2452, 1995.

Gamerdinger, A.P., R.J. Wagenet, and M.T. van Genuchten. Application of two-site/two-region models for studying simultaneous nonequilibrium transport and degradation of pesticides. *Soil Sci. Soc. Am. J.*, 54:957–963, 1990.

Gerke, H.H. and M.T. van Genuchten. A dual-porosity model for simulating the preferential movement of water and solutes in structured porous media. *Water Resour. Res.*, 29:305–319, 1993a.

Gerke, H.H. and M.T. van Genuchten. Evaluation of a first-order water transfer term for variably saturated dual-porosity flow models. *Water Resour. Res.*, 29:1225–1238, 1993b.

Germann, P. and K. Beven. A distribution function approach to water flow in soil macropores based on kinematic wave theory. *J. Hydrol.*, 83:173–183, 1986.

Glass, R.J., T.S. Steenhuis, and J.-Y. Parlange. Mechanism for finger persistence in homogeneous unsaturated porous media: theory and verification, *Soil Sci.*, 148:60–70, 1989.

Glass, R.J. and M.J. Nicholl. Physics of gravity fingering of immiscible fluids within porous media: An overview of current understanding and selected complicating factors. *Geoderma*, 70:133–163, 1996.

Gottesbüren, B., W. Mittelstaedt, and F. Führ. Comparison of different models to simulate the leaching behaviour of quinmerac predictively. In *Proceedings of the BCPC Symposium 'Pesticide Movement to Water,'* Walker, A., R. Allen, S.W. Bailey, A.M. Blair, C.D. Brown, P. Günther, C.R. Leake, and P.H. Nicholls (Eds.), Warwick, 1995, pp. 155–160.

Gwo, J.P., P.M. Jardine, G.V. Wilson, and G.T. Yeh. A multiple-pore-region concept to modeling mass transfer in subsurface media. *J. Hydrol.*, 164:217–237, 1995.

Hall, D.G.M. An amended functional leaching model applicable to structured soils. I. Model description. *J. Soil Sci.*, 44:579–588, 1993.

Hendrickx, J.M.H., L.W. Dekker, and O.H. Boersma. Unstable wetting fronts in water repellent field soils. *J. Environ. Qual.*, 22:109–118, 1993.

Hendrickx, J.M.H. and T.-M. Yao. Prediction of wetting front stability in dry field soils using soil and precipitation data. *Geoderma*, 70:265–280, 1996.

Hill, D.E. and J.-Y. Parlange. Wetting front instability in layered soils. *Soil Sci. Soc. Am. Proc.*, 36:697–702, 1972.

Hillel, D. Unstable flow in layered soils: a review. *Hydrological Processes*, 1, 143–147, 1987.

Hutson, J.L. and R.J. Wagenet. A multiregion model describing water flow and solute transport in heterogeneous soils. *Soil Sci. Soc. Am. J.*, 59:743–751, 1995.

Jamison, V.C. The penetration of irrigation and rain water into sandy soils of central Florida. *Soil Sci. Soc. Am. Proc.*, 10:25–29, 1946.

Jarvis, N.J. CRACK—A model of water and solute movement in cracking clay soils. Technical description and user notes. *Report 159, Division of Agricultural Hydrotechnics, Department of Soil Sciences, Swedish University of Agricultural Sciences*, Uppsala, 1989.

Jarvis, N.J. The MACRO model (Version 3.1)—Technical description and sample simulations. *Reports and Dissertations, 19, Department of Soil Sciences, Swedish University of Agricultural Sciences*, Uppsala, Sweden, 1994.

Jarvis, N.J., M. Stähli, L. Bergström, and H. Johnsson. Simulation of dichlorprop and bentazon leaching in soils of contrasting texture using the MACRO model. *J. Environ. Sci. Health*, A29(6):1255–1277, 1994.

Jarvis, N.J. Simulation of soil water dynamics and herbicide persistence in a silt loam soil using the MACRO model. *Ecol. Modeling*, 81:97–109, 1995a.

Jarvis, N.J. The implications of preferential flow for the use of simulation models in the registration process, In *Proc. 5th Int. Workshop 'Environmental Behaviour of Pesticides and Regulatory Aspects,'* Brussels, April 1994, 1995b, pp. 464–469.

Jarvis, N.J. and I. Messing. Near-saturated hydraulic conductivity in soils of contrasting texture as measured by tension infiltrometers. *Soil Sci. Soc. Am. J.*, 59:27–34, 1995.

Jarvis, N.J., M. Larsson, P. Fogg, A.D. Carter. Validation of the dual-porosity model MACRO for assessing pesticide fate and mobility in soil. In *Proc. BCPC Symposium 'Pesticide movement to water,'* Walker, A., R. Allen, S.W. Bailey, A.M. Blair, C.D. Brown, P. Günther, C.R. Leake, and P.H. Nicholls (Eds.), Warwick, U.K., 1995a, pp. 161–170.

Jarvis, N.J., L. Bergström, and C.D. Brown. The use of pesticide leaching models for management purposes. In *Environmental Behaviour of Agrochemicals,* Roberts, T.R. and P.C. Kearney (Eds.). John Wiley & Sons, Ltd. 1995b, pp. 185–220.

Jarvis, N.J., P. Nicholls, J.M. Hollis, T. Mayr, and S.P. Evans. Pesticide exposure assessment for surface waters and groundwater using the decision-support tool MACRO_DB. In Proceedings of the X Symposium of Pesticide Chemistry, *The Environmental Fate of Xenobiotics,* Del Re, A.A.M., E. Capri, S.P. Evans, and M. Trevisan (Eds.), Piacenza, Italy, 1996, pp. 381–388.

Johnsson, H., L. Bergström, P.-E. Jansson, and K. Paustian. Simulated nitrogen dynamics and losses in a layered agricultural soil. *Agric., Ecosystems Environ.*, 18:333–356, 1987.

Jury, W.A. and H. Flühler. Transport of chemicals through soil: Mechanisms, models and field applications. *Adv. Agron.*, 47:141–201, 1992.

Kung, K.-J.S. Preferential flow in a sandy vadose zone. 1. Field observation. *Geoderma,* 46:51–58, 1990a.

Kung, K.-J.S. Preferential flow in a sandy vadose zone. 2. Mechanism and implications. *Geoderma,* 46:59–71, 1990b.

Lafolie, F. and Ch. Hayot. One-dimensional solute transport modelling in aggregated porous media. Part 1. Model description and numerical solution. *J. Hydrol.*, 143:63–83, 1993.

Larsson, M.H., N.J. Jarvis, and H. Johnsson. A dual-porosity model for nitrogen turnover and nitrate leaching from arable soil. *Annales Geophysicae, Supplement to Volume 14, Part II,* C323, 1996.

Leij, F.J., N. Toride, and M.T. van Genuchten. Analytical solutions for non-equilibrium solute transport in three-dimensional porous media. *J. Hydrol.*, 151:193–228, 1993.

Luxmoore, R.J., P.M. Jardine, G.V. Wilson, J.R. Jones, and L.W. Zelazny. Physical and chemical controls of preferred path flow through a forested hillslope. *Geoderma,* 46:139–154, 1990.

Mallawatantri, A.P., B.G. McConkey, and D.J. Mulla. Characterization of pesticide sorption and degradation in macropore linings and soil horizons of Thatuna silt loam. *J. Environ. Qual.*, 25:227–235, 1996.

Matthews, A.M., A.C. Armstrong, P.B. Leeds-Harrison, G.L. Harris, and J.A. Catt. Development and testing of a model for predicting tillage effects on nitrate leaching from a cracked clay soil, submitted to *Soil and Tillage Research,* 1997.

Mayr, T., N.J. Jarvis, P.B. Leeds-Harrison, and C. Kechavarzi. Pedotransfer functions to estimate soil matrix hydraulic properties. Submitted to *Geoderma,* 1997.

Monteith, J.L. Evaporation and environment. In *The State and Movement of Water in Living Organisms,* Fogg, G.E. (Ed.), 19th Symposium of the Society of Experimental Biology, Cambridge, 1965, pp. 205–234.

Mualem, Y. A new model for predicting the hydraulic conductivity of unsaturated porous media. *Water Resour. Res.,* 12:513–522, 1976.

Nieber, J.L. and D. Misra. Modeling flow and transport in heterogeneous, dual-porosity drained soils. *Irrig. Drain. Syst.,* 9:217–237, 1995.

Nieber, J.L. Modeling finger development and persistence in initially dry porous media. *Geoderma,* 70:207–229, 1996.

Othmer, H., B. Diekkrüger, and M. Kutilek. Bimodal porosity and unsaturated hydraulic conductivity. *Soil Sci.,* 152:139–150, 1991.

Ritsema, C.J., L.W. Dekker, J.M.H. Hendrickx, and W. Hamminga. Preferential flow mechanism in a water repellent sandy soil. *Water Resour. Res.,* 29:2183–2193, 1993.

Roth, K., H. Flühler, and W. Attinger. Transport of a conservative tracer under field conditions: Qualitative modeling with random walk in a double-porous medium. In *Field-scale water and solute flux in soils, Proceedings Centro Stefano Franscini, Monte Verita, Ascona,* Roth, K., H. Flühler, W.A. Jury, and J.C. Parker (Eds.), Birkhäuser, Basel, 1990, pp. 239–249.

Russell, M.H., R.J. Layton, and P.M. Tillotson. The use of pesticide leaching models in a regulatory setting: An industrial perspective. *J. Environ. Sci. Health,* A29:1105–1116, 1994.

Saxena, R., N.J. Jarvis, and L. Bergström. Interpreting non-steady state tracer breakthrough experiments in sand and clay soils using a dual-porosity model. *J. Hydrol.,* 162:279–298, 1994.

Smettem, K.R.J. and C. Kirkby. Measuring the hydraulic properties of a stable aggregated soil. *J. Hydrol.,* 117:1–13, 1990.

Steenhuis, T.S., J.-Y. Parlange, and M.S. Andreini. A numerical model for preferential solute movement in structured soils. *Geoderma,* 46:193–208, 1990.

Thoma, S.G., D.P. Gallegos, and D.M. Smith. Impact of fracture coatings on fracture/matrix flow interactions in unsaturated, porous media. *Water Resour. Res.,* 28:1357–1367, 1992.

van Dam, J.C., J.M.H. Hendrickx, H.C. van Ommen, M.H. Bannink, M.T. van Genuchten, and L.W. Dekker. Water and solute movement in a coarse-textured water-repellent field soil. *J. Hydrol.,* 120:359–379, 1990.

van Genuchten, M.T. and P.J. Wierenga. Mass transfer in sorbing porous media. I. Analytical solutions. *Soil Sci. Soc. Am. J.,* 40:473–480, 1976.

van Genuchten, M.T. A closed-form equation for predicting the hydraulic conductivity of unsaturated soils. *Soil Sci. Soc. Am. J.,* 44:892–898, 1980.

van Genuchten, M.T. A general approach for modeling solute transport in structured soils. In Proceedings of the 17th International Congress IAH, *Hydrogeology of Rocks of Low Permeability.* Memoires IAH, 17, 1985, pp. 513–526.

van Genuchten, M.T. and R.J. Wagenet. Two-site/two-region models for pesticide transport and degradation: theoretical development and analytical solutions. *Soil Sci. Soc. Am. J.,* 53:1303–1310, 1989.

Wagenet, R.J. and J.L. Hutson. LEACHM: Leaching estimation and chemistry model. A process based model of water and solute movement, transformations, plant uptake and chemical reactions in the unsaturated zone. *Continuum 2,* Water Resources Institute, Cornell University, Ithaca, NY, 1987.

White, I., M.J. Sully, and K.M. Perroux. Measurement of surface-soil hydraulic properties: disk permeameters, tension infiltrometers, and other techniques. In *Advances in Measurement of Soil Physical Properties: Bringing Theory into Practice,* Topp, G.C., et al. (Eds.), SSSA Special Publication no. 30, SSSA, Madison, WI, 1992, pp. 69–103.

Youngs, E.G. and P.B. Leeds-Harrison. Aspects of transport processes in aggregated soils. *J. Soil Sci.,* 41:665–675, 1990.

Zurmühl, T. and W. Durner. Modeling transient water and solute transport in a biporous soil. *Water Resour. Res.,* 32:819–829, 1996.

Chapter Eight

Modeling the Interaction Between Leaching and Intraped Diffusion

T.M. Addiscott, A.C. Armstrong, and P.B. Leeds-Harrison

INTRODUCTION

This volume is concerned with nonequilibrium processes in soils. Leaching is a nonequilibrium process, like all forms of flow, simply because there are no flows in systems at equilibrium. Diffusion, which interacts with leaching, is a nonequilibrium process for the same reason, and preferential flow is far removed from any form of equilibrium. This chapter is concerned with the interaction between leaching and the diffusion within aggregates and other structural units in clay soils. We are not concerned in detail here with the within-flow diffusion that occurs as one aspect of dispersion during leaching, and we do not review models for leaching, this having been done by Wagenet (1983) and Addiscott and Wagenet (1985a). We begin with a brief introduction to diffusion theory and the way in which it has been integrated with the classical approach to modeling leaching. Both these approaches are limited to systems in which a steady state is established, something that does not happen when intermittent rain falls on unsaturated soil, as usually happens in rainfed agriculture. We go on to describe a diffusion model that was designed to cope with these circumstances and the PEDAL and CRACK-NP leaching models, both of which were developed for these circumstances and incorporated the diffusion model. Simulations of leaching data obtained with these models are presented and discussed.

The classical approach to modeling solute leaching presumes completely uniform flow of solute and water through a perfectly homogeneous medium. This is a concept unlikely to be found in reality, and even the flow of tritiated water in a laboratory column of carefully packed sand was found by De Smedt et al. (1986) to show evidence that mobile and immobile categories of water developed. Uniform flow is very unlikely to happen in soils in the field, and the more clay the soil contains, the less likely it becomes. The breakthrough of chloride in a Rothamsted soil with more than 30% of clay (Figure 8.1) was far removed from the smooth bell-shaped breakthrough curve given by a uniform soil. Clay soils contain a very wide range of pore sizes and rates of water movement. Water is for many purposes immobile in the very small pores within aggregates, but during heavy rain it may move down rapidly through large cracks and other macropores, a process described as *channelling* (Beven, 1981) or *preferential flow* (Bouma, 1981). Such soils often have field drains, and if a preferential

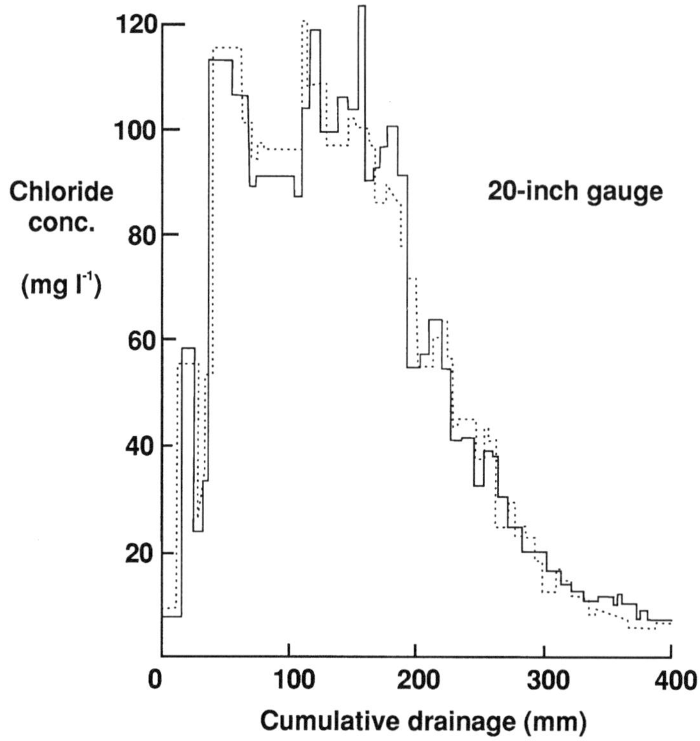

Figure 8.1. Measured (solid line) and simulated (broken line) concentrations of chloride plotted against cumulative drainage for the 20-inch (500 mm) drain gauge at Rothamsted from 9 October 1974 (Application of chloride) until drainage ceased. From Addiscott et al. (1978).

flow pathway intersects a drain, solutes applied to the soil surface can be carried very rapidly through these macropores into the drainage system and then into a ditch or stream without interacting appreciably with the bulk of the soil matrix. Preferential flow can be the cause of a rapid pollution event should unexpected heavy rain follow the application of an agrochemical and carry it rapidly downward, as shown by Hallard and Armstrong (1992) who recorded movement of a dye used as a tracer from the soil surface to a mole drain within about 30 minutes. It can, however, help to retain solute that is already in the soil matrix, because the water passes through the profile without putting the solute at risk and the latter remains safe unless it diffuses to a water pathway. Both these aspects of preferential flow are important because of concerns about losses from agricultural land to water of nitrate and pesticides (e.g., Addiscott et al., 1991; Cheng, 1990), and a more comprehensive review of preferential flow is given elsewhere in this volume.

Lawes and his colleagues at Rothamsted recognized the phenomenon of preferential flow more than 100 years ago (Lawes et al., 1882), but it received little attention until the late 1970s and early 1980s. This period saw the seminal work of Bouma and his colleagues in the Netherlands on the evaluation of the size and function of macropores (e.g., Bouma and Dekker, 1978; Bouma, 1981) and also the development of leaching models that assume the soil to contain both mobile and immobile water. This period also saw some useful theoretical development of the topic by Beven and Germann (1981, 1982), notably in the field of representative elementary volumes for leaching.

Mobile and Immobile Water and Related Concepts

The concepts of preferential flow and mobile and immobile water are related to each other and to that of the holdback of solute against leaching. These concepts all depend on our ideas about the structure of the soil. Pore sizes in soils vary enormously and rates of water flow reflect this variation. One simplifying assumption that has been made, particularly for modeling purposes, is that soil water can be divided into mobile and immobile categories. The implication for solute leaching is that solute in immobile water is held back from leaching until it moves from the immobile water into the mobile category. This holdback depends greatly on the mechanism by which the solute moves into or out of the immobile water. This in turn depends on the location of the immobile water and then on the type of soil involved.

Aggregated soils provide the simplest understanding of mobile and immobile water; the immobile water is within the aggregates and the mobile water flows around them. Soils do not need to be aggregated to contain mobile and immobile water, but we are concerned here with soils that are aggregated in nature. Solute can enter the aggregates in two ways. If water flows around dry aggregates, they will imbibe it, and solute may be carried into them by convection. This does not happen with moist aggregates, but solute can enter them by diffusion, provided there is a concentration gradient from the outside to the inside of each aggregate. Pores within aggregates are very small, so water flows from them are rarely large enough to carry appreciable amounts of solute. Solute can, however, move out of aggregates by diffusion if the concentration gradient is in the right direction.

If diffusion is the main mechanism controlling the movement of solute, the size of the aggregates determines the degree of holdback. More solute diffuses out of small aggregates than from an equivalent volume of larger aggregates in the same time, and diffusion theory suggests that the holdback will be proportional to the square of the diameter or side length of the aggregates.

One question that arises is whether preferential flow is to be identified with flow in the mobile category of water. Because of the wide range of pore sizes in soils, particularly clay soils, it may be appropriate to consider them as separate phenomena. In some soils, mobile water may carry solute within the soil matrix, but not very rapidly so that the solute is able to move between the mobile and immobile categories, while fast preferential flow takes place in the largest pores and does not interact very much with the matrix. This scenario is adopted in the SLIM model (Addiscott and Whitmore, 1991). In heavy cracking and swelling clay soils, however, the only downward movement occurs in large cracks and is effectively preferential flow. The water and solute in this flow interact with the clay peds, but only to a rather limited extent. This is the scenario of the CRACK and CRACK-NP models (Jarvis and Leeds-Harrison, 1987a; Armstrong et al., 1995, 1996; Matthews et al., in press a and b).

Early "Mobile-Immobile" Models

The best known mobile-immobile model is that of van Genuchten and Wierenga (1976, 1977). At the time of its development it was used and evaluated at the scale of the laboratory column, although Barraclough (1989a,b) subsequently produced a variant that was applicable at the field scale. A simpler mobile-immobile leaching model was developed by Addiscott (1977) for use in the field; this allowed for preferential flow when rain exceeded a critical value. This model provided good simulations (Figure 8.1) of the complex pattern of leaching that resulted from the application of ^{36}Cl to the surfaces of the Rothamsted Drain Gauges (Addiscott et al., 1978), despite the fact that it did not simulate explicitly the diffusion within the aggregates and structural units of the soil. These data are far removed from the smooth bell-shaped breakthrough curve given by a uniform soil.

THEORY AND MODELS

Classical Approaches to Leaching and Diffusion

Solute flow in a homogeneous porous medium is described by the following general equation (e.g., Youngs and Leeds-Harrison, 1990).

$$\frac{\partial C}{\partial t} = \frac{D}{r^n}\frac{\partial}{\partial r}\left(r^n \frac{\partial C}{\partial r}\right) = \frac{1}{r^n}\frac{\partial}{\partial r}(r^n vC) \tag{8.1}$$

with: C the solute concentration at time t
D the dispersion coefficient when there is convective flow, or the diffusion coefficient when there is none
v the average pore water velocity
r the space coordinate
n a shape factor, zero for planar slabs, one for cylinders and two for spheres

The whole equation is applicable when convective-dispersive flow occurs, but in water-filled aggregates v can be assumed to be zero so that the term containing it vanishes, leaving an expression of Fick's second law for diffusion.

Diffusion

Diffusion is described by Fick's first and second laws. The first law states that the diffusive flux, F, is proportional to the concentration gradient, the coefficient of proportionality being the diffusion coefficient, D.

$$F = -D\frac{\partial C}{\partial x} \tag{8.2}$$

An expression for the second law is obtained by omitting the term containing v in Equation 8.1. Solutes obviously do not diffuse as rapidly in soils as they do in free solution, because of the obstruction to movement caused by the soil matrix. The "effective" diffusion coefficient, D_{eff}, that operative within the volume of soil of interest, is obtained by multiplying the diffusion coefficient in free solution, D_l, by the volumetric water content, θ, and an impedance factor, f_l, (Nye and Tinker, 1977).

$$D_{eff} = D_l \theta f_l \tag{8.3}$$

Fick's laws are relatively simple in principle. The main problem in using them lies in defining appropriate boundary conditions for particular applications and finding easily usable solutions. Soil aggregates almost never have the simple shapes required by diffusion theory, and where a definite geometry has had to be assigned, they have been treated as spheres or cubes. Rao et al. (1982) showed that nonspherical aggregates can be represented by equivalent spheres. For a sphere, the expression for Fick's second law has to be solved in radial coordinates with suitable boundary conditions. One solution that has been found (Crank, 1956) and used fairly often is that for the uptake of solute from a stirred solution of limited volume, V, by an initially solute-free sphere of radius r. It is

$$\frac{M_t}{M_\infty} = 1 - \sum_{n=1}^{\infty} \frac{6\alpha(1+\alpha)\exp(-Dq_n^2 t/r^2)}{9+9\alpha+q_n^2\alpha^2} \tag{8.4}$$

where M_t and M_∞ are the uptakes of solute at time t and infinite time, D is the diffusion coefficient, the q_n's are the nonzero roots of $\tan q_n = 3q_n/(3 + \alpha q_n)$ and $\alpha = 3V/(4r^3)$. The same equation can be used for diffusion from a sphere into an initially solute-free solution, and Rao et al. (1980a) obtained satisfactory fits when they used it to obtain values of the effective diffusion coefficient in porous ceramic spheres of 55 mm and 75 mm radius. They also obtained values for the time-averaged mass transfer coefficient used in the mobile-immobile leaching model of van Genuchten and Wierenga (1976). Identical spheres were placed in leaching columns that were used in breakthrough experiments to evaluate two leaching models (Rao et al., 1980b), as described in the next section.

Equation 8.2 described these laboratory experiments quite satisfactorily, but the boundary conditions attached to it greatly limit its use in unsaturated soils subject to intermittent rain. The solution surrounding the aggregate in such soils is not constant in either its volume or its concentration, because both may change when rain occurs, and neither the solution inside the aggregate nor that outside it is initially zero in most cases. It was for these reasons that a more robust but less exact numerical diffusion model able to cope with irregularly imposed changes in the volume and concentration of the external solution was developed (Addiscott, 1982a).

Simultaneous Leaching and Diffusion

The problem of simulating solute transport in aggregated porous media has been on the soil physics agenda for 25 years or more. Passioura (1971) and Passioura and Rose (1971) developed an approach based on the classical hydrodynamic dispersion theory, which they explored in laboratory experiments with small porous glass beads in columns. Rao et al. (1980b) made some interesting column experiments in which they compared leaching in aggregated and nonaggregated porous media. The aggregated medium comprised the porous ceramic spheres of 55 mm and 75 mm radius described in the previous section, separated from each other by fine sand or small glass beads, which also made up the nonaggregated medium. Both media were saturated for the experiments, and the diffusion characteristics of the porous ceramic spheres had been determined in the separate diffusion experiments described above. Rao et al. used these experiments to evaluate two models. The first assumed convective-dispersive solute transport in the inter-aggregate water and treated diffusion between the inter- and intra-aggregate water explicitly, using Fick's second law written in radial coordinates. The second model was that of van Genuchten and Wierenga (1976) which does not take specific account of diffusion or the geometry of the spheres, and uses the transfer coefficient determined in the diffusion experiment to regulate the transfer of solute between mobile and immobile water. Rao et al. (1980b) concluded that both models were able to simulate the breakthrough curves obtained in the experiments. They also found that the transfer coefficient needed by the second model could be estimated from an equation incorporating the volumes of mobile and immobile water and the radius of the spheres. They discussed whether it would be worth extending the first model to handle distributions of aggregate sizes as well as single size aggregates but concluded that it was not necessary, given the satisfactory performance of the second model. In support of this conclusion they cited Passioura (1971) and Passioura and Rose (1971), who considered that for this purpose a distribution of sizes could be represented adequately by the root mean square radius.

Computer Models for Leaching and Diffusion Based on Soil Structural Concepts

Diffusion

The simple diffusion model (Addiscott, 1982a) considers cubic aggregates bathed by a solution. It assumes the aggregates to be cubic rather than spherical for two reasons. One is that aggregates in cultivated soils are frequently described as blocky, particularly in clay soils. The other is simply that cubes pack more conveniently than spheres.

The model divides each cube (of side length a) into n concentric volumes, each of which contains a variable amount S_i of solute ($1 \leq i \leq n$), and computes the diffusive flux of solute between adjacent concentric volumes and between the outermost concentric volume and the external solution. Either the volumes or the interface distances can be made equal, but making the volumes equal is the more effective. The volumes are then all a^3/n, and the (nonequal) distances, b_i, between the interfaces are

$$b_i = a\left[i^{1/3} - (i-1)^{1/3}\right]/n^{1/3} \tag{8.5}$$

where i is counted from the center outward. The area, A_i, across which diffusion between adjacent volumes occurs, increases with the square of the distance from the center. To get an appropriate value, it is taken to be the mean surface area between the surfaces lying halfway between the interfaces delineating the adjacent volumes. These halfway surfaces are at distances x_1 and x_2 from the center, where, for diffusion between volumes i and $(i + 1)$,

$$x_1 = a\left[i^{1/3} + (i-1)^{1/3}\right]/4n^{1/3} \tag{8.6}$$

$$x_2 = a\left[i^{1/3} + (i+1)^{1/3}\right]/4n^{1/3} \tag{8.7}$$

and for diffusion between the outermost volume and the external solution,

$$x_1 = a\left[n^{1/3} + (n-1)^{1/3}\right]/4n^{1/3} \tag{8.8}$$

$$x_2 = a/2 \tag{8.9}$$

The distance for diffusion between adjacent volumes is $(x_2 - x_1)$.

Diffusional solute transfer is computed using Fick's first law. The change, ΔS_i, during time Δt in the quantity of solute in layer i resulting from diffusion into layer $(i - 1)$ is given by

$$\frac{1}{A_i}\frac{\Delta S_i}{\Delta t} = -D_l \theta f_l \frac{(S_i - S_{i-1})}{w(x_X - x_1)} \tag{8.10}$$

where $w = a^3\theta/n$. The model can be used without modification for spherical aggregates of diameter a, because the ratio of the volume of a sphere to that of a cube is the same as the ratio of the surface areas; both ratios are $\pi/6$. It is also potentially applicable to a regular tetrahedron and a rhombic octahedron (Addiscott, 1982a). Its most important feature, however, is its ability to simulate diffusion when the volume or concentration of the external solution undergoes externally imposed changes.

The model gave better simulations of diffusion as the number of concentric volumes was increased to 10, but no further, and when tested against data for diffusion of bromide from cubes of natural chalk, it gave satisfactory simulations (Addiscott, 1982a). Tests against the data of Rao et al. (1980a) for diffusion of $^{36}Cl^-$ from porous ceramic spheres again gave satisfactory simulations, especially when allowance was made for the removal of aliquots of the external solution. The model also simulated nitrate diffusion from natural aggregates of three size ranges well, but did less well for diffusion from two other size ranges (Addiscott et al., 1983). The model's main disadvantage is that it is not as exact as analytical solutions of the diffusion equations.

Simultaneous Leaching and Diffusion

Two models are discussed in this section. The PEDAL model was the result of integrating the diffusion model described in the previous section with the simple mobile-immobile leaching model of Addiscott (1977). It has been published in outline (Addiscott, 1984) but not in detail. The original CRACK model was a water flow model (Jarvis and Leeds-Harrison, 1987a) into which Jarvis (1989) introduced the movement of a conserved solute. The current model is CRACK-NP, so called because it incorporates routines for the mineralization of the soil's organic nitrogen and the sorption and degradation of pesticides (Armstrong et al., 1995, 1996; Matthews et al., 1997a,b). This latter model also incorporates the simple diffusion model described above.

The PEDAL Model

The simple mobile-immobile leaching model of Addiscott (1977) was a layer model in which the amounts of mobile and immobile water in each layer were fixed parameters. Rain caused downward movement of water and solute in the mobile water only, and a "fast leaching routine" allowed movement through two or more layers when heavy rain occurred. After the downward flow, solute moved laterally between the two categories of water. In the original version of the model, this movement simply equalized the concentrations of solute between the two categories, but subsequent versions included a "holdback factor." The rationale of this factor was that the lateral movement was by diffusion and the concentrations were never equalized (in finite time). Calculations based on the aggregate size distribution in Rothamsted topsoil (Addiscott et al., 1983) suggested a holdback factor of 0.1 was appropriate for this soil, implying that the lateral solute movement went 90% of the way to complete equalization of the concentrations. The holdback factor was larger in the subsoil, because the structural units were larger.

The PEDAL model was similar to the model described above except that the lateral movement of solute was simulated explicitly as diffusion-limited, using the diffusion model for cubic aggregates described in the section above (Eqs. 8.6–8.10). Addiscott et al. (1983) found that the diffusion of bromide from mixed 20, 30, and 40 mm cubes of chalk could be simulated better with separate computations for each size than by taking a cube root mean cube size; soil aggregates have a much wider range of sizes than the chalk cubes, so the model was set up to accept normal or lognormal distributions of cube sizes. Lognormality is likely when the distribution of sizes results from shattering (Epstein, 1948) and has been observed in cultivated surface soils (Gardner, 1956; Smith, 1977). The model uses the Sectioning Method of Addiscott and Wagenet (1985b) to define a set of aggregate sizes representing the distribution, which can be either normal or lognormal, and runs the diffusion model for each of these sizes after each flow event in the mobile water around the aggregates.

Aggregate size distribution is a soil property that is changed by cultivation. The PEDAL model was therefore developed more as a means of exploring the effects of aggregate size and size distribution on the holdback against leaching of solute within the aggregates than as a management tool. These effects were explored (Addiscott, 1984) in terms of the holdback, H, against leaching of the solute initially in the profile; that is, the proportion still in the profile at the end of leaching. We saw above that diffusion theory shows that holdback should depend on the square of the aggregate diameter or side length, so the size-dependent holdback, H_{sd}, and the proportion, P_{sd}, of the potential size-dependent holdback, H_p, which it constituted were computed as follows for various aggregate sizes and amounts and patterns of rainfall.

The scenario for the investigation was a 500 mm "soil profile" comprising 10 layers each with the same mean aggregate size. The profile was assumed to be fully recharged and the aggregates not to shrink or swell. The aggregates had a porosity of 0.35, and air and nonporous matter made up a proportion of 0.25 of each layer. The computer program was written so that the amounts of mobile and immobile water in each layer were the same for all mean aggregate sizes, so that only the direct effects of aggregate size on holdback were simulated. The volumetric water content was 0.31, and the solute was assumed to be nonsorbed and to be uniformly distributed throughout the profile at the start. The diffusion parameters D_l and f_l were 1.2×10^{-9} m^2 s^{-1} and 0.3, respectively, values appropriate to calcium nitrate or a calcium halide. This profile was subjected to 50 or 100 mm of rain falling in 2.5 or 10 mm aliquots or alternating 10 and 2.5 mm aliquots.

The minimum holdback, H_{min}, was computed with the version of the model that allows full equilibration between the mobile and immobile categories of water and takes no account of the constraint on lateral transfer imposed by diffusion. The maximum holdback, H_{max}, was calculated on the assumption that no solute initially in the immobile water was leached. This implies aggregates of infinite size, and running PEDAL with an aggregate size (cube side length) of 1000 mm gave holdbacks similar to H_{max}. The potential size-dependent holdback, H_p, was $(H_{max} - H_{min})$, and if the holdback found for an aggregate size or size distribution was H, the size-dependent holdback, H_{sd}, was $(H - H_{min})$. The proportion, P_{sd}, of the potential size-dependant holdback achieved is then

$$P_{sd} = H/H_p = (H - H_{min})/(H_{max} - H_{min}) \tag{8.11}$$

With uniformly sized aggregates, the proportion P_{sd} depended greatly on the aggregate size (Table 8.1), being proportional to its square, as expected from diffusion theory. It was also influenced strongly by the rainfall pattern, being smallest with the 2.5 mm aliquots and greatest with the 10 mm aliquots. The overall quantity of rain had most effect with the small aliquots and least with the large. Also shown in Table 8.1 is the aggregate size needed for discernible size-dependent holdback, a_{sd},

Table 8.1. Proportion (P_{sd}) of the Potential Size-Dependent Holdback Achieved. Effects of Aggregate Size, a (Uniform), and Amount and Pattern of Rainfall.[a]

	Amount of Rain (mm)					
	50			100		
	Aliquot Size (mm)					
	2.5	Alt	10	2.5	Alt	10
a (mm)	P_{sd}					
20	0.000	0.000	0.001	0.000	0.000	0.001
40	0.000	0.006	0.015	0.004	0.008	0.017
60	0.001	0.038	0.081	0.021	0.043	0.082
80	0.008	0.107	0.194	0.052	0.108	0.183
100	0.026	0.203	0.322	0.092	0.189	0.289
1000	0.955	0.985	0.986	0.942	0.974	0.983
	a_{sd}					
	37	18	15	18	17	15

[a] Also, aggregate size, a_{sd}, needed to give discernible size-dependent holdback. Alt = alternating 2.5 and 10 mm aliquots. Based on data in Table 1 of Addiscott (1984).

which was defined as that at which P_{sd} first exceeded 0.0005. This also varied appreciably with the rainfall regime.

The relationship between P_{sd} and the aggregate size (cube side), a, was influenced strongly by the nature of the aggregate size distribution, there being substantial differences between uniformly, normally, and lognormally distributed aggregates (Figure 8.2). The reason for this was that the smallest aggregates released their solute more rapidly than larger ones (Addiscott et al., 1983). Where aggregates are mixed in size, this rapid release from the small ones will increase the concentration around the larger ones, lessen the concentration gradient, and delay diffusion from them. Whether the distribution is normal or lognormal will influence the ratios of large to small aggregates. The relationship between P_{sd} and a was also influenced by the amount and pattern of rainfall, with appreciable differences between the 2.5 and 10 mm aliquots; the regime in which these aliquots alternated gave an intermediate result (omitted for clarity in the figure).

The CRACK-NP Model

The CRACK-NP model is based on soil physical concepts. Transport of solute depends on convection in pores around the soil aggregates and diffusion within them. Additionally, if the aggregates are initially dry, there may be convective flow of solute into the aggregates as water is drawn into them by the hydraulic gradient between their surfaces and their interiors. How much convective flow and diffusion occur depends on the proportion of the surface of the aggregates bathed by the water in the pores around them, and this will be determined by conditions in both the aggregates and the pores.

Water infiltrating into the soil will run off the aggregates if the rate of infiltration into them is exceeded and repeats the process further down the profile until all the water has been taken up by the aggregates (Jarvis and Leeds-Harrison, 1987b). During this process, by no means all the aggregate surfaces will be bathed. Dye tracing studies by Bouma and Dekker (1978) suggested that only about 5% of the surfaces in heavy clay soils were bathed at any one time. This can change if there is a neck in a key

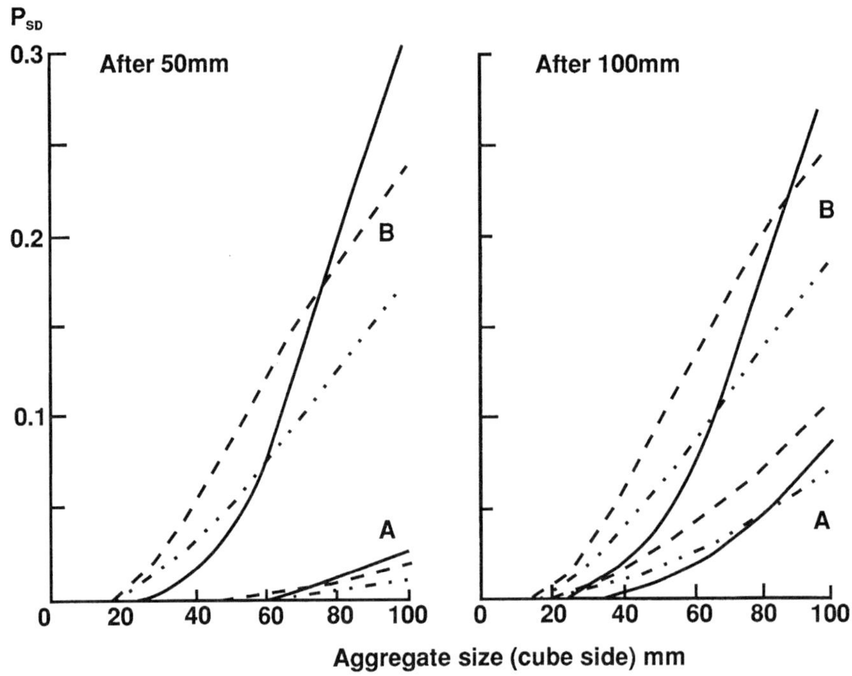

Figure 8.2. Proportion (P_{sd}) of size-dependent holdback achieved following 50 or 100 mm of rain. Effects of aggregate size (cube side) and size distribution, uniform (solid line), normal (broken line) or lognormal (dash-dotted line), with 2.5 mm (A) or 10 mm (B) rainfall aliquots. From Addiscott (1984) with permission.

downward pore, because water will accumulate, cause local saturation, and bathe a much larger proportion of the surfaces.

In a heavy clay soil during winter, the peds are usually nearly saturated with effectively immobile water, and water flows only in the cracks between them. This suggests that the concept underlying the CRACK-NP model, which is illustrated in Figure 8.3, gives a fair description of the behavior of such soils. It assumes the soils to be made up of a series of layers of cubic peds separated by planar cracks. The change in terminology from "aggregates" to the related term "peds" is deliberate and reflects their nature and structure, particularly at depth in these soils.

Water falling on the soil surface infiltrates the surface layer of peds shown in Figure 8.3 such that the rate of infiltration, I_p, at time t is described by the first term of Philip's (1957) infiltration equation,

$$I_p = S/2t^{0.5} \tag{12}$$

where S is the ped sorptivity, which can be measured without the macropore component as described by Leeds-Harrison et al. (1994). Infiltration in excess of that able to infiltrate the peds in the surface layer initiates macropore flow and flows down the cracks in accordance with the Hagan-Poiseuille equation (Childs, 1969),

$$Q = d^3 g\rho \, (\text{grad } \psi)/12 \, \mu \tag{13}$$

where Q (volume per unit time per unit depth of crack) is the flow rate in a vertical planar crack of width d under a hydraulic gradient grad ψ, g is the acceleration due to

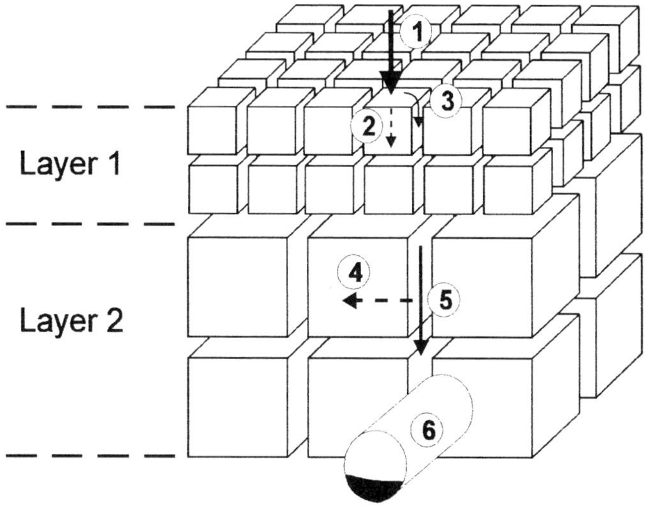

Figure 8.3. Diagrammatic representation of soil structure and hydrological processes in the CRACK-NP model. Rain may infiltrate the surface, but if its intensity (1) exceeds the infiltration capacity of the peds at the surface (2), it enters the cracks (3). Sorption by peds deeper in the soil (4) may lessen the amount of water flowing in the cracks (5). Flow will occur in the drain (6) if the local water table in the cracks rises.

gravity, ρ is the density of water, and μ the viscosity. g, ρ, and μ are constants and grad ψ can be assumed to be 1. The importance of macropores and other large pores in the transmission of water and solute is emphasized by the power to which d is raised. The cracks are assumed to form a grid, and the equation is implemented for the crack between each pair of nodes in the grid. All the downward flow is assumed to be in the cracks, with water remaining immobile in the peds. The equation is solved by an iterative relaxation technique that maintains continuity of flow between each pair of nodes. Because of the power to which it is raised, the crack width, d, is an important parameter, and it is determined by the shrinking and swelling of the peds. The initial crack width is a function of the difference between the initial water content of each layer and that at "field capacity," when the peds are fully swollen. Subsequently, individual peds are assumed to shrink or swell according to a swelling function defined by the slope of the line relating the void ratio to the volumetric water content. CRACK-NP assumes that there is a stable crack volume and width, and therefore some fissuring, in a fully swollen soil.

Solute movement is by mass flow in the water moving in the cracks or by diffusion into and out of the peds. The quantity of solute transported by mass flow is computed from the amount of water flowing in the cracks, computed as above and its solute concentration, which is determined by events at the surface and by the intraped diffusion. The diffusion into and out of the peds is simulated using the simple diffusion model described above.

Like the PEDAL model, CRACK-NP has been used to investigate the effect of ped size on solute losses from the soil. It simulated greater solute concentrations in water moving through a soil with small aggregates than a soil with larger ones (Figure 8.4a), as observed by Leeds-Harrison et al. (1992) and expected from Table 8.1, and smaller concentrations when the diffusion coefficient was lessened (Figure 8.4b). Matthews et al. (1997a) used mean crack spacings of 100 mm and 200 mm to simulate plowing and direct drilling, respectively, obtaining N losses by leaching, 31 and 27 kg N/ha/yr, that compared well with the measured losses of 33 and 27 kg N/ha/yr found by Catt et al. (1992) in the Brimstone Experiment, suggesting that the model may be useful for simulating the effects of cultivation practice.

EVALUATING THE MODELS

PEDAL

The PEDAL model, as noted above, was developed more for investigating the effects of aggregate size on solute losses than for management or similar purposes, and it has not been subjected to formal evaluation. However, two tests at very different scales are worth mentioning, one against a laboratory column of chalk cubes and the other against a 25-m field profile comprising topsoil, clay, and chalk.

In the laboratory experiment (Addiscott, 1984), chalk cubes from the batch used to test the diffusion model were packed into a column 400 mm high with a square cross-section, separated by thin layers of 1–2 mm quartz chips that enabled water to flow between them. The solutions within and between the cubes were initially 0.1*M* in calcium bromide. Distilled water equivalent to 2–12 mm aliquots of rain were applied at the top of the column, usually at 15-minute intervals but with much longer intervals overnight and at the weekend. Immediately after each aliquot was applied, an equivalent volume of solution was withdrawn at the bottom of the column, giving virtually instantaneous drainage, and analyzed for bromide using a previously calibrated bromide-specific electrode. Experiments were conducted with identical "rainfall" regimes for (a) 56 20-mm chalk cubes arranged in 14 layers separated by 0.5 mm of quartz chips, and (b) seven 40-mm cubes, each constituting a layer, separated by 1.25 mm of quartz chips.

The bromide concentration in the solution withdrawn at the bottom of the column decreased steadily during much of the experiment for both sizes of the cubes (Figure 8.5). There was a slight increase in bromide concentration (Point A) as bromide diffused out of the cubes during a 13-hour overnight break in the applications of water, and a very much larger increase during the 42- and 31-hour breaks at the weekend (Points B and C), when bromide had much longer to diffuse out. The larger cubes had the same total volume as the smaller ones, but their size held the bromide back longer and resulted in greater increases in concentration during the breaks from water application.

When PEDAL was used to simulate these results, the diffusion coefficient in free calcium bromide solution was set at 1.2×10^{-9} m^2 s^{-1} and the impedance factor at 0.4, the values used in the simulation of the experiment on bromide diffusion in the chalk cubes. The simulations for both sizes of cube represented the measured points quite well (Figure 8.5). The small increase in concentration at Point A and the much larger one at Point B were simulated with reasonable accuracy for both sizes. The columns were in effect experimental analogs of the model, so the agreement ought to have been satisfactory, but the results suggest that the conclusions drawn from the model about holdback in aggregates were probably soundly based.

During a collaborative exercise, a team from the Water Research Centre at Medmenham used the technique described by Young and Gray (1978) to take three 25-m cores through topsoil, heavy clay, and chalk at a site at Rothamsted plowed out of long-term grass 18 years previously. Although the cores were taken only a short distance from each other, the depths to the clay-chalk interface varied from 3 to 11 m. The PEDAL model was used to simulate the nitrate profiles found in all three cores (Addiscott, 1982b), but only the core with the clay-chalk interface at 3 m is shown here.

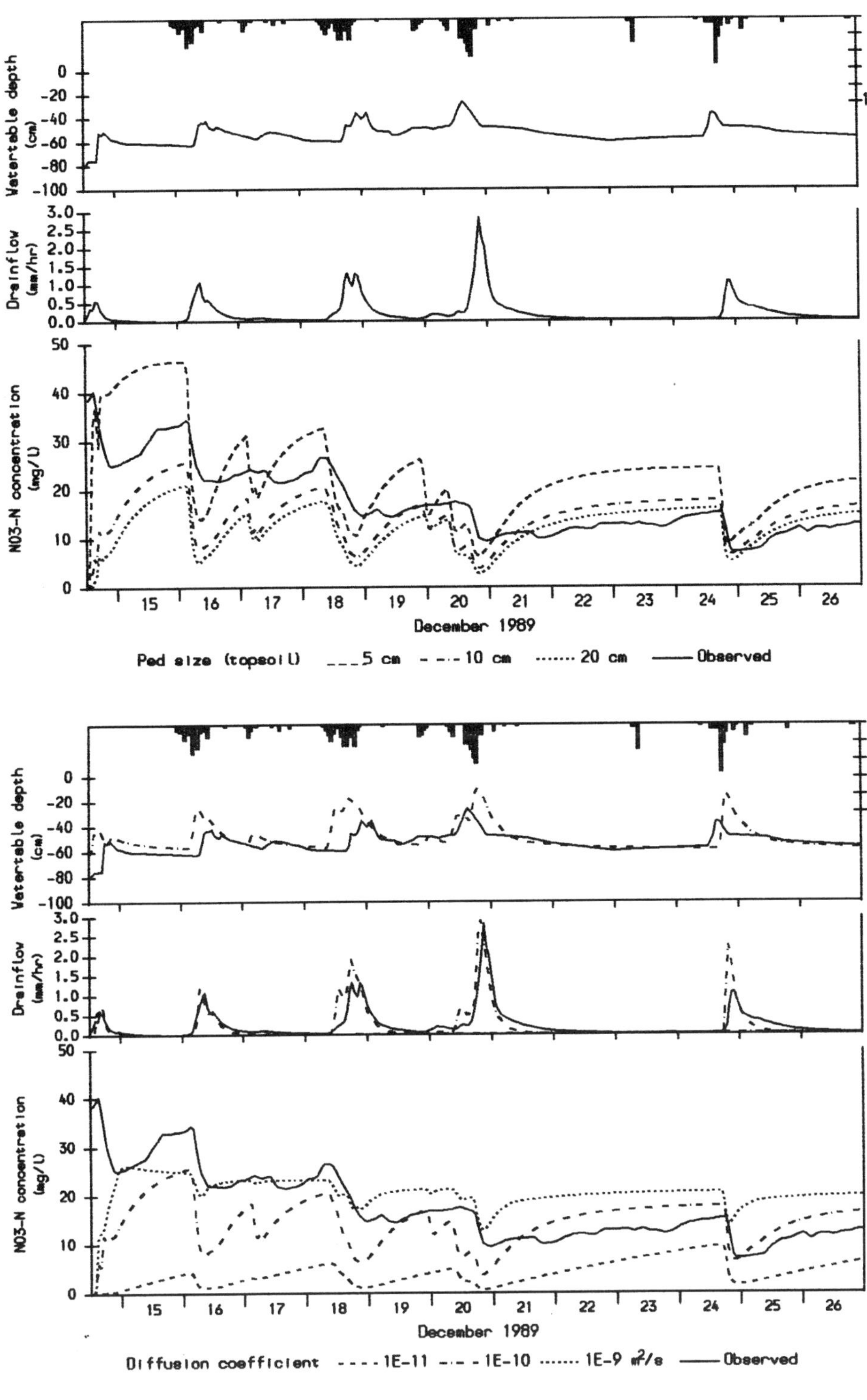

Figure 8.4. (a) Effect of different (uniform) topsoil ped sizes on simulated nitrate concentrations in drainage from the Brimstone Experiment during 12 days of December 1989. 50 mm (broken line), 100 mm (dash-dotted line), 200 mm (dotted line), Observed (solid line). (b) Effect of different intraped diffusion coefficients. 10^{-11} (broken line), 10^{-10} (dash-dotted line), 10^{-9} (dotted line) $m^2 s^{-1}$, Observed (solid line).

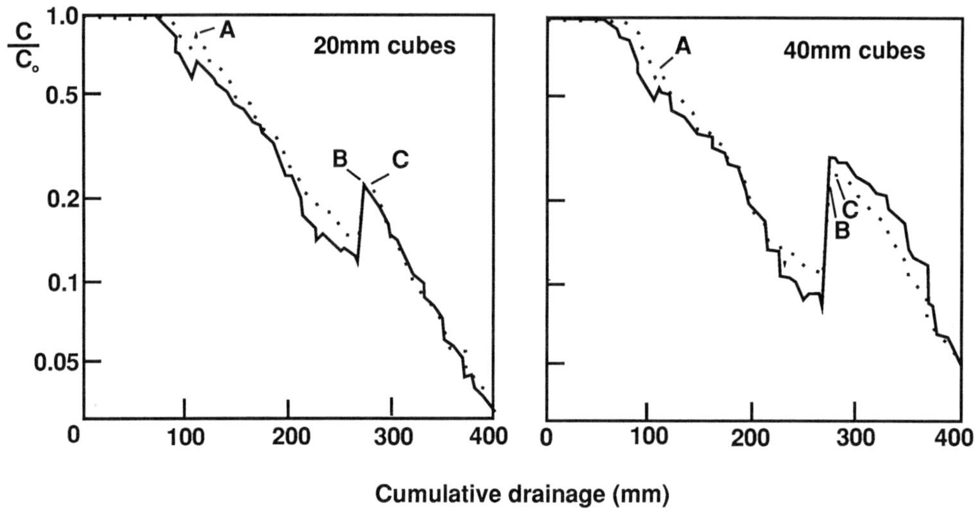

Figure 8.5. Ratio of bromide concentration in solution drained from the chalk cube columns to that initially in the columns (C/C_0). Measured (points) and simulated (line) ratios for 20 and 40 mm cubes plotted against cumulative drainage. A, B and C are the first measurements taken after pauses in the drainage of water from the column of 16, 42, and 22 hours, respectively. From Addiscott (1984) with permission.

The top meter of the core was mainly topsoil, and its aggregates were known to be likely to have a lognormal size distribution with a mean of 17 mm. There was no information about the size distributions further down, other than that the clay-chalk interface was at 3 m, so the second and third meters were assumed to have normally distributed 90 mm peds and the third and fourth meters (clay-chalk interface) lognormally distributed 10 mm peds. The remaining 21 meters were assigned normally distributed 200 mm structural units. These assumptions were in the nature of "best guesses" made in the light of observations on chalk cliffs in Kent, UK, and they were not manipulated (or "tweaked" or "fiddled") to improve the fit.

The clay and chalk were assigned porosities of 0.35 and 0.5 (Williams, 1978 and personal communication). The diffusion coefficient for calcium nitrate was assumed to be similar to the value of 1.2×10^{-9} m^2 s^{-1} used for calcium bromide. The impedance factor was assumed to be numerically similar to the value taken for the porosity. Rain and evaporation were measured about 200 m from the site, and the input of nitrate-N from the plowed-out grassland was estimated from the regression of the loss of total N from the topsoil ($N_0 - N_t$) on the square root of time, t.

$$(N_0 - N_t) = 922t^{1/2};\ r^2 = 0.9997 \tag{8.14}$$

The simulation of the nitrate profile in the core overestimated the nitrate concentration in the 5 m beneath the clay-chalk interface but was otherwise reasonably good (Figure 8.6). The same was true of the two profiles not shown. Because the sizes of the structural units could not be assessed *in situ*, there is a question as to whether the fit shown in Figure 8.6 constitutes a validation of the model, but it certainly shows that the model can be used reasonably successfully at widely differing scales. One reason for this is possibly that the size of the structural unit is a property which retains its meaning regardless of the scale at which the model is used.

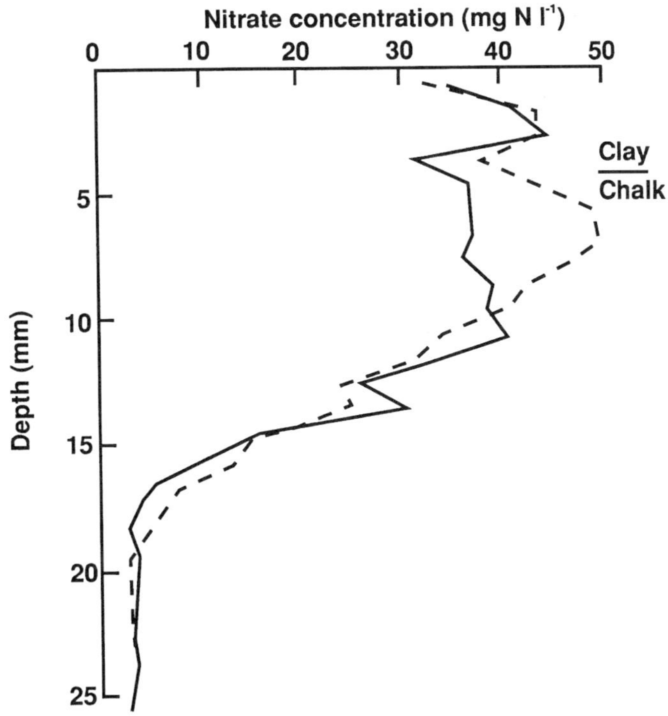

Figure 8.6. Measured (solid line) and simulated (broken line) nitrate concentrations in the interstitial water of a profile at Rothamsted comprising topsoil, clay and chalk. From Addiscott (1982b) with permission. Measurements made by the Water Research Centre, Medmenham.

CRACK-NP

The Brimstone Farm Experiment near Wantage in Oxfordshire has a heavy clay soil of the Denchworth Series. This soil cracks and swells, providing an excellent site for evaluating the CRACK-NP model. The plots are hydrologically isolated by barriers of heavy polyethylene, and are drained artificially by a system of mole and pipe drains. The subsoil is impermeable, and water leaves the plots either through the drains, as lateral flow at the base of the plow layer, or as surface runoff. These flows are measured and sampled for analysis automatically. Much the largest proportion of the water passes through the drains. The plots have a total area of 0.22 ha and an experimental area of 0.1 ha. The length of the representative area for leaching is 1 to 10 m, when macropore flow is likely, compared with 10 mm for micropore flow (Beven and Germann, 1982), so these plots are more likely than lysimeters to provide representative drainage samples in macroporous soils.

CRACK-NP has been evaluated against data for nitrate and pesticide leaching from several plots, and Figure 8.7 gives an example. Measurements of the depth to the water table, drainflow, and nitrate concentration in drainage were simulated for a two-week period in 1989. The drainflow was simulated quite well and the water-table depth somewhat less well. The simulations of the nitrate concentration underestimated the concentration at the start of the period, mainly because the amount of nitrate initially in the soil profile was not known. However, the simulations improved as the run progressed, and the timings of the increases and decreases in concentration were simulated well. The decreases in concentration coincided with drainflow events, as observed by Armstrong and Burt (1993).

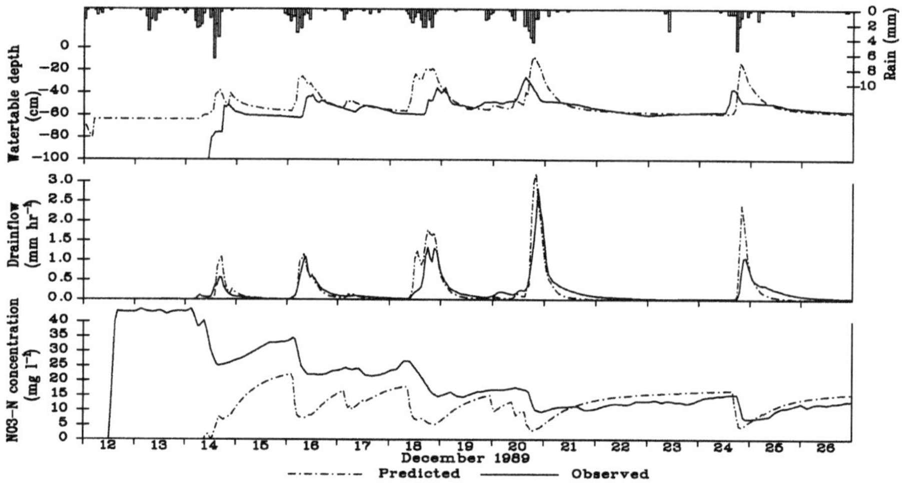

Figure 8.7. Evaluation of the CRACK-NP model against nitrate leaching data from the Brimstone Experiment during 12 days in 1989. Rain (histogram) and measured (solid line) and simulated (dash-dotted line) depth to water table, drainflow, and nitrate concentrations in drainage.

DISCUSSION

Validation

Neither of the leaching models has been subjected to statistically-based validation, but we were able to check whether the behavior they simulated tallied with reality. The PEDAL model suggested *inter alia* that aggregates had to have a cube side of at least 15 mm to give any discernible holdback. When Addiscott et al. (1983) measured the diffusion of nitrate out of aggregates from Rothamsted topsoil, they found that for 4–5 mm aggregates, the largest tested, M_t/M_∞ took 1600 seconds (less than half an hour) to reach 0.93. For periods of leaching measured in tens of days, 5 mm aggregates would give negligible holdback. These authors also investigated bromide diffusion from clods of the heavier Saxmundham soil ranging from 25 to 57 mm in equivalent cube side. In the smallest of these cubes, M_t/M_∞ took six hours to reach 0.8, while in the largest it reached only about 0.4 during that time. The sizes suggested in Table 8.1 to be needed for appreciable holdback thus look reasonable and give hope that the model is efficacious.

The CRACK-NP model's simulation of the Brimstone Experiment showed drops in concentration that coincided with those measured and with the peaks in drainflow suggesting that the model is reasonably efficacious. A similar conclusion can be drawn from the simulation of the effects of plowing and direct drilling mentioned in the section describing the model. One residual problem is that it is difficult to evaluate the model at more than one scale. The model has been shown to work reasonably well at the plot scale, but we do not know how well it simulated processes at the ped scale, because these processes could not be measured *in situ*.

Ped Sizes and Size Distributions

It has been possible to draw some conclusions about the effect of ped size on the holdback of solute against leaching. One is that peds need to have a size (cube side) of at least 15 mm to give appreciable holdback. Another is that conclusions cannot be drawn without reference to the amount, and particularly the pattern, of rainfall. With 20 2.5-mm aliquots of rain, the aggregates needed to be about twice as large to give appreciable holdback as with five 10-mm aliquots. Whether the peds were uniformly sized or had normal or lognormal size distributions also seemed important, for the reason outlined earlier.

The very large degree of holdback given by a 1000-mm structural unit raises an interesting question. Heavy clay subsoils with fissures may have structural units approaching this order of magnitude. It will take a very long time for solutes to diffuse into or out of them. If the subsoil is unsaturated so that water flows over only a small part of them, there will be very little solute exchange, but if the subsoil is saturated for part of the year, there will be much more diffusion. It could be that pollutants are accumulating slowly in these structural units. If so, they will eventually diffuse out, and because of the size of the units, they will go on diffusing out for a very, very long time. This may, of course, already be happening.

ACKNOWLEDGMENTS

We are grateful to colleagues and collaborators whose ideas and efforts helped in the preparation of this chapter. N.J. Jarvis kindly made available his version of the CRACK model and provided help and encouragement. A.M. Matthews and A.M. Portwood made valuable contributions to its development to CRACK-NP, and the former drew Figure 8.3. V.H. Thomas helped with the development and use of the PEDAL model. G.L. Harris and J.A. Catt provided data from the Brimstone Experiment. The chapter drew on research funded by the UK Ministry of Agriculture, Fisheries and Food and the Natural Environment Research Council, whose support is acknowledged gratefully. IACR-Rothamsted receives grant-in-aid from the Biotechnology and Biological Sciences Research Council of the United Kingdom.

REFERENCES

Addiscott, T.M. A simple computer model for leaching in structured soils. *J. Soil Sci.,* 28:554–563, 1977.

Addiscott, T.M. Simulating diffusion within soil aggregates: A simple model for cubic and other regularly shaped aggregates. *J. Soil Sci.,* 33:37–45, 1982a.

Addiscott, T.M. Modelling nitrate movement in profiles containing soil, heavy clay and chalk, In *Modelling Agricultural-Environmental Processes in Crop Production,* Golubev, G. and I. Schvytov (Eds.), Laxenburg: IIASA, 1982b.

Addiscott, T.M. Modelling the interaction between solute leaching and intra-ped diffusion in heavy clay soils, In *Proceedings of the ISSS Symposium on Water and Solute Movement in Heavy Clay Coils,* Bouma, J. and P.A.C. Raats (Eds.), Wageningen: ILRI (ILRI Publication No 37). pp. 279–292, 1984.

Addiscott, T.M. and R.J. Wagenet. Concepts of solute leaching in soils: A review of modelling approaches. *J. Soil Sci.,* 36:411–424, 1985a.

Addiscott, T.M. and R.J. Wagenet. A simple method for combining soil properties that show variability. *Soil Sci. Soc. Am. J.,* 49:1365–1369, 1985b.

Addiscott, T.M. and A.P. Whitmore. Simulation of solute leaching in soils of differing permeabilities. *Soil Use Manage.,* 7:94–102, 1991.

Addiscott, T.M., D.A. Rose, and J. Bolton. Chloride leaching in the Rothamsted Drain Gauges: Influence of rainfall pattern and soil structure. *J. Soil Sci.,* 29:305–314, 1978.

Addiscott, T.M., V.H. Thomas, and M.A. Janjua. Measurement and simulation of anion diffusion from natural soil aggregates and clods. *J. Soil Sci.,* 34:709–721, 1983.

Addiscott, T.M., A.P. Whitmore, and D.S. Powlson. *Farming, Fertilizers and the Nitrate Problem.* CAB International, Wallingford, 1991.

Armstrong, A.C. and T.P. Burt. Nitrate losses from agricultural land, in *Nitrate: Processes, Patterns and Management,* Burt, T.P., A.L. Heathwaite, and S.T. Trudgill (Eds.), Wiley and Sons, Chichester, pp. 239–267, 1993.

Armstrong, A.C., T.M. Addiscott, and P.B. Leeds-Harrison. Methods for Modelling Solute Movement in Structured Soils, In *Solute Modelling in Catchment Systems,* Trudgill, S.T. (Ed.), Wiley and Sons, Chichester, pp. 134–161, 1995.

Armstrong, A.C., A.M. Matthews, A.M. Portwood, N.J. Jarvis, and P.B. Leeds-Harrison. *CRACK-NP: A Model to Predict the Movement of Water and Solutes from Cracking Clay Soils.* Version 2.0. Technical Description and User's Guide. ADAS Land Research Centre, Gleadthorpe, Meden Vale, UK, 1996.

Barraclough, D. A usable mechanistic model of nitrate leaching. I. The model. *J. Soil Sci.,* 40:543–554, 1989a.

Barraclough, D. A usable mechanistic model of nitrate leaching. II. Applications. *J. Soil Sci.,* 40:555–562, 1989b.

Beven, K. Micro-, meso- and macro-porosity and channelling phenomena in soils. *Soil Sci. Soc. Am. J.,* 45:1245, 1981.

Beven, K. and P.F. Germann. Water flow in macropores. II. A combined flow model. *J. Soil Sci.,* 32:115–129, 1981.

Beven, K. and P.F. Germann. Macropores and water flow in soils. *Water Resour. Res.,* 18:1311–1325, 1982.

Bouma, J. Soil morphology and preferential flow along macropores. *Agric. Water Manage.,* 3:235–250, 1981.

Bouma, J. and L.W. Dekker. A case study on the infiltration into dry clay soil. I. Morphological observations. *Geoderma* 20:27–40, 1978.

Catt, J.A., D.G. Christian, M.J. Goss, G.L. Harris, and K.R. Howse. Strategies to reduce nitrate leaching by crop rotation, minimal cultivation and straw incorporation in the Brimstone Farm Experiment, Oxfordshire. *Ann. Appl. Biol.* 30:255–262, 1992.

Cheng, H.H. Pesticides in the soil environment—An overview, In *Pesticides in the Soil Environment: Process, Impacts and Modelling,* Cheng, H.H. (Ed.), Soil Science Society of America, Madison, WI, 1990.

Childs, E.C. *An Introduction to the Physical Basis of Soil Water Phenomena.* Wiley, London, 1969.

Crank, J. *The Mathematics of Diffusion.* Clarendon Press, Oxford, 1956, pp. 88–91.

De Smedt, F., F. Wauters, and J. Sevilla. Study of tracer movement through unsaturated sand. *Geoderma,* 38:223–236, 1986.

Epstein, B. Logarithmico-normal distribution in breakage of solids. *Indust. Eng. Chem.,* 40:2289–2291, 1948.

Gardner, W.R. Representation of soil aggregate size distribution by a logarithmic-normal distribution. *Soil Sci. Soc. Am. Proc.,* 20:151–153. 1956.

Hallard, M. and A.C. Armstrong. Observations of water movement to and within mole drainage channels. *J. Agric. Eng. Res.,* 52:309–315, 1992.

Jarvis, N.J. *CRACK: A Model of Water and Solute Movement in Cracking Clay Soils.* Report 159 Swedish University of Agricultural Sciences. Department of Soil Sciences, Uppsala, 1989.

Jarvis, N.J. and P.B. Leeds-Harrison. Modelling water movement in drained clay soil. I. Description of the model, sample output and sensitivity analysis. *J. Soil Sci.,* 38:487–498, 1987a.

Jarvis, N.J. and P.B. Leeds-Harrison. Modelling water movement in drained clay soil. II. Application of the model in Evesham series clay soil. *J. Soil Sci.,* 38:499–509, 1987b.

Lawes, J.B., J.H. Gilbert, and R. Warington. On the amount and composition of drainage water collected at Rothamsted. III. The quantity of nitrogen lost by drainage. *J. Royal Agric. Soc. England, 2nd Series* 18:43–71, 1882.

Leeds-Harrison, P.B., B.J. Vivian, and T.C.H. Chamen. Tillage effects on drained clay soils. *ASAE Paper* 92-2648, 1992.

Leeds-Harrison, P.B., E.G. Youngs, and B. Udin. A device for measuring the sorptivity of soil aggregates. *Eur. J. Soil Sci.,* 45:269–272, 1994.

Matthews, A.M., A.M. Portwood, A.C. Armstrong, P.B. Leeds-Harrison, G.L. Harris, J.A. Catt, and T.M. Addiscott. CRACK-NP. Development of a model for predicting pollutant transport at the Brimstone Farm site, Oxfordshire, UK. *Soil Use and Management,* 1997a.

Matthews, A.M., A.C. Armstrong, P.B. Leeds-Harrison, G.L. Harris, and J.A. Catt. Development and testing of a model for predicting tillage effects on nitrate leaching from a cracked clay soil. *Soil Tillage Res.,* 1997b.

Nye, P.H. and P.B.H. Tinker. *Solute Movement in the Soil-Root System.* Blackwell, Oxford. 1977.

Passioura, J.B. Hydrodynamic dispersion in porous media. I. Theory. *Soil Sci.,* 111:339–344, 1971.

Passioura, J.B. and D.A. Rose. Hydrodynamic dispersion in porous media. II. Effects of velocity and aggregate size. *Soil Sci.,* 111:345–351, 1971.

Philip, J.R. The theory of infiltration. 4. Sorptivity and algebraic infiltration equations. *Soil Sci.,* 84:257–264, 1957.

Rao, P.S.C., R.E. Jessup, and T.M. Addiscott. Experimental and theoretical aspects of solute diffusion in spherical and nonspherical aggregates. *Soil Sci.,* 133:342–349, 1982.

Rao, P.S.C., R.E. Jessup, D.E Rolston, J.M. Davidson, and D.P. Kilcrease. Experimental and mathematical description of nonabsorbed solute transfer by diffusion in spherical aggregates. *Soil Sci. Soc. Am. J.,* 44:684–688. 1980a.

Rao, P.S.C., D.E. Rolston, R.E. Jessup, and J.M. Davidson. Solute transport in aggregated porous media: Theoretical and experimental evaluation. *Soil Sci. Soc. Am. J.,* 44:1139–1146, 1980b.

Smith, K.A. Soil aeration. *Soil Sci.,* 123:284–291, 1977.

Van Genuchten, M.Th. and P.J. Wierenga. Mass transfer studies in porous sorbing media. I. Analytical solutions. *Soil Sci. Soc. Am. J.,* 40:473–480, 1976.

Van Genuchten, M.Th. and P.J. Wierenga. Mass transfer studies in porous sorbing media. II. Experimental evaluation with tritium (^3H_2O). *Soil Sci. Soc. Am. J.,* 41:272–278, 1977.

Wagenet, R.J. Principles of salt movement in soils, In *Chemical Mobility and Reactivity in Soil Systems,* Nelson, D.W. et al. (Eds.), Special Publication 11, American Society of Agronomy, Madison, WI, 1983.

Whitmore, A.P. A method for assessing the goodness of computer simulation of soil processes. *J. Soil Sci.,* 42:289–299, 1991.

Williams, R.J.B. Effects of management and manuring on physical properties of some Rothamsted and Woburn soils. *Report of the Rothamsted Experimental Station for 1977,* Part 2, pp. 37–54, 1978.

Young, C.P. and E.M. Gray. *The Distribution of Nitrate in the Chalk and Triassic Sandstone Aquifers.* WRC Technical Report No. 69. WRC, Medmenham, Bucks, 1978.

Youngs, E.G. and P.B. Leeds-Harrison. Aspects of transport processes in aggregated soils. *J. Soil Sci.,* 41:665–675, 1990.

CHAPTER NINE

Experimental Techniques for Confirming and Quantifying Physical Nonequilibrium Processes in Soils

P.M. Jardine, R. O'Brien, G.V. Wilson, and J.-P. Gwo

INTRODUCTION

Local equilibrium during solute transport in soils suggests that the interchange or transformation of mass in a heterogeneous system is instantaneous, or that the residence time of infiltrating water and solutes is sufficiently large so that hydraulic and concentration gradients are negligible within the system. In terms of physical processes, the validity of the local equilibrium assumption depends on the degree of interaction between macroscopic transport properties (i.e., water flux, hydrodynamic dispersion) and microscopic physical properties (i.e., diffusional mass transfer, aggregate size, length scales between pore classes). Often, microscopic physical properties impose a rate-limiting constraint on the transport of solutes in heterogeneous soil systems and deviations from local equilibrium occur. These conditions of physical nonequilibrium can arise when (1) the diffusion of mass into the Neurst film of water that surrounds soil particles is slow relative to the bulk solute flux and (2) the soil aggregate size or pore-class heterogeneity is sufficiently large to impose concentration and hydraulic gradients on the system (Valocchi, 1985; Parker and Valocchi, 1986).

The occurrence of physical nonequilibrium in soils was evidenced more than a century ago (Schumacher, 1864; Lawes et al., 1882); however, its significance was overlooked until about 20 to 30 years ago (Whipkey, 1965; 1969; Elrick and French, 1966; Kissel et al., 1973; McMahon and Thomas, 1974; Shuford et al., 1977; DeVries and Chow, 1978; Shaffer et al., 1979). Based on studies dealing with water and solute movement in undisturbed soils, it became increasingly apparent that the physical structure of the soil media often controlled solute fate and transport. The soils literature has become inundated with both laboratory and field investigations that document the preferential movement of solutes coupled with time-dependent mass exchange within the heterogeneous structure of media. At the laboratory scale, observable clues of such physical nonequilibrium conditions are evidenced from column studies showing the asymmetric breakthrough of solute concentration with time (Ritchie et al., 1972; Bouma and Wosten, 1979; Nkedi-Kizza et al., 1982, 1983; German et al., 1984; Smettem, 1984; Selim et al., 1987; Seyfried and Rao, 1987; Jardine et al., 1988; 1993a,b; Anamosa et al., 1990). Observations of physical nonequilibrium at the field scale are noted from the irregular distribution of solutes discharged from and "stored" within the media (Shuford et al., 1977; Shaffer et al., 1979; Johnston et al., 1983; Jardine et al., 1989; 1990a,b; McDonnell, 1990; Luxmoore et al., 1990; Wilson et al., 1991a,b; 1993; Roth et al., 1991; Hornberger et al., 1991). More often than not, experiments involving physical nonequilibrium are performed under a single set of conditions making it difficult, if not impossible, to accurately mathematically quantify the observations. A single experimental condition by itself is typically not sensitive enough to definitively document or quantify the condition of physical nonequilibrium. In this chapter we present a variety of laboratory and field experimental techniques that lend credibility to the observation of physical nonequilibrium processes in heterogeneous soil systems. The use of these techniques not only improves our conceptual understanding of solute migration in soils, it also provides the necessary constraints that are needed for the accurate numerical quantification of the physical nonequilibrium processes.

CONTROLLING FLOW-PATH DYNAMICS

Variations in Pore-Water Flux

A relatively simple technique for confirming and quantifying physical nonequilibrium in soil systems involves displacement experiments performed at a variety of experimental fluxes using a single representative tracer. Alteration of the experimental flux or specific discharge through a soil system perturbs the rate of approach toward equilibrium by changing the hydraulic or concentration gradient. In heterogeneous systems that exhibit a large distribution of pore sizes, an increase in the overall pore-water flux should result in greater system nonequilibrium due to a decrease in solute residence time within the porous media. This is usually the case since solute movement into the matrix is a combination of advective and diffusive processes, and is typically the rate-limiting step as the system approaches equilibrium.

In homogeneous systems, such as a well-packed bed of silica sand, where pore size heterogeneities are limited, physical nonequilibrium may still control the rate of solute interaction with the solid phase. This results from diffusion limitations through the film of water that surrounds particles in solution (Neurst film). The thickness of the film is dependent on the bulk fluid flux through the system, with conditions of more rapid flux resulting in thinner films surrounding the particles. Thus, the significance of physical nonequilibrium in a homogeneous system decreases with increased flux; exactly opposite of what is observed in heterogeneous soil systems. Often for homogeneous systems solute diffusion through particle water films is rapid, and changes in solute breakthrough as a function of pore-water flux are difficult to detect macroscopically.

For conditions of very low solute flux, axil diffusion (movement in response to a longitudinal concentration gradient) can become the predominant source of dispersion in both heterogeneous and homogeneous systems. However, in heterogenous systems competing processes such as matrix diffusion may limit the dominance of axil diffusion.

The technique of varying pore-water flux for defining physical nonequilibrium processes in soil systems has been used by a number of investigators (Lai and Jurinak, 1972; Nkedi-Kizza et al., 1982; Akratanakul et al., 1983; Hornberger et al., 1991; Schulin et al., 1987; Hu and Brusseau, 1995; Reedy et al., 1996). Elrick and French (1966) used the technique on small undisturbed columns and repacked columns containing air-dried sieved soil. The effluent breakthrough curves for Cl^- in the undisturbed soil differed with changes in pore-water flux, whereas little difference in tracer breakthrough was noted for the disturbed soil under the same conditions. The pore-water velocity effect noted for the undisturbed media can be attributed to physical nonequilibrium processes since the structure of the soil media was preserved and the pore-class heterogeneity was maintained. Similarly, Nkedi-Kizza et al. (1982) found that ^3H_2O breakthrough in an aggregate oxisol soil became increasingly asymmetric as the bulk pore-water flux increased. The rapid initial breakthrough and extended tailing of this nonreactive tracer was an indication of physical nonequilibrium between advection-dominated pore domains and the soil matrix pores within the aggregates.

O'Brien (1994, Oak Ridge National Laboratory, unpublished data) also used variations in pore-water flux to assess the significance of physical nonequilibrium processes in large undisturbed columns of weathered fractured shale (Figure 9.1a,b). Using Br^- as a nonreactive tracer at conditions of constant hydraulic gradient, the authors varied the experimental flux nearly 30 orders of magnitude which was consistent with in situ measurements for groundwater infiltration and recharge in the area of study. As is typical of heterogeneous media, tracer displacement was characterized by an initial rapid solute breakthrough which was followed by extended tailing to longer times (Figure 9.1a). In this system, the fracture network of weathered shale controlled the advective transport of solutes which was coupled with diffusion into the surrounding matrix blocks. The largest and smallest flux experiments were conducted over periods of 0.25 d and 94 d, respectively, with the relative amount of tracer mass remaining in the column at the end of each pulse ranging from 22% (fast flux) to 38% (slow flux). These results indicated that the system became increasingly removed from equilibrium as the pore-water flux was increased. At the conditions of faster flux, the tracer residence times in the mobile fracture regions were significantly decreased, and thus not as much mass was lost to the matrix. However, mass loss to the matrix as a function of pore-water flux was highly nonlinear, suggesting that the rate of mass transfer from fractures to the matrix is greater at larger fluxes. The velocity dependence of mass-transfer processes in various porous media and soils has been shown by others (Nkedi-Kizza et al., 1983; Akratanakul et al., 1983; Jensen, 1984; Schulin et al., 1987; Anamosa et al., 1990; Kookana et al., 1993; Reedy et al., 1996).

The results of O'Brien (1994) which show extensive physical nonequilibrium during solute transport (Figure 9.1 a,b) are compared to experiments by Hu and Brusseau (1994) who used a homogeneous nonreactive silica sand to demonstrate the effect of pore-water flux on tracer transport (Figure 9.1c). Pentafluorobenzoic acid (PFBA) was used as a nonreactive tracer, and the pore-water flux was varied over 100 orders of magnitude. For conditions where axil diffusion was negligible, pore-water flux had no effect on the transport of PFBA, which suggested that the system was not affected by physical nonequilibrium processes. This is consistent with the fact that the silica bed was macroscopically homogeneous with limited pore-size variability. Where the pore-water flux became exceedingly slow (i.e., <0.1 cm/h), axil diffusion became a significant contributor to the overall dispersion in the system and the PFBA breakthrough curves showed earlier breakthrough and a delayed approach to equilibrium (Figure 9.1c).

Quantifying physical nonequilibrium processes at the field scale using variations in pore-water flux are rare due to the difficulty of providing enough water to drive solute mobility within the system. Hornberger et al. (1991) used variations in pore-water flux to assess the significance of physical nonequilibrium in a forested field soil in Orono, ME. The authors isolated a 3x9x1 m deep soil block and applied a Br^- tracer

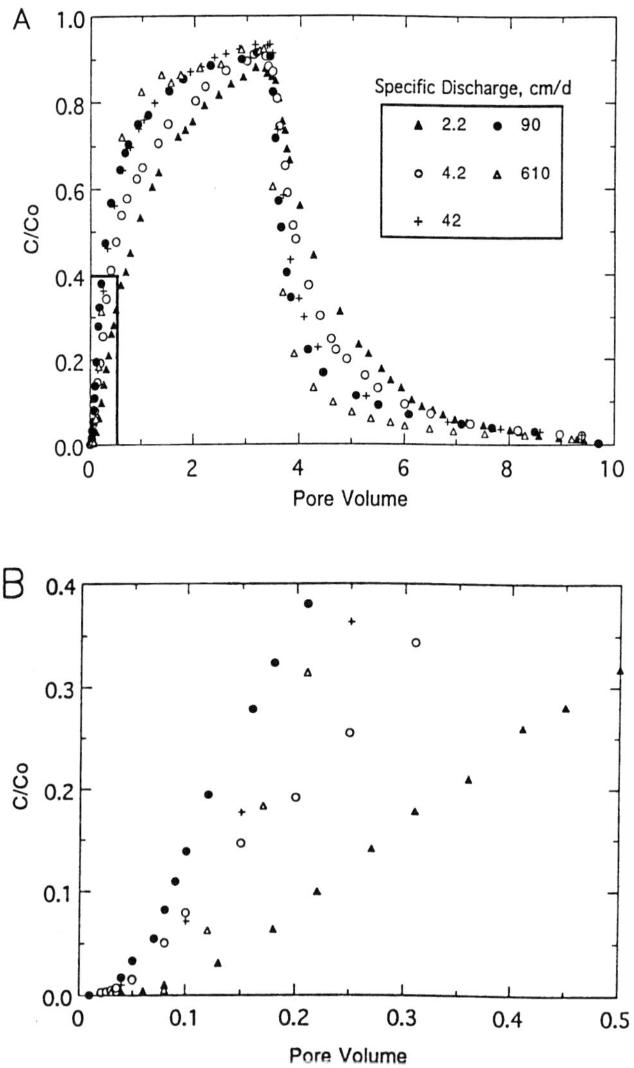

Figure 9.1. (a) Bromide breakthrough curves for a series of steady-state specific discharges in an undisturbed column of weathered, fractured shale. The largest and slowest flux experiments were conducted over periods of 6 and 2200 hr, respectively. The rectangle in the lower corner defines the expanded portion of the plot shown in (b). From O'Brien, R., 1994, Oak Ridge National Laboratory, unpublished data. With permission.

using sprinkler irrigation at three application rates (2.5, 5, and 10 cm/h). The total mass of Br^- applied in each experiment was constant. The flux-averaged breakthrough curves of Br^- for the entire soil block were significantly more asymmetric at the higher infiltration rates, suggesting enhanced preferential flow of solutes and marked physical nonequilibrium (Figure 9.2). Solute breakthrough curves were dominated by a single peak, with solute travel times being inversely proportional to the application rate. A secondary peak in the outflow curve was observed in all cases and is conceptually consistent with flow and transport occurring in at least two distinct pore size classes (Figure 9.2).

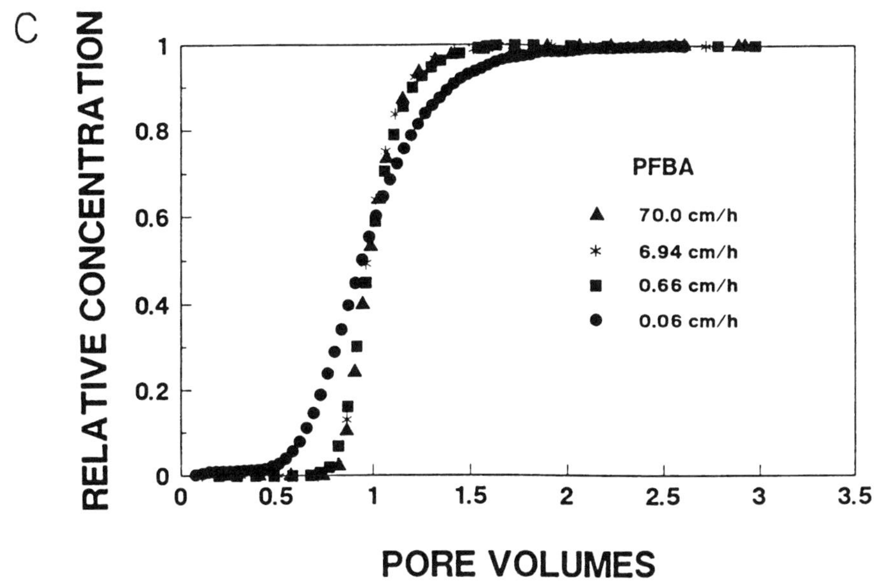

Figure 9.1. (c) Pentafluorobenzoic acid (PFBA) breakthrough curves for a series of steady-state specific discharges in a column of homogeneous nonreactive silica sand. From Hu, Q. and M.L. Brusseau, *J. Hydrol.,* 158:305–317, 1994. With permission.

Controlling flow-path dynamics through variations in pore-water flux is a relatively simple technique for assessing the significance of physical nonequilibrium in soil systems. However, when used by itself the technique is semiquantitative since it is difficult to know how system variables change in response to flux variations (e.g., are the proportion of advective flow paths constant with changes in flux?). When this technique is combined with others described below, it can become a powerful means of quantifying physical nonequilibrium processes in soil systems (see Hu and Brusseau, 1995, and Reedy, 1996 as examples).

Variations in Pressure-Head

Another technique that is used for controlling flow-path dynamics for the purpose of quantifying physical nonequilibrium processes, involves the manipulation of the soil water content with pressure-head variations. The basic concept of the technique is to collect water and solutes from select sets of pore-classes in order to determine how each set contributes to the bulk flow and transport processes that are observed for the whole system. In heterogeneous systems, a decrease in pressure-head (more negative) will cause larger pores to drain and become nonconductive during solute transport. Since advective flow processes tend to dominate in large pore regimes, a decrease in pressure-head, which will restrict flow and transport to smaller pores, will limit the disparity of solute concentrations among pore groups. By minimizing the concentration gradient in the system, the extent of physical nonequilibrium is decreased.

There are several techniques that are used in laboratory and field investigations for collecting water and solutes from select sets of pore classes. In the laboratory, pressure-heads are imposed upon soil samples or columns so that pore class sets can be either included in, or excluded from the tracer displacement experiment. Fritted glass or stainless steel endplates are typically attached to both ends of the soil column and a

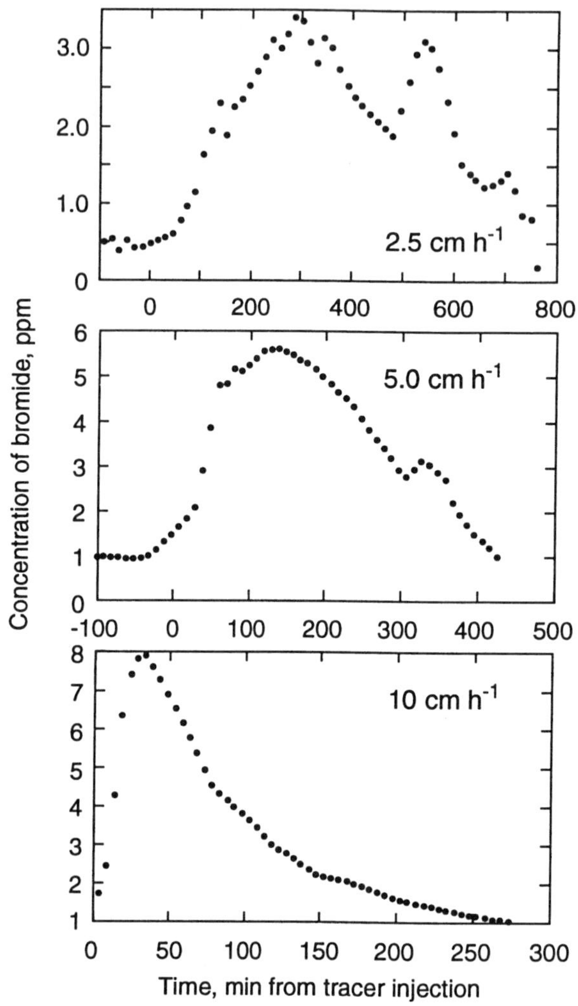

Figure 9.2. Flux-averaged bromide breakthrough curves for a series of steady-state infiltration rates on an undisturbed *in situ* pedon. Note the change in the x- and y-axes scales for the three figures. Modified from Hornberger, G.M., et al., *J. Hydrol.,* 124:81–99, 1991. With permission.

pressure-head is imposed at the top using a constant head marriott device, and a pressure-head is imposed at the bottom using either a hanging water head or a vacuum source (Figure 9.3). It is important that the bubbling pressure of the frit is not so high as to control the conductivity of the drainage profile within the column. For most structured media where preferential flow controls physical nonequilibrium processes, frits should be used that allow water to freely pass without suction. These frits are of the coarse variety (25–50 μm maximum pore diameter) and generally have bubbling pressures <–30cm (less negative). If only the flow characteristics of the matrix porosity are of interest, then finer frits can be used that have higher bubbling pressures.

Seyfried and Rao (1987) quantified physical nonequilibrium processes in an aggregated tropical soil by controlling ^3H_2O dynamics in undisturbed columns using pressure-head variations. The authors used a similar setup as that shown in Figure 9.3 except a hanging water column was used to apply tension at the column exit. Tracer breakthrough curves were found to be asymmetric in shape when pressure heads be-

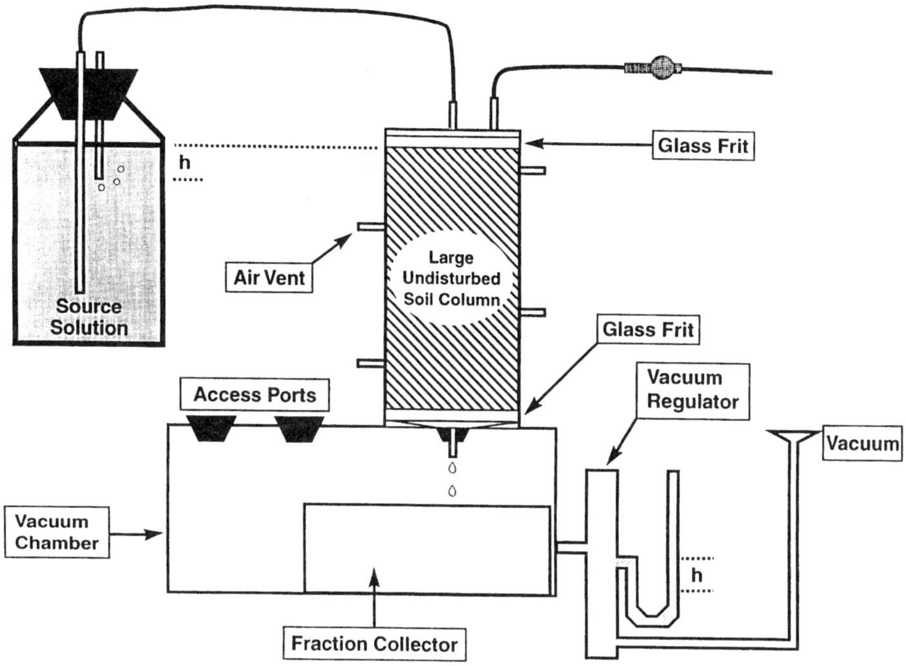

Figure 9.3. Schematic diagram of a saturated-unsaturated column flow apparatus illustrating a constant pressure-head (h) design on the column inlet and outlet boundaries. From Jardine, P.M., et al., *Soil Sci. Soc. Am. J.*, 57:945–953, 1993a. With permission.

tween 0 and –10 cm were applied (Figure 9.4a,b). Breakthrough curves became much more symmetric when displacement experiments were performed using pressure heads smaller than –10 cm (more negative). Similar results were found by Jardine et al. (1993a) for solute transport processes in undisturbed columns of weathered fractured shale (Figure 9.4c). Both of these studies conducted displacement experiments under a unit hydraulic gradient so that physical nonequilibrium processes could be independently addressed. The increasing asymmetry of the breakthrough curves with increasing saturation (less negative pressure head) was indicative of enhanced preferential flow. As the soils became increasingly unsaturated, breakthrough curve tailing became less significant because of a decrease in the participation of larger pores involved in the transport process (Figures 9.4a,b,c) . Effluent tracer concentrations were modeled with the classical convective-dispersive (CD) equation. This model could not describe data for experiments conducted between 0 and –10 cm pressure head, suggesting significant solute mass-transfer limitations between pore classes during these conditions. Seyfried and Rao (1987) used the mobile-immobile model to describe these data, and Gwo et al. (1995) used a multiregion flow and transport model to simulate the data of Jardine et al. (1993a). The CD equation was found to adequately describe observed effluent concentrations during conditions <–10 cm pressure-head (Figures 9.4b,c). This suggested that mass transfer limitations (physical nonequilibrium) became increasingly negligible for these unsaturated conditions because macropore flow was eliminated. It is interesting to note in both soil systems described above, the application of tension resulted in a 5- to 40-fold decrease in the mean pore water flux with relatively little change in the soil-water content. These results suggest that most of the water and solute flux is channeled through pores that hold water with tensions >10 cm (macropores; see Luxmoore, 1981), even though their surface area and contribution

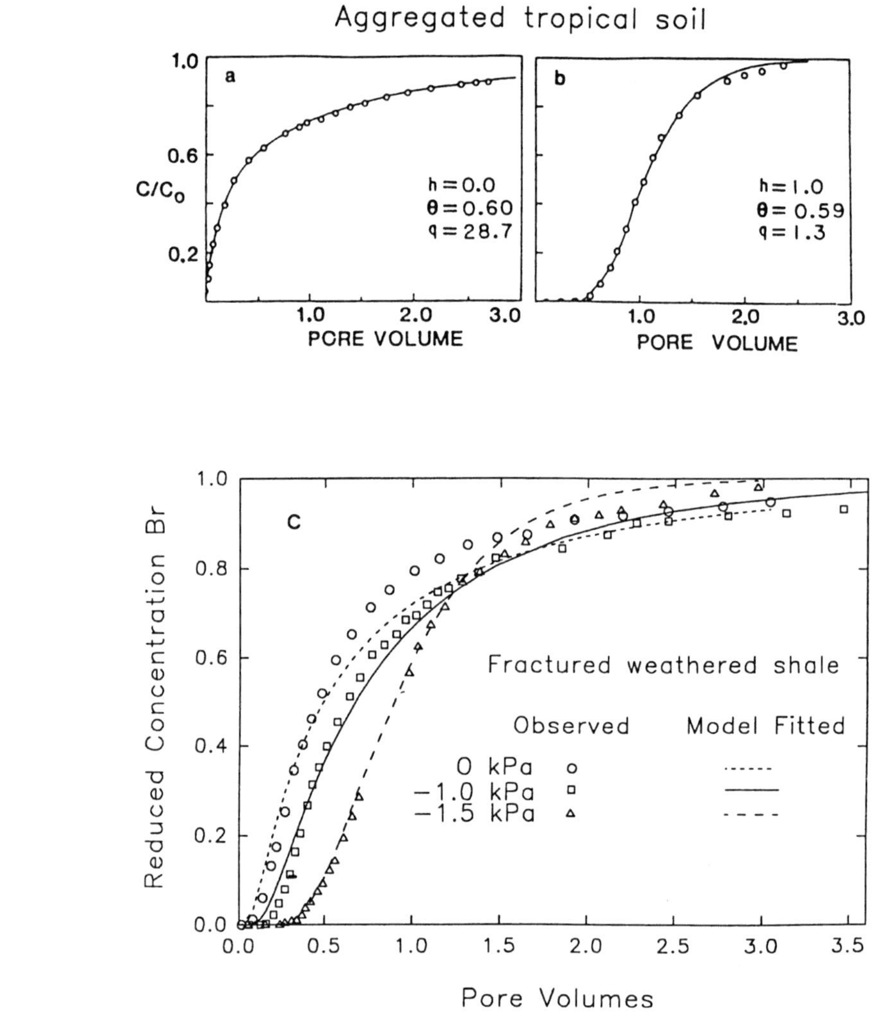

Figure 9.4. (a and b) Breakthrough curves for a nonreactive ^3H_2O tracer as a function of soil water tension (h) in an aggregated tropical soil. The solid lines represent the best-fit curves from the mobile-immobile model at h=0 and the classical convective-dispersive model at h=1.0 kPa. Modified from Seyfried, M.S. and P.S.C. Rao, *Soil Sci. Soc. Am. J.*, 51:1434–1444, 1987. With permission. (c) Breakthrough curves for a nonreactive Br^- tracer as a function of pressure-head in a fractured weathered shale. The model-fitted curves used the classical convective-dispersive model with optimization of the dispersion coefficient to the observed data. Modified from Jardine, P.M. et al., *Soil Sci. Soc. Am. J.*, 57:945–953, 1993. With permission.

to the total system porosity is very small (Wilson and Luxmoore, 1988; Wilson et al., 1989, 1992).

At the field-scale, tension infiltrometers have been used to assess infiltration rates as a function of soil porosity (Watson and Luxmoore, 1986; Wilson and Luxmoore, 1988; Clothier et al., 1992; Jaynes et al., 1996). The technique is similar to that described above for columns, where new water infiltrates into the soil surface under tension, thereby controlling the flow-path dynamics of the system (Figure 9.5). Increasing tensions (more positive values) exclude increasingly smaller pore class sets, where the largest pores drain first. This technique is useful for estimating macroporosity

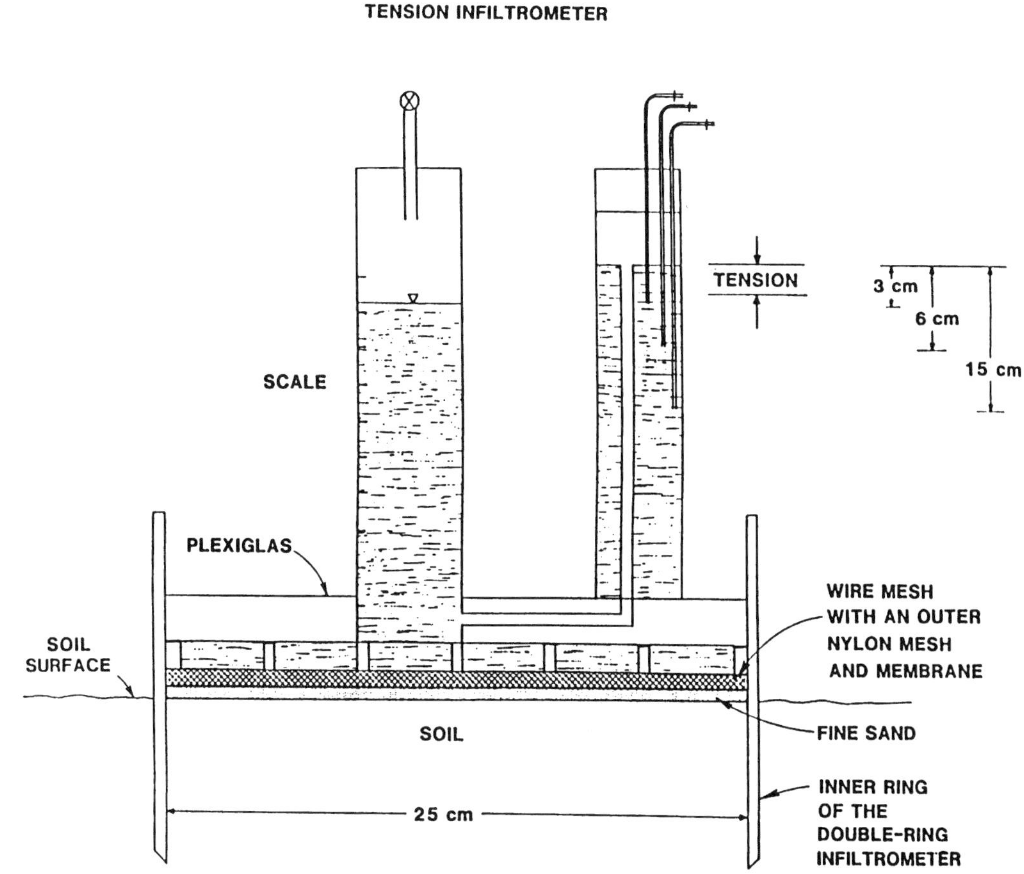

Figure 9.5. Schematic diagram of a tension infiltrometer showing designated tensions of 3, 6, and 15 cm water which exclude pores > 0.1, 0.05, and 0.02 cm in diameter from the transport process. From Watson, K.W. and R.J. Luxmoore, *Soil Sci. Soc. Am. J.*, 50:578–582, 1986. With permission.

and mesoporosity distributions in the field and the prevalence of preferential flow processes. In most instances, tracer-free water is used as the infiltrating solution, and only qualitative estimates can be made of physical nonequilibrium processes. Recently, however, Clothier et al. (1992) and Jaynes et al. (1995) added single and multiple nonreactive tracers, respectively, to the infiltrating solutions of the tension infiltrometer in order to quantify physical nonequilibrium processes in the field. The technique of Jaynes et al. (1995) is designed to provide a field-relevant measure of the mobile and immobile water fractions and the first-order mass transfer coefficient that accounts for mass exchange between the pore regions. The technique is described in detail in Chapter 11 by Jayne and Horton in this book, and it is a good method for obtaining transport parameters by manipulating flow-path dynamics.

It is also possible to assess physical nonequilibrium at the field-scale using a negative pressure-head or tension (opposite sign of pressure head) in order to withdraw water and solutes from the soil. Unlike the previous method in this section where flow paths are turned off and on, the field-scale sampling strategy is designed to extract water and solutes from different pore class sets using a range of suction heads. Porous ceramic, fritted stainless-steel, and fritted glass have been used by researchers to design samplers for extracting soil water. Again, the porosity of the sampler is critical as to what set of soil pore classes is being analyzed. Large pores that are dominated by advective processes need to be sampled with zero-tension pan lysimeters or fritted glass that has a low bubbling pressure. Porous ceramic, fine fritted glass, and most stainless steel frits are unable to sample water and solutes that move by advection. The pore diameter of these samplers is very small and they can only be used to sample the soil matrix porosity (Severson and Grigal, 1976; Shaffer et al., 1979; Haines et al., 1982; Barbee and Brown, 1986; Litaor, 1988; Jardine et al., 1989, 1990a). Under certain conditions, however, small pore samplers can extract large pore water and solutes (e.g., stagnate water in dead-end preferential flow-paths).

Jardine et al. (1990a) and Wilson and Jardine (unpublished data) utilized this technique to assess the significance of physical nonequilibrium on solute transport in a structured clay soil and a weathered fractured shale, respectively. Jardine et al. (1990a) instrumented a 2 × 2 × 3 m deep soil block with coarse and fine fritted glass solution samplers maintained at –20 and –100 cm pressure-head, respectively. The coarse samplers collected soil solution held at pressure-heads of 0 to –20 cm and were believed to have monitored predominately macropore and mesopore channels (large pores). The fine samplers were thought to have collected soil water from predominately matrix mesopores (small pores) which were assumed to approximate the lower pore size limit for gravitationally mobile water in the soil. Wilson and Jardine (unpublished data) later refined this technique by using three types of fritted glass solution samplers: very coarse, medium, and fine. These authors instrumented a 2 × 2 × 3 m deep block of weathered fractured shale with the various samplers, where (1) the very coarse frits, held at zero pressure, collected free-flowing advective pore water, (2) the medium frits, held at –10 cm pressure, collected soil water from mesopores, and (3) the fine frits, held at –250 cm pressure, collected soil water from micropores. Tracer experiments conducted at these facilities revealed large differences for solute mobility within the various pore regimes (see Chapter 2 by Wilson et al. in this book). An example is shown in Figure 9.6 illustrating the storm-driven transport of Br^- in large and small pore on the structured clay soil. The tracer was applied slowly to the surface soil matrix in order to establish a concentration gradient between the soil pore regimes. These data suggested that during the vertical flux of storm water through the soil profile, solutes were transported by advection and diffusion from small-pore regions to large-pore regions via hydraulic and concentration gradients, respectively, with small pores being a major source for solute transport in large pores. Solutes that were rapidly transported to greater depths through large pores increased solute reserves in adjacent small pores at the same depths through a reversal of the mechanism described above. Thus, hydraulic and physical nonequilibrium processes in this soil were driven by hydraulic and concentration gradients, respectively, that were established as a result of the heterogeneous distribution of different sized pores within the system (see Chapter 2 in this book).

FLOW INTERRUPTION

An additional technique for confirming and quantifying physical nonequilibrium in soil systems involves flow interruption during a portion of a tracer displacement experiment. The technique involves inhibiting the steady-state flow process during an experiment for a designated period of time, thereby eliminating hydraulic nonequilibrium, and allowing a new physical or chemical equilibrium state to be approached. The flow-interrupt method has predominantly been implemented during reactive tracer studies that focus on the determination of chemical nonequilibrium (Murali and Aylmore, 1980; Hutzler et al., 1986; Brusseau et al., 1989; Ma and Selim,

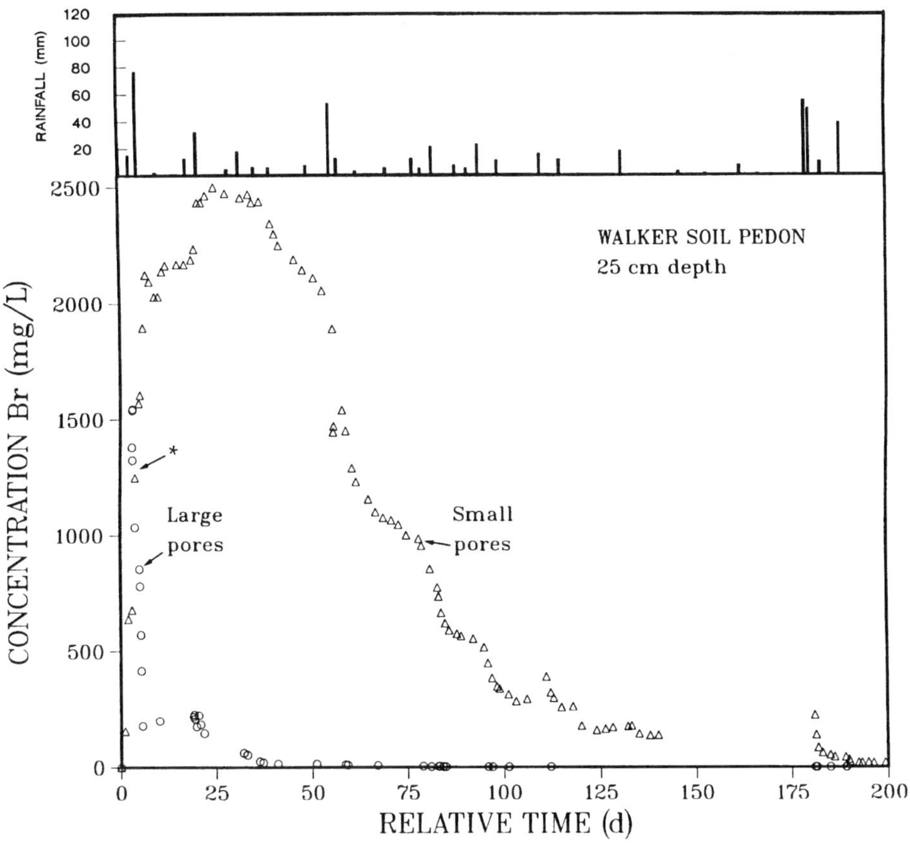

Figure 9.6. Unsaturated transport of Br^- through large and small pores of an *in situ* pedon using coarse and fine fritted glass solution samplers, respectively, as solute collection devices. The symbol * designates the point where a shift in the concentration gradient occurs between large and small pores. From Jardine, P.M. et al., *Geoderma,* 46:103–118, 1990. With permission.

1994). Recently, the technique has been employed to quantify physical nonequilibrium in glass beads (Koch and Flühler, 1993), porous ceramic spheres (Hu and Brusseau, 1995), and undisturbed fractured weathered shale (Reedy et al., 1996; O'Brien, 1994, Oak Ridge National Laboratory, unpublished data).

When physical nonequilibrium processes are significant in a soil system, the flow-interruption method will cause an observed concentration perturbation for a conservative tracer when flow is resumed. Interrupting flow during tracer injection will result in a decrease in tracer concentration when flow is resumed, whereas interrupting flow during tracer displacement (washout) will result in an increase in tracer concentration when flow is resumed. The concentration perturbations that are observed after flow interrupt are indicative of solute diffusion between pore regions of heterogeneous media. Conditions of preferential flow create concentration gradients between pore domains (physical nonequilibrium), resulting in diffusive mass transfer between the regions. Therefore, during injection, tracer concentrations within advection dominated flowpaths (i.e., fractures, macropores) are higher than those within the matrix. Upon flow interruption, the relative concentration decrease that is observed indicates that solute diffusion is occurring from larger, more conductive pores, into the smaller pores. During tracer displacement or washout, the concentrations within the preferred flow-

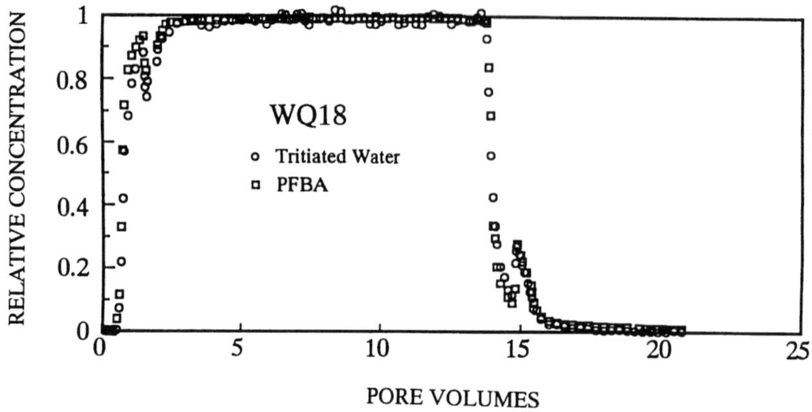

Figure 9.7. Breakthrough curves with flow interruption for the nonreactive tracers, ^3H_2O and PFBA, in columns of mixed porous ceramic spheres and glass beads. Flow interruption was initiated for several hours after displacement of approximately 2 and 14.5 pore volumes of the two tracers. Modified from Hu, Q. and M.L. Brusseau, *Water Resour. Res.,* 31:1637–1646, 1995. With permission.

paths are lower than those within the matrix. Thus, solute diffusion is occurring from smaller pores into larger pores, and a concentration increase is observed when flow interruption has been imposed.

The position of the flow interrupt along the course of the tracer breakthrough curve is important, such that longitudinal molecular diffusion (axil dispersion) within a particular pore region is minimal (Brusseau et al., 1989). Therefore, the flow interrupt method should only be performed in the latter stages of solute injection and displacement to minimize longitudinal concentration gradients within a pore domain. When used appropriately, the flow-interruption method is a sensitive technique for assessing physical nonequilibrium in soil systems. If physical nonequilibrium is significant in a soil system, flow interruption will cause a concentration perturbation of a conservative tracer when flow is resumed. If physical nonequilibrium is *not* significant, a concentration perturbation will not be observed.

The utility of the flow interrupt method for confirming and quantifying physical nonequilibrium in heterogeneous media is best shown by the work of Hu and Brusseau (1995), and Reedy et al. (1996). Hu and Brusseau (1995) fabricated physical heterogeneities by mixing porous ceramic spheres with glass beads and used ^3H_2O and PFBA to confirm physical nonequilibrium through flow interrupt (Figure 9.7). Reedy et al. (1996) used undisturbed fractured weathered shale and Br^- to show the effect of time and mean flux on the flow interrupt technique for quantifying physical nonequilibrium (Figure 9.8). These authors showed that concentration perturbations that were induced by flow interruption were significantly more pronounced at larger fluxes. This was because the system was further removed from equilibrium at the larger fluxes since a greater concentration gradient was established between advection-dominated flow paths and the soil matrix. Using a two-domain flow and transport model, the solute diffusive mass transfer rate between pore domains was quantified and found to be a linear function of flux (Reedy et al., 1996). The authors noted that the flow interrupt placed a constraint on the modeling effort that was advantageous for defining a more accurate set of parameter values for describing the observed transport data.

The results of Hu and Brusseau (1995) and Reedy et al. (1996) are compared with those of Ma and Selim (1994) who used uniform sieved soil particles for displacement experiments involving atrazine and ^3H_2O. Using the flow interrupt method, Ma and

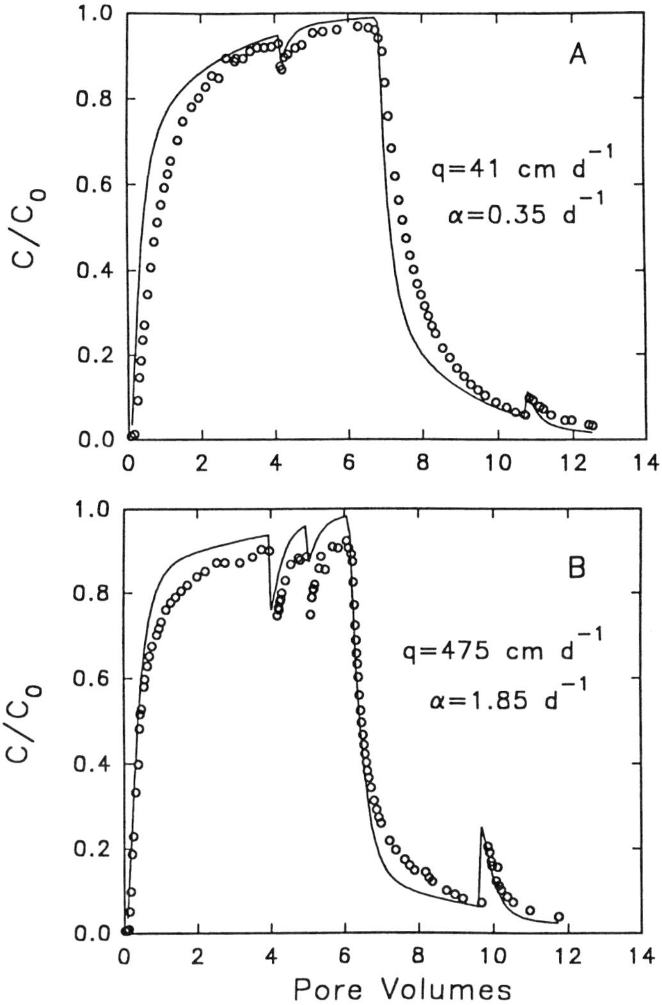

Figure 9.8. Breakthrough curves with flow interruption, at two specific discharges for a nonreactive Br^- tracer in undisturbed columns of weathered, fractured shale. Flow interruption was initiated for 7 days after (a) approximately 4 and 11 pore volumes of tracer were displaced at a flux of 41 cm d^{-1} and (b) approximately 4, 5, and 10 pore volumes of tracer were displaced at a flux of 475 cm d^{-1}. The solid lines represent simulations using a two-region model with optimization of the mass transfer coefficient, α, that accounts for mass exchange between different pore regions. Modified from Reedy, O.C. et al., *Soil Sci. Soc. Am. J.*, 60:1376–1384, 1996. With permission.

Selim (1994) showed dramatic chemical nonequilibrium sorption behavior for atrazine but did not observe a concentration perturbation for the nonreactive tracer ^3H_2O (Figure 9.9). These results indicated that physical nonequilibrium processes were insignificant in their system and this was consistent with the use of uniform soil particles that had minimal structural heterogeneities.

The flow interruption technique is difficult to implement in the field since this scale is most often representative of an open system that is influenced by an extensive number of coupled processes. However, when forced gradients are imposed upon an open system and then quickly terminated, a flow-interrupt scenario develops. This is a common procedure in groundwater remediation efforts that use pulsed pump-and-

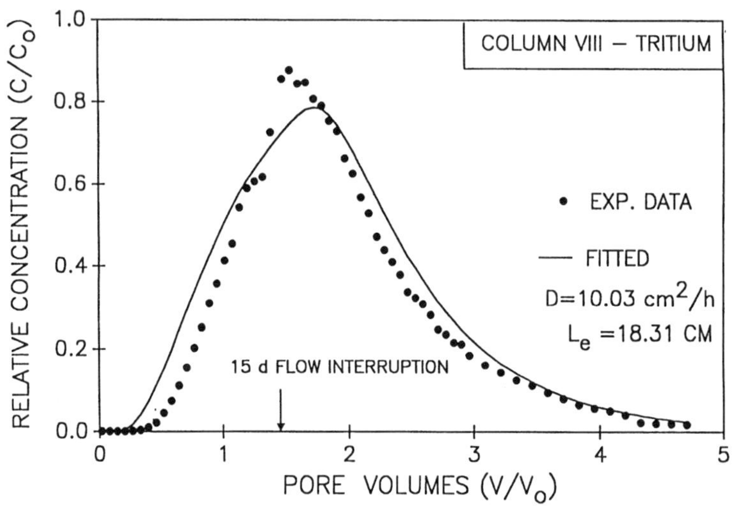

Figure 9.9. Breakthrough curve with a 15-day flow interruption for a nonreactive ^3H_2O tracer in a column of uniform sieved soil particles. Solid lines represent model fits to the convective-dispersive equation with optimization of the dispersion coefficient, D, and the effective solute path length, L_e. From Ma, L. and H.M. Selim, *Water Resour. Res.,* 30:3489–3498, 1994. With permission.

treat technologies to remove contaminants from heterogeneous subsurface environments. Typically, forced advective flow fields are imposed on a contaminated site in an effort to withdraw waste from the subsurface. More often than not, the rate-limiting step in waste removal is the extraction of the contaminants from the soil or rock matrix. Increasing the advective flow rate will help to accelerate the removal of contaminants from the matrix; however, the diffusional barrier that exists between the advective flow domain and the matrix is often far too dominant to allow extensive pump-and-treat strategies. The rate limitations imposed by physical and chemical nonequilibrium processes in subsurface systems often limit the success of classical pump-and-treat techniques for site remediation. By using a pulsed pump-and-treat technique where the forced advective flow fields are intermittently imposed on the system, the removal of contaminants from the soil and rock becomes much more cost-effective. Essentially, a flow interruption scenario occurs during the pulsed pump-and-treat technique. When the advective flow field of the system is returned to its natural condition, both concentration and hydraulic gradients will drive contaminant mass exchange from the soil and rock matrix into the advective flow domain. As the mass exchange process nears equilibrium, the pump and treat technique can be reintroduced for the clean-up of contaminants residing within the advective flow domain. By repeating this technique a number of times, the removal of contaminants from the soil and rock matrix will become evident. The pulsed pump-and-treat strategy for groundwater remediation may not accelerate the time required for cleanup relative to classical, continuous pump-and-treat techniques. The pulsed strategy, however, does minimize the amount of groundwater that must be treated, and this will significantly reduce cleanup cost.

MULTIPLE TRACERS WITH DIFFERENT DIFFUSION COEFFICIENTS

Of all the techniques used for quantifying physical nonequilibrium in soils, one of the most powerful is the simultaneous use of multiple tracers with different diffusion coefficients during a displacement experiment. Maloszewski and Zuber (1993) can be quoted as saying "...double tracing should be applied as a rule in order to check the assumptions related to matrix diffusion and to distinguish the apparent dispersion from the intrinsic one." When multiple tracers are used in conjunction with flow interruption and controlled flow path dynamics, the former technique becomes even more indispensable for assessing and quantifying chemical or physical nonequilibrium (see Hu and Brusseau, 1995 as an example).

In general, when the technique is used to quantify physical nonequilibrium processes in soils, two or more conservative tracers that have different diffusion coefficients are simultaneously displaced through the porous media. Tracers such as Br^-, Cl^-, fluorobenzoates, ^3H_2O, and dissolved gases such He, Ne, SF_6, and Kr (for water-saturated conditions) are frequently suitable for assessing physical nonequilibrium processes in soil using the multiple tracer technique (Carter et al., 1959; Raven et al., 1988; Bowman and Gibbens, 1992; Wilson and MacKay, 1993; Gupta et al., 1994; Jaynes, 1994; Linderfelt and Wilson, 1994; Clark et al., 1996; Sanford et al., 1996; Sanford and Solomon, 1996). Colloidal tracers and tracers that are size-excluded from the matrix porosity of soils are not included in this type of experimental technique. When physical nonequilibrium processes are significant in porous media, tracers with larger molecular diffusion coefficients will be preferentially lost to the matrix porosity relative to tracers with smaller molecular diffusion coefficients. When advective processes are dominant in a system and matrix diffusion is negligible, multiple tracer breakthrough profiles will not differ considerably.

These scenarios can be demonstrated in the laboratory by comparing the results of O'Brien (1994, Oak Ridge National Laboratory, unpublished data) with those of Brusseau (1993). The former authors investigated the simultaneous transport of Br^- and PFBA through a large undisturbed column of fractured weathered shale (Figure 9.10a). The molecular diffusion coefficient for PFBA is 40% smaller than the diffusion coefficient for Br^- (Bowman, 1984). Differences in the breakthrough curves for these solutes could be solely attributed to diffusive (versus advective) transport. Since the PFBA diffused more slowly into the weathered shale matrix, its breakthrough at the column exit was initially more rapid than Br^-, but required longer times to approach equilibrium (Figure 9.10a). Thus, Br^- had a larger mass loss to the matrix at any given time, and will therefore reach equilibrium more rapidly than PFBA. In contrast, Brusseau (1993) showed that ^3H_2O and PFBA exhibited nearly identical breakthrough curves in a column of nonstructured sandy soil (Figure 9.10b). Even though the molecular diffusion coefficient of ^3H_2O is ~40% larger than for PFBA, the two tracers were not affected by physical nonequilibrium processes in this media so their mobility was essentially identical. In the same study, Brusseau (1993) showed the separation of ^3H_2O and PFBA breakthrough curves in an aggregated soil where the structure of media was partially maintained. Recent studies by Hu and Brusseau (1995) have further shown the utility of the multiple tracer technique where four tracers with different diffusion coefficients (^3H_2O; PFBA; 2,4-D, or 2,4-dichlorophenoxyacetic acid; and HPCD, or hydroxypropyl-β-cyclodextrix) were used to assess physical nonequilibrium in a mixture of glass beads and porous ceramic spheres. The molecular weight of HPCD is 1500 g/mole which is small enough to remain unaffected by pore exclusion mechanisms.

Studies by Shropshire (1995) also revealed the utility of dissolved gas tracers in saturated systems for assessing physical nonequilibrium (Figure 9.11). Helium was simultaneously injected with Br^- into a column of fractured weathered shale for about 150 min. and then followed by tracer-free solution for nearly 600 min. The breakthrough and displacement (washout) of He was significantly more sluggish relative to Br^-. This results from the fact that the molecular diffusion coefficient for He is ~3x

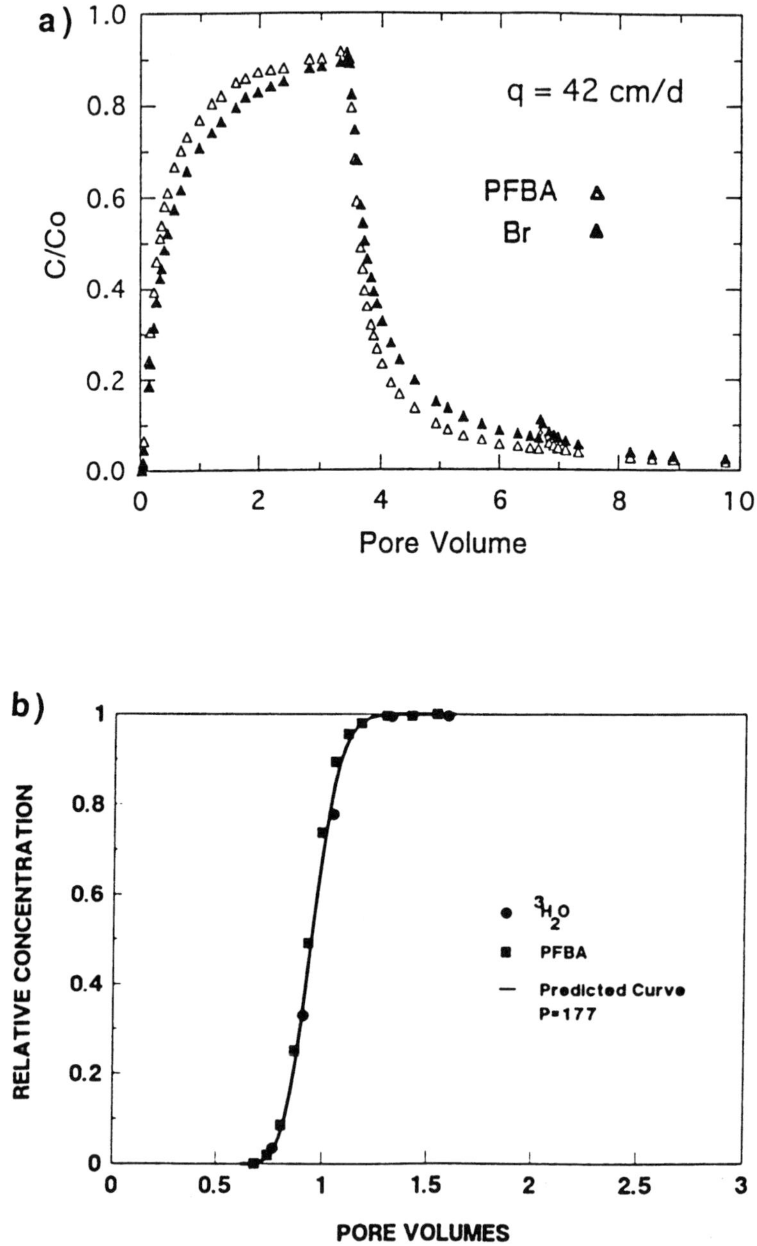

Figure 9.10. (a) Breakthrough curves for the simultaneous injection of two nonreactive tracers, Br^- and PFBA, at a flux (q) of 42 cm d^{-1} in an undisturbed column of weathered, fractured shale. Modified from O'Brien, R. (1994, Oak Ridge National Laboratory, unpublished data). With permission. (b) Breakthrough curves for the simultaneous steady-state injection of two nonreactive tracers, ^3H_2O and PFBA, in a column packed with a structureless sandy soil. From Brusseau, M.L. *Water Resour. Res.,* 29:1071–1080, 1993. With permission.

greater than that for Br^-. Therefore, the He tracer is preferentially lost to the matrix relative to Br^-. Modeling results by Shropshire (1995) indicated that the multiple tracer technique was an excellent method for approximating fracture aperture and network geometry in a weathered shale media.

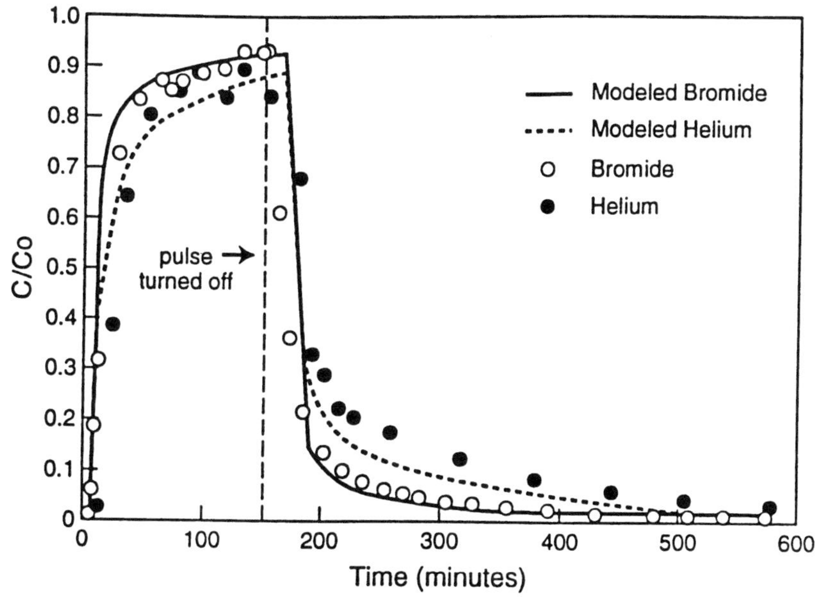

Figure 9.11. Breakthrough curves for the simultaneous, steady-state injection of two nonreactive tracers, He an Br^-, in an undisturbed column of weathered, fractured shale. A discrete fracture flow and transport code was used to simulate the data as described by Shropshire (1995). From Sanford, W.E. et al., *Water Resour. Res.,* 32:1635–1642, 1996. With permission.

Jardine et al. (1998) also showed the utility of multiple dissolved gas and solute tracers for assessing physical nonequilibrium processes in fractured shale bedrock (Figures 9.12a,b,c). The natural gradient injection of Br^-, He, and Ne was initiated for 6 months, followed by a 12-month washout, within an isolated fracture zone of a shallow aquifer at the WAG 5 site on the U.S. Department of Energy's Oak Ridge Reservation in east Tennessee. Spatial and temporal monitoring of the tracers was performed in the matrix and fracture regimes of the bedrock using multilevel sampling wells instrumented downgradient from the injection source. Tracers migrated preferentially along strike, and their concentrations in the fracture regime quickly reached a consistent steady-state value. Thus, observed differences in tracer breakthrough into the matrix were a function of their molecular diffusion coefficients. The breakthrough of the three tracers 6 m from the source and 0.8 m into the matrix relative to the fracture is shown in Figure 9.12a. The movement of He and Ne into and from the matrix was more rapid than Br^-, and this is consistent with the larger molecular diffusion coefficients of the dissolved gases relative to Br^-. These results confirmed the contribution of matrix diffusion to the overall physical nonequilibrium process that controls solute transport in shale bedrock.

At greater distances from the source, the contribution of matrix interactions are still prominent, and tracer breakthrough profiles remain suggestive of a diffusion mechanism, although at first glance this may not be apparent (Figure 9.12b,c). At 13 m from the source and 0.6 m into the matrix, the three tracers break through nearly simultaneously, with the concentration of the gas tracers eventually surpassing Br^- (Figure 9.12b). This is followed by tracer washout after the input pulse was terminated at 180 d. At 23 m from the source and 0.1 m into the matrix, the movement of Br^- into the matrix is actually more rapid than that of the noble gas tracers, which is exactly opposite of what was observed 6 m from the source. This apparent paradox is

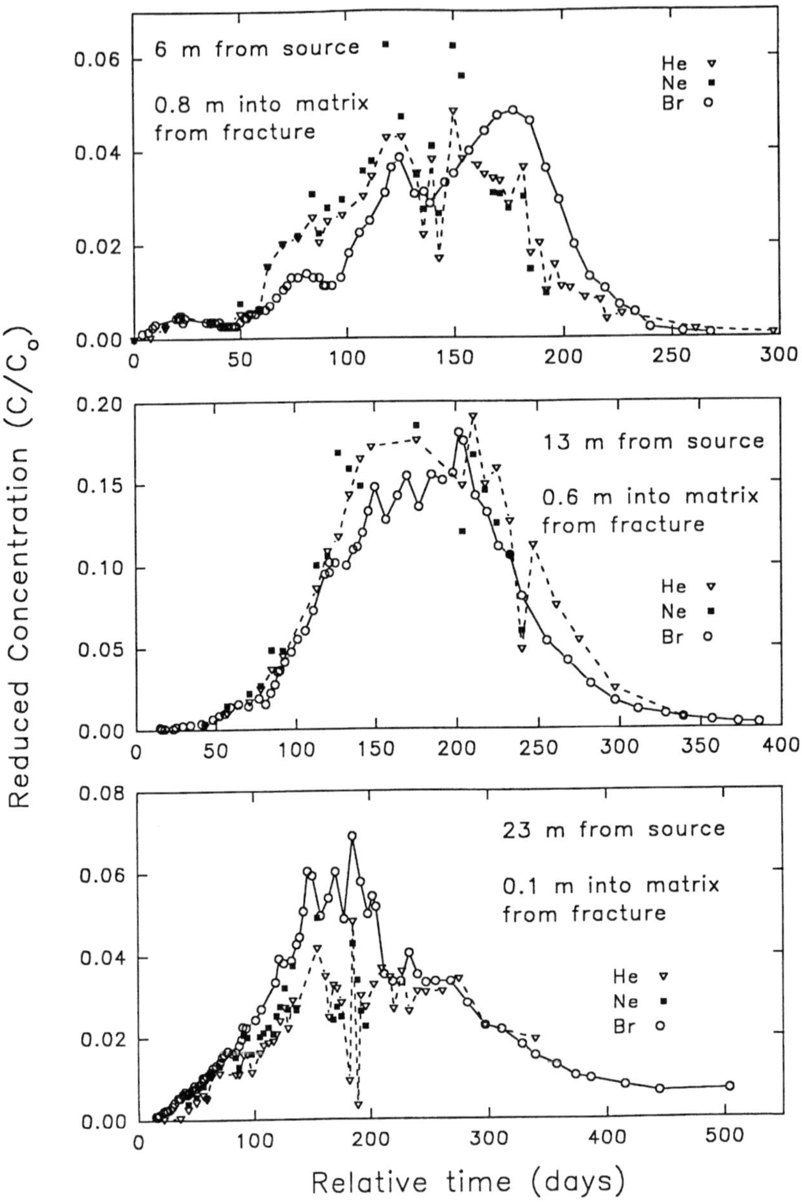

Figure 9.12. Breakthrough curves of He, Ne, and Br⁻ in the matrix porosity of shale bedrock located (a) 6, (b) 13, and (c) 23 meters down gradient of, and along strike from a tracer injection well. From Jardine, P.M. et al. (*Water Resour. Res.*, pending 1998).

caused by the preferential loss of gas tracers to the rock matrix closer to the source. Thus, Br^- remains within the advective flow field (fracture regime) for a longer time period, allowing it to be transported greater distances. Having been transported farther downgradient, Br^- experiences the first opportunity to begin diffusing into the matrix at distances further from the source. Eventually, He and Ne arrive at the same locations and also begin to diffuse into the matrix, lagging behind that of Br^- (Figures 9.12b,c). Because of the faster diffusion rate of the noble gases, the movement of He and Ne into the matrix is more rapid and, if given time, the He and Ne breakthrough curves will eventually cross over and surpass the Br^- breakthrough curves (see Figure 9.12b as an example).

Sanford and Solomon (1996) also used the noble gas tracers He and Ne to quantify physical nonequilibrium processes in a saturated shale saprolite. Separation of the gas breakthrough curves was attributed to their different molecular diffusivities. The authors stress that the advantage of using noble gases as tracers are (1) they are chemically inert, (2) the gases behave nearly identically except for differing diffusion coefficients, (3) the injection water can be saturated several orders of magnitude above background levels without significantly altering the physical properties of the fluid, and (4) the samples can be detected over a concentration range of five orders of magnitude.

Garnier et al. (1985) also performed multiple tracer studies in a fissured chalk formation using four tracers, D_2O, I^-, uranine (disodium fluorescein), and $H^{13}CO_3^-$, that were instantaneously injected, and their concentrations measured in a pumping well 10 m away from the injection well. Maloszewski and Zuber (1990) replotted the data of Garnier et al. (1985) and found that a larger quantity of uranine was mobilized relative to the other tracers, most likely because uranine has a smaller diffusion coefficient, which limits its movement into and out of the matrix. Conversely, both D_20 and I^- had significantly smaller mass recoveries relative to uranine, and this can be attributed to their larger diffusion coefficients which enhances their movement into the matrix.

MULTIPLE TRACERS WITH GROSSLY DIFFERENT SIZES

Another sensitive technique for confirming and quantifying physical nonequilibrium in heterogeneous soil systems is the use of multiple tracers with distinctly different sizes. Specifically, this technique uses both dissolved solutes and colloidal tracers so that flow-path accessibility can be controlled. Colloidal particles are typically large enough to be excluded from the matrix porosity of most soils and geologic material. If they are not chemically retarded by the porous media, colloidal particles serve as excellent tracers for quantifying advective flow velocities in systems conducive to preferential flow. When colloidal tracers are coupled with dissolved solutes that can interact with the matrix porosity, a unique technique emerges for assessing physical nonequilibrium processes in subsurface media.

The choice of colloidal particles for use as tracers in structured subsurface media depends on a number of factors (McKay et al., 1993a); namely, (1) loss due to matrix diffusion, (2) loss due to sorption, (3) loss due to decay or inactivation, and (4) detection limits. Viruses, bacteria, fluorescent microspheres, DNA-labeled microspheres, radiolabeled Fe-oxide particles, and synthetic polymers have all been used as colloidal tracers in various subsurface media (Barraclough and Nye, 1979; Gerba et al., 1981; Smith et al., 1985; Bales et al., 1989; Harvey et al., 1989, 1993, 1995; Toran and Palumbo, 1991; McKay et al., 1993a,b; Hinsby et al., 1996; Yang et al., 1996; O'Brien et al., 1996; Reimus, 1996). Smith et al. (1985) showed that *Escherichia coli* were preferentially transported through undisturbed soil columns relative to Cl^-. The bacteria were effectively excluded from pores that had a diameter <15 μm and therefore arrived in the column effluent much more rapidly than Cl^-; the latter being affected by diffusion into the matrix porosity of the soil. Similarly, Champ and Schroeter (1988) showed that colloidal-sized bacteria traveled more rapidly than Na fluorescein and $^{82}Br^-$ during field experiments in fractured crystalline rock. Bales et al. (1989) observed that colloidal-sized bacteriophage (virus) migrated three times faster through columns of fractured porous tuff relative to nonreactive solute tracers. The difference in breakthrough characteristics for the tracers was attributed to size exclusion of the phage from the matrix porosity of the tuff. McKay et al. (1993a) point out that certain bacteriophage can serve as excellent tracers since (1) many are sufficiently hydrophobic to discourage sorption on mineral surfaces, (2) phage inactivation is slow at normal groundwater temperatures, thereby permitting use in long-term field experiments, and (3) the detection limits are reasonably low to allow several orders of magnitude dilution during a tracer study.

O'Brien (1994, Oak Ridge National Laboratory, unpublished data) simultaneously injected two strains of bacteriophage (PRD-1 and MS-2) and two dissolved solutes (Br^- and PFBA) into a column of fractured weathered shale in order to assess physical nonequilibrium processes (Figure 9.13). The bacteriophage travel times were much more rapid than the dissolved solutes and exhibited significantly less total dispersion in their transport, as evidenced by steeper breakthrough characteristics relative to PFBA and Br^-. The larger bacteriophage were preferentially transported through a smaller range of flow-paths and were minimally affected by diffusion into the matrix porosity. The dissolved tracers, on the other hand, are influenced by diffusive mass transfer processes between fractures and the matrix (Figure 9.13). Besides providing visual evidence of physical nonequilibrium processes in structured media, the experiments provided advective flow velocities that were used to parameterize numerical models designed to simulate the observed data. Similar results were noted by Hinsby et al. (1996) and Harton (1996), who investigated bacteriophage and dissolved solute transport in fractured clay tills and weathered shales, respectively.

Recently, McKay et al. (1993a,b; 1995) used bacteriophage in field experiments to quantify advective flow velocities and physical nonequilibria phenomena in a fractured clay till and a fractured shale saprolite. The transport velocity of the phage was several orders of magnitude greater than nonreactive tracers at both field facilities (Table 9.1a,b). The large contrast in transport velocities between the phage and the conservative tracers was due to extensive diffusion of the dissolved solutes into the matrix porosity. McKay et al. (1995) noted large losses of phage, on the order of 1–2 1og cycles per meter of travel, under natural gradient flow conditions within the shale saprolite. However, the phage remained a usable tracer because of the high injection concentration relative to the detection limit.

Yang et al. (1996) developed a unique set of colloidal tracers that consisted of silica microspheres to which DNA segments had been attached. This technique offers the advantage of providing a large number of uniquely labeled tracers that have identical chemical and physical properties. The tracers are colloidal-sized (0.5 μm) and are ideal for delineating fracture interconnectivity and advective flow rates in subsurface media that is complicated by physical nonequilibrium (Moline et al., 1995). The DNA-labeled silica microsphere tracers are physically and chemically more uniform than DNA-labeled clay particles tracers, and the former association is a stronger covalent attachment relative to DNA-labeled clay particles. Further, the DNA-labeled microspheres are more flexible than fluorescent microspheres, since a limited number of fluorescent varieties are available.

TRACERS FOR FLOW-PATH MAPPING

An additional method for confirming the significance of physical nonequilibrium processes in soils involves flow-path mapping with tracers such as dyes and fluorescent microspheres. The basic concept behind this technique is to label flow-paths with chemical tracers that are readily observable by the naked eye. This approach permits the observation of conducting pathways which are related to the structural features in the soil that transport water and solutes. The technique is usually qualitative since

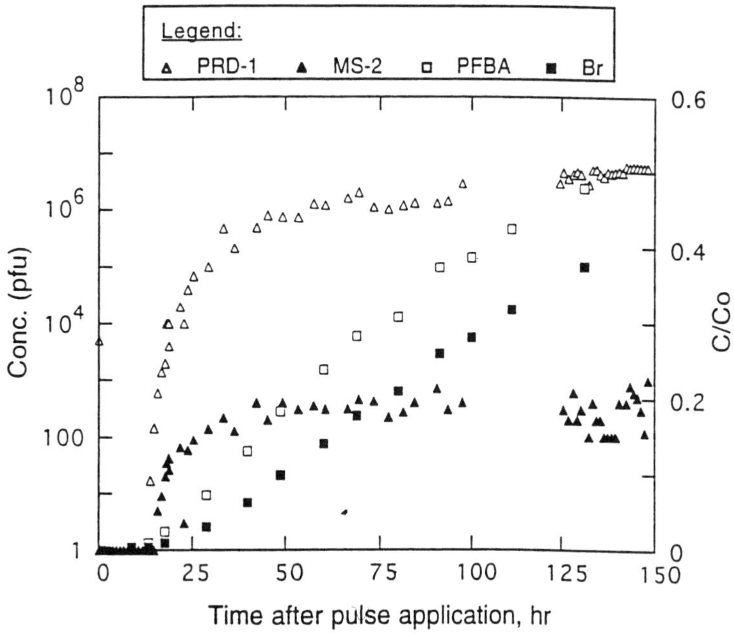

Figure 9.13. Effluent concentrations of two bacteriophage strains (PRD-1 and MS-2) and reduced concentrations of the dissolved tracers PFBA and Br⁻, that were simultaneously injected at 2.2 cm d^{-1} onto an undisturbed column of weathered fractured shale. Note the different concentration axes used for each of the two tracer types. From O'Brien, R. (1994, Oak Ridge National Laboratory, unpublished data). With permission.

Table 9.1. Comparison of Colloid and Solute Transport.[a]

Collector[b] RT2	Depth (m)	Transit Time, Day: Colloids, C > 0.1 pfu mL^{-1}	Transit Time, Day: Bromide, C/Co > 0.01	Fracture Aperture,[c] µm
1C	1.6	0.74–1.76	130–135	43
1D	1.6	0.74–1.76	78–82	40
1E	1.6	0.74–1.76	<110	9
1F	1.6	0.74–1.76	80–83	9
2E	2.1	0.74–1.76	<110	20
2F	2.1	<0.74	50–70	16
3F	2.6	0.74–1.76	50–115	13

[a] From McKay et al., Environ. Sci. Technol., 27:1075–1079, 1993. With permission.
[b] Collectors were 4.0 m from source trench.
[c] Fracture aperture based on measured flow rate and fracture spacing.

it is difficult to quantify these tracers when they are adsorbed or associated with the soil surface. However, the method is a useful tool for assessing the significance of physical nonequilibrium processes in soils.

One of the most common types of tracers used for flow-path mapping are dyes (Ritchie et al., 1972; Smart and Laidlaw, 1977; Starr et al., 1978; Omoti and Wild, 1979; Johnston et al., 1983; Davis et al., 1985; Jardine et al., 1988; Richard and Steenhuis, 1988; Ghodrati and Jury, 1990; Kung, 1990; Hornberger et al., 1991). The basic assumption made in interpreting dye patterns is that the more dye solution that con-

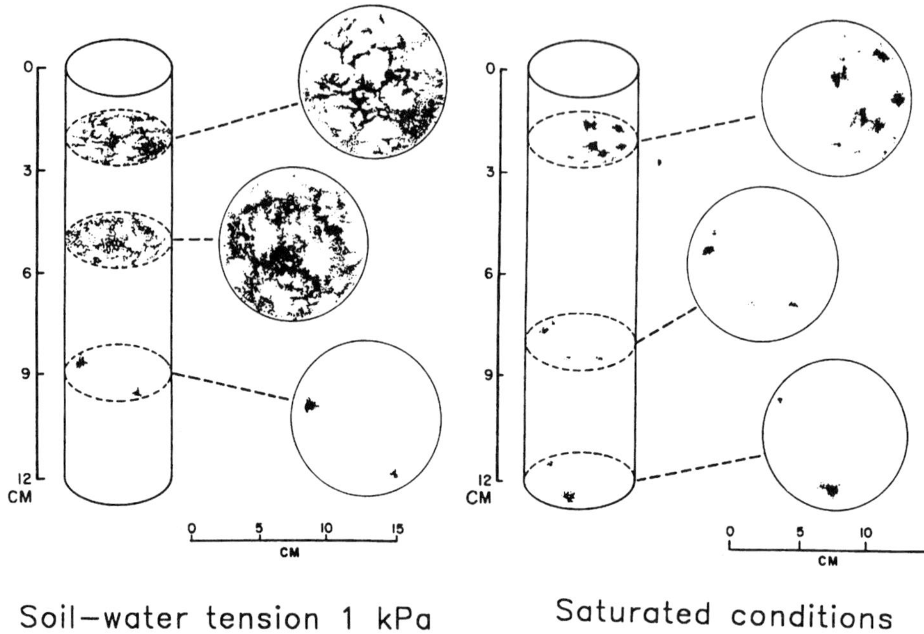

Figure 9.14. Rhodamine B dye staining patterns in undisturbed columns of an aggregated tropical soil resulting from dye displacement under soil-water tensions of 1 kPa and 0 kPa (saturated). Modified from Seyfried, M.S., and P.S.C. Rao, *Soil Sci. Soc. Am. J.*, 51:1434–1444, 1987. With permission.

tacts a surface, the more darkly stained that surface will be (Seyfried and Rao, 1987). Thus, preferential flow-paths would have a tendency to become more darkly stained relative to flow-paths within the matrix, since the former carry the majority of the dye flux and their surface area is small. Seyfried and Rao (1987) used Rhodamine B as a strongly sorbing dye to confirm physical nonequilibrium processes in a structured tropical soil (Figure 9.14). Using saturated conditions they showed that preferential advective flow was confined to small regions within their undisturbed columns, and was associated with a series of relatively large pores, or continuous pore sequences. When the dye tracer was applied during unsaturated conditions, it was observed that the most rapidly conducting pore sequences became disconnected, thus compressing the dye "solute front" (Figure 9.14). These observations were in qualitative agreement with tracer studies using ^3H_2O which quantified the nonequilibrium processes in these soils. Similar findings were reported by Buchter et al. (1995) who used an iodine starch method to assess heterogeneous flow paths in an undisturbed column of soil. The qualitative assessment of physical nonequilibrium using the dye test was consistent with the more quantitative solute tracer tests that were also performed. Likewise, Omoti and Wild (1979) used fluorescein and pyranine to map solute flow-paths in an agricultural field soil. They noted that the vertical distribution of pyranine matched that of a nonreactive $^{36}Cl^-$ tracer, with both tracers following preferential flow paths. Ghodrati and Jury (1990) used the negatively charged amine red dye on a field soil and showed preferential flow of solution which was consistent with observations by Jury et al. (1986a,b) of deep leaching of pesticides in this same soil.

Since strongly adsorbing dyes introduce a filtering effect by staining only preferential flow-paths, it could be beneficial to simultaneously apply a nonsorbing dye with a sorbing dye so that physical nonequilibrium processes could be more quantitatively assessed (Wildenschild et al., 1994). By using dyes with different fluorescent and sorption tendencies (Smart and Laidlaw, 1977), it may be possible to distinguish pore-class groups that are separated by physical nonequilibrium processes. The authors are not familiar with any studies that may exist on this topic.

Image analysis has been applied to dye-stained soil surfaces to obtain more quantitative information on soil porosity. Using this approach, thin-sections or other polished, stained surfaces are converted from a visual to a digital image using electronic scanners. The digital image can be analyzed for the number, size, and shape of pore spaces in a soil (see Singh et al., 1991; Vermeul et al., 1993; and Ringrosevoase, 1996 as examples of this technique). While improved measures of macroporosity are provided with this approach, it is limited by the two-dimensional nature of the analysis as well as the destructiveness of the technique. Observations of the connectivity and tortuosity of pore spaces—critical information to predicting solute transport in heterogeneous soils—is lost without an analysis of the third dimension (Glasbey et al., 1991; Singh et al., 1991; Eggleston and Peirce, 1995).

Several nondestructive, three-dimensional imaging techniques are being used in laboratory vadose and/or saturated flow studies. Computed tomography (Warner et al., 1990; Anderson and Gantzer, 1989; Heijs et al., 1996) and magnetic resonance imagery (Hoffman et al., 1996; Posadas et al., 1996) have successfully identified and characterized preferential water and solute movement in soils. Repeated observations of the spatial and temporal variability in flow and transport experiments may provide the necessary data set to quantify physical nonequilibrium in a soil system.

Fluorescent microspheres have also been used as tracers of advective flow processes in soils and groundwater (Harvey et al., 1989, 1993, 1995, Lenczewski, 1993; Reimus, 1996; Crumbie, 1997). There are only a few studies that the authors are aware of where fluorescent microspheres have been used to map flow-paths in structured media (O'Brien, 1994, Oak Ridge National Laboratory, unpublished data; Crumbie, 1997). In the studies of O'Brien (1994, Oak Ridge National Laboratory, unpublished data) and Crumbie (1997), fluorescent microspheres were infiltrated into columns of shale saprolite in order to provide insight into the flow-path network and structure of the media. Columns were physically dismantled and gently pried open along natural planes of weakness, and exposed surfaces were examined under UV light. The fluorescent microsphere tracers provided direct visual evidence that the hydraulically active fractures within the media followed a tortuous path and comprised only a small portion of the sample. Further, it was visually apparent that the intersection of fracture sets appeared to govern advective flow. Such observations are critical when developing conceptual models regarding the influence of physical nonequilibrium on solute transport in subsurface media.

SUMMARY

This chapter has described a number of techniques that are useful for assessing physical nonequilibrium processes in heterogeneous subsurface environments. The techniques involve systematic experimental manipulations of the physical processes governing the flow and transport of water and solutes in subsurface media. The experimental manipulations create system perturbations so that the rate-limiting physical processes can be isolated and quantified. The techniques described in this chapter for assessing physical nonequilibrium during solute transport include: (1) controlling flow path dynamics with manipulations of pore-water flux and soil-water tension, (2) isolating diffusion processes with flow interruption, (3) using multiple tracers with different diffusion coefficients, (4) using multiple tracers with grossly different sizes, and (5) using tracers that visually map flow-path dynamics. All of the techniques are useful for confirming the significance of physical nonequilibrium processes in heterogeneous systems, with some techniques being more quantitative than others. Techniques such as flow interruption and multiple tracers with different diffusion coefficients and sizes are more quantitative since known alterations in system properties are invoked to produce an expected behavior that is consistent with the theoretical property constraint. For example, if diffusion is a process contributing to physical nonequilibrium in a system, then the use of multiple tracers with different diffusion coefficients will result in solute travel time differences that *must* be described by the difference of the tracer molecular diffusion coefficients. Likewise, if diffusion is a significant contributor to physical nonequilibrium during transport, then flow interruption *must* result in an observable concentration perturbation when flow is reinitiated.

When the techniques described in this chapter are combined, they become an extremely powerful means of quantifying physical nonequilibrium processes in heterogeneous systems. For example, when flow interruption is coupled with the use of multiple tracers, the system must behave in accordance with known tracer diffusion coefficients, and the concentration perturbations that result during the interrupt must reflect this. Similarly, when the use of multiple pore-water velocities is coupled with both dissolved and colloidal tracers, the system must behave in accordance with known tracer sizes, and any change in flow-path dynamics must reflect this condition. The combined use of the described techniques not only improves our conceptual understanding of time-dependent solute migration in subsurface media, it also provides the necessary experimental constraints that are needed for the accurate numerical quantification of the physical nonequilibrium processes.

REFERENCES

Akratanakul, S., L. Boersma, and G.O. Klock. Sorption processes in soils as influenced by pore water velocity: 2. Experimental results. *Soil Sci.,* 135:331–341, 1983.

Anamosa, P.R., P. Nkedi-Kizza, W.G. Blue, and J.B. Sartain. Water movement through an aggregated, gravelly oxisol from Cameroon. *Geoderma,* 46:263–281, 1990.

Anderson, S.H. and C.J. Gantzer. Determination of soil water content by X-ray computed tomography and magnetic resonance imaging, *Irrigation Sci.,* 10(1):63–71, 1989.

Bales, R.C., C.P. Gerba, G.H. Grondin, and S.L. Jensen. Bacteriophage transport in sandy soil and fractured tuff. *Appl. Environ. Microbiol.,* 55:2061–2067, 1989.

Barbee, G.C. and K.W. Brown. Comparison between suction and free-drainage soil solution samplers. *Soil Sci.,* 141:149–154, 1986.

Barraclough, D. and P.H. Nye. The effect of molecular size on diffusion characteristics in soil. *J. Soil Sci.,* 30:29–42, 1979.

Bouma, J. and J.H.M. Wosten. Flow patterns during extended saturated flow in two, undisturbed swelling clay soils with different macrostructures. *Soil Sci. Soc. Am. J.,* 43:16–22, 1979.

Bowman, R.S. and J.F. Gibbens. Difluorobenzoates as nonreactive tracers in soil and groundwater. *Ground Water,* 30:8–14, 1992.

Bowman, R. Evaluation of some new tracers for soil water studies. *Soil Sci. Soc. Am. J.,* 48:987–993, 1984.

Brusseau, M.L., P.S.C. Rao, R.E. Jessup, and J.M. Davidson. Flow interruption: A method for investigating sorption nonequilibrium. *J. Contamin. Hydrol.,* 4:223–240, 1989.

Brusseau, M.L. The influence of solute size, pore water velocity, and intraparticle porosity on solute dispersion and transport in soil. *Water Resour. Res.,* 29:1071–1080, 1993.

Buchter, B., C. Hinz, M. Flury, and H. Flühler. Heterogeneous flow and solute transport in an unsaturated stony soil monolith. *Soil Sci. Soc. Am. J.,* 59:14–21, 1995.

Carter, R.C., W.J. Kaufman, G.T. Orlob, and D.K. Todd. Helium as a ground-water tracer. *J. Geophys. Res.,* 64:2433–2439, 1959.

Champ, D.R. and J. Schroeter. Bacterial transport in fractured rock—A field-scale tracer test at the Chalk River Nuclear Laboratories. *Water Sci. Technol.,* 20:81–87, 1988.

Clark, J.F., P. Schlosser, M. Stute, and H. J. Simpson. SF_6–^3He tracer release experiment: A new method of determining longitudinal dispersion coefficients in large rivers. *Environ. Sci. Technol.,* 30:1527–1532, 1996.

Clothier, B.E., M.B. Kirkham, and J.E. McLean. In situ measurements of the effective transport volume for solute moving through soil. *Soil Sci. Soc. Am. J.,* 56:733–736, 1992.

Crumbie, D.H. Laboratory Scale Investigations into the Influence of Particle Diameter on Colloid Transport in Highly Weathered and Fractured Saprolite, Thesis presented to the University of Tennessee, in fulfillment of the requirements for the degree of Master of Science, 1997.

Davis, S.N., D.J. Campbell, H.W. Bentley, and T.J. Flynn. Ground Water Tracers. National Water Well Assoc., Worthington, OH, 1985, p. 200.

De Vries, J. and T.J. Chow. Hydrologic behavior of a forested mountain soil in Coastal British Columbia. *Water Resour. Res.,* 5:935–942, 1978.

Eggleston, J.R. and J.J. Peirce. Dynamic programming analysis of pore space. *J. Soil Sci.,* 46:581–590, 1995.

Elrick, D.E. and L.K. French. Miscible displacement patterns on disturbed and undisturbed soil cores. *Soil Sci. Soc. Am. Proc.,* 30:153–156, 1966.

Garnier, J.M., N. Crampon, C. Preaux, G. Porel, and M. Vreulx. Tracage par ^{13}C, ^2H, I et uranine dans la nappe de la craie sénonienne en écoulement radial convergent (Béthune, France), *J. Hydrol. Amsterdam,* 78:379–392, 1985.

Gerba, C.P., S.M. Goyal, I. Cech, and G.F. Bogdan. Quantitative assessment of the adsorptive behavior of viruses to soils. *Environ. Sci. Technol.,* 15:940–944, 1981.

Germann, P.F., W.M. Edwards, and L.B. Owens. Profiles of bromide and increased soil moisture after infiltration into soils with macropores. *Soil Sci. Soc. Am. J.,* 48:237–244, 1984.

Ghodrati, M. and W.A. Jury. A field study using dyes to characterize preferential flow of water. *Soil Sci. Soc. Am. J.,* 54:1558–1563, 1990.

Glasbey, C.A., G.W. Horgan, and J.F. Darbyshire. Image analysis and three-dimensional modelling of pores in soil aggregates. *J. Soil Sci.,* 42:479–486, 1991.

Gupta, S.K., L.S. Lau, and P.S. Moravcik. Ground-water tracing with injected helium. *Ground Water,* 32:96–102, 1994.

Gwo, J.P., P.M. Jardine, G.V. Wilson, and G.T. Yeh. A multiple-pore-region concept to modeling mass transfer in subsurface media. *J. Hydrol.,* 164:217–237, 1995.

Haines, B.L., J.B. Waide, and R.L. Todd. Soil solution nutrient concentrations sampled with tension and zero-tension lysimeters: Report of discrepancies. *Soil Sci. Soc. Am. J.,* 46:658–661, 1982.

Harton, A.D. Influence of Flow Rate on Transport of Bacteriophage in a Column of Highly Weathered and Fractured Shale, Thesis presented to the University of Tennessee in fulfillment of the requirement for the degree of Master of Science, 1996.

Harvey, R.W., L.H. George, R.L. Smith, and D.L. LeBlanc. Transport of microspheres and indigenous bacteria through a sandy aquifer: Results of natural- and forced-gradient tracer experiments. *Sci. Technol.,* 23:51–56, 1989.

Harvey, R.W., N.E. Kinner, D. MacDonald, D.W. Metge, and A. Bunn. Role of physical heterogeneity in the interpretation of small-scale laboratory and field observations of bacteria, microbial-sized microsphere, and bromide transport through aquifer sediments. *Water Resour. Res.,* 29:2713–2721, 1993.

Harvey, R.W., N.E. Kinner, D. MacDonald, D.W. Metge, and A. Bunn. Transport behavior of groundwater protozoa and protozoan-sized microspheres in sandy aquifer sediments. *Appl. Environ. Microbiol.,* Jan, 209–217, 1995.

Heijs, A.W.J., C.F. Ritsema, and L.W. Dekker. Three-dimensional visualization of preferential flow patterns in two soils. *Geoderma,* 70:101–116, 1996.

Hinsby, K., L.D. McKay, P. Jørgensen, M. Lenczewski, and C.P. Gerba. Fracture aperture measurements and migration of solutes, viruses and immiscible creosote in a column of clay till. *Ground Water,* 34:1065–1075, 1996.

Hoffman, F., D. Ronen, and Z. Pearl. Evaluation of flow characteristics of a sand column using magnetic resonance imaging. *J. Contam. Hydrol.,* 22(1–2):95–107, 1996.

Hornberger, G.M., P.F. Germann, and K.J. Beven. Throughflow and solute transport in an isolated sloping soil block in a forested catchment. *J. Hydrol.,* 124:81–97, 1991.

Hu, Q. and M.L. Brusseau. The effect of solute size on diffusive-dispersive transport in porous media. *J. Hydrol.,* 158:305–317, 1994.

Hu, Q. and M.L. Brusseau. Effect of solute size on transport in structured porous media. *Water Resour. Res.,* 31:1637–1646, 1995.

Hutzler, N.J., J. C. Crittenden, J.S. Gierke, and A.S. Johnson. Transport of organic compounds with saturated groundwater flow: Experimental results. *Water Resour. Res.,* 22:285–295, 1986.

Jardine, P.M., G.V. Wilson, and R.J. Luxmoore. Modeling the transport of inorganic ions through undisturbed soil columns from two contrasting watersheds. *Soil Sci. Soc. Am. J.,* 52:1252–1259, 1988.

Jardine, P.M., G.V. Wilson, R.J. Luxmoore, and J.F. McCarthy. Transport of inorganic and natural organic tracers through an isolated pedon in a forested watershed. *Soil Sci. Soc. Am. J.,* 53:317–323, 1989.

Jardine, P.M., G.V. Wilson, and R.J. Luxmoore. Unsaturated solute transport through a forest soil during rain storm events. *Geoderma,* 46:103–118, 1990a.

Jardine, P.M., G.V. Wilson, J.F. McCarthy, R.J. Luxmoore, D.L. Taylor, and L.W. Zelazny. Hydrogeochemical processes controlling the transport of dissolved organic carbon through a forested hillslope. *J. Contamin. Hydrol.,* 6:3–19, 1990b.

Jardine, P.M., G.K. Jacobs, and G.V. Wilson. Unsaturated transport processes in undisturbed heterogeneous porous media: I. Inorganic contaminants. *Soil Sci. Soc. Am. J.,* 57:945–953, 1993a.

Jardine, P.M., G.K. Jacobs, and J.D. O'Dell. Unsaturated transport processes in undisturbed heterogeneous porous media: II. Co-contaminants. *Soil Sci. Soc. Am. J.,* 57:954–962, 1993b.

Jardine, P.M., W.E. Sanford, O.C. Reedy, J.P. Gwo, D.S. Hicks, J.S. Riggs, and N.D. Farrow. Quantifying diffuse mass transfer processes in fractured shale bedrock using multiple tracers. *Water Resour. Res.,* 1998, pending.

Jaynes, D.B. Evaluation of fluorobenzoate tracers in surface soils. *Ground Water,* 32:532–538, 1994.

Jaynes, D.B., S.D. Logsdon, and R. Horton. Field method for measuring mobile/immobile water content and solute transfer rate coefficient. *Soil Sci. Soc. Am. J.* 59:352–356, 1995.

Jensen, J.R. Potassium dynamics in soil during steady flow. *Soil Sci.,* 138:285–293, 1984.

Johnston, C.D., D.H. Hurle, D.R. Hudson, and M.J. Height. Water Movement through Preferred Paths in Lateritic Profiles of the Darling Plateau, Western Australia. Ground Water Res. Tech. Pap. No. 1, pp. 1–34, Commonw. Sci. Ind. Res. Organ. (CSIRO), Melbourne, Australia, 1983.

Jury, W.A., H. El Abd, and M. Resketo. Field study of napropamide movement through unsaturated soil. *Water Resour. Res.,* 22:749–755, 1986a.

Jury, W.A., H. El Abd, L.D. Clendening, and M. Resketo. Evaluation of pesticide transport models under field conditions. *ACS Symp. Ser.* 315, pp. 384–395, 1986b.

Kissel, D.E., J.T. Ritchie, and E. Burnett. Chloride movement in undisturbed swelling clay soil. *Soil Sci. Soc. Am. Proc.,* 37:21–24, 1973.

Koch, S. and H. Flühler. Non-reactive solute transport with micropore diffusion in aggregated porous media determined by a flow-interruption method. *J. Contam. Hydrol.,* 14:39–54, 1993.

Kookana, R.S., R.D. Schuller, and L.A.G. Aylmore. Simulation of simazine transport through soil columns using time-dependent sorption data measured under flow conditions. *J. Contam. Hydrol.,* 14:93–115, 1993.

Kung, K.-J.S. Preferential flow in a sandy vadose zone. 1. Field observation. *Geoderma,* 46:51–58, 1990.

Lai, S.-H. and J.J. Jurinak. The transport of cations in soil columns at different pore velocities. *Soil Sci. Soc. Am. Proc.,* 36:730–733, 1972.

Lawes, J.B., J.H. Gilbert, and R. Warington. On the amount and composition of the rain and drainage waters collected at Rothamsted. III. The drainage water from land cropped and manured. *J. R. Agric. Soc. Engl.,* 18:1–17, 1882.

Lenczewski, M. Comparative Transport of Bacteriophage and Microspheres in an Aquifer under Forced-Gradient Conditions, Thesis presented to the University of Arizona in fulfillment of the requirement for the degree of Master of Science, 1993.

Linderfelt, W.R. and J.L. Wilson, Field study of capture zones in a shallow sand aquifer, in *Transport and Reactive Processes in Aquifers,* Dracos and Stauffer, Eds., Balkema, Rotterdam, 1994.

Litaor, M.I. Review of soil solution samplers. *Water Resour. Res.,* 24:727–733, 1988.

Luxmoore, R.J. Micro-, meso- and macroporosity of soil. *Soil Sci. Soc. Am. J.,* 45:671, 1981.

Luxmoore, R.J., P.M. Jardine, G.V. Wilson, J.R. Jones, and L.W. Zelazny. Physical and chemical controls of preferred path flow through a forested hillslope. *Geoderma,* 46:139–154, 1990.

Ma, L. and H.M. Selim. Predicting the transport of atrazine in soils: Second-order and multireaction approaches. *Water Resour. Res.,* 30:3489–3498, 1994.

Maloszewski, P. and A. Zuber. Mathematical modeling of tracer behavior in short-term experiments in fissured rocks. *Water Resour. Res.,* 26:1517–1528, 1990.

Maloszewski, P. and A. Zuber. Tracer experiments in fractured rocks: Matrix diffusion and the validity of models. *Water Resour. Res.,* 29:2723–2735, 1993.

McDonnell, J.J. A rationale for old water discharge through macropores in a steep, humid catchment. *Water Resour. Res.,* 26:2821–2832, 1990.

McKay, L.D., J.A. Cherry, R.C. Bales, M.T. Yahya, and C.P. Gerba. A field example of bacteriophage as tracers of fracture flow. *Environ. Sci. Technol.,* 27:1075–1079, 1993a.

McKay, L.D., R.W. Gillham, and J.A. Cherry. Field experiments in a fractured clay till. 2. Solute and colloid transport. *Water Resour. Res.,* 29:3879–3890, 1993b.

McKay, L.D., W.E. Sanford, J.M. Strong-Gunderson, and V. De Enriquez. Microbial Tracer Experiments in a Fractured Weathered Shale near Oak Ridge, Tennessee. International Assoc. Hydrogeologists Congress, Edmonton, Alberta, Canada, June 5–10, 1995.

McMahon, M.A. and G.W. Thomas. Chloride and tritiated water flow in disturbed and undisturbed soil cores. *Soil Sci. Soc. Am. Proc.,* 38:727–732, 1974.

Moline, G.R., W.E. Sanford, R.S. Burlage, J.M. Strong-Gunderson, D. Crumbie, A. Harton, and L.D. McKay. Assessing fracture interconnectivity and fracture flow dynamics using multiple groundwater tracers. *EOS, Trans. Am., Geophys. Union,* 76(46):F239, 1995.

Murali, V. and L.A.G. Aylmore. No-flow equilibration and adsorption dynamics during ionic transport in soils. *Nature,* 283:467–469, 1980.

Nkedi-Kizza, P., P.S.C. Rao, R.E. Jessup, and J.M. Davidson. Ion exchange and diffusive mass transfer during miscible displacement through an aggregated oxisol. *Soil Sci. Soc. Am. J.,* 46:471–476, 1982.

Nkedi-Kizza, P., J.W. Biggar, M. Th. van Genuchten, P.J. Wierenga, H.M. Selim, J.D. Davidson, and D.R. Nielsen. Modeling tritium and chloride 36 transport through an aggregated oxisol. *Water Resour. Res.,* 19:691–700, 1983.

Omoti, U. and A. Wild. Use of fluorescent dyes to mark the pathways of solute movement through soils under leaching conditions: 2. Field experiments. *Soil Sci.,* 128:98–104, 1979.

Parker, J.C. and A.J. Valocchi. Constraints on the validity of equilibrium and first-order kinetic transport models in structured soils. *Water Resour. Res.* 22:399–407, 1986.

Posadas, D., A. Tannus, H. Panepucci, and S. Crestana. Magnetic resonance imaging as a non-invasive technique for investigating 3-D preferential flow occurring with stratified soil samples. *Comp. Electr. Agric.,* 14:255–267, 1996.

Raven, K.G., K.S. Novakowski, and P.A. Lapcevic. Interpretation of field tracer tests of a single fracture using a transient solute storage model. *Water Resour. Res.,* 24:2019–2032, 1988.

Reedy, O.C., P.M. Jardine, G.V. Wilson, and H.M. Selim. Quantifying the diffusive mass transfer of nonreactive solutes in columns of fractured saprolite using flow interruption. *Soil Sci. Soc. Am. J.,* 60:1376–1384, 1996.

Reimus, P.W. The Use of Synthetic Colloids in Tracer Transport Experiments in Saturated Rock Fractures, Thesis presented to the University of New Mexico in fulfillment of the requirements for the degree of Doctor of Philosophy, LA-13004-T, 1996.

Richard, T.L. and T.S. Steenhuis. Tile drain sampling of preferential flow on field. *J. Contam. Hydrol.,* 3:307–325, 1988.

Ringrosevoase, A.J. Measurement of soil macropore geometry by image analysis of sections through impregnated soil. *Plant Soil,* 183:27–47, 1996.

Ritchie, J.T., D.E. Kissel, and E. Burnett. Water movement in undisturbed swelling clay soil. *Soil Sci. Soc. Am. Proc.,* 36:874–879, 1972.

Roth, K., W.A. Jury, H. Flühler, and W. Attinger. Field scale transport of chloride through an unsaturated field soil. *Water Resour. Res.,* 27:2533–2541, 1991.

Sanford, W.E., R.G. Shropshire, and D.K. Solomon. Dissolved gas tracers in groundwater: Simplified injection, sampling, and analysis. *Water Resour. Res.,* 32:1635–1642, 1996.

Sanford, W.E. and D.K. Solomon. Site characterization and containment assessment with dissolved gases. *J. Environ. Engineering,* (in press), 1996.

Schulin, R., P.J. Wierenga, H. Flühler, and J. Leuenberger. Solute transport through a stony soil. *Soil Sci. Soc. Am. J.,* 51:36–42, 1987.

Schumacher, W. *Die Physik des Bodens,* Berlin, 1864.

Selim, H.M., R. Schulin, and H. Flühler. Transport and ion exchange of calcium and magnesium in an aggregated soil. *Soil Sci. Soc. Am. J.,* 51:876–884, 1987.

Severson, R.C. and D.F. Grigal. Soil solution concentrations: effect of extraction time using porous ceramic cups under constant tension. *Water Resour. Bull.,* 12:1161–1170, 1976.

Seyfried, M.S. and P.S.C. Rao. Solute transport in undisturbed columns of an aggregated tropical soil: Preferential flow effects. *Soil Sci. Soc. Am. J.,* 51:1434–1444., 1987.

Shaffer, K.A., D.D. Fritton, and D.E. Baker. Drainage water sampling in a wet, dual-pore soil system. *J. Environ. Qual.,* 8:241–246, 1979.

Shropshire, R.G. Dual-Tracers: A Tool for Studying Matrix Diffusion and Fracture Parameters, Thesis presented to the University of Waterloo, Ontario, Canada, in fulfillment of the thesis requirement for the degree of Master of Science in Earth Science, 1995.

Shuford, J.W., D.D. Fritton, and D.E. Baker. Nitrate-nitrogen and chloride movement through undisturbed field soil. *J. Environ. Qual.,* 6:255–259, 1977.

Singh, P., R.S. Kanwar, and M.L. Thompson. Macropore characterization for two tillage systems using resin-impregnation technique. *Soil Sci. Soc. Am. J.,* 55:1674–1679, 1991.

Smart, P.L. and I.M.S. Laidlaw. An evaluation of some fluorescent dyes for water tracing. *Water Resour. Res.,* 13:15–33, 1977.

Smettem, K.R.J. Soil-water residence time and solute uptake. III. Mass transfer under simulated winter rainfall conditions in undisturbed soil cores. *J. Hydrol. (Amsterdam),* 67:235–248, 1984.

Smith, M.S., G.W. Thomas, R.E. White, and D. Ritonga. Transport of *Escherichia coli* through intact and disturbed soil columns. *J. Environ. Qual.,* 14:87–91, 1985.

Starr, J.L., H.C. DeRoo, C.R. Frink, and J.-Y. Parlange. Leaching characteristics of a layered field soil. *Soil Sci. Soc. Am. J.,* 42:386–391, 1978.

Toran, L. and A.V. Palumbo. Colloid transport through fractured and unfractured laboratory sand columns. *J. Contam. Hydrol.,* 9:289–303, 1991.

Valocchi, A.J. Validity of the local equilibrium assumption for modeling sorbing solute transport through homogeneous soils. *Water Resour. Res.,* 21:808–820, 1985.

Vermeul, V.R., J.D. Istok, A.L. Flint, and J.L. Pikul, Jr. An improved method for quantifying soil macroporosity. *Soil Sci. Soc. Am. J.,* 57:809–816, 1993.

Warner, G.S. Characterization of Soil Macropores by Computed Tomography, Ph.D. thesis, University of Minnesota, 1990.

Watson, K.W. and R.J. Luxmoore. Estimating macroporosity in a forest watershed by use of a tension infiltrometer. *Soil Sci. Soc. Am. J.,* 50:578–582, 1986.

Whipkey, R.Z. Subsurface stormflow from forested slopes. *Bull. Int. Assoc. Sci. Hydrol.,* 10:74–85, 1965.

Whipkey, R.Z. Storm runoff from forested catchments by subsurface routes. In: *Floods and Their Computation,* Vol. II. I.A.S.H. Publ. 85, 1969.

Wildenschild, D., K.H. Jensen, K. Villholth, and T.H. Illangasekare. A laboratory analysis of the effect of macropores on solute transport. *Ground Water,* 32:381–389, 1994.

Wilson, G.V. and R.J. Luxmoore. Infiltration, macroporosity, and mesoporosity distributions on two forested watersheds. *Soil Sci. Soc. Am. J.,* 52:329–335, 1988.

Wilson, G.V., J.M. Alfonsi, and P.M. Jardine. Spatial variability of saturated hydraulic conductivity of the subsoil of two forested watersheds. *Soil Sci. Soc. Am. J.,* 53:679–685, 1989.

Wilson, G.V., P.M. Jardine, R.J. Luxmoore, L.W. Zelazny, D.A. Lietzke, and D.E. Todd. Hydrogeochemical processes controlling subsurface transport from an upper subcatchment of Walker Branch watershed during storm events. 1. Hydrologic transport processes. *J. Hydrol.,* 123:297–316, 1991a.

Wilson, G.V., P.M. Jardine, R.J. Luxmoore, L.W. Zelazny, D.E. Todd, and D.A. Lietzke. Hydrogeochemical processes controlling subsurface transport from an upper subcatchment of Walker Branch watershed during storm events. 2. Solute transport processes. *J. Hydrol.,* 123:317–336, 1991b.

Wilson, G.V., P.M. Jardine, and J.P. Gwo. Modeling the hydraulic properties of a multiregion soil. *Soil Sci. Soc. Am. J.,* 56:1731–1737, 1992.

Wilson, G.V., P.M. Jardine, J.D. O'Dell, and M. Collineau. Field-scale transport from a buried line source in variable saturated soil. *J. Hydrol.,* 145:83–109, 1993.

Wilson, R.D. and D.M. MacKay. The use of sulfur hexafluoride as a conservative tracer in saturated sandy media. *Ground Water,* 31:719–724, 1993.

Yang, Z., R.S. Burlage, W.E. Sanford, and G.R. Moline. DNA-Labeled Silica Microspheres for Groundwater Tracing and Colloid Transport Studies. Proc., 212th National Meeting, Am. Chem. Soc., Orlando, FL, Aug. 25–30, 1996.

CHAPTER TEN

Transport in Unsaturated Soil: Aggregates, Macropores, and Exchange

B.E. Clothier, I. Vogeler, S.R. Green, and D.R. Scotter

INTRODUCTION

Soil, the thin and fragile skin of Planet Earth, plays a critical role in the sustainability of land management systems. As a biologically active and fertile realm, soil provides us with the base means of producing food, and it also offers us the possibility to safely dispose of certain benign waste products. Furthermore, soil lies astride the main thoroughfare along which rainwater enters our subterranean and surface water reserves. The water quality of these receiving bodies is thus strongly influenced by the transport processes and mechanisms of exchange that prevail during the rainwater passage through the biologically-active surficial layer of the globe—the medium we call soil. Development of sustainable systems of primary production and land protection demands a proper understanding of flow and transport processes in soil.

Physically, soil is a porous medium comprising a complex arrangement of the three phases of solid, liquid, and air. Biological and chemical processes in the soil require that there be an unimpeded passage through the medium of water and gases. Such transport is greatest when the soil possesses a pedological structure characterized by aggregate stability and macroporosity. Indeed, pedological structure is a key indicator of the health of soil.

Unfortunately, when the structure of soil is not an isotropically ordered arrangement of these phases, with a small representative elementary volume, the classical techniques of mathematical analysis struggle to describe water flow and chemical transport processes since they are characterized by nonequilibrium processes. Whereas these partial differential equations have provided us with underpinning descriptions of flow and transport in uniform and isotropic soil with short-range order, their successful application to the structured soils of the field has been hampered by an ability to treat adequately processes that display local physical nonequilibria.

In structured soils, not all of the soil water appears to be equally involved in actively transporting solute. Many strategies have been used for modeling the preferential transport of solute by the more-rapidly moving water, as the wide range of chapters in this book indicates. These approaches range, *inter alia,* from the use of convective-stochastic transfer functions, pore network schemes, and cellular-automaton fluids, through to the traditional convective-dispersion equation. As in previous chapters, we are concerned with the problem of parameterizing the mobile-immobile water version of the convection-dispersion equation during inert solute transport engendered by unsaturated water flow. We conclude by considering the impact of the initial water content on physical nonequilibrium conditions, as well as reactive solute transport during flow characterized by physical nonequilibrium.

THE MOBILE-IMMOBILE MODEL

We present here some observations of the role of aggregates and macropores in providing flow-paths of differing mobilities which serve to create physical nonequilibrium conditions between the invading solution and that initially resident in the soil. We revisit data obtained with the strongly aggregated Ramiha silt loam (Magesan et al., 1995; Clothier et al., 1996), and compare them with transport data obtained with the weakly structured Manawatu fine sandy loam which is known to have a matrix-macropore dichotomy (Clothier and Smettem, 1990). Attempts to model these observations in order to parameterize the mobile-immobile water version of the convection-dispersion equation are described.

Aggregates and Macropores

Aggregates and macropores can create a wide spectrum of pore-water velocities associated with a solution that is either invading the soil, or draining from it. In an aggregated soil, the more rapid water flow-paths will be those that are connected throughout the inter-aggregate domain. So inter-aggregate water will preferentially wend its way around the aggregates, which would tend to possess smaller, yet connected intra-aggregate pores. These would not seem to be as effective at transport.

Pedogenic and biogenic processes can impart upon a soil a macroporous structure characterized by a longitudinal connectedness. The enhanced mobility of flow down these variously sized macropores, as predicted by Poiseuille's law, is abetted by macropore connectivity to produce preferential infiltration of the invading solution, or enhanced drainage of some part of the resident solution. So, in both aggregated soil and macroporous soil, local heterogeneity in the soil pedological structure leads to conditions of physical nonequilibrium between the invading or draining solution, and that initially resident in the soil.

One of the most popular mechanistic models that is currently used to model the physical nonequilibrium that results from either the presence of aggregates or macropores was presented by van Genuchten and Wierenga (1976), which was based on the scheme developed in petroleum engineering by Coats and Smith (1964). Occam's razor is simply used to cut the soil volumetric water content θ (m^3 m^{-3}) in two. The great simplicity of this model is that one fraction of the soil water, θ_{im} (m^3 m^{-3}), is considered, for transport purposes, to be stagnant. The mind's eye would see this immobile water as residing in one case in the intra-aggregate domain, or being resident within the micropores of the matrix in the other. The complementary portion of the soil water, θ_m (m^3 m^{-3}), is deemed to be mobile, as might be imagined of the water that is between the aggregates, or flowing in the macropores. Thus at any time t, and place z, it is simply considered that

$$\theta(z,t) = \theta_m(z,t) + \theta_{im}(z,t) \tag{10.1}$$

Such a binary split of course will not be realistic, yet the simplicity of the separation is appealing because of its analytical convenience. Formally, this conceptual model allows solute exchange between the two domains only by a diffusion-like processes.

Meanwhile the convective part of the solute transport model permits hydrodynamic dispersion, but only in the mobile domain.

During invasion of the soil by an infiltrating solution, or drainage of some part of the resident solution, the mobile-immobile model will predict a smearing process to occur as a result of physical nonequilibrium between the two domains. Inert solute contained in the infiltrating solution at concentration C_i (mol m^{-3}), will appear preferentially at depth as C_o (mol m^{-3}) in aggregated soil because of more rapid transport with the inter-aggregate water (Figure 10.1). The shape of the breakthrough of inert solute is in this case most likely to be dominated by diffusive exchange processes due to physical nonequilibrium between the inter-aggregate and intra-aggregate solutions, and only partly by pore-water velocity in the inter-aggregate space. Given the sphere-like nature of the aggregates, and the nonlinear nature of spherical diffusion, the shape of the solute front will be further modified by the range in the radii of the aggregates. The theoretical and experimental studies of Passioura (1971) and Rao et al. (1980a, 1980b) indicated that for artificially-aggregated media in the laboratory, spherical diffusion typically begins to dominate when the aggregate diameter exceeds about 2 mm.

For a soil ramified by macropores that are connected within a microporous matrix, an earlier-than-expected arrival of solute will occur due to preferential transport along the larger and more-connected macropores (Figure 10.1). This higher velocity transport, coupled with mechanical mixing between the variously-sized macropore networks, will result in a breakthrough of inert solute in which the smearing is dominated by hydrodynamic dispersion. The relative role of diffusion from the macropores will depend upon the comparative size of the macropores. In the lower limit the pore sizes will be of the same order, and the flow will become uniform since local equilibrium will prevail. Scotter (1978) observed solute breakthrough in soil containing well-defined channels or cracks. A dramatic jump in preferential solute flow was shown when channels exceeded 0.2 mm in diameter, and when cracks were wider than 0.1 mm.

Theory

Solute Transport

Diffusion-Dispersion Modeling

The simplest deterministic model that accounts analytically for preferential solute flow is the steady-state mobile-immobile water version of the convection-dispersion equation (van Genuchten and Wierenga, 1976). Solute exchange between the two domains is only allowed by diffusion. So this model can *sensu stricto* only apply to flow in a soil when the water content θ exceeds θ_{im}. At any depth, the steady pore-water velocity in the mobile domain $v_m(z)$ (m s^{-1}) can simply be found as the steady invading flux density q(z) (m^3 m^{-2} s^{-1}) divided by $\theta_m(z)$, being $\theta(z) - \theta_{im}(z)$. This two-domain formulation for solute transport can be written as

$$\theta_m \frac{\partial C_m}{\partial t} + \theta_{im} \frac{\partial C_{im}}{\partial t} = \theta_m D_s \frac{\partial^2 C_m}{\partial z^2} - v_m \theta_m \frac{\partial C_m}{\partial z} \tag{10.2}$$

where C_m and C_{im} are the concentrations of inert chemical in the mobile and immobile domains, and D_s is the solute dispersion coefficient (m^2 s^{-1}). It is assumed that molecular diffusion in the direction of flow in the mobile domain is negligible, and

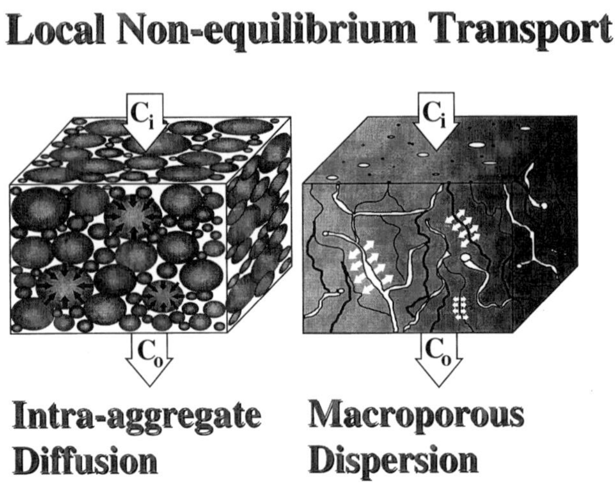

Figure 10.1. Local nonequilibrium transport through an aggregated soil in which the preferential transport is dominated by intra-aggregate diffusion (left); and in a macroporous soil in which the solute breakthrough is controlled by longitudinal hydrodynamic dispersion (right).

that hydrodynamic dispersion is in proportion to the average pore-water velocity in the mobile domain,

$$D_s = \lambda \, v_m \tag{10.3}$$

where the constant of proportionality λ (m) is the dispersivity. The only mechanism to transfer solute between domains is, for analytical convenience, limited to a diffusion-like process that can be described by

$$\theta_{im} \frac{\partial C_{im}}{\partial t} = \alpha \, (C_m - C_{im}) \tag{10.4}$$

where α (s^{-1}) is the first-order mass transfer coefficient. For an aggregated soil, Rao et al. (1980a, 1980b) predicted α to vary with aggregate radius, the molecular diffusion coefficient, and the fraction of the soil water that resides in the intra-aggregate pores. Kookana et al. (1993) even found a strong coherence between reported values of this diffusion parameter α and pore water velocity v_m.

Application of this mobile-immobile model requires binary separation of the soil water content into two complementary domains of θ_{im} and θ_m, plus a knowledge of α, the rate of solute diffusion between them, as well as an understanding of λ in the mobile domain. van Genuchten and Dalton (1986) lamented the "...obscure dependency of the mass transfer coefficient on the diffusion properties of...aggregate geometry and the diffusion coefficient." The prime objective of the mobile-immobile model was to account for the physical nonequilibrium brought on by the structured nature of the pore system of soil, yet α does not allow a clear modeling of this. However, if an assumption is made about the geometrical structure of soil, then appropriate forms of the diffusion equation can be used to examine solute exchange between a domain containing an inert solute, and the complementary pore space that is devoid of it.

Lumped Parameter Modeling

Passioura (1971) developed a theory for hydrodynamic dispersion in aggregated media by considering the microscopic flow geometry of viscous transport through the inter-aggregate pore space, and molecular diffusion within the spherical aggregates of radius a (m). In the mobile-immobile notation of above, Passioura's (1971) theory predicts the dispersed concentration in the inter-aggregate pore space at the surface of the spherical aggregates, *viz.* $C_m(z,a,t)$. But the aggregates are themselves a sink for part of this solute that is passing over the aggregates' surface, so that the mass transfer can be found by solving the spherical diffusion equation

$$\frac{\partial C(r)}{\partial t} = \frac{D_a}{r^2}\frac{\partial}{\partial r}\left(r^2 \frac{\partial C(r)}{\partial r}\right) \qquad 0 \le r \le a \tag{10.5}$$

where r is the radial distance from the center of the aggregate, and D_a (m^2 s^{-1}) is the molecular diffusion coefficient for solute within the aggregates. The average solute concentration within the aggregate is given as

$$C_{im}(z,t) = \frac{3}{a^3}\int_0^a r^2\, C(z,r,t)\,\mathrm{d}r \tag{10.6}$$

Thus the impact of the diffusive sink for solute within the aggregates is to increase the smearing of the passing solute front, over and above that which would be expected solely from hydrodynamic dispersion in the inter-aggregate space. For structured soils in which there are linked dispersive and diffusive processes there will be difficulties in isolating the effects due to either transport mechanism. Passioura (1971) proceeded to present a lumped dispersion parameter, $\bar{D}$, that would add to D_s a correction that, to a first approximation, accounts for the effects of intra-aggregate diffusion. For spherical aggregates,

$$\bar{D} = \left(\frac{\theta_m}{\theta}\right) D_s + \frac{\left(1 - {}^{\theta_m}\!/_{\theta}\right) a^2\, v^2}{15\, D_a} \tag{10.7}$$

subject to a criterion that

$$\frac{\left(1 - {}^{\theta_m}\!/_{\theta}\right) D_a\, L}{a^2\, v} > 0.3 \tag{10.8}$$

where L (m) is the depth into the soil, and here v (m s^{-1}) is q (Darcy's flux) divided by the entire water content θ. With this lumped dispersion parameter (Eq. 10.7), a standard, local-equilibrium version of the convection-dispersion equation can be used to account for the enhanced solute dispersion due to interdomain diffusion during transport through an aggregated medium (Rao et al., 1980a, 1980b; Parker and Valocchi, 1986).

Van Genuchten and Dalton (1986) extended this type of approach to consider other well-defined geometries, such as rectangular aggregates, cylindrical aggregates, and hollow cylindrical macropores. In the latter case, they considered a macropore of radius b (m) embedded within a surrounding matrix of radius c (m). They showed the dispersive add-on for diffusion, in this lumped flow system, to be approximately

$$\overline{D} = \left(\frac{\theta_m}{\theta}\right) D_s + \frac{\left[2\ln(c/b) - 1\right]\left(1 - {}^{\theta_m}\!/_{\theta}\right) c^2\ v^2}{4\ D_a} \tag{10.9}$$

These analyses hint at the difficulty of being able to separate out the role of diffusion and of dispersion. They also suggest that it is going to be difficult to find independent ways of measuring appropriate values of α and λ for use in the mobile-immobile transport model of Eqs. 10.2 and 10.3. Given the unrealistic binary split of the pore space either into mobile (β) or stagnant water (1-β), and given the unitary α and λ's that purport to represent flows between and within these domains, it is then not surprising that these parameters are themselves found to possess complex dependencies upon geometric and flow characteristics (Kookana et al., 1993; Rao et al., 1980a, 1980b).

Water Flow

Infiltration experiments on prewet soil that is structured, often during steady flow, have been used to examine the utility of this mobile-immobile model (Eqs. 10.2 and 10.3). Frequently the predictions have been an improvement over straightforward use of the local equilibrium model. Because this mobile-immobile form only considers physical nonequlibrium in terms of solute transport rather than water movement, difficulties will arise when trying to use it to describe solute invasion during transient infiltration into a soil which is initially at a low water content (Gaudet et al., 1977; Bond and Wierenga, 1990). Here the pores in the so-called immobile domain might be empty and possess a capillary attractiveness that will result in direct convective invasion of solute-bearing water. In Eqs. 10.2 and 10.3 there is no means by which such convective water transfer can occur.

Transient water flow into unsaturated soil is normally described by Richards' equation which accounts both for the diffusion-like capillary attraction of water into soil, as well as direct convection by gravity. With water content as the dependent variable, and including a sink for root water uptake S ($m^3\ m^{-3}\ s^{-1}$), Richards' equation can be written as

$$\frac{\partial\theta}{\partial t} = \frac{\partial}{\partial z}\left[D_W(\theta)\frac{\partial\theta}{\partial z}\right] - \frac{\partial K_W(\theta)}{\partial z} - S(\theta, z) \tag{10.10}$$

where z (m) is the vertical ordinate, D_w the soil water diffusivity function ($m^2\ s^{-1}$), and K_w the hydraulic conductivity function ($m\ s^{-1}$). Boundary and initial conditions that would be appropriate for unsaturated flow into a column of soil from a disk permeameter set at a head h_o (m), are

$$\begin{aligned} &\theta(z,t) = \theta_n(z,t) && z \geq 0, && t = 0 \\ &\theta_o = \theta(h_o) && z = 0, && t > 0 \end{aligned} \tag{10.11}$$

Infiltration using this approach involves imbibition of the wetted pore space θ comprising θ_m and θ_{im}. Thus if the soil initially has $\theta_n < \theta_{im}$, then solution of the water flow equation should provide for convection of water into the so-called immobile domain, which poses more problems than just the semantic concern of what to call any water that moves into the immobile domain! Laboratory experiments with undisturbed columns (Kluitenberg and Horton, 1990), and also with repacked media (Gaudet, 1978; Bond and Wierenga, 1990), have demonstrated the impact that the mode of water entry into soil can have on the preferential nature of flow, due to physical nonequilibrium between the invading solution and some part of the porosity that does not fully partake in flow. So whereas over the last 20 years we seem to have readily accepted a two-domain model for solute transport, only recently have we begun to model more realistically the water vehicle in like fashion (Gerke and van Genuchten, 1993; Jarvis, 1994). Macropore flow and soil dry-down by roots conspire to create conditions of physical nonequilibrium during infiltration that demand proper attention.

Nonetheless, given the simplicity of the mobile-immobile model, and because of the pressing need to model preferential solute transport over time periods longer than just the first wetting of soil, Eqs. 10.2, 10.3, 10.4, and 10.10 have been cobbled together forcibly in an unholy alliance to provide some useful predictions of solute transport under conditions of physical nonequilibrium. Tillman et al. (1991) allowed water and solute to be directly convected into the immobile domain upon first wetting if $\theta < \theta_{im}$. Otherwise solute transport between the domains was by diffusion only. In their numerical scheme, Vanclooster et al. (1992) took a different tack. They solved in unison the transient forms of Eqs. 10.2, 10.3, 10.4, and 10.10 by simply assuming that, irrespective of the absolute value of θ, only some fixed fraction β of the wetted pore space is mobile. Somewhat incongruously then, $\theta_m(z,t) = \beta.\theta(z,t)$, so that the immobile domain is not associated with any fixed pore class. The numerical solution we later use here for illustrative purposes also adopts this approach.

Parameterization

Hydraulic Properties

Permeametry has, over the last two decades, provided us with better measurements of the hydraulic character of unsaturated soil in the critical region close to saturation (Clothier and White, 1981; Perroux and White, 1988; Ankeny et al., 1991). These data have revealed the matrix macropore dichotomy of field soils. The two functions required to describe water transport are those of the diffusivity D_w and the conductivity K_w (Eq. 10.10), which are linked via the water characteristic $h(\theta)$: $D_w = K_w\, dh/d\theta$ where h is the soil water pressure head (m). We present here in Figure 10.2, an accumulation of disk permeameter data and laboratory results for the hydraulic conductivity function $K_w(h)$ of Manawatu fine sandy loam (Clothier and Smettem, 1990; Clothier et al., 1995). This Dystric Fluventic Eutrochrept is a recent alluvial soil. The knee in the data around –50 mm is typical of that found for many soils by Jarvis and Messing (1995) indicating a quite-distinct separation between matrix pores and macropores.

White and Sully (1987) developed a macroscopic capillary length scale that they termed the flow-weighted mean pore size l_m (m). This mean pore size is proportional to the slope of the logarithmic plot of $K_w(h)$. The data of Figure 10.2 would indicate that in the region close to saturation, at about h_o= –20 to –40 mm, the hydraulically inferred mean pore size would be about 0.5 mm. At this head, the hydraulic conductivity of the soil is around 5–10 mm/hr, which happens to be about the rainfall rate of

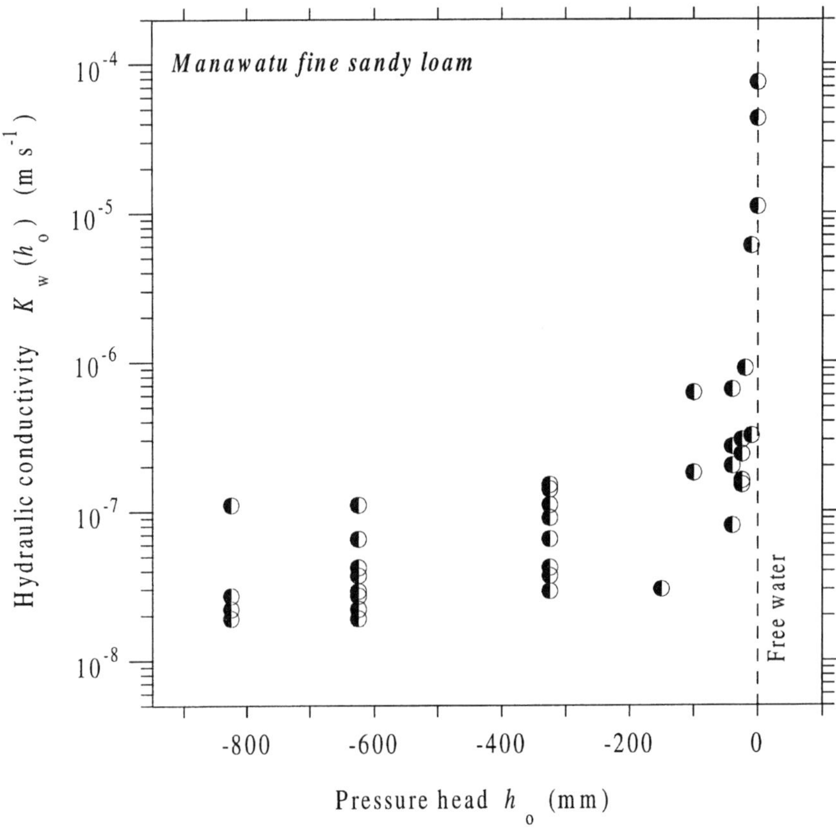

Figure 10.2. Hydraulic conductivity K_w of Manawatu fine sandy loam as a function of pressure head h_o. In the range close to saturation the data derive from disc permeametry, and at heads drier than –200 mm they were obtained in the laboratory on undisturbed cores. (Combined data from Clothier and Smettem, 1990; Clothier et al., 1995).

an average storm, but less than any rate likely to be used for irrigation. At heads wetter than this, K_w climbs dramatically, with l_m being of the order of 1–2 mm. This indicates a prevalence of significant macropores. So for heavy rainfall and irrigation, matrix ponding is likely to occur on the surface of this soil, creating free water which would have the opportunity to flow preferentially down any surface-vented macropores and create subterranean conditions of physical nonequilibrium.

Chemical Transport Properties

A great deal of effort has been invested in hydraulically characterizing soil, and the results of this work highlight the role that local variability in macroporous networks can play in generating preferential water transport in soil. Whereas there have been many observations of the impact of this preferential flow on solute transport, there has not been a commensurate effort to measure independently the properties governing solute transport. In relation to the mobile-immobile model, van Genuchten and Dalton (1986) noted the circularity that "...in general, values for α, and to some extent the immobile water content θ_{im}, must be fitted to observed data before the model can be used. Because of the diffuse spatial location and configuration of immobile water pockets and associated sorption sites, α and θ_{im} for such soils are difficult to quantify by other than curve-fitting techniques."

In order to measure θ_{im} independently, Clothier et al. (1992) proposed using a tracer-filled disk permeameter to allow infiltration into the soil of a solution at some concentration C_m. They ignored any effects that might accrue from interdomain diffusion. After an infiltration of depth I (m) of this solution, it was hoped that hydrodynamic dispersive effects would have disappeared so that the flux concentration in the mobile domain in the soil just under the disk would be C_m. Thus by removing soil samples from under the disk, and determining the resident concentration of tracer solute C^*, the mobile fraction could be determined using the partition formula

$$\theta C^* = \theta_m C_m + \theta_{im} C_{im} \tag{10.12}$$

If a tracer not initially present in the soil were used, and if diffusive exchange (Eq. 10.4) is considered sufficiently weak, then C_{im} should remain essentially zero over the timescale of the experiment. Clothier et al. (1995) presented results from some field experiments that suggested the value of α might indeed be sufficiently small to allow use of Eq. 10.12 in the form,

$$\theta_m = \theta\left(\frac{C^*}{C_m}\right) \tag{10.13}$$

via which θ_m can be obtained independently from measurements immediately under the disk of just C^* and θ. For a soil having $\lambda \cong 20$ mm, as might the Manawatu fine sandy loam, they showed that an infiltration of about $I \cong 25$ mm is needed to allow dispersive effects to be eliminated from the flux concentration just under the disk. Their results for this soil suggested that θ_m/θ is about 0.7 to 0.75 at a head of $h_o = -40$ mm.

Subsequently, Jaynes et al. (1995) added a novel aspect to this technique. Through the use of the sequential invasion into the same soil under the permeameter of i inert tracers, they showed how differences in the respective resident concentrations of C_i^* would reflect the rate of diffusion into the immobile domain. The resident concentration in the soil of any tracer i would reflect the diffusion that had taken place during the time t_i it had been present in the mobile domain. From the regression of the measurements using

$$\ln\left[1 - \frac{C_i^*}{C_{m,i}}\right] = \frac{-\alpha}{\theta_{im}} t_i + \ln\left[\frac{\theta_{im}}{\theta}\right] \tag{10.14}$$

they could deduce both θ_m and α. From many experiments, they found α to range between about 5×10^{-7} and 5×10^{-6} s^{-1}. Necessarily, they had to assume that the flux concentration in the mobile domain of the ith tracer was immediately $C_{m,i}$, so that any changes in the resident concentration would only be due to diffusion (Eq. 10.4). This then assumes an instantaneous rise in the flux concentration under the disk to $C_{m,i}$, which implies there is no hydrodynamic dispersion, that is $\lambda \cong 0$.

Simulations

The technique of Clothier et al. (1992) accounts for λ, but assumes $\alpha \cong 0$ to realize the value of θ_m (Eq. 10.13). Conversely, the multiple tracer technique of Jaynes et al.

(1995) demands that hydrodynamic dispersion be ignored, $\lambda \cong 0$, so that θ_{im} and α can be determined. Is it possible to devise simple field experiments, such as those described by either Clothier et al. (1992) or Jaynes et al. (1995) to resolve independently the properties required by the mobile-immobile model of Eqs. 10.2 and 10.3?

To explore this possibility we use a numerical scheme that solves simultaneously transient forms of Eqs. 10.2, 10.3, 10.4, and 10.10, subject to condition (Eq. 10.11). These 1-D simulations are carried out using properties appropriate to a disk permeameter set at h_o=–40 mm, sitting atop a 1-D pedestal of Manawatu fine sandy loam initially at θ_n=0.3. At this head, the soil will wet to θ_o=0.42, whereas θ_s=0.43. The saturated K_w was set at $K_s = 3 \times 10^{-3}$ mm s^{-1}, with a power function exponent of 15 used to reduce K_w from K_s as a function of the normalized water content. A sorptivity of 0.2 mm s$^{-1/2}$ was used, and 4 was taken as the exponential slope of the D_w function with respect to the normalized water content (Clothier and White, 1981). The simulation was run for 15,000 s so that an infiltration I of 26 mm entered the soil. These conditions and properties were used to mimic conditions that were observed in the field on pedestals, just under the disk with permeameters under tension. The time of infiltration would most likely represent the interest span of field technologists who might consider using permeameters to measure the soil chemical transport characteristics, as well as its hydraulic properties, especially given the obvious need to replicate observations.

The water transport equation of Eq. 10.10, subject to Eq. 10.11, was solved by finite differencing using Newton-Raphson iteration over a column length of 350 mm divided into 35 intervals. The simulated profile of water content at the end of this infiltration event is shown in Figure 10.3a, such that the front of water invasion is seen to be diffused around a depth of 225 mm.

The prime purpose of the exercise however, is to determine the respective roles of θ_m, λ, and α in generating profiles of the resident concentration, θC^* under the permeameter. We will also explore the potential that measurements of this θC^* might be used to obtain independent measures of those inert solute transport parameters that are a consequence of the physical nonequilibrium that prevails during infiltration and drainage in aggregated and macroporous soil. A transient form of the mobile-immobile model (*cf.* Eqs. 10.2, 10.3, and 10.4) was solved for a given value of the mobility fraction, β, by using finite differencing with a Crank-Nicolson scheme that used a numerical dispersion correction similar to that of Gerke and van Genuchten (1993). The value of $v_m(z, t)$ used in Eq. 10.2 came from numerical solution of the water transport equation. The concentration of the invading solution, C_m, was for simplicity set to unity. Hence in the absence of an immobile fraction, or after interdomain diffusion had ironed out any local differences, θC^* would eventually attain a proximal value of $\theta_o C_m = 0.42$.

Diffusion by α

The invading front of inert solute pivots about $z \cong 55$ mm for various values of the interdomain mass transfer coefficient α (Figure 10.3a). If α were lower than 10^{-6} s^{-1}, then interdomain diffusion could be considered ineffective over the 4 hr 10 min of this simulated disk measurement, and proximally the total resident concentration θC^* would remain close to 0.273 which here is $\theta_m C_m$. Separation in the profiles of resident concentration is only evident if $10^{-6} < \alpha$ (s^{-1}) $< 5\times10^{-5}$. Of course, Jaynes et al. (1995) used the time rise in θC^* close to the disk to resolve α, rather than depthwise observations of $\theta(z)C^*(z)$. Nonetheless, this simulation serves to show the possible range within which a value of α might be deduced. Over a two-day period, Clothier et al. (1995) thought their observations of the changing θC^* implied an α of 6×10^{-6} s^{-1}.

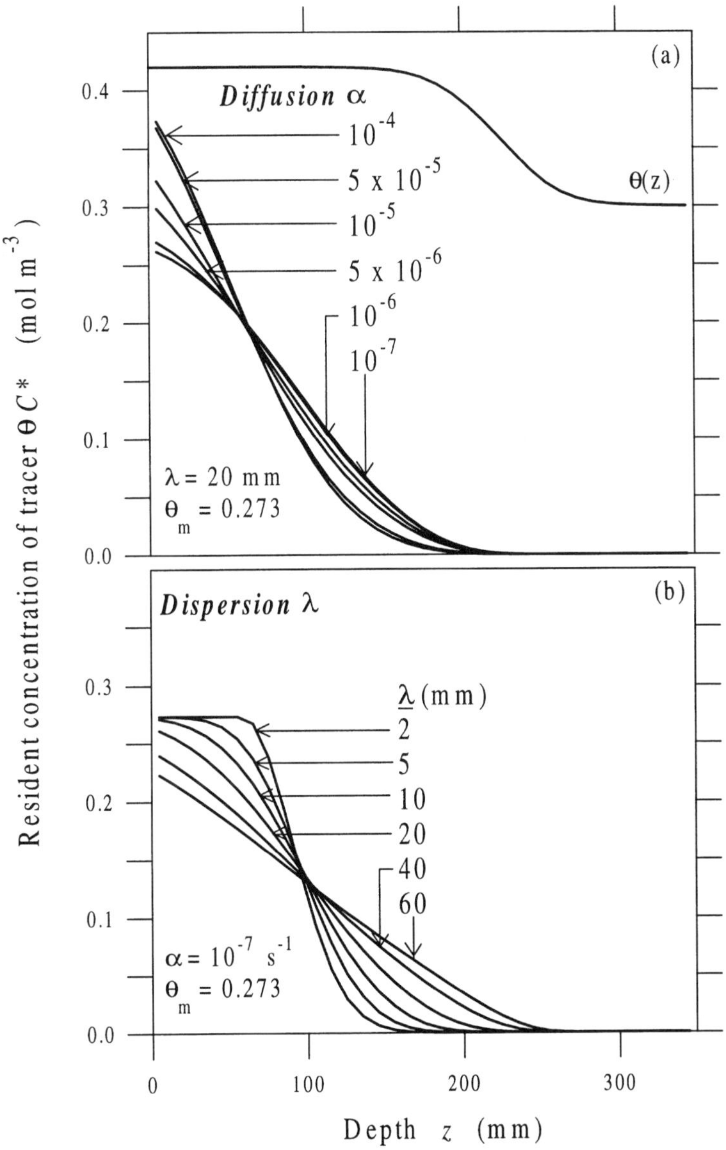

Figure 10.3. (a) The impact of variation in the interdomain mass transfer coefficient, α, on profiles of the resident concentration, θC*, under the disk permeameter. Here λ was set at 20 mm, and β at 0.65, so that $\theta_m C_m$ would, in the absence of diffusive exchange, only rise to 0.273. **(b)** The impact of variation in the longitudinal dispersion coefficient, λ, on profiles of the resident concentration, θC*, under the disk permeameter. Here α was set at 10^{-7} s^{-1}, hence interdomain diffusion was negligible, so given that β=0.65, the maximum value of θC* would be around 0.273.

Yet their measurements at 14 days did not bear this out, for interdomain differences in *C* still appeared to remain. Values of α from 10^{-8} s^{-1} to 10^{-5} s^{-1} have been reported (Vanclooster et al., 1992; Andreu et al., 1994). As Vanclooster et al. (1992) noted "...α can play a key role in predicting the center of mass [and] parameter identification in the field needs additional investigation." However, separation of such diffusive effects, from dispersion-induced changes in either depth- or timewise changes in θC*, needs to be considered when attempting to measure α in the field.

Dispersion by λ

If, in the same simulated experiment, a fixed value of $\alpha = 10^{-7}$ s^{-1} were assumed, then the depthwise profiles in θC^* that would result due to variations in the dispersivity λ are shown in Figure 10.3b. If the diffusion-free technique of Clothier et al. (1992) were used, then after this infiltration of $I = 26$ mm of solution, a reliable value of θ_m (=0.273) could only be obtained from the proximal θC^* if the soil had a $\lambda < 10$ mm. A dispersivity of this order would seem to be the lower limit expected for field soils (Magesan et al., 1995). Indeed, the latter authors suggested that the longitudinal dispersion of solute via the macro- and mesopore networks might be used as an index of soil structure. For the well-structured Ramiha silt loam, an aggregated soil lacking in macropores, Magesan et al. (1995) found λ to be 15–20 mm. For the Manawatu fine sandy loam, an essentially apedal soil, they found that the well-connected macropore network (Figure 10.2) resulted in significant solute dispersion such that λ was 30–70 mm.

Comparison of Figures 10.3a and 10.3b highlight the difficulty of using inverse procedures from *in situ* observations of θC^* to characterize the diffusive component (α) from the dispersive effect (λ) during transport in a mobile-immobile system of fixed mobility fraction β. The confusion mounts when the value of β is also sought using the same information.

Mobility Fraction β

If the values of α and λ happened to be known *a priori*, then the value of θ_m could be deduced from observations of θC^*. The impact of variation in β (=θ_m/θ) is shown in Figure 10.4 for a case where diffusion dominates (α=10^{-5} s^{-1}, Figure 10.4a), and for a case where dispersion accounts for the shape in the solute profile (λ=20 mm, Figure 10.4b). The profiles of Figures 10.4a and 10.4b possess shapes characteristic of either the interdomain diffusion and dispersion dominated profiles of Figures 10.3a or 10.3b. Logically, the penetration of solute is greatest when the mobility fraction is least. The proximal value of the resident concentration rises with an increase in the mobility fraction β such that if the respective roles of diffusion and dispersion were known, measurements of θC^* right under the disk could be used to infer θ_m.

However in the absence of *a priori* information, comparison of Figures 10.3 and 10.4 reveals that it is not possible directly to isolate the multifarious processes that act in unison to shape the invading profile of solute. Figures 10.3a, 10.3b, 10.4a, and 10.4b indicate that there would be an indistinguishable difference in profiles that have either α=10^{-6}, β=0.65 with λ=20 (Figure 10.3a); or α=10^{-7}, β=0.65 with λ=5 (Figure 10.3b); or α=10^{-5}, β=0.35 with λ=20 (Figure 10.4a); or even α=10^{-6}, β=0.55 with λ=20 (Figure 10.4b). Given the range of choice, there will undoubtedly be difficulty in the use of inverse procedures to infer collectively *α, β,* and *λ*, based on observations of $\theta(z,t)\,C^*(z,t)$: even if unitary values of α, β, and λ were truly to exist! Having explored numerically these difficulties, we now turn to observations in the field that might allow some insight into the relative importance, or even relevance, of an immobile-mobile, diffusive-dispersive representation of solute transport.

AGGREGATES

The Ramiha silt loam, a soil of aeolian origin, contains some variable charge due to the presence of windblown volcanic ash in the soil (Cowie, 1978). It is described pedologically as being strongly aggregated. Upon breaking open a clod of this soil, spherical crumbs of consistent radius of $a \cong 2.5$ mm break off the fracture face. So as to

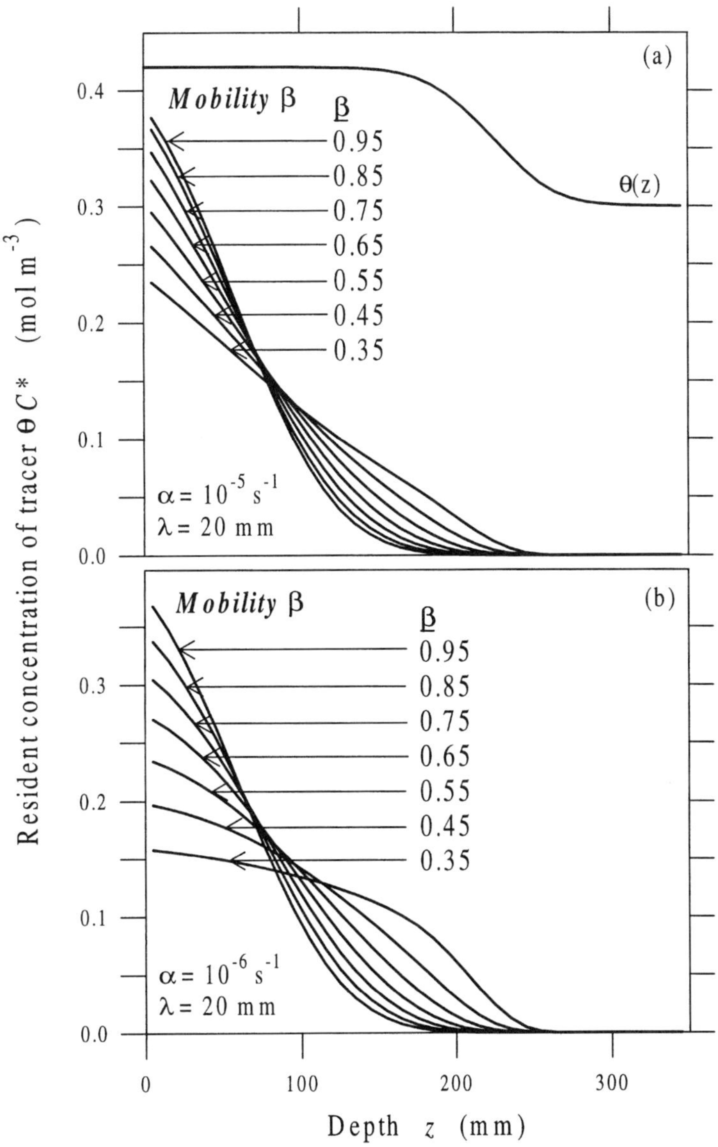

Figure 10.4. **(a)** The impact of variation in the mobility fraction β in shaping the profile in the resident concentration θC* under the permeameter when $\alpha=10^{-5}$ s^{1} and λ=20 mm. Here diffusion is relatively important and α controls the profile shape. **(b)** As for above, but for $\alpha=10^{-6}$ s^{-1} so that dispersion is relatively more important than diffusion.

avoid problems of hydrodynamic dispersion, Clothier et al. (1996) allowed the infiltration of some 90 mm of dual-tracered solution into four undisturbed columns of this soil. All solutions contained a strong concentration of 0.1 *M* KBr such that any anion adsorption effects might be overcome so that this tracer could be considered inert. Two cores were infiltrated with low concentration radioactive sulfate (20 μmol L^{-1} $K_2{}^{35}SO_4$), whereas the other two received a higher strength solution (2 mmol L^{-1} $K_2{}^{35}SO_4$). Results from Clothier et al. (1996) are combined here and reinterpreted in terms of mobile-immobile transport of inert solute into an aggregated soil. Comment is made later with regard to the reactive anion transport of sulfate.

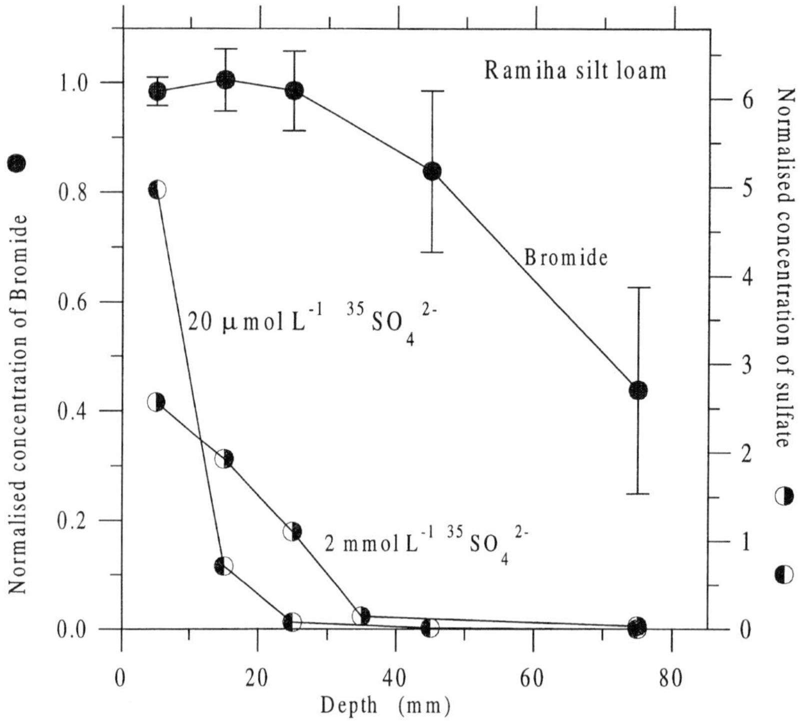

Figure 10.5. Combination of the data presented by Clothier et al. (1996) for the inert transport of bromide and sulfate into the strongly aggregated Ramiha silt loam after an infiltration of some 90 mm of solution. The data for bromide represent the mean from all four cores, whereas the sulfate data each relate to two of the four experiments. The standard deviations are shown as the error bars.

In Figure 10.5, the bromide data from all four cores are presented, where at each depth in every core, a row of five samples was removed. Thus each point of normalized bromide concentration is, in Figure 10.5, the mean and standard deviation of 20 samples. Despite the soil being strongly aggregated, the pore space invasion of bromide is complete, and the 20 proximal values have a very small variance around unity. The variance in the measured resident concentrations of bromide increases with depth of penetration, as might suggest unrequited diffusive transfer.

It could simply be that all the water is effectively mobile ($\theta_m = \theta_o$, Eq. 10.2). Alternatively, it could be that this aggregated soil has a substantial amount of immobile water θ_{im} and that during the course of this 12-hour experiment, α is sufficiently large that interdomain transport is so effective that the resident concentration can become θC_m, because C_{im} rises to the influent value of C_m. On the basis of this single but replicated observation, one cannot choose between either proposition. Irrespective, it is fascinating that flow in this soil of complex pedological structure should appear to be capable of simple description with $\theta_m = \theta_o$, either because truly β=1, or apparently so because α is large enough to make it appear this way.

The aggregate analysis of Passioura (1971), presented as Eq. 10.7, allows some insight as to which horn of the dilemma we reside upon. The Ramiha silt loam crumbles easily to form spherical aggregates of a = 2.5 mm, and finally the average flow through the cores under a surface h_o = –40 mm, was at about 3.3 µm s^{-1}. This average infiltration rate of around 7 mm/hr, as might happen in the field with this soil, is about half the slowest rate in the experiments of Rao et al. (1980b) with artificially aggregated media. If we accept that the hydrodynamic dispersivity for this soil is λ=20 mm, then should this soil have, hypothetically, an mobile fraction of $\beta=\theta_m/\theta_o=0.75$, the first term in Eq. 10.7 would be 0.05 mm^2 s^{-1}. The second term that is due to aggregate diffusion, given that $D_a = 3.33 \times 10^{-4}$ mm^2 s^{-1} (Rao et al., 1980a), would just be 0.0035 mm^2 s^{-1}, even given this $\beta=\theta_m/\theta_o=0.75$. Thus, rather than use a λ of 20 mm in a mobile-immobile formulation for this Ramiha silt loam, the same result could be achieved in a local-equilibrium, convection-dispersion model with just λ=21.4 mm to account for intra-aggregate diffusion. So in this strongly aggregated soil, it is possible to use a local equilibrium model such that we need not consider intra-aggregate diffusion. The flow rate is characteristically low, and the aggregates are typically quite small. So, despite being pedologically complex, solute transport in this soil can be treated quite simply.

MACROPORES

Unlike the complex pedological structure of the Ramiha silt loam, the Manawatu fine sandy loam lacks distinct pedal structure, being a young (*ca.* 500 years) alluvial soil. However, this soil displays a distinct matrix-macropore dichotomy (Figure 10.2). The nexus occurs at an h of around –20 to –50 mm, where K_w is of the order of $K_w \cong 10^{-6}$ m s^{-1}, which happens to be about the common rate of rainfall (5 mm hr^{-1}), yet lower than any expected rate of irrigation. Thus, incipient ponding of the matrix to allow entry of free water into the macropore network will be possible, and will depend critically on the local flux density q at the soil surface.

The results of Clothier et al. (1995) are extended here to explore the role of macropores and the matrix in transporting solute into this pedologically simple soil. These results are compared with the results of other studies carried out on this soil. In Figure 10.6 are shown five normalized profiles of bromide (C^*/C_m) for various values of infiltration I. Dispersive effects can be seen to cause the proximal concentration of bromide to rise as I increases. However, even after some 80 mm of infiltration, a large rainstorm or excessive irrigation, the value of C^*/C_m is still a long way from unity. Indeed, there is a consistency in the 5 near-surface values when I >20 mm, with $C^*/C_m \cong 0.7$. These five points, comprising 30 samples of the resident concentration, possess a low variance, despite representing measurements made after 5 hours through to 3 days of solute contact with the soil surface. Thus it would appear that this soil does indeed possess some immobile water, and/or that inter-domain transfer is ineffective. The mobile water would here appear to be that conducted in the macropore network, and the immobile water that residing in the interstitial volume of the matrix. The flow interruption method devised for field use by Clothier et al. (1995) would lend credence to the inefficacy of inter-domain diffusion over such inter-macropore distances. The dispersion correction of van Genuchten and Dalton (1986), Eq. 10.9, is used illustratively to explore this further.

Intensive sampling was used to determine the spatial pattern of bromide penetration under one disk containing bromide tracer (Figure 10.7). The lateral pattern in C^*/C_m derived from cheek-by-jowl sampling across a diameter under the disk reveals the presence of macropore pathways of preferential flow. A central conduit around the middle of the disk can be seen to take bromide preferentially down to beyond 70 mm. There is a characteristic transverse wavelength in the pattern, suggesting that the annular radius of the surrounding immobile ped (c in Eq. 10.9) could be about 15 mm. The $K_w(h)$ data of Figure 10.2 suggest a flow-weighted mean macropore size l_m of around 0.5 to 2 mm, so that in Eq. 10.9 we can take b as being 1 mm. If we take the same values for β, λ, and D as before, since the velocity here is much lower at 0.625

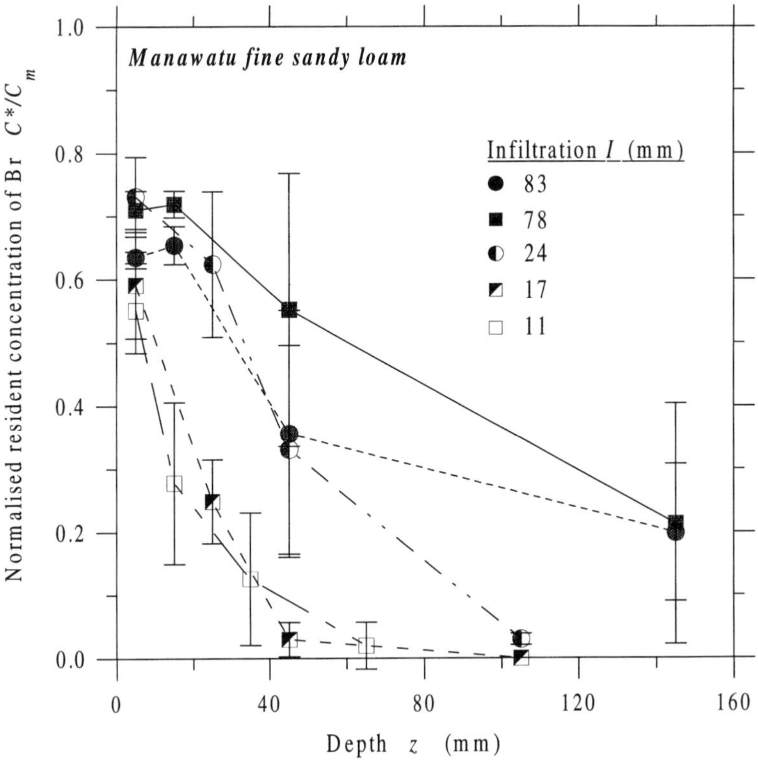

Figure 10.6. A compilation of normalized bromide profiles, C^*/C_m, measured under 5 disk permeameters set at either h_o=–20 or –40 mm and resting on Manawatu fine sandy loam. Each data point comprises 5–6 samples with the standard deviation shown as an error bar.

μm s^{-1}, the hydrodynamic dispersion term in Eq. 10.9 is just 0.0094 $mm^2\ s^{-1}$. However, the second term on the right-hand side, which accounts for the role of matrix diffusion from the macropore, is an order of magnitude bigger at 0.073 $mm^2\ s^{-1}$. The mean dispersion coefficient is thus $\bar{D}$ = 0.082 $mm^2\ s^{-1}$, which is similar to that found for the aggregated Ramiha silt loam, but for very different reasons. For the Manawatu fine sandy loam, this simple geometric consideration suggests that matrix diffusion from the mobile flow in isolated macropores will dominate dispersion during solute transport into this dual-porosity soil. Because of the much slower pore water velocity in this soil, the mean dispersion length $\bar{\lambda}$ that would apply, given this corrected $\bar{D}$ in a local equilibrium model (van Genuchten and Wierenga, 1976), is 175 mm. Such a large dispersivity, due to the correction for simple macropores throughout the matrix, would predict C^*/C_m to rise to just 0.72 after an I of 80 mm (*cf.* Figure 10.6), while after some 135 mm of infiltration C^*/C_m would still only be predicted to be 0.85. Vogeler et al. (1996) found measured λ's to range from just 30–80 mm, stressing that the above analysis is just illustrative, for it only allows convective flow in a series of separated macropores, each of radius b embedded in an annular diffusive matrix of radius c.

Nonetheless, here we find that the proximal value of C^*/C_m rises to just 0.7 after an I of about 80 mm, and using undisturbed cores in the laboratory, Magesan et al. (1995) found C^*/C_m to be 0.85 after I=135 mm. Vogeler et al. (1996) found that after some 1000 mm of infiltration, there still remained a volume fraction of about 0.08 (or 0.04 $m^3\ m^{-3}$) that was apparently excluded from anion transport. This they attributed

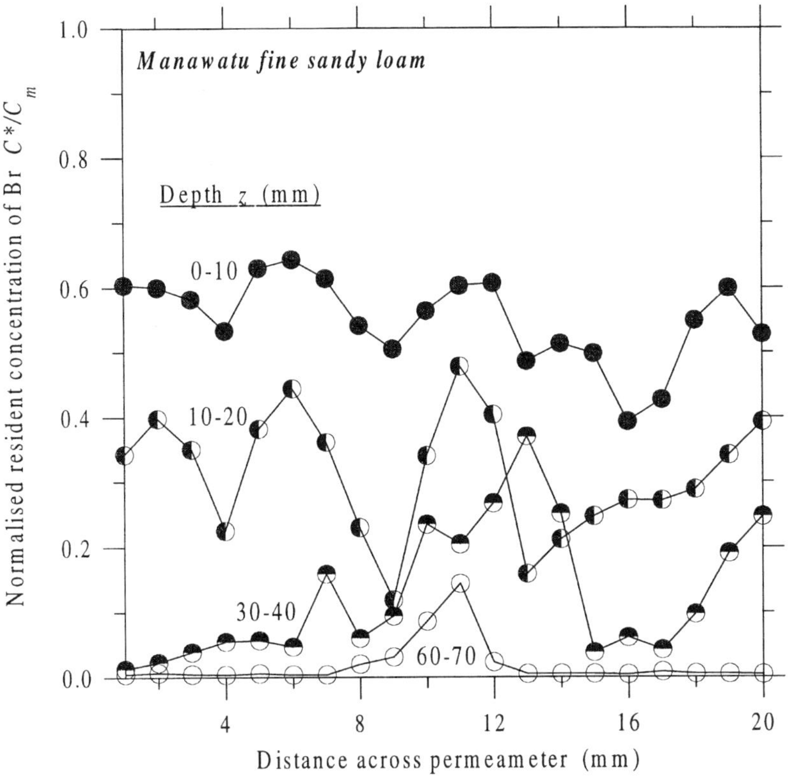

Figure 10.7. The individual measurements of the normalized resident concentration of bromide, C^*/C_m, made under one of the permeameters of Figure 10.6 (l=17 mm). Four transects each comprising twenty 10 mm cores were extracted from underneath the disc of diameter 20 cm. Profiles were taken at depths 0–10, 10–20, 30–40, and 60–70 mm.

to double layer exclusion. Of the remaining fraction, they suggested that θ_{im}/θ_o was 0.31, as found by Clothier et al. (1995), and shown in Figure 10.6. On the basis of the flux concentration coming from the base of their unsaturated cores, Vogeler et al. (1996) found by trial-and-error that the breakthrough shape could be accounted for by dispersion characterized by λ in the mobile phase of 20 mm, and by flow interruption that diffusion could be represented by $\alpha=1.1 \times 10^{-5}\ s^{-1}$. The resident concentration data of Figure 10.6 would suggest a higher λ and a lower α, otherwise the proximal concentration would have risen to unity after the 3 days. This conflict serves to reinforce the difficulty of trying to use inverse procedures to infer independent measures of α, β, and λ.

Thus, flow in this pedologically simple soil is quite complex because of the matrix-macropore dichotomy whose break point occurs in the region of flows that can be expected to occur naturally. In our structured soil, the Ramiha silt loam, the aggregates were sufficiently small, and the flow rates were quite fast, so that it was possible to ignore mobile-immobile water considerations, for either they didn't exist or they were hidden because of effective diffusive exchange. However, in the macroporous Manawatu fine sandy loam, the slower unsaturated flow into this soil, wherein there is a significant matrix separation between flow-paths, means that we either need to consider that the soil has significant immobile water with ineffective diffusive exchange, or that the soil has no immobile water yet possesses a large dispersivity that is corrected for matrix exchange of solute from the macropores. Irrespective, for the likely size of infiltration events shown in Figure 10.5, it is obviously important for the Manawatu fine sandy loam that an apposite form of a nonequilibrium flow model be chosen, or that an apt form of a local equilibrium description be selected.

Use of the disk permeameter to supply tracer to the soil, on the time scale, and with the I that is likely to occur naturally can be used to infer from measured $\theta(z,t)$ $C^*(z,t)$ which of these propositions to adopt. However, independent parameterization from these data of unitary values of α, β, and λ does not appear feasible.

The role of water flow in controlling solute transport is shown to be critical (Figures 10.7 and 10.2), yet the currently popular form of the mobile-immobile water model does not properly acknowledge this. Rather, the steady form adopted (Eq. 10.2), and its use only of inter-domain diffusion (Eq. 10.4), appears popular mainly because of its analytical convenience, rather than because of its proper representation of exchange mechanisms. The dye staining observations under a disk permeameter of Lin and McInnes (1995) suggest that such preferential transport and ineffective interdomain exchange in macroporous soil is widespread, especially in silt and clay-textured soils.

EXCHANGE

The water-borne transport of only inert solutes has so far been considered. Reaction of invading chemicals with field soil needs to be better addressed. The role of convective exchange of the invading solution into the empty pores in the initially dry soil has been ignored. Such an exchange process is likely to be important in carrying nutrients into the root zone, and passing pollutants on to groundwater.

Chemical Exchange

The mobile-immobile transport model (Eqs. 10.2 and 10.4) can be simply written to account for reactive transport if a linearized adsorption isotherm is assumed. In this way the retarded movement of any chemical, relative to an inert tracer, is given by

$$R = 1 + \frac{\rho K_D}{\theta} \tag{10.15}$$

where ρ (kg m^{-3}) is the soil bulk density, and K_D is the distribution coefficient linking the sorbed chemical S (mol kg^{-1}) to that in solution C (mol m^{-3}) via $S = K_D C$. It is possible to use batch procedures to measure K_D in the laboratory. Commonly, the $S(C)$ data are fitted to some nonlinear function, such as the Freundlich equation,

$$S = mC^n \tag{10.16}$$

where m and n are the Freundlich parameters. However, despite there being an industry based on measurement and interpretation of laboratory isotherms, there has been little effort in measuring reactive chemical exchange in the field, especially on soils that are either aggregated or macroporous, wherein there might be sequestered exchange sites due to physical nonequilibrium, or other nonequilibria.

Clothier et al. (1996) proposed using a permeameter laden with both an inert tracer and reactive chemical to resolve the adsorption isotherm in the field. Unlike the proposal to measure θ_m using the resident concentration of inert tracer (Eq. 10.13), they showed that dispersive effects would persist so long for reactive chemical as to render that approach impractical. Rather, they suggested that measurement of the respective invasion fronts of inert and reactive chemical could be used. If the retardation of reactive chemical were measured at two concentrations (R_1, C_1; R_2, C_2), then the exponent of the nonlinear isotherm could be found as

$$n = 1 + \frac{\ln\left(\frac{R_1 - 1}{R_2 - 1}\right)}{\ln\left(\frac{C_1}{C_2}\right)} \tag{10.17}$$

Given this n, the retardation for a Freundlich isotherm, $R=1+(\rho/\theta)mnC^{n-1}$, could be used to resolve m from any one of the (R, C) values. Their Ramiha silt loam data using $^{35}SO_4$ are represented here in Figure 10.5 such that from the two retardations it was found that for sulfate $S=2.3\ C^{0.7}$. This dynamically deduced form of the sulfate adsorption isotherm is only about 80% that found using batch procedures, despite their being complete invasion of the aggregated pore space in this soil by the invading inert tracer. This apparent sequestration of exchange sites merits further investigation.

Using both their observations of resident concentration and flux concentration from undisturbed cores of Manawatu fine sandy loam undergoing unsaturated flow, Vogeler et al. (1997a) found that reactive exchange of Ca^{2+} and Mg^{2+} could only be predicted if 80% of the batch-measured cation exchange capacity were considered. They considered their ammonium acetate method for measuring the exchange capacity in the laboratory was in error due to an ionic strength effect.

When considering reactive chemical transport in field soil it will be necessary that new and appropriate means of measuring the exchange isotherm be developed. For soils possessing some form of dual (or multiple) porosity, it will be necessary to apportion the exchange sites between the various domains.

Convective Exchange

In many circumstances where there is an interest in describing chemical transport through soil, the soil is growing plants. The near-surface soil is thus ramified by roots such that between infiltration events, the soil can be dried down to quite low water contents. Hence, should there hypothetically be a dual porosity soil whose initial water content θ_n becomes less than the previously determined θ_{im}, then obviously there is difficulty in using a simple mobile-immobile model like Eq. 10.2 subject only to the diffusion of Eq. 10.4.

For the Manawatu silt loam, Tillman et al. (1991), Snow et al. (1994), and Vogeler et al. (1997a) have all explored in the field the impact of vegetation and antecedent dryness on the passage of chemicals during unsaturated flow. For another macroporous soil, Kluitenberg and Horton (1990) demonstrated the impact of the method of solute application on the preferential transport of solutes in soil. In a field study, over 60 days with a lysimeter of Manawatu fine sandy loam growing pasture, Snow et al. (1994) had great difficulty in describing the transport of solute in the top 250 mm of soil due to small-scale heterogeneity in the incident water fluxes, and local variation in the soil properties that were exacerbated by cycles of wetting and drying. Subsurface solute transport was found capable of description using a local equilibrium version of the convective-dispersion equation if just two-thirds of the soil wetted porosity was considered mobile during the drainage of about 400 mm of water. This result supports the permeameter observations of Figure 10.6, although their measures of λ were only 10–20 mm.

In a controlled plot experiment on the same site, Tillman et al. (1991) used a watering can to apply 5-mm of KBr to plots of either a wet soil or dry soil. Using the same can, in 5 mm pulses every 30 minutes at an instantaneous q of greater than 50 mm hr^{-1} under the wetted "footprint" of the watering can, an additional 50 mm of water was applied to wash the Br into the soil of both plots. The penetration of solute under the dry soil was much less (50 mm) than that for the wet (>300 mm). This indicates that for this surface boundary condition, the soil antecedent water content can play a dramatic role in the transport of solute. Under even more controlled flow conditions in the laboratory, Vogeler et al. (1997a) used large undisturbed cores under a rain simulator set at a steady $q = 10$ mm hr^{-1}. They found there to be no difference in the convective-diffusive transport of solute due to the antecedent moisture conditions. These conflicting results can be explained by consideration of the Manawatu fine sandy loam's $K_w(h)$ function, relative to the applied flux. At an instantaneous rate of 50 mm hr^{-1}, the matrix of the surface soil would certainly pond incipiently, so that free water could access any surface-vented macropores and carry with it Br (Tillman et al., 1991). In the dry soil case, this macropore transport would be convectively curtailed by capillary absorption of the solution into the dry walls of the macropore. However, under a flow rate of just 10 mm hr^{-1} it is likely, given $K_w(h)$, that there would be little, or no, incipient ponding (Vogeler et al., 1997). Without the confusion generated by macropore flow, both convective-diffusive water flow and solute transport could proceed in a more orderly fashion, as they found.

Modeling of the dual roles of convection and diffusion in water flow is now being coupled with dual-porosity solute transport models to describe better transport in aggregated and macroporous soils. The dual-porosity deterministic model of Gerke and van Genuchten (1993) involves two linked continua: a macropore system, and a matrix pore system. For aggregated soils, Jarvis (1994) presented a numerical model that allows convection between the inter-aggregate mobile zone, and intra-aggregate pores. Field measurement of these parameters under typical field conditions will require the development of novel techniques and new methodologies. Inverse procedures, possibly using this model for guidance, may, however, allow determination of the key parameters that control solute transport during unsaturated flow in aggregated or macroporous soils.

PROSPECTIVE

Since the 1970s, there has been a significant improvement in our ability to observe, and our capacity to describe physical nonequilibrium phenomena in soils. New devices are providing us with vision of greater acuity concerning the role that aggregates and macropores play in controlling chemical transport through unsaturated soil.

Presently, however, our understanding of certain critical nonequilibrium phenomena is not well advanced. In modeling waterborne chemical transport we tend only to consider steady flow in mobile-immobile water soils, with a crude diffusion-like mechanism linking the two domains. The inability to separate independently the mechanisms of mobility, dispersion, and diffusion renders troublesome the application of this analytically convenient model. The description of water flow in biporous soil is less advanced, although the analyses of Jarvis (1991) and Gerke and van Genuchten

(1993) have signposted the way ahead. There is seen a need to describe better the convective exchange that can occur between the various pore classes. During chemical transport in soil that was antecedentally moist, it may be that only interdomain diffusion is required to describe exchange between the mobility classes. Transport might well be dominated by flow in the mesopore and macropore domains, and so be amenable to mobile-immobile domain description, if we were able to adequately parameterize the model. The simpler local-equilibrium model might even prove as effective in certain circumstances

However, plant roots can extract water from across a wide range of pore sizes, so that after a period of dry-down, smaller pores may well possess a capillary attractiveness to invading water such that its passenger chemicals will be drawn in to take up sheltered residency within these small pores. Once located there, they may well be rendered somewhat immobile, being resistant to subsequent leaching should the water flow be macropore-dominated (Tillman et al., 1991). In the absence of incipient ponding, with the attendant lack of generation of macropore flow, the role of antecedent soil moisture disappears, and solute transport appears well-behaved and capable of convective-diffusive description (Vogeler et al., 1997a). Thus there is seen a need to improve our hydraulic understanding of the generation of macropore water and its impact of interdomain convective exchange of water. The physical role that the initial and soil surface conditions play on chemical transport is of paramount importance (Kluitenberg and Horton, 1990). Physical nonequilibrium phenomena, especially in vegetated surface soil, play a critical role in controlling solute transport.

Such nonequilibrium phenomena are probably the field rule, rather than the exception, and this demands that we achieve better observations of flow and transport mechanisms, and that we acknowledge Joseph Fourier's bromide that "...Nature is indifferent to the difficulties it causes mathematicians."

REFERENCES

Andreu, L., F. Moreno, N.J. Jarvis, and G. Vachaud. Application of the model MACRO to water movement and salt leaching in drained and irrigated marsh soil, Marismas, Spain *Agric. Water Manag.*, 25:71–88, 1994.

Ankeny, M.D., M. Ahmed, T.C. Kaspar and R. Horton. Simple field method for determining unsaturated hydraulic conductivity. *Soil Sci. Soc. Am. J.*, 55:467–470, 1991.

Bond, W.J. and P. Wierenga. Immobile water during solute transport in unsaturated sand columns. *Water Resour. Res.*, 26:2475–2481, 1990.

Clothier, B.E. and I. White. Measurement of sorptivity and soil water diffusivity in the field. *Soil Sci. Soc. Am. J.*, 45:241–245, 1981.

Clothier, B.E. and K.R.J. Smettem. Combining laboratory and field measurements to define the hydraulic properties of soil. *Soil Sci. Soc. Am. J.*, 54:299–304, 1990.

Clothier, B.E., M.B. Kirkham, and J.E. MacLean. *In situ* measurement of the effective transport volume for solute moving through soil. *Soil Sci. Soc. Am. J.*, 56:733–736, 1992.

Clothier, B.E., L. Heng, G.N. Magesan, and I. Vogeler. The measured mobile-water content of an unsaturated soil as a function of hydraulic regime. *Austr. J. Soil Res.*, 33:397–414, 1995.

Clothier, B.E., G.N. Magesan, L. Heng, and I. Vogeler. *In situ* measurement of the solute adsorption isotherm using a disc permeameter. *Water Resour. Res.*, 32:771–778, 1996.

Coats, K.H. and B.D. Smith. Dead-end pore volume and dispersion in porous media. *Soc. Pet. Eng. J.*, 4:73–84, 1964.

Cowie, J.D. Soils and agriculture of Kairanga County, *N.Z. Soil Bur.* 33, DSIR, 1978.

Gaudet, J.-P. Transferts d'eau et de soluté dans le sol non-saturé—Mesures et simulation. Thesis (Docteur d'Etat), University de Grenoble, France, 1978, p. 246.

Gaudet, J.-P., H. Legat, G. Vachaud, and P.J. Wierenga. Solute transfer, with exchange between mobile and stagnant water, through unsaturated soil. *Soil Sci. Soc. Am. J.,* 41:665–671, 1977.

Gerke, H.H. and M.T. van Genuchten. Evaluation of a first-order water transfer term for variably saturated dual-porosity flow models. *Water Resour. Res.,* 29:1225–1238, 1993.

Jarvis, N.J. MACRO—A model of water movement and solute transport in macroporous soils. *Reports & Dissertations* 9, Swedish University of Agricultural Science, Uppsala, 1991.

Jarvis, N.J. The MACRO Model (Version 3.1)—Technical description and sample simulations. *Reports & Dissertations* 19, Swedish University of Agricultural Science, Uppsala, 1994.

Jarvis, N.J. and I. Messing. Near-saturated hydraulic conductivity in soils of contrasting texture measured by tension infiltrometers. *Soil Sci. Soc. Am. J.,* 59:27–35, 1995.

Jaynes, D.B., S.D. Logsdon, and R. Horton. Field method for measuring mobile/immobile water content and solute transfer rate coefficient. *Soil Sci. Soc. Am. J.,* 59:352–356, 1995.

Kluitenberg, G.J. and R. Horton. Effect of solute application method on preferential transport of solutes in soil. *Geoderma,* 46:283–297, 1990.

Kookana, R.S., R.D. Schuller, and L.A.G. Aylmore. Simulation of simazine transport through soil columns using time dependent sorption data measured under flow conditions. *J. Contaminant Hydrol.,* 14:93–115, 1993.

Lin, H.S. and K.J. McInnes. Water flow in clay soil beneath a tension infiltrometer. *Soil Sci.,* 159:375–382, 1995.

Magesan, G.N., I. Vogeler, D.R. Scotter, B.E. Clothier, and R.W. Tillman. Solute movement through two unsaturated soils. *Aust. J. Soil Res.,* 33:585–596, 1995.

Parker, J.C. and A.J. Valocchi. Constraints on the validity of equilibrium and first-order kinetic transport models in structures soils. *Water Resour. Res.,* 22:399–407, 1986.

Passioura, J.B. Hydrodynamic dispersion in aggregated media. I. Theory. *Soil Sci.,* 111:339–344, 1971.

Perroux, K.M. and I. White. Designs for disc permeameters. *Soil Sci. Soc. Am. J.,* 52:1205–1215, 1988.

Rao, P.S.C., R.E. Jessup, D.E. Rolston, J.M. Davidson, and D.P. Kilcrease. Experimental and mathematical description of nonadsorbed solute transfer by diffusion in spherical aggregates. *Soil Sci. Soc. Am. J.,* 44:684–688, 1980a.

Rao, P.S.C., D.E. Rolston, R.E. Jessup, and J.M. Davidson. Solute transport in aggregated porous media: Theoretical and experimental evaluation. *Soil Sci. Soc. Am. J.,* 44:1139–1146, 1980b.

Scotter, D.R. Preferential solute movement through larger soil voids. I Some computations using simple theory. *Austr. J. Soil Res.,* 16:257–267, 1978.

Snow, V.O., B.E. Clothier, D.R. Scotter, and R.E. White . Solute transport in a layered field soil: Experiments and modeling using the convection-dispersion approach. *J. Contaminant Hydrol.,* 16:339–358, 1994.

Tillman, R.W., D.R. Scotter, B.E. Clothier, and R.E. White. Solute movement during intermittent water flow in a field soil and some implications for irrigation and fertiliser application. *Agric. Water Manage.,* 20:119–133, 1991.

Vanclooster, M., H. Veerecken, J. Diels, F. Huysmans, W. Verstaete, and J. Feyen. Effect of mobile and immobile water in predicting nitrogen leaching from cropped soils. *Modeling Geo-Biosphere Processes,* 1:23–40, 1992.

Van Genuchten, M.Th. and F.N. Dalton. Models for simulating salt movement in aggregated field soils. *Geoderma,* 38:165–183, 1986.

Van Genuchten, M. Th. and P.J. Wierenga. Mass transfer studies in sorbing porous media. I. Analytical solutions. *Soil Sci. Soc. Am. J.,* 40:473–480, 1976.

Vogeler, I., D.R. Scotter, B.E. Clothier, and R.W. Tillman. Anion transport through intact soil columns during intermittent unsaturated flow. *Soil Technol.* (in press), 1996.

Vogeler, I., D.R. Scotter, S.R. Green, and B.E. Clothier. Solute movement through undisturbed soil columns under pasture during transient and steady unsaturated flow. *Austr. J. Soil Res.,* 35:1153–1163, 1997a.

White, I. and M.J. Sully. Macroscopic and microscopic capillary length and time scales from field infiltration. *Water Resour. Res.,* 23:1514–1522, 1987.

CHAPTER ELEVEN

Field Parameterization of the Mobile/Immobile Domain Model

D.B. Jaynes and R. Horton

INTRODUCTION

It is clear from numerous field and laboratory experiments that solute movement is often poorly described by the classical advection-dispersion model. Rather, solute breakthrough curves frequently exhibit earlier arrival and more pronounced tailing than predicted by this model. These observations have spurred the development of conceptual models that specifically include physical nonequilibrium to more accurately depict solute movement. In the simplest version of these models, the water-filled pore space is partitioned into two domains, a mobile domain, where water is free to move and solute movement is by advection and dispersion, and an immobile domain, where water is stagnant and solute moves only by diffusion.

Coats and Smith (1964) published an early version of this mobile/immobile domain (MIM) model which was later popularized by the work of van Genuchten and Wierenga (1976, 1977). For one-dimensional transport of a noninteracting, conservative solute, the MIM model can be written:

$$\theta_m \frac{\partial C_m}{\partial t} + \theta_{im} \frac{\partial C_{im}}{\partial t} = \theta_m D_m \frac{\partial^2 C_m}{\partial x^2} - q \frac{\partial C_m}{\partial x} \tag{11.1}$$

where θ_m and θ_{im} are the mobile and immobile volumetric water contents, the sum of which equals the total volumetric water content, θ; C_m and C_{im} are the solute concentrations in the mobile and immobile domains; t is time; x is distance; D_m is the hydrodynamic dispersion coefficient for the mobile domain; and q is Darcy flux. Exchange between the mobile and immobile domains is described by:

$$\theta_{im} \frac{\partial C_{im}}{\partial t} = \alpha (C_m - C_{im}) \tag{11.2}$$

where α is a first-order mass transfer coefficient. Thus, the MIM model can account for more rapid solute transport because flow occurs in only a fraction of the water-filled pore space and the model can account for the tailing observed in breakthrough curves via slow exchange of solute between the two domains (van Genuchten and Wierenga, 1977).

Equations 11.1 and 11.2 have been successfully applied to many laboratory column leaching studies. These studies have shown that both θ_m and α vary with the average pore water velocity, v (= q/θ), water content, and soil aggregate size (van Genuchten and Wierenga, 1977; Gaudet et al., 1977; Nkedi-Kizza et al., 1983; Kookana et al., 1993). Clothier et al. (1995) drew parallels between the apparent change in these transport parameters with the apparent change in the characteristic length determined from hydraulic conductivity measurements. Characteristic length typically decreases as the soil water pressure head decreases, presumably due to the emptying of the macro- and mesopores in the soil (Parlange, 1972; Jarvis and Messing, 1995).

Little information exists, however, as to the magnitude and behavior of MIM model parameters in field soils. In field tracer studies, the ratio θ_m/θ has been observed to vary from 1 (i.e., no preferential flow; Cassel, 1971) to 0.45–0.65 (Smettem, 1984; Gvirtzman and Magaritz, 1986), and even to as small as 0.25 in a weakly structured tropical soil (Seyfried and Rao, 1987) and a massive desert soil (Rice et al., 1986). Methods have been proposed for estimating θ_{im} by equating it to the water remaining at some soil water pressure such as –33 kPa (Addiscott, 1977) or –202 kPa (Addiscott et al., 1986), or equating it to the residual water content from the water retention function (Jaynes et al., 1988b). However, these estimates appear to be soil-specific and have not been tested over an extensive range of soils.

Typically, θ_m and α are estimated from solute breakthrough curves where extended measurements of effluent concentration versus time are required. An inverse method such as described by Parker and van Genuchten (1984) is then used to estimate the model parameters giving the best fit (least sum-of-squares) to the data. Breakthrough curves are time-consuming to measure and are very difficult to conduct in the field. To fully characterize how α and θ_{im} vary by soil, management, and initial and boundary conditions, robust, easy-to-use measurement methods are required.

FIELD METHOD

As a first approach, we can use Equation 11.2 in conjunction with a simple tracer leaching study to get an estimate of θ_m in field soils. If we infiltrate a solution containing a conservative, noninteracting tracer such as tritiated water into soil and sample the soil for tracer after a period of leaching, the average concentration in the soil, C, will be a combination of the tracer concentration in both the mobile and immobile domains. Assuming no immobile water and piston displacement of tracer within the mobile domain, the tracer concentration in the mobile domain, C_m, will equal the input concentration, C_o, and C will equal C_o behind the tracer front. If, however, θ_{im} > 0, C will be less than the input concentration for soil initially tracer-free. By further assuming that α is small compared to the leaching time for the tracer and that no tracer is initially present in the immobile water, from conservation of mass we find:

$$\theta_{im} = \theta\left(1 - \frac{C}{C_o}\right) \tag{11.3}$$

This is the approach proposed by Clothier et al. (1992). They used a tension infiltrometer to apply a Br^- tracer to soil. This allowed them to vary the infiltration rate and resulting water content and quantify the effect on θ_{im}. Using this method they found θ_m/θ to be 0.49 in a Manawatu fine sandy loam at an application pressure head of –0.02 m. Further experiments by Clothier et al. (1995) showed that θ_m/θ was a function of the pressure head at which the tracer was applied, increasing from 0.41 at –0.02 m to 0.64 at –0.15 m of water head.

While a promising first approximation, assuming the $\alpha = 0$ does not appear to be a realistic assumption in many cases. For example, Clothier et al. (1995) estimated α to be 0.02 h^{-1} over the first few days of tracer application. At this rate, measurable diffusion of solute into the immobile domain will occur in only a few hours. This estimate of α is within the range of 0.001 to 10 h^{-1} found from surveying a number of studies (Kookana et al., 1993). Thus, assuming $\alpha \approx 0$ in most cases where tracer application exceeds about an hour will systematically over estimate θ_m/θ using Eq. 11.3.

Alternatively, we can assume α is not negligible while still assuming piston displacement of tracer. A piston displacement assumption simplifies further analysis by removing the need to evaluate Eq. 11.1. Dispersion is ignored and for soil near the inlet boundary we assume that $C_m = C_o$. These assumptions allow for a focus on Eq. 11.2 which is solved to give:

$$\ln\left(1 - \frac{C}{C_o}\right) = \ln\left(\frac{\theta_{im}}{\theta}\right) - \frac{\alpha}{\theta_{im}} t^* \tag{11.4}$$

where $t^* = t - \ell/v$ and is defined as the time required for the tracer front to reach the depth of sampling, ℓ. Substituting for t^* in Eq. 11.4 gives:

$$\ln\left(1 - \frac{C}{C_o}\right) = \ln\left(\frac{\theta_{im}}{\theta}\right) + \frac{l\alpha\theta_m}{\theta_{im}q} - \frac{\alpha}{\theta_{im}} t \tag{11.5}$$

Thus by regressing ln(1–C/Co) versus time, both α and θ_{im} can be found from the resulting intercept and slope.

This is the approach first proposed by Jaynes et al. (1995), where they infiltrated a sequence of conservative, noninteracting tracers over time and measured the resident concentrations in a shallow soil sample. By again using a tension infiltrometer to apply the tracers, the dependence of α and θ_{im} on the infiltrometer tension (and thus v and θ) can be determined.

In their approach, Jaynes et al. (1995) used four different tracers rather than repeated measurements with a single tracer. An advantage to using multiple tracers is that only one soil sample need be taken, eliminating sample-to-sample variability. A disadvantage of the technique is that the transport properties of the different tracers may not be identical. Jaynes et al. (1995) used Br^- and the fluoridated benzoates, pentafluorobenzoate, *o*-trifluoromethylbenzoate, and 2,6-difluorobenzoate, which are known to have near identical transport properties in many soils (Jaynes, 1994) and similar aqueous diffusion coefficients (Bowman and Gibbens, 1992; Benson and Bowman, 1994).

Jaynes et al. (1995) found good linearity for the data (Figure 11.1) when resident tracer concentration was plotted vs. time as given in Eq. 11.5. Discrepancy between the measured and predicted behavior may have been due to experimental limitations such as difficulty in measuring the small differences in tracer concentration or nonidentical transport behavior of the individual tracers. This latter possibility can be compensated for by alternating the order of tracer application during replicate determinations. Values determined for α and θ_{im} by Jaynes et al. (1995) were well within the range observed by others in laboratory column experiments.

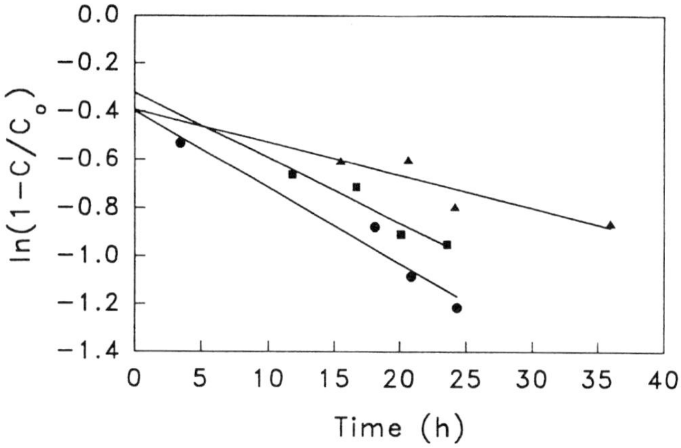

Figure 11.1. Examples of resident solute concentrations from three replicate leaching experiments plotted as per Eq. 11.5 and the best fit lines to the data. From Jaynes, D.B., *Soil Sci. Soc. Am. J.*, 59:352–356, 1995. With permission.

The sequential tracer technique can potentially be used for measuring the distribution of α and θ_{im} in field soils. While not as simple to use as the single tracer method of Clothier et al. (1992), it retains many of the advantages (a single soil sample is taken and total experimental time is relatively short), while giving estimates of the additional parameter, α. In its simplest form, only two tracers need be applied, although the use of multiple tracers allows for the confirmation of log-linear behavior predicted by Eq. 11.5. If accurate, the method will make possible the determination on how α and θ_{im} vary as functions of soil type, crop, and tillage over a range of wetting heads.

EVALUATION OF MULTIPLE TRACER APPROACH

While Jaynes et al. (1995) showed that Eq. 11.5 gave reasonable estimates of α and θ_{im}, they did not compare these estimates to independent estimates determined by other means such as curve fitting of breakthrough curves (inverse method). Nor did they evaluate the range over which the assumptions leading to Eq. 11.5 are valid. A direct approach would be to compare Eq. 11.5 to the analytical solution of Eqs. 11.1 and 11.2 such as given by Parker and van Genuchten (1984). However, we have been unable to reformulate their solution into a form directly comparable to Eq. 11.5. As an alternative, we can compare the estimates of α and θ_{im} given by Eq. 11.5 when applied to data generated from the complete analytical solution with known input values.

Figure 11.2 shows the results of three simulations using the analytical solution to Eqs. 11.1 and 11.2 (Parker and van Genuchten, 1984). In these simulations, v was set to 1 cm hr^{-1}, θ_m/θ to 0.66 and the resident concentration vs. time at a depth of 2 cm below the surface was calculated. Three combinations of dispersivity, γ (= D/v), and α were used. In the first simulation, γ = 0.01 cm and α = 0.005 h^{-1}. After about 1.7 h, ln(1–C/C_o) vs. time is well represented by a straight line as described by Eq. 11.5 (Figure 11.2). In the second simulation, α was again set to 0.005 h^{-1} and γ was increased to 1 cm. Plotting the results of the simulations in Figure 11.2 results in a concave-upward curve that is poorly represented by a straight line. At this higher dispersivity, the assumption of piston displacement of solute is not valid and transport is poorly described by Eq. 11.5. Finally, Figure 11.2 shows the results of a simulation where γ = 0.01 cm and α = 0.5 h^{-1}. Again, the resulting curve is poorly described by Eq. 11.5, being

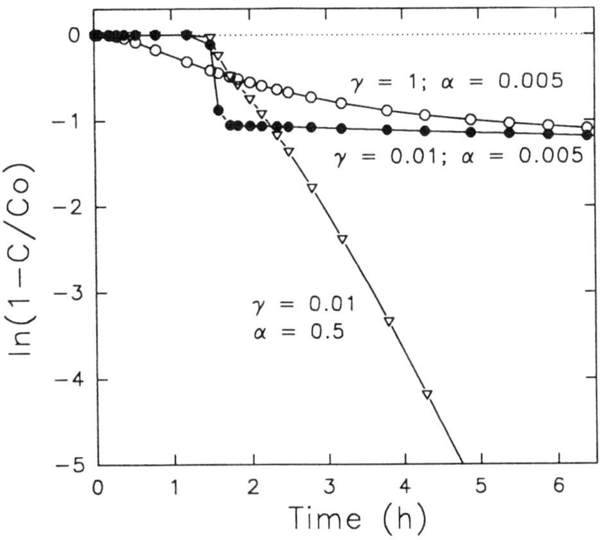

Figure 11.2. Examples of simulated resident concentration data for a depth of 2 cm generated from the analytical solution to the MIM model for a θ_m/θ ratio of 0.66 and a water flux of 0.5 cm h^{-1} and plotted as per Eq. 11.5. Values on figure are for dispersivity, γ, in cm and the inter-domain solute transfer coefficient, α, in h^{-1} used in each simulation.

concave downward. Thus even at a low dispersivity, the rapid exchange between the mobile and immobile domains caused by setting α high negates the assumption that $C_m = C_o$ made in developing Eq. 11.5.

Using the analytical solution to compute soil solution concentration data for a range of parameter values and applying Eq. 11.5 to the generated data, we can determine the range in conditions for which calculated values of α and θ_{im} are accurate. We compute a relative error for the calculated parameters by dividing the absolute difference between the known parameter value used in the simulation and the value calculated from Eq. 11.5, by the known value. Figure 11.3a shows the relative error in calculated values of θ_{im} over a range of log(γ) and log(α/v) values and a θ_{im}/θ ratio of 0.33. Relative errors in the computed value of θ_{im} are less than 0.2 over a wide range of γ and α/v values. However, as γ increases above 1 cm the relative error increases rapidly for all values of α/v. Likewise, as α/v increases beyond 0.01 cm^{-1}, the relative error increases rapidly. For low values of γ and values of α/v greater than about 0.3 cm^{-1}, Eq. 11.5 actually gives values of θ_{im} that are greater than the value of θ.

Figure 11.3b is the graph of the relative error in α calculated from Eq. 11.5. Low relative errors are found for values of γ less than 0.1 cm and for α/v values less than 0.03 cm^{-1}. Unlike the results for θ_{im}, reasonable values of α are still calculated for low γ values and α/v ratios greater than 0.03 cm^{-1}, the values never being worse than about a factor of 2 in error. Decreasing the θ_m/θ ratio used in the simulations, enlarges the regions in Figure 11.3 giving reasonable parameter values; increasing the ratio reduces these regions.

Based on comparisons with the complete solution to the MIM model, Eq. 11.5 would appear to give reasonable estimates of α and θ_{im} over a restrictive range of conditions. However, a survey of the literature shows that most reported values of α/v are ≤ 0.1 cm^{-1} (Kookana et al., 1993), which corresponds to the lower half of Figure 11.3. Thus, it would appear that expected α/v ratios will not appreciably limit the applicability of Eq. 11.5.

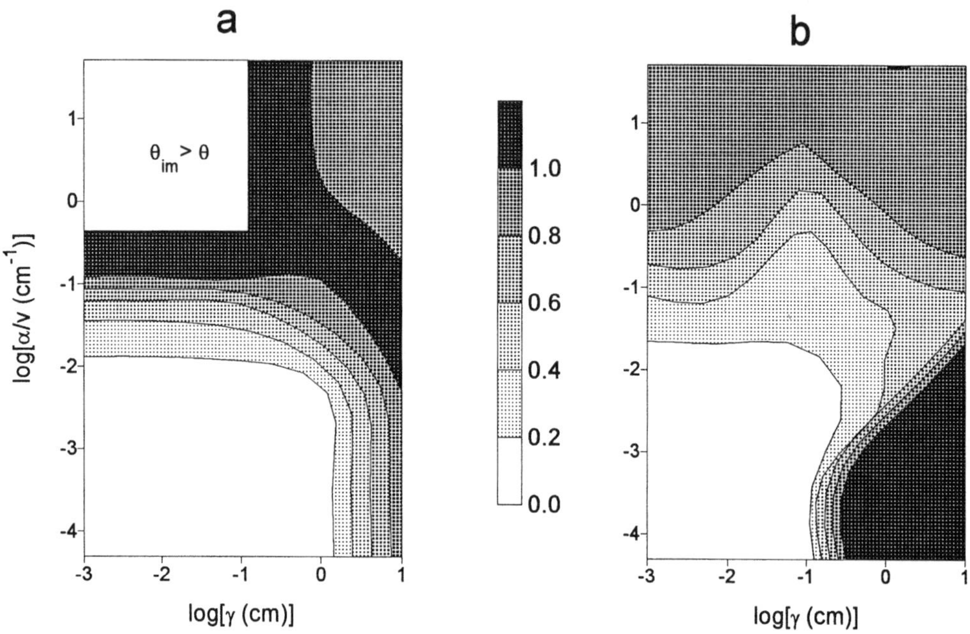

Figure 11.3. Map of the relative error in the estimated value of (a) θ_{im} and (b) α as functions of log(γ) and log(α/v). Data were generated from the analytical solution of the MIM model and known model parameters with a θ_m/θ ratio of 0.66 and a simulation depth of 2 cm. Parameter estimates were obtained from applying Eq. 11.5 to the generated data. Blanked space in upper left corner of (a) represents values where computed θ_{im} values exceed θ.

Representative values for γ are less certain. In laboratory column studies, reported values for γ have ranged from 0.16 to 2.0 cm (van Genuchten and Wierenga, 1977; Gamerdinger et al., 1990; Kookana et al., 1993). Over this range of γ values, Eq. 11.5 should give reasonable estimates of α and θ_{im} (Figure 11.3). However, estimates of γ based on field-scale experiments are usually 10 to 100 times larger (Jaynes et al., 1988a; Yasuda et al., 1994), well outside the range of applicability for Eq. 11.5. These large γ values may in part be due to the greater heterogeneity of soil over the larger spatial scales used, and will limit the accuracy of Eq. 11.5. However, the sequential tracer approach may still be helpful since no other practical method exists for estimating the MIM transport parameters in the field.

LABORATORY EVALUATIONS

The applicability of Eq. 11.5 can be evaluated from leaching studies using laboratory soil columns by comparing estimates based on Eq. 11.5 vs. estimates obtained by applying inverse methods. By applying a sequence of tracers to a soil column over time, α and θ_{im} can be calculated using Eq. 11.5 and resident solute concentrations from within the soil column. Similarly, independent estimates for α and θ_{im} can be made by applying inverse methods to the solute concentrations in the column outflow vs. time (Parker and van Genuchten, 1984). Inverse methods applied to solute breakthrough curves are typically used to estimate the parameter values in Eqs. 11.1 and 11.2 (van Genuchten and Wierenga, 1977; Seyfried and Rao, 1987).

Table 11.1. Soil, Bulk Density (ρ), Water Content (θ), and Pore-Water Velocity (v) for Each Column.

Column #	Soil Material	ρ	θ	v
		Mg m^{-3}	kg kg^{-1}	cm h^{-1}
1	Florida beach sand[a]	1.54	0.40	32.8
2	Clarion Ap, sicl[b]	0.94	0.56	41.4
3	Clarion C, sicl[b]	0.94	0.49	16.2
4	Tama Ap, l[b]	0.96	0.59	45.0
5	Tama C1g, l[b]	0.96	0.49	35.3

[a] 0.5–1 mm sieve fraction.
[b] 1–2 mm sieve fraction.

Five 12.5-cm long columns were packed with different soil materials (Table 11.1). Details of these and other column experiments can be found in Lee et al. (1998). In each experiment, the soil was first saturated with a 0.4 m*M* $CaCl_2$ solution. The columns were then leached with the same solution under a slight ponded head until steady flow conditions were achieved. A sequence of solutions were then introduced at the top of each column. The first solution contained 0.3 mM $CaCl_2$ and 0.1 m*M* of either pentafluorobenzoate (PFBA), *o*-trifluoromethylbenzoate (TFMBA), 2,6-difluorobenzoate (DFBA), or 2,3,6-trifluorobenzoate (TFBA) tracer. After leaching the column with about 1 pore volume of the first solution, a second solution was applied containing 0.2 m*M* of $CaCl_2$, 0.1 mM of the first benzoate tracer, and 0.1 mM of a second benzoate tracer. The process was repeated until the final solution contained no $CaCl_2$ and the four benzoate tracers at a concentration of 0.1 mM each. Outflow containing the tracers was collected from the columns with a fraction collector. Following infiltration of the last tracer, the columns were sectioned in 1 cm increments and the tracers extracted and measured to calculate the resident tracer concentration vs. depth.

Figure 11.4 shows the concentrations of the four tracers in outflow from column 4 vs. time, normalized by the input concentrations. Breakthrough curves for the other columns were similar. Each breakthrough curve had early arrival of tracer indicative of physical nonequilibrium processes. The breakthrough curve for each tracer was normalized by the input concentration and adjusted so that t=0 when the individual tracer was first applied to the column. The four breakthrough curves were then combined and the best fit values for the parameters D, α, and θ_{im} in Eqs. 11.1 and 11.2 found by inverse methods using the program CXTFIT (Parker and van Genuchten, 1984).

Resident concentrations for the four tracers in column 4 are shown in Figure 11.5. Tracer profiles in the other columns were similar. Relative concentrations were all less than 1, indicating the presence of immobile water in each column. At most depths, relative concentrations were greater for tracers applied the longest, indicative of transfer processes occurring between the mobile and immobile domains. Finally, the last tracer applied showed considerable dropoff in concentration with depth and clearly shows a zone where dispersion processes are important.

Equation 11.5 was applied to the resident concentration data. Because insufficient leaching of the last applied tracer would cause interference from dispersion processes, only the concentrations from the top layers were used. We also discarded data from the top 1-cm layer in each column since this layer had bulk densities considerably lower than the rest of the column. Figure 11.6 shows the resident concentrations in the four layers of column 4 plotted as $\ln(1-C/C_o)$ vs. time and the resulting best fit line to each layer. For each soil layer straight lines fit the data reasonably well.

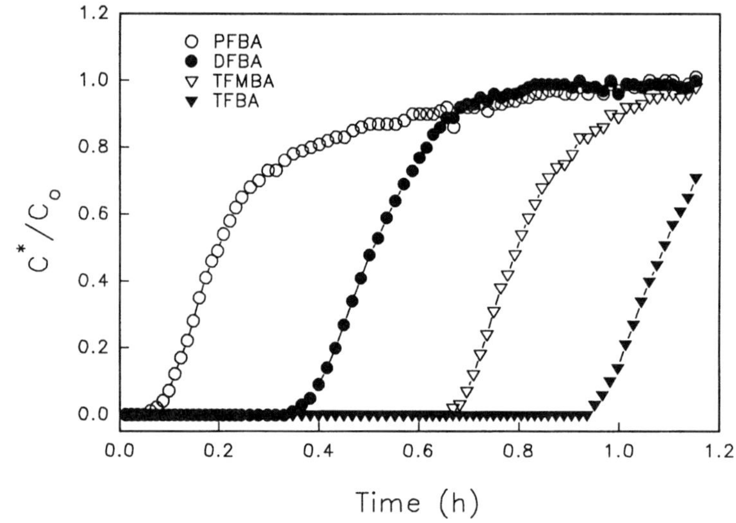

Figure 11.4. Effluent concentrations, C*, normalized by input concentrations, C_o, vs. time of 4 sequentially applied benzoate tracers in column 4.

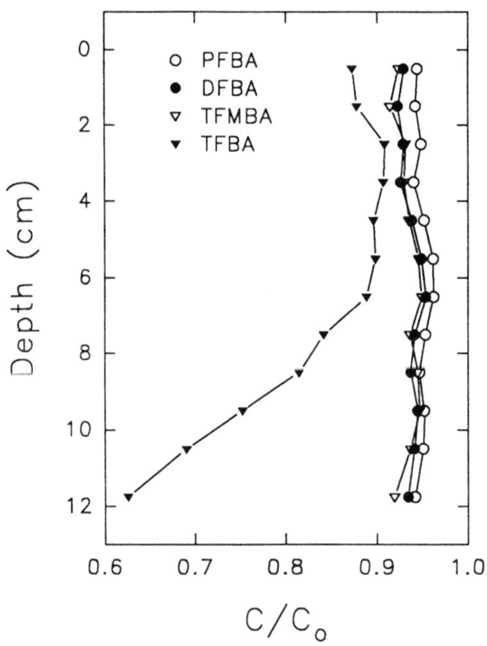

Figure 11.5. Average pore-water concentrations vs. depth normalized by the input concentrations of 4 benzoate tracers sequentially applied to column 4.

We can compare values of α and θ_{im} calculated using an inverse method (CXTFIT) and the outflow data to values calculated using Eq. 11.5 and the resident solute concentration data (Figure 11.7). For the inverse method, the best fit value and its computed 95% confidence limit is plotted. For the regression method, the four estimates from the top four layers are plotted. Calculating γ from the D values estimated by the

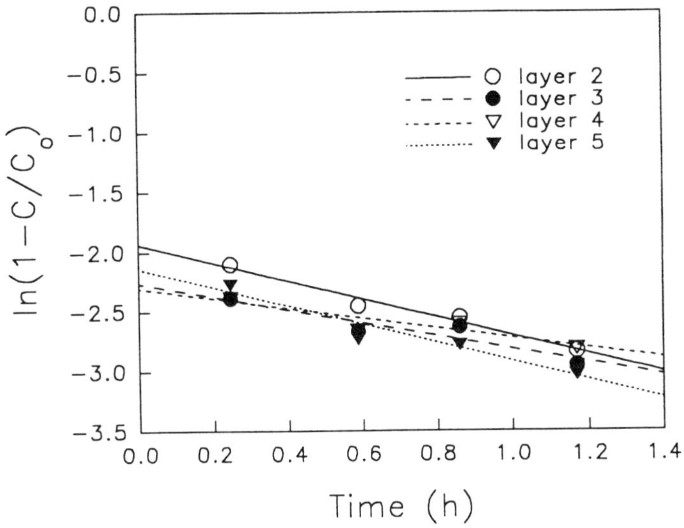

Figure 11.6. Resident concentrations of tracers within layers 2–5 of column 4 plotted as per Eq. 11.5 and the best fitting lines to the data.

inverse method, we find that the experimental conditions for all columns except 1 were marginal for using Eq. 11.5 to estimate α and θ_{im} (Figure 11.3). Even so, the two estimation methods gave similar estimates of α and θ_{im}/θ for each column except column 5 where the inverse estimate of θ_{im}/θ is greater than that found using Eq. 11.5.

FIELD APPLICATIONS

Eq. 11.5 was applied to two different field sites in central Iowa. Tension infiltrometers were used in both experiments to introduce a sequence of tracer solutions at the soil surface. The first experiment used only a single pressure head, –0.03 m. The second experiment included infiltration at four pressure heads, 0.01, –0.03, –0.06, and –0.15 m. The first experiment included a transect of 47 infiltration sites. The second experiment was set as a grid of 40 infiltration sites with 10 at each tension randomly assigned. Full details of the experiments can be found in Casey (1996), Casey et al. (1997), and Casey et al. (1998).

Figure 11.8a presents the spatial distribution of the θ_{im}/θ values obtained from the first field experiment. The median of θ_{im}/θ was 0.627, the mean 0.646, and standard deviation 0.065. Based upon semivariogram analysis, no spatial correlation of (θ_{im}/θ) was detected (i.e., a pure nugget semivariogram). The values of θ_{im}/θ determined in this experiment fall within the range of values reported by other investigators (van Genuchten et al., 1977; Nkedi-Kizza et al., 1983, 1984; Smettem, 1984; Gvirtzman and Magaritz, 1986; Rice et al., 1986). However, the average of these field values is on the high side of earlier reported values.

Figure 11.8b shows the spatial distribution of the α values obtained from the first field experiment. The median α was 0.078 h^{-1}, the mean 0.09 h^{-1}, and the standard deviation 0.054 h^{-1}. Semivariogram analysis indicated no spatial correlation. The values of α obtained in this experiment were similar to those reported from earlier laboratory studies (Kookana et al., 1993).

Figure 11.9 shows a relationship between α and pore water velocity. The graph contains data from earlier laboratory studies as well as the Casey et al. (1997) field

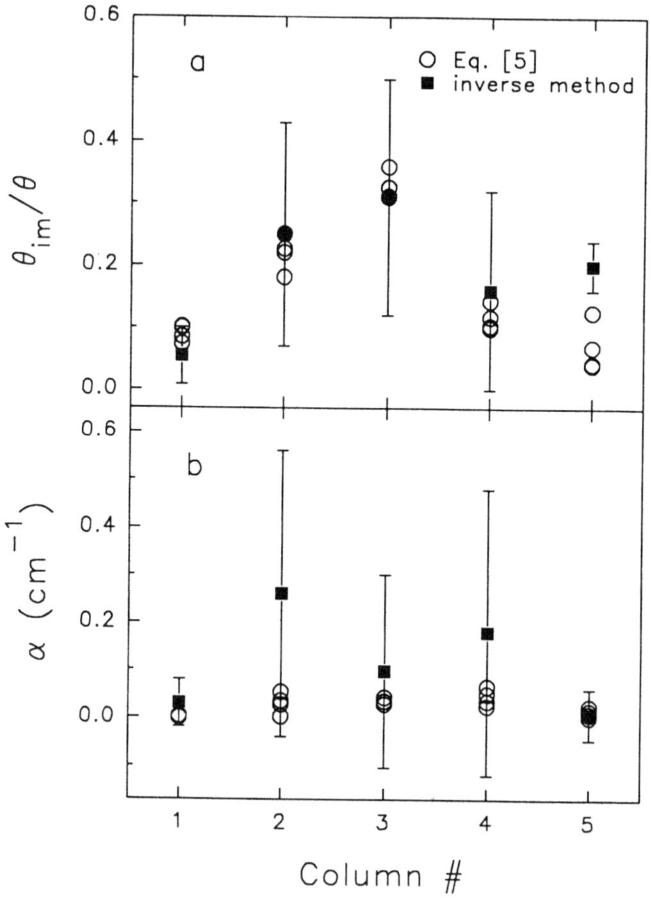

Figure 11.7. Comparison of solute transport parameters (a) θ_{im}/θ and (b) α estimated from both solute breakthrough curves and inverse methods, and the resident concentrations in layers 2–5 and Eq. 11.5, for 5 column experiments. Error bars on values from inverse method represent the 95% confidence limits of the estimates.

data. The laboratory and field data indicate a similar relationship between α and pore water velocity. The field estimates of α based upon Eq. 11.5 are of similar range and of similar relationship to pore-water velocity as reported by others from lab studies. This finding supports the usefulness of the sequential tracer method for determining field transport parameters.

Figure 11.10a shows values of θ_{im}/θ as a function of pressure head from the second field experiment. Mean and standard deviation of θ_{im}/θ for pressure heads of 0.01, –0.03, –0.06, and –0.15 m were 0.40 (0.17), 0.27 (0.11), 0.22 (0.04), and 0.35 (0.20), respectively. Angulo-Jaramillo et al. (1996) also reported fluctuations in θ_{im}/θ values with pressure head. θ_{im} was largest at a pressure head of 0.01 m and was nearly constant for the other three pressure heads. The measured total water contents for pressure heads of 0.01, –0.03, –0.06, and –0.15 m were 0.41, 0.35, 0.34, and 0.34, respectively. Thus, total water content and immobile water content had similar trends with pressure head.

Figure 11.10b shows values of α as a function of pressure head for the second field experiment. Mean and standard deviation of α for pressure heads of 0.01, –0.03, –0.06, –0.15 m were 1.3 (1.5), 0.036 (0.054), 0.0044 (0.0041), and 0.0033 (0.0024) h^{-1}, respec-

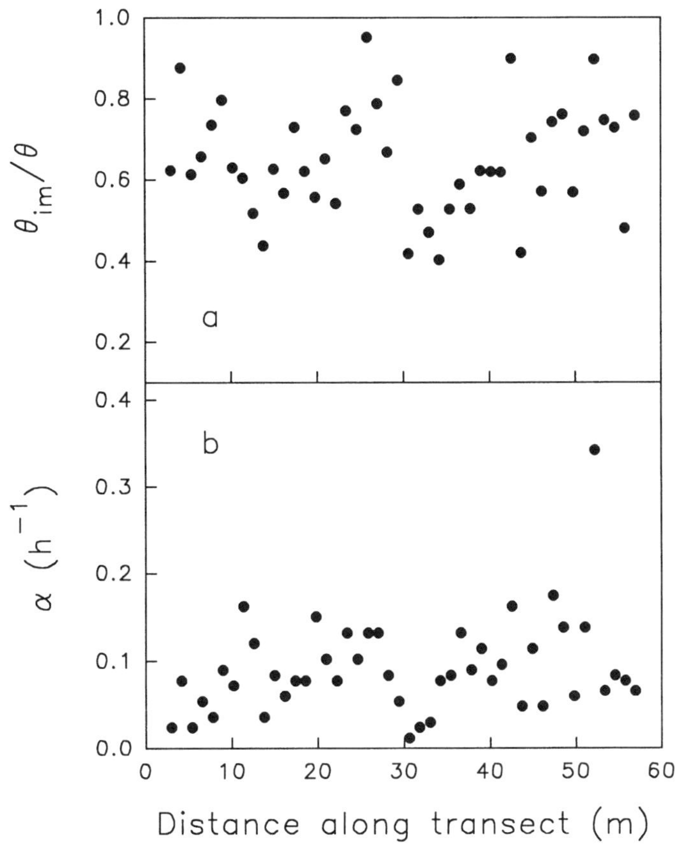

Figure 11.8. Values of (a) θ_{im}/θ and (b) α measured along a field transect. From Casey, F.X.M. et al., *Soil Sci. Soc. Am. J.*, 61:1030–1036, 1997. With permission.

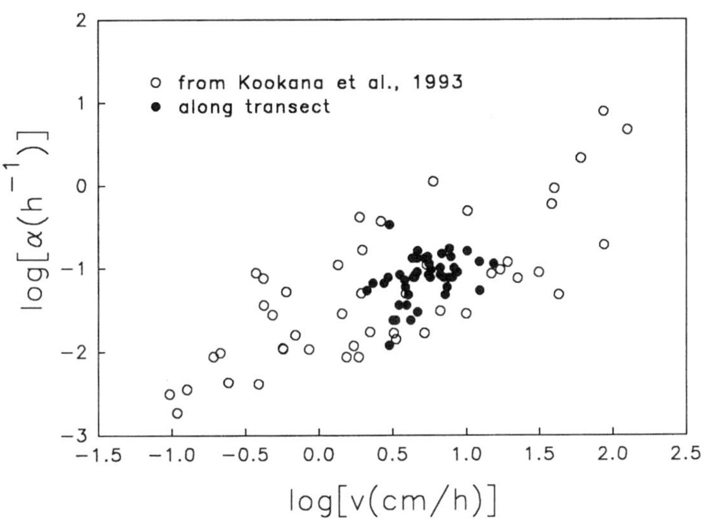

Figure 11.9. α versus v values measured along a transect compared to values summarized from earlier column experiments by Kookana et al., 1993. From Casey, F.X.M. et al., *Soil Sci. Soc. Am. J.*, 61:1030–1036, 1997. With permission.

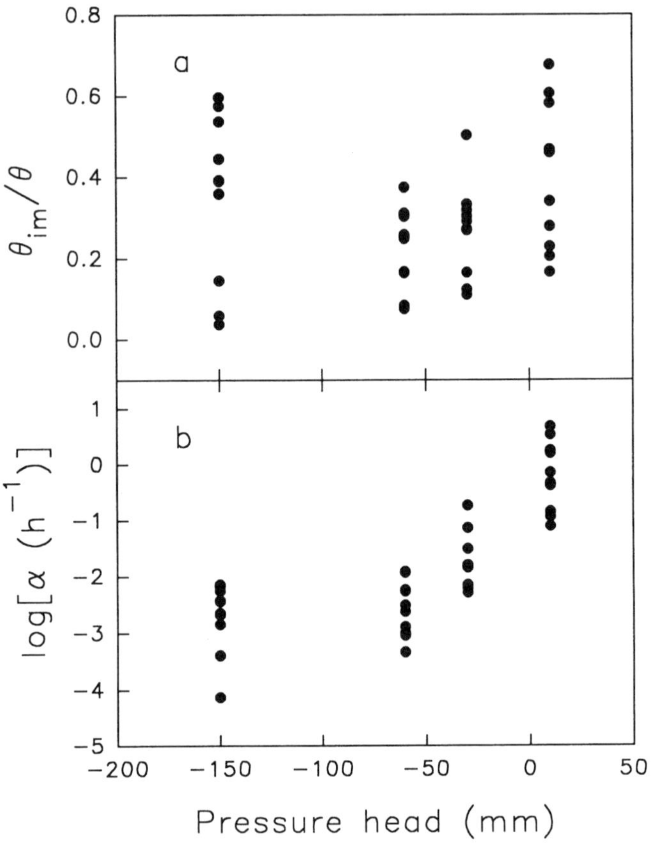

Figure 11.10. Values of (a) θ_{im}/θ and (b) α versus infiltrometer pressure head measured at 40 locations in a field (Casey, 1996).

tively. As found by others (Kookana et al., 1993), the α values were correlated with infiltration flux and thus pore water velocity and pressure head. The average ratios for α/v for these same pressure heads were 0.015, 0.012, 0.003, and 0.006 cm^{-1}. Thus for dispersivities less than about 1 cm, the calculated values for α and θ_{im} should be no worse than 20% in error (Figure 11.3).

We can compare the values of α and θ_{im}/θ found at a pressure head of –0.03 m in the two field experiments directly. At a probability of 0.05, the α values in the two field studies were not significantly different. However, the θ_{im}/θ values were significantly different at a probability of 0.05. The two field sites differed in a couple of important ways. Tillage differed, no-till versus ridge-till. Corn growth stage also differed at the time of measurements. Both soils were glacial till, but soil types differed. Similar α values implies similar microporosity makeup of the two field soils. Differing θ_{im}/θ values implies differences in the macroporosity of the field soils. Transport parameter comparisons between the two sites imply that α is somewhat independent of soil management practices while θ_{im}/θ responds to soil management practices.

SUMMARY

A simple method for estimating immobile water content and mass exchange coefficient for the MIM model has been presented and evaluated. The method is useful for laboratory and field applications. A numerical study indicates the experimental conditions to which the method can be best applied. Laboratory tests indicate that the simple method provides values of α and θ_{im} similar to those obtained from solute breakthrough curve analysis. This method is the only practical method available for determining these solute transport parameters in the field. Finally, because tension infiltrometers are used, field soil hydraulic properties, water retention and unsaturated hydraulic conductivity, also can be determined at the same time that the mass exchange coefficient and immobile water content are determined.

REFERENCES

Addiscott, T.M. A simple computer model for leaching in structured soils. *J. Soil Sci.,* 28:554–563, 1977.

Addiscott, T.M., Ph.J. Heys, and A.P. Whitmore. Application of simple leaching models in heterogeneous soils. *Geoderma,* 38:185–194, 1986.

Angulo-Jaramillo, R., J.-P. Gaudet, J.-L. Thony, and M. Vauclin. Measurement of hydraulic properties and mobile water content of a field soil. *Soil Sci. Soc. Am. J.,* 60:710–715, 1996.

Benson, C.F. and R.S. Bowman. Tri- and tetrafluorobenzoates as nonreactive tracers in soil and groundwater. *Soil Sci. Soc. Am. J.,* 58:1123–1129, 1994.

Bowman, R.S. and J.F. Gibbens. Difluorobenzoates as nonreactive tracers in soil and ground water. *Ground Water,* 30:8–14, 1992.

Casey, F.X.M. Determining Solute Transport Parameters in Field Soil, Thesis presented to the Iowa State University, Ames, IA, in partial fulfillment of the requirements for the degree of Master of Science, 1996.

Casey, F.X.M., S.D. Logsdon, R. Horton, and D.B. Jaynes. Immobile water content and mass exchange coefficient of a field soil. *Soil Sci. Soc. Am. J.,* 61:1030–1036, 1997.

Casey, F.X.M., S.D. Logsdon, R. Horton, and D.B. Jaynes. Field soil hydraulic and solute transport measurement of parameters as a function of water pressure head. *Soil Sci. Soc. Am. J.,* (in press), 1998.

Cassel, D.K. Water and solute movement in Svea loam for two water management regimes. *Soil Sci. Soc. Am. J.,* 35:859–866, 1971.

Clothier, B.E., L. Heng, G.N. Magesan, and I. Vogeler. The measured mobile-water content of an unsaturated soil as a function of hydraulic regime. *Aust. J. Soil Res.,* 33:397–414, 1995.

Clothier, B.E., M.B. Kirkham, and J.E. McLean. In situ measurements of the effective transport volume for solute moving through soil. *Soil Sci. Soc. Am. J.,* 56:733–736, 1992.

Coats, K.H. and B.D. Smith. Dead-end pore volume and dispersion in porous media. *SPE J.,* 4:73–84, 1964.

Gamerdinger, A.P., R.J. Wagenet, and M.Th. van Genuchten. Application of two-site/two-region models for studying simultaneous nonequilibrium transport and degradation of pesticides. *Soil Sci. Soc. Am. J.,* 54:957–963, 1990.

Gaudet, J.P., H. Jégat, G. Vachaud, and P.J. Wierenga. Solute transfer, with exchange between mobile and stagnant water, through unsaturated sand. *Soil Sci. Soc. Am. J.,* 41:665–671, 1977.

Gvirtzman, H. and M. Magaritz. Investigation of water movement in the unsaturated zone under an irrigated area using environmental tritium. *Water Resour. Res.,* 22:635–642, 1986.

Jarvis, N.J. and I. Messing. Near-saturated hydraulic conductivity in soils of contrasting texture measured by tension infiltrometers. *Soil Sci. Soc. Am. J.,* 59:27–34, 1995.

Jaynes, D.B. Evaluation of nonreactive tracers for mid-Iowa soils. *Ground Water,* 32:532–538, 1994.

Jaynes, D.B., R.S. Bowman, and R.C. Rice. Transport of a conservative tracer in the field under continuous flood irrigation. *Soil Sci. Soc. Am. J.*, 52:618–624, 1988a.

Jaynes, D.B., S.D. Logsdon, and R. Horton. Field method for measuring mobile/immobile water content and solute transfer rate coefficient. *Soil Sci. Soc. Am. J.*, 59:352–356, 1995.

Jaynes, D.B., R.C. Rice, and R.S. Bowman. Independent calibration of a mechanistic-stochastic model for field-scale solute transport under flood irrigation. *Soil Sci. Soc. Am. J.*, 52:1541–1546, 1988b.

Kookana, R.S., R.D. Schuller, and L.A.G. Aylmore. Simulation of simazine transport through soil columns using time-dependent sorption data measured under flow conditions. *J. Contamin. Hydrol.*, 14:93–115, 1993.

Lee, J., D.B. Jaynes, and R. Horton. Mobile/immobile model parameter estimation: laboratory column evaluation. *Soil Sci. Soc. Am. J.*, (in press), 1998.

Nkedi-Kizza, P., J.W. Biggar, H.M. Selim, M.Th. van Genuchten, P.J. Wierenga, J.M. Davidson, and D.R. Nielsen. On the equivalence of two conceptual models for describing ion exchange during transport through an aggregated oxisol. *Water Resour. Res.*, 20:1123–1130. 1984.

Nkedi-Kizza, P., J.W. Biggar, M.Th. van Genuchten, P.J. Wierenga, H.M. Selim, J.M. Davidson, and D.R. Nielsen. Modeling tritium and chloride 36 transport through an aggregated oxisol. *Water Resour. Res.*, 19:691-700, 1983.

Parker, J.C. and M.Th. van Genuchten. Determining transport parameters from laboratory and field tracer experiments. *Virginia Agric. Exp. Sta. Bull.*, 84, 1984, p. 96.

Parlange, J.-Y. Theory of water movement in soils: 4. Two and three dimensional steady infiltration. *Soil Sci.*, 113:96–101, 1972.

Rice, R.C., R.S. Bowman, and D.B. Jaynes. Percolation of water below an irrigated field. *Soil Sci. Soc. Am. J.*, 50:855–859, 1986.

Seyfried, M.S. and P.S.C. Rao. Solute transport in undisturbed columns of an aggregated tropical soil: Preferential flow effects. *Soil Sci. Soc. Am. J.*, 51:1434–1444, 1987.

Smettem, K.R.J. Soil-water residence time and solute uptake. *J. Hydrol.*, 67:235–248, 1984.

van Genuchten, M.Th. and P.J. Wierenga. Mass transfer studies in sorbing porous media: I. Analytical solutions. *Soil Sci. Soc. Am. J.*, 40:473–480, 1976.

van Genuchten, M.Th. and P.J. Wierenga. Mass transfer studies in sorbing porous media: II. Experimental evaluation with tritium (^3H_2O). *Soil Sci. Soc. Am. J.*, 41:272–278, 1977.

Yasuda, H., R. Berndtsson, A. Bahri, and K. Jinno. Plot-scale solute transport in a semiarid agricultural soil. *Soil Sci. Soc. Am. J.*, 58:1052–1060, 1994.

CHAPTER TWELVE

Transfer Function Approaches to Modeling Solute Transport in Soils

R.E. White, L.K. Heng, and R.B. Edis

INTRODUCTION

Definition of the Problem

Variability of Solute Inputs in Soil-Plant-Animal Systems and Heterogeneity of Soil Processes at Different Scales

The distribution and flux of solutes in the field varies considerably with time and space. This variability develops through differences in soil, climate, vegetation, animals, management, and their interactions. It affects solutes applied to the surface and those resident in the soil. Heterogeneity occurs at a hierarchy of spatial and temporal scales, ranging from microscopic scales involving time-dependent chemical sorption and precipitation/dissolution reactions, to intermediate scales involving the preferential movement of water and chemicals through macropores, to a much larger scale involving spatial variability of soils in the landscape, and changes in soil properties with season. Differences between the laboratory and field in the degree of heterogeneity and characteristic solute mixing times can lead to errors when laboratory results are applied to the field. For example, the lateral dimension of a soil column is normally between 10 and 100 cm. On the other hand, although the root zone of an agricultural field may have the same length scale in the direction of flow as a soil column, its lateral dimension is of the order of 100 m to 1 km.

Soils are not spatially homogeneous. For example, Schulin et al. (1987) observed significant spatial heterogeneity of bromide concentrations even 400 days after application of a uniform bromide pulse to a field soil in Switzerland. A similar observation was also reported by Flury et al. (1994) from a dye-tracing experiment where remarkable differences were observed in the spatial flow patterns. There is often a large change from the scale of intensive measurements to the scale of practical application. Similarly, there is a need to extrapolate from sites where soil properties and their variability have been intensively studied, to sites where much less is known. Ideally, we seek to identify and quantify differences between one site and the next that affect the distribution and flux of solutes, but spatial variability makes such quantification difficult. Models provide a means of extrapolating from short-term, site-specific observations to longer times and larger scales.

The Role of Simulation Modelling

Models are essentially hypotheses and no model is perfect. They are useful not because they reproduce reality but because they simplify reality and enable the most important processes to be identified, studied, and simulated, and for outcomes to be predicted in advance (Addiscott, 1993). When used appropriately, a model allows extrapolation from a limited set of data so that the amount of repetitive and time-consuming experimentation can be reduced. In addition, a model should help to identify gaps in our knowledge. Computer-based models are potentially capable of helping us advance our knowledge of soil processes in unexpected ways. We may be challenged to reexamine our conceptual foundations and describe the world more precisely than has hitherto been possible (Narasimhan, 1995). However, models are often used inappropriately, outside the context in which they were developed and beyond the parameter range for which they have been tested.

Stochastic Versus Deterministic Approaches

Addiscott and Wagenet (1985) classified solute transport models as either deterministic or stochastic. But the distinction is becoming less clear because there is increasing use of both approaches within a single model (Miralles-Wilhelm and Gelhar, 1996). Deterministic models assume that the occurrence of a given set of events leads to a uniquely definable outcome which can be described mathematically. These models are usually centered around derivations of the convective-dispersive equation (CDE) which may be solved analytically or numerically. Stochastic models were developed to overcome spatial and temporal heterogeneity in soil properties and processes in the field, on the premise that the functions governing the many coupled processes involved in solute transport are unquantifiable in a deterministic sense. The outcome of any one realization of the transport process is uncertain, but if enough sample realizations are observed, the probability density function (pdf) of the underlying process can be defined. The challenge is to characterize that function, and manipulate the probability distribution by coupling additional processes for which the quantifying expressions are determinable. This chapter reviews the stochastic approach to modeling solute transport in soil systems ranging from large intact cores to unconfined fields, based on a transfer function model (TFM), for a variety of solutes and initial and boundary conditions.

METHODOLOGY

Solutes of Importance in Natural and Agricultural Systems

Solutes in natural and agricultural systems can be categorized by their role in the system. For example, there are nutrients, pesticides, waste compounds including heavy metals, organic chemicals, bacteria and viruses, and salts in the case of salinity. The movement of solutes from the soil to surface waters or groundwater represents a degradation of the water resource. In the case of nutrients, leaching represents degradation in the form of loss of fertility, which can be expressed in direct economic terms. In addition to loss/gain degradation, nitrification followed by nitrate leaching is a major cause of accelerated soil acidification in many agricultural soils. It is also useful to categorize solutes by their chemical stability and reactivity with soil constituents. Thus, solutes can be classified as conservative or labile, and as reactive or nonreactive.

Conservative solutes are those that undergo no irreversible reactions, and are unchanged chemically or physically when removed from, or remaining in, the soil system. The total amount of solute (solute in soil plus that removed) remains unchanged with time. Chloride and Br are examples of conservative solutes which have commonly been used as tracers in solute transport studies (see below).

Labile solutes can undergo reversible or irreversible reactions which change the chemical form of the solute, or the physical phase it occupies in the soil. These reactions may be physicochemical or biochemical and the amount of solute may increase or decrease with time. Nitrate, SO_4^{2-}, $H_2PO_4^-$, and NH_4^+ are examples of labile solutes which may be involved in microbial mineralization/immobilization and redox reactions in soil. Many nonindigenous solutes such as pesticide molecules are labile, undergoing decomposition which can be described by first-order decay reactions, and their lability quantified in terms of their half-life.

Reactive solutes are those which undergo reversible or irreversible reactions with soil constituents. Adsorption on surfaces is the most common reaction, but precipitation and dissolution can also occur (e.g., the reversible precipitation of Ca^{2+} as $CaSO_4$ or $CaCO_3$). Cations and anions behave to varying degrees as reactive solutes in soil. Their transport through soil is therefore affected relative to the movement of water, the general effect being that of retardation of the reactive solute. Uncharged molecules such as N_2O (in solution) and urea, which do not react with soil surfaces, or precipitate in insoluble compounds, are called *nonreactive*. Anions such as NO_3^-, Cl^-, and Br^-, which are not strongly adsorbed even at positively charged sites nor form insoluble salts, are also classed as nonreactive.

Tracers in Solute Transport Studies

Tracers are usually conservative, nonreactive solutes that are used to "trace" the movement of water through soils. They are most useful for elucidating the characteristics of the soil solute transport system, providing breakthrough curves (BTCs) and depth concentration profiles, from which the parameters of transport models can be calculated. If the reactions of target solutes that are labile and/or reactive in a soil are known, they can be superimposed on the transport characteristic derived from a tracer experiment. Tracers, in particular dyes, can also be used to identify flow pathways and the distribution of soil pore space (Bouma and Wösten, 1979; Kung, 1990; McBratney et al., 1992; Flury et al., 1994), and to measure the transport volume of a soil (Clothier et al., 1992; Jaynes et al., 1995). Examples of tracers used in laboratory and field studies include Cl^-, Br^-, pentafluorobenzoic acid (PFBA) (Pearson et al., 1996), tritiated water (White et al., 1984) and $^{35}SO_4^{2-}$ (Bolan et al., 1986). Anions such as Cl^- and Br^- may be excluded by charge repulsion from very small pores, which has the effect of accelerating their movement relative to water molecules. In some strongly weathered soils with high sesquioxide content, Cl^- and Br^- may be retarded due to anion exchange. For example, Porro et al. (1993) reported retardation coefficients of 0.79 and 0.76 for Cl^- and Br^-, respectively, relative to tritium in a loamy fine sand, and 0.84 for bromide in a column with layers of sand and calcareous silty clay loam. Plant uptake of tracers also needs to be considered.

Origin of Solutes

Surface-Applied and Indigenous Solutes

Surface-applied solutes enter the soil by mass flow in water and by diffusion. In most cases, the infiltration of water is the dominant factor for the movement of solute to depth, and the concentration of solute transported below the surface will tend to be

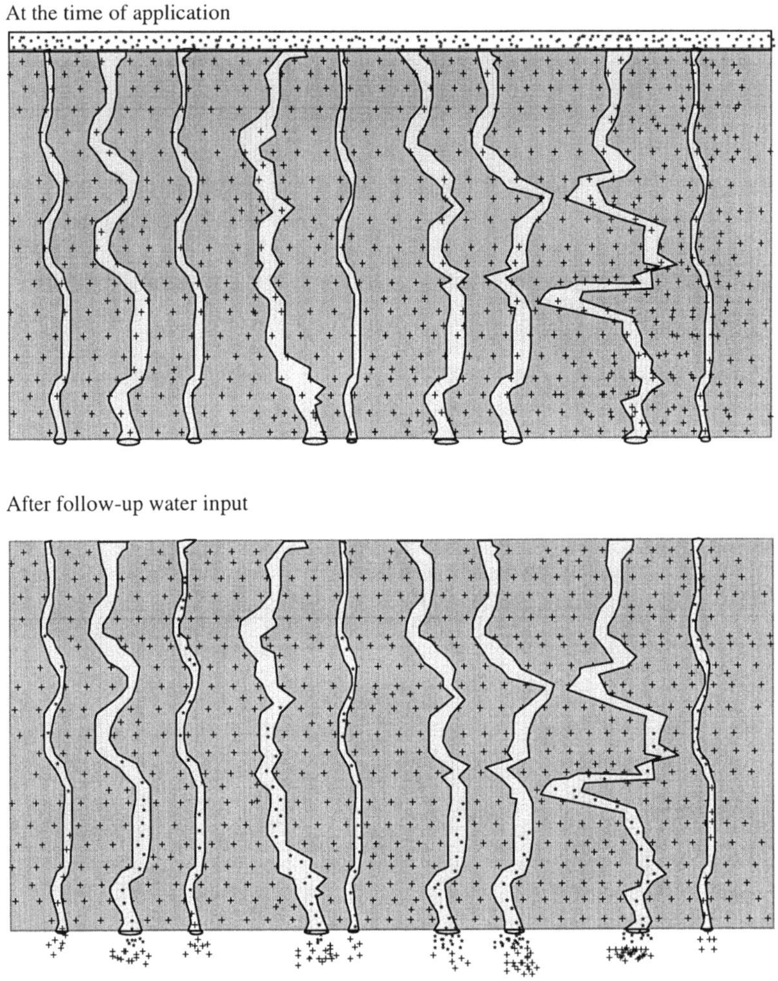

Figure 12.1. Conceptual diagram showing applied (*) and indigenous (+) solutes in a soil with a clearly defined flow tubes, at the time of application and after follow-up water input. Note that the replacement of the indigenous solute by the applied solute, and then by solute-free water, occurs at different rates in separate flow tubes of the transport volume.

highest in the transport volume (see below). Examples of surface-applied solutes are fertilizers, pesticides, and contaminant spills.

Indigenous solutes are those present in the soil before the period of observation. They are generally distributed throughout the water-filled porosity, regardless of the potential flux of water through that porosity. With an influx of solute-free water, the concentration of indigenous solute will be highest in the pores experiencing the least water flux. Indigenous solutes may be generated within the soil, such as through the mineralization of organic matter or mineral weathering. They may also be derived from external sources over a long period (for example, from atmospheric deposition, long-term fertilizer application, or saline groundwater intrusion); or from sources long enough ago to allow diffusion equilibrium in all the soil water (e.g., ancient marine intrusions). A diagrammatic representation of the behavior of surface-applied and indigenous solutes during transport through soil is shown in Figure 12.1.

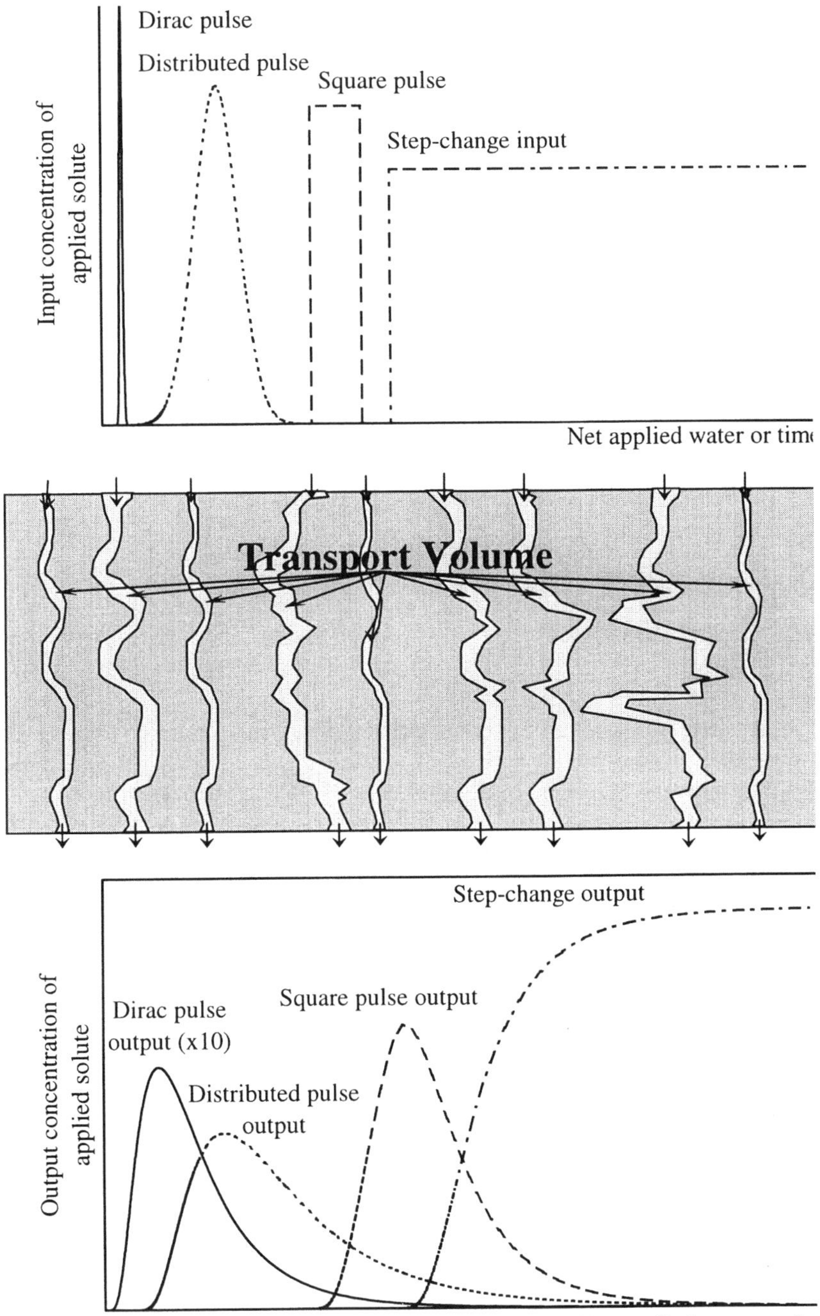

Figure 12.2. Conceptual diagram showing different types of input conditions and typical shapes of breakthrough curves after passing through the soil transport volume.

Pulse and Step-Change Inputs

An input of solute over an infinitesimally short period is known as a Dirac pulse, and in practice is the surface application of a solute in a period which is very brief relative to the period of leaching. Examples which approximate a Dirac pulse include the application of highly soluble or dissolved fertilizers in a single application, and application of soluble herbicides and other pesticides. A step-change input relates to an ongoing change to the input rate of the solutes. This may be an increase in solute input, such as may occur under a permanent, unsealed slurry lagoon, or a decrease, such as through the leaching of a saline soil with good quality irrigation water. A square pulse can be considered as a step-up change, followed after an interval by a step-down change, as might occur with wastewater disposal by irrigation or through occasional fertigation. A diagrammatic representation of the different forms of solute input to a soil, in terms of concentration, and their transformation to output concentrations, is shown in Figure 12.2.

Resident (Volume-Averaged) and Flux-Averaged Concentrations of Solutes

The resident concentration C^r (kg solute m^{-3} soil) is the volume-averaged concentration in the soil, measured, for example, by extracting a known volume of soil in water. It may also be expressed as the mass of solute per unit volume of soil water, to make it more readily compared to a flux concentration (Toride et al., 1993). The flux concentration C^f (kg solute m^{-3} solution) is the solute concentration in the water flowing through the soil. It is defined as the flux density of solute mass divided by the flux density of water (Kreft and Zuber, 1978; Jury and Roth, 1990). The concentration of solutes in drainage water from the bottom of lysimeters or in agricultural pipe drains is an unequivocal measure of the flux concentration at that depth. However, there is evidence that the resident concentration of solutes may vary over a microscale, in particular between the inside and outside of soil aggregates (White, 1985a). Such variations may be caused by the natural clustering of nitrifying bacteria, and therefore the production of nitrate, within aggregates (Darrah et al., 1987). In the case of resident NO_3^-, the inter- and intra-aggregate pore spaces probably correlate with regions of the flowing and nonflowing water, respectively. Thus when Booltink (1995) measured nitrate leaching in a strongly structured clay soil, with a high incidence of macropores through which drainage occurred rapidly after storms, he found that nitrate leached from the finely aggregated topsoil to the groundwater, bypassing almost all the pore space in the subsoil. White et al. (1983) observed a similar effect in a well-structured clay soil as it wet up after a dry summer, and Dekker and Bouma (1984) observed this effect for cores of a clay soil under unsaturated conditions.

If the distinction between flux and resident concentrations is not made, leaching losses of soil-generated solutes such as nitrate can be overestimated, while the losses of surface-applied solutes can be underestimated (Magesan et al., 1994b). For example, White (1989) measured the leaching of NO_3^- from a sample set of 13 large intact cores, and modeled NO_3^- leaching using a TFM with a pdf derived from the leaching of surface-applied Cl^-. If all the nitrate in each core was assumed to be available for leaching, the predicted mean mass of nitrate leached was four times greater than the mean-measured N leached. However, a good approximation of the observed data was obtained by assuming the soil nitrate was distributed throughout the length of each core, and using the concentration in the first small sample of drainage as an estimate of the nitrate concentration in the transport volume that was available for leaching.

The Transport Volume

The occurrence of preferential flow of water and solutes, and the apparent "protection" of solutes inside aggregates from leaching (Addiscott and Cox, 1976), have led to the postulation of a "transport volume" θ_{st} (Jury et al., 1986), which is defined as the fractional fluid volume effective in transporting solute during the period of observation. It is an operationally defined parameter, highly irregular in shape and may vary in size as the rate of water flow through the soil changes in space and time (White, 1987) (see Figures 12.1 and 12.2). Solute can enter or exit via the surface or base of the transport volume V_{st}. Within that volume, dissolution/precipitation or adsorption/desorption, diffusion, and biological and/or chemical transformations can take place.

The transport volume θ_{st} is analogous to the mobile water θ_m used in mechanistic models, which is the fraction of the soil volume through which water containing dissolved solute is moving (Coats and Smith, 1964; van Genuchten and Wierenga, 1976). White (1985b) found that the operational transport volume obtained from a TFM experiment bore a 1:1 relationship to the mechanistic mobile water volume. As with θ_{st}, θ_m has been shown in both laboratory and field studies to change with time and to depend on the concentration of soil solution, initial soil water content, soil water flux and aggregate size (Nkedi-Kizza et al., 1983; Smettem, 1984; Seyfried and Rao, 1987; Jaynes et al., 1988; Clothier et al., 1992, 1995). Values of θ_m have been deduced as the unknown in model validation experiments (Jarvis et al., 1991a,b), or by a curve-fitting procedure in mechanistic models (van Genuchten and Wierenga, 1976), or chosen as an arbitrary fraction of the soil's resident water that is bypassed by the invading solution (Addiscott, 1977; Corwin et al., 1991), or from the relationship between mobile water and pressure head (Angulo-Jaramillo et al., 1996).

In the TFM approach, measurements of θ_{st} have been obtained from the pdf of solute travel times. If the parametric form of the pdf is known or can be assumed, the center of location of solute distribution can be expressed in terms of the median or mean travel time or pathway. The median is the time by which a solute molecule applied to the soil surface has a 50% probability of appearing in the drainage at the depth of interest. The mean travel time defines the time by which half the solute applied to the surface has been recovered in the drainage at the depth of interest. The advantages of using the median travel time are discussed by Dyson and White (1987). For a lognormal pdf of travel times, the transport volume can be obtained from (White et al., 1986b)

$$\theta_{st} = \frac{q}{\bar{v}} = \frac{q\bar{t}}{L} \qquad (12.1)$$

where q is the soil water flux density, $\bar{v}$ is the median pore water velocity, L is the calibration depth, and $\bar{t}$ is the median travel time, given as

$$\bar{t} = \exp \mu \qquad (12.2)$$

where μ is the mean of the lognormal pdf.

Under steady flow conditions, the cumulative drainage $I = qt$ is directly proportional to time so that the pdf of travel times can be expressed in terms of I. Where drainage cannot be measured, net applied water (rainfall plus irrigation minus evaporation) can be used as the driving variable. Since I has the dimensions of length, the pdf of I can be considered as a distribution of travel pathlengths for solute molecules. If the drainage distribution is lognormal, the transport volume is estimated from the median by

$$\theta_{st} = \frac{\exp \mu}{L} \qquad (12.3)$$

and from the mean by

$$\theta_{st} = \frac{\exp(\mu + \sigma^2)}{L} \tag{12.4}$$

where μ and σ^2 are the mean and variance, respectively, of the lognormal distribution of the normalized outflow concentration against cumulative drainage. When flow is not steady state, the mean travel pathlength is the preferred estimator for the calculation of the transport volume (Jury et al., 1986; Heng and White, 1996). An estimate of mean travel time can also be made simply by using the location of the solute peak in a breakthrough curve (Zhang et al., 1996).

While this approach allows a transport volume to be calculated from a solute breakthrough curve, the method is slow and does not allow prediction of solute movement in the absence of breakthrough data (Addiscott, 1994). A simple, routine method of direct measurement of the transport volume is therefore needed. Recently, Clothier et al. (1992, 1995) developed a method for measuring the effective transport volume in the field using a disk-permeameter, where the mobile water was calculated after infiltration with a tracer such as bromide. A similar approach was adopted by Jaynes et al. (1995), whereby rather than a single tracer, a sequence of tracers was used.

Measurement of Solute Transport

Size of the Experimental Unit in Relation to the Scale of Variation in Soil Properties

Most experimental systems for measuring solute transport in soil consider only one-dimensional flow. For this assumption to be valid, the net effect of vertical and lateral flow at the site of measurement must be measured. Zones of high conductivity may cover only a small proportion of the area of the land and be easily missed in a point measurement survey, yet dominate vertical flow of water. Even if the full range of vertical hydraulic conductivities is encompassed, significant horizontal flow over an area larger than the calibration or measurement area may invalidate the functions for solute transport derived from those measurements. This problem presents a major challenge for solute transport studies at the field, farm, and catchment scale. Van Wesenbeeck and Kachanoski (1994) found the recovery and movement of solutes was largely controlled by the spatial pattern of B horizon depth, with lateral movement concentrating vertical flow through "tongues" in the B horizon. Geophysical techniques may be of use in identifying zones of potentially different conductivity at scales larger than the calibration unit, such as ground penetrating radar, mineral magnetics, airborne gamma spectrometry, and other electromagnetic survey methods (Wilford et al., 1992).

To avoid the vagaries of heterogeneity associated with open-system fields, measurements and model testing may be most productively conducted in closed systems such as intact cores and lysimeters where the variability of flow can be retained but effectively integrated.

Large Intact Cores

Intact soil cores, in which the structure is retained and heterogeneity preserved, allow solute transport to be studied under controlled conditions, which is less easily achieved in the field. Cores from 10 to 30 cm diameter and from 20 to 40 cm long are collected from the field with minimal disturbance. Water and solute are applied to the top of the cores in various ways, including needle-array rainfall simulators without surface ponding (a flux-controlled boundary condition), constant head ponding (potential-controlled boundary condition) and disk permeameters which, by controlling the suction at which the solution is applied, allow either boundary condition to be applied. The bottom of the core may be free-draining (zero suction at the lower boundary) or under a controlled suction. Measurements of moisture content, electrical conductivity, and temperature can be made along the length of the cores, and solution samples collected using porous suction cups. On completion, the core is often destructively sampled for measurement of soil properties, and the distribution of solutes remaining within the core. The potential intensity of study and control of conditions make cores indispensable in developing and testing models. For example, Vanclooster et al. (1995) used TDR probes to determine θ and bulk EC, in an intact column of sandy soil, during quasi-steady-state water flow following a pulse input of KCl. The bulk EC measurements were calibrated for the column conditions using suction cup samples taken between probe pairs. The TDR enabled solute resident concentrations to be measured quickly, nondestructively and without disturbing flowpaths, at detailed time and space intervals. The results were used to evaluate whether solute transport was stochastic-convective or convective-dispersive in behavior down to several depths.

Intact cores have been widely used to study the effect of preferential flow in structured soils. There is concern about compression of soil during coring, and that laboratory procedures may enhance or suppress the efficacy of macropore channels. Murphey et al. (1981) used layers of polyester resin and fiberglass cloth to prevent soil breakage, smearing, or compaction when handling soil cores. The confining walls of the core may terminate nonvertical flow channels. Alternatively, cores of finite length may create an exit surface that is not present in the field soil, and flow may occur preferentially between the soil core and enclosing walls, particularly under conditions of surface ponding (Seyfried and Rao, 1987; Jardine et al., 1993). Smith et al. (1985) tested the effect of edge flow in enhancing or diminishing solute transport by comparing the effluent collected from the inner and outer regions of intact soil cores under nonponding conditions. They found no significant effect either way. Buchter et al. (1984) and Cameron et al. (1990) sought to prevent edge flow by injecting vaseline or paraffin between the soil core and casing. However, any effect of edge flow or preferential flow from surface applications can be eliminated by the application of suction at both the infiltration and exit surfaces (Magesan et al., 1995). By carefully controlling the boundary conditions and head gradient in the core, specific behavior predicted by various models can be tested.

Lysimeters

Lysimeters are commonly used for leaching studies. They are essentially larger versions of intact soil cores, having dimensions up to ca. 1 m diameter and 1 to 1.5 m depth. They offer the advantage of being set in the field and are therefore amenable to realistic soil and crop management practices. Measurements, including drainage, can be made on a defined volume of soil and generally the large variations associated with field studies can be avoided. Lysimeters may experience similar problems to intact cores of edge flow, and they are unable to reproduce the effects of lateral flow which can occur in field soils having a marked textural contrast between the A and B horizons. Some lysimeters have sand layers at the base to allow suctions to be applied, but if not, the maximum suction that can be applied to the lysimeter is equal to the difference in gravitational potential between the soil surface and the base.

Drained Fields

In soils with shallow groundwater or seasonally perched water tables, drains are sometimes installed to remove excess water and improve drainage. Fields with underdrainage, such as a network of mole and tile (pipe) drains, offer the opportunity to capture field-scale influences on solute transport and to integrate the effects of natural variability in both water flow and solute concentration. Leaching losses are normally calculated from the flow rate measured using small flumes or V-notch weirs with continuous flow recording, and the measured drainage concentration. Studies have shown that intermittently sampled pipe drains may not provide a good estimate of nutrient losses (Kolenbrander, 1969; Cooke and Williams, 1970) because the concentration can vary considerably with flow rate (Wild and Cameron, 1980; White, 1988). However, when samples are taken at a frequency proportional to flow rate, the drainage concentration provides the flux concentration required to calculate solute loss. Not all the excess water may leave the soil via the drains, so that leaching losses may be underestimated (Thomas and Barfield, 1974). This shortcoming can be overcome by having closely spaced mole drains over pipe drains set in an impermeable subsoil, which creates an almost watertight system (Haigh and White, 1986; Scotter et al., 1990). As such, the system is excellent for monitoring leaching losses because the spatial variability in soil solute concentrations and water flow over a relatively large area is integrated in the drainage. Several workers have used underdrainage systems to characterize subsurface water quality and quantity, including Baker et al. (1975), Smith et al. (1983), White (1987), Richard and Steenhuis (1988), Heng et al. (1991, 1994) and Magesan et al. (1994a,b).

Undrained Fields

In undrained fields, there is a compounding difficulty in estimating both the solute flux concentration and the flux of water. Normally, an estimate is made of the flux or resident concentration at a number of points, and this is multiplied by the area-average water flux. If there is no covariance in the water flux and concentration variables, the product of the arithmetic means gives the most reliable result. However, it is difficult to unambiguously estimate the flux concentration without sampling actual drainage. The concentration of solutes in the soil water has been measured in several ways, including by repetitive soil sampling, suction cup samplers, minilysimeters, and ion exchange resins. Suction cups, which are widely used to sample soil water, are generally assumed to provide a flux concentration. However, to provide a representative flux concentration, a suction cup system should sample the complete spectrum of solute flow paths, in proportion to the contribution that each makes to the overall flux concentration. In some cases this condition does not apply, and the suction cups do not provide a representative sample of either a flux or resident concentration. Magesan et al. (1994b) found this to be true of resident soil NO_3^-, but not for surface-applied Br^- for which concentrations estimated from suction cups were comparable to flux concentrations measured in the drainage water. Conversely, Ellsworth et al. (1996) found that suction cup samplers consistently yielded low estimates of the leaching of an applied solute. The consistent bias suggested that, apart from the problem of representative sampling spatially, either the flux of the water sampled by cups was greater than the average simulated flux, or the solute concentration in the water sampled by the cups was less than the flux average. Whereas resident concentrations in this study were measured on soil samples after complete excavation of the soil block, thus eliminating any effects of spatial variability, the flux concentrations from the suction cup samples incorporated such variability. Thus, the fact that mean travel times estimated from suction cup samples were only ca. 10% greater than those estimated from soil sampling is an encouraging result. Adams and Gelhar (1992) used suction cups to monitor the leaching of bromide and obtained mass recoveries ranging from 45 to 300%. The major problem was probably the estimation of water flux past individual cups.

Influence of Hydraulic Conditions on Solute Transport

Flow Rate

Preferential movement of solutes through macropores appears to be initiated above a threshold water input rate (Bouma et al., 1981; Beven and Germann, 1982). This suggests that a change in the transport characteristics of solutes (and hence the transport volume) should occur once that threshold is exceeded. In studying the effect of water flux density through soil on solute transport, Dyson and White (1989b) divided soils into two classes: those for which solute transport is insensitive to the rate of water throughput (Class 1), and those for which solute transport is significantly affected by changes in water throughput (Class 2). BTCs (normalized concentration against cumulative drainage) for surface-applied Cl^- leached through intact cores of a well-structured clay soil were very similar for irrigation rates from 3 to 30 mm hr^{-1}. This soil exhibited a pronounced asymmetric leaching pattern, suggestive of flow through macropores, yet the BTCs were almost identical for all flow rates. Flow was unsaturated but the soil moisture content, once near steady state was attained, showed little change for the different rates. Thus, this Class 1 soil had an approximately constant θ and transport volume θ_{st} with varying irrigation rate. Similarly, Richter and Jury (1986) found no noticeable effect of irrigation rate (0.55 and 0.95 cm h^{-1}) on the characteristics of bromide transport through microlysimeters of an undisturbed sandy loam soil. Hornberger et al. (1990) found that, despite the importance of macropore flow, the transport volume of a soil block in a forested catchment irrigated at near steady-state at three rates decreased only slightly with increasing flow rate. Results from Costa et al. (1994), recalculated in terms of cumulative drainage, show that the center of location of the pdf for Br^- transport through a sandy loam soil changed little as flow rate varied from 4.6 to 21.4 mm h^{-1}.

Equation 12.1 shows that θ_{st} will remain constant if the mean pore-water velocity increases in the same proportion as the water flux density. This is possible, according to Poiseuille's equation for flow of liquid with viscosity η, through a cylindrical tube of radius r and length L at a rate Q, i.e.,

$$Q = \frac{\pi r^4 \Delta P}{8 \eta L} \tag{12.5}$$

provided the head gradient $\Delta P/L$ remains constant. However, in soils for which the pore space is very heterogeneous, the resistance to flow is very dependent on the size, regularity, and continuity of the pores conducting water, so the proportionality between flux density and mean pore-water velocity is likely to break down once very large pores begin to fill with water.

Variable flow rates occur in the field under rainfall when water movement is episodic, with an unpredictable variation in both duration of flow and the interval between flows. Under such transient conditions, for soils whose transport characteristics do not change appreciably with flow rate (Class 1), the pdf of I is the appropriate function to describe solute transport (Jury and Roth, 1990).

Initial Water Content

There is likely to be an interaction between the effects of initial water content and flow rate which is determined by the balance between gravity-driven convective flow, mainly in the large pores, and convective flow due to matric suction gradients into the small pores. When the unsaturated conductivity of dense aggregates is very low, the rate of water and hence solute movement into the intra-aggregate pores will be slow even though the suction gradient is large. Thus, flow of water and solutes that bypasses much of the intra-aggregate pore space can occur in well-structured clay soils over a range of initial soil water contents (Kneale and White, 1984; White et al., 1986b). Consistent with these observations, Quisenberry et al. (1994) found that the replacement of resident water by applied water was greater at lower application rates in a well-structured silt loam soil. Booltink (1995) reported drainage occurring after less water had been applied than was required to fill the available storage capacity in a heavy clay soil.

Flury et al. (1994) found a strong interaction between initial soil moisture content and flow rate, particularly for soils that cracked on drying and swelled when wet. Ponding on a dried and cracked soil was likely to cause more bypass flow than slow rewetting. White et al. (1986a) found that the median travel time for surface-applied Cl^-, reflecting the skewness of the BTC, depended on the initial water content at a given flow rate in a structured clay soil. Leaching of surface-applied chemicals, under unsaturated conditions, was more pronounced in this clay soil when it was initially dry compared to initially wet. At high initial soil water contents, applied solutes can diffuse faster into the intra-aggregate pore space and so be less vulnerable to leaching. However, Tillman et al. (1991) found that when Br^- was applied to a field soil in a small volume of water (5 mm) before leaching with 50 mm of water, less Br^- was leached when the soil was initially dry than when it was wet. In this case, solute was drawn by capillary flow into the dry soil aggregates and became less vulnerable to leaching. Similarly, solutes initially resident, or generated, in the intra-aggregate pore space tend to be protected from leaching (Addiscott and Cox, 1976).

Hysteresis in the soil water characteristic may affect the size of the transport volume, and therefore solute transport, particularly in poorly structured coarse textured soils. Depending on whether the drying or the wetting boundary curve is used as a reference, solute transport was, respectively, faster or slower relative to the nonhysteretic case (Vereecken et al., 1995). Russo et al. (1989) concluded that hysteresis was likely to retard the downward transport of solutes compared to nonhysteretic conditions.

The "fingering" phenomenon in hydrophobic soils is sensitive to initial water content (Ritsema and Dekker, 1995), with the transport volume in initially dry hydrophobic soils being much smaller than in initially wet soils. The accumulation of organic matter at the surface, which becomes increasingly hydrophobic with prolonged drying, may increase preferential flow down pathways such as worm burrows (Edwards et al., 1993).

MODELING APPROACHES

Choice of a Simulation Model

The one-dimensional CDE has been shown to work successfully at the laboratory scale and on small scale in the field where soils are homogeneous (Miller et al., 1965; Warrick et al., 1971). However, many large field-scale studies of solute transport through the unsaturated zone have shown enormous variability in pore-water velocities and very large hydrodynamic dispersion coefficients (Quisenberry and Phillips, 1976; Schulin et al., 1987; Andreini and Steenhuis, 1990). Such variability, coupled with imprecision in the quantification of other processes relevant to solute behavior, poses a serious problem for the simulation of heterogeneous field-scale transport. This has led to the adoption of stochastic models and geostatistical methods to assist in the monitoring and prediction of solute transport in the unsaturated zone.

A number of stochastic approaches are available; e.g., scaling theories, Monte Carlo simulations, stochastic-continuum models (Russo and Dagan, 1991) and stochastic-convective models (Dagan and Bresler, 1979; Jury, 1982; Jury and Roth, 1990). Jury and Flühler (1992) discussed the appropriateness of these models for simulating solute transport in the unsaturated zone. Any process model, whether it is deterministic such as the CDE or stochastic, such as the stochastic-convective stream tube model, conforms to a transfer function model, provided it maintains mass balance (Sposito et al., 1986). However, of all these stochastic models, the stochastic-convective stream tube model appears to be most appropriate, when solutes are applied over a wide area and observations are made at early times, or shallow depths (Butters and Jury, 1989). This model assumes that solute moves by convection at different velocities in individual flow tubes without mixing between adjacent tubes. Adopting such a model means that, once the pdf of solute travel times or pathlengths (the transfer function) between input and output surfaces separated by a depth L has been defined, the transport of the solute to other depths can be predicted. It also permits the relationship between travel time and travel distance pdfs to be identified (see below).

Properties of the Transfer Function

Stationarity and Ergodic Properties of the Transfer Function

The assumptions of stationarity and ergodicity underlie the application of the TFM. Stationarity assumes that the pdf for solute travel does not depend on the time or spatial position at which the transport process is observed. Thus, the moments of the pdf should be applicable over a local area, and the values obtained in one year applicable to the same area in other years when seasonal conditions may be different. Stationarity of the transfer function has been observed in studies by Heng et al. (1994) and Magesan et al. (1994a), where the transport parameters derived from the leaching of Cl^- over one winter season were successfully used to predict its movement during two other years and also extended to the leaching of SO_4^{2-}. Similarly, White et al. (1986b) observed stationarity in the pdf of travel times for surface-applied Br^- moving to a single mole drain in a clay loam soil.

For ergodicity to apply, a statistical or ensemble average should be equal to the time average of individual realizations of the stochastic process. This assumption has rarely been tested in solute transport studies. Mallawatantri and Mulla (1996) compared the mean μ and standard deviation σ for Br^- travel, calculated as an average for 40 individual plots, with the average of these parameters for the whole field (57 ha).

There was good agreement between the two sets of parameters. However, a true test of ergodicity would be to compare the statistical average of the solute travel moments derived from a number of point measurements in a field, with the independently measured moments of the pdf for solute travel through the *whole* field. This could best be achieved using a mole and pipe-drained field which was continuously monitored for leaching, and from which several large intact soil cores were taken for measurement of the transport of a conservative, nonreactive tracer.

Travel Time and Travel Distance pdfs

The transfer function for a soil is most readily determined from the response of a defined soil volume to a narrow pulse input of conservative, nonreactive solute to the surface. Consider that the solute pulse is leached with water at a constant flux and the concentration in the drainage at depth z below the surface is monitored with time. The fraction of the applied solute molecules leached below depth z between drainage I and $I + \Delta I$ gives the probability of a molecule applied to the surface reaching depth z in this amount of drainage. For steady flow, the change in the normalized concentration in the drainage as I builds up, defines the pdf for solute travel to depth z, or the travel time pdf $f^f(z,I)$. Because the pdf is expressed in terms of flux concentrations, measured in drainage from a soil core or lysimeter, or from a drained field, it is given the superscript f. The normalization constant is the mass of solute applied to the surface and, provided the observation time is long enough (or the cumulative drainage very large), the area under the breakthrough curve should be one. The general form of the transfer function equation for such a system, given a time-varying flux concentration $C^f(0,I)$ at the input surface, is given by (Jury and Roth, 1990; Jury and Scotter, 1994)

$$C^f(z,I) = \int_0^I C^f(0, I - I')f^f(z,I)dI' \tag{12.6}$$

where $C^f(z, I)$ = the flux concentration in the drainage at depth z.

Alternatively, we may consider a soil volume in which a known amount of solute is present at z = 0 and time t = 0, and monitor the redistribution of this solute pulse through the soil as it is leached with a steady flux of water. The resident concentration in the soil can be measured at different depths after an amount of drainage has occurred, and the distribution of normalized solute concentration with depth then gives the travel distance pdf $f^r(z,I)$. The transfer function equation for this initial condition, for a semi-infinite system, is given by the equation

$$C^r(z,I) = \int_0^z C^r(z - z', 0)f^r(z', I)dz' \tag{12.7}$$

where $C^r(z, I)$ is the normalized resident concentration after drainage I has redistributed the initial concentration $C^r(z, 0)$. The relationship between the travel distance pdf and the drainage flux pdf, assuming stochastic-convective transport, is (Jury and Roth, 1990)

$$f^r(z,I) = \frac{I}{z} f^f(I,z) \tag{12.8}$$

In the flux pdf, I is a variable and z a parameter, whereas in the travel distance pdf, z is a variable and I a parameter. Although the soil water content θ and flux q may be assumed constant within a single stream tube, they are likely to vary between stream tubes in a field soil. For this situation, Jury and Scotter (1994) derived the relationship between the travel distance and drainage flux pdfs as

$$f_i^r(z,I) = \frac{LI^2}{E[I,L]z^2} f_b^f(I,z) \tag{12.9}$$

where L is the calibration depth for the drainage flux pdf, and $E\{I,L\}$ is the first moment of the drainage flux pdf. The subscript b refers to the "boundary value problem" where the pdf is the normalized flux concentration in the drainage in response to a narrow pulse input at the surface when there is no solute in the soil initially (see Equation 12.6). The subscript i refers to the "initial value problem" where the pdf is the normalized resident concentration in the soil at different depths in response to the presence of a pulse of solute in the soil at $z = 0$ and $I = 0$ (see Equation 12.7). The advantage of the relationship in Equation 12.9 is that once one pdf has been parameterized by experimentation (see below), the other can be estimated without further measurements.

Depth-Scaling

Another property that the stochastic-convective assumption confers on the transfer function is that of depth-scaling of solute transport according to the equation

$$f^f(z,I) = \frac{L}{z} f^f\left(L, \frac{IL}{z}\right) \tag{12.10}$$

This means that, for a vertically homogeneous soil, the probability that solute applied to the surface at $z = 0$ and $I = 0$ will appear at depth z after cumulative drainage $\leq I$ is the same as the probability it will reach depth L after drainage $\leq LI/z$. The testing of solute transport models in naturally heterogeneous soils is poorly developed. A correct identification of the governing flow concept requires depth-dependent measurements of solute behavior (Jury and Flühler, 1992). For example, the depth-dependence of solute dispersion is central to discriminating between the CDE and stochastic-convective TFM approaches. This is the result of different fundamental assumptions in the two models about mixing between regions of different flow velocity. The CDE assumes that lateral mixing of solute between such regions occurs over a time period which is much smaller than the time required for solute to be moved from the inlet to outlet end of the transport volume (Taylor, 1953). In the CDE, the effective dispersion coefficient D is constant and the spread of the solute pulse as it is leached through the soil is directly proportional to depth at which it is observed. In the TFM, however, solute moving in individual stream tubes is assumed to be isolated from that in neighboring tubes and spreading of the pulse is entirely due to the effect of variations in pore-water velocity; that is, to hydrodynamic dispersion. This leads to D appearing to increase linearly with depth and the spreading of the BTC to increase proportionally to z^2.

Parameterizing the pdf

To parameterize a pdf, an appropriate function is fitted to a solute BTC, or to a resident concentration depth profile, using an optimization procedure to obtain the best estimates of the first and second moments of the pdf. The most common procedure is sum of squares optimization, but for some applications other estimation procedures may be more appropriate (Jury and Sposito, 1985). A number of functions have been used to parameterize pdfs, the more common of which are discussed below.

The Fickian pdf

Sposito et al. (1986) showed that the two component CDE can be presented as a special case of the TFM. The pdf of the CDE is known as a Fickian pdf (Jury and Sposito, 1985) and in terms of I is written as

$$f(I) = \frac{z\exp[-(z - vI)^2 / (4DI)]}{2\sqrt{\pi D I^3}} \tag{12.11}$$

where D is the dispersion coefficient and v the pore-water velocity. The Fickian pdf implies that a solute pulse spreads symmetrically about the center of location as it is leached through soil. In practice, because the distribution of pore-water velocities is markedly skewed, especially in well-structured soils, the Fickian pdf is not usually appropriate for the transfer function.

The Lognormal pdf

In many field studies and with large intact soil cores in the laboratory, a lognormal pdf has been found to be most appropriate (Biggar and Nielsen, 1976; Jury et al., 1982; White et al., 1984; 1986a; Dyson and White, 1987; Butters and Jury, 1989). Jury and Roth (1990) described this pdf as a convective lognormal transfer function (CLT) which has the form

$$f(I) = \frac{\exp[-(\ln(I) - \mu)^2 / 2\sigma^2]}{\sqrt{2\pi}\sigma I} \tag{12.12}$$

where μ and σ are the mean and standard deviation of the lognormal function. When the lognormal function is parameterized at a depth L, it may be evaluated at a depth z other than the calibration depth by using Equation 12.10, which gives

$$f(z, I) = \frac{\exp\left[-(\ln\left(\frac{IL}{z}\right) - \mu_L)^2 / 2\sigma_L{}^2\right]}{\sqrt{2\pi}\sigma_L I} \tag{12.13}$$

The pdf can be modified to estimate flux as a function of time or cumulative drainage, or resident concentrations as a function of depth. The pdf can also be modified to accommodate alternative initial and boundary conditions, as discussed below.

Narrow Pulse of Applied Tracer

Consider a quantity of dissolved solute M_0 that is applied per unit area of soil surface in a small volume of leaching water. This can be treated as a Dirac pulse input to the transport volume, and the flux concentration at a particular depth L following cumulative drainage I is calculated from the equation

$$\frac{C^f}{M_o} = f^f(z, I) \tag{12.14}$$

where $f^f(z,I)$ is the drainage flux pdf (see Equation 12.13) evaluated at the calibration depth L. The depth concentration profile for the solute, after a given amount of drainage I, can be obtained from the travel distance pdf according to (Jury and Sposito, 1985)

$$\frac{C^r}{M_o} = f^r(z, I) \tag{12.15}$$

where

$$f^r(z,I) = \frac{\exp\left(\mu - \frac{\sigma^2}{2}\right)}{L} \frac{\exp\left[-(\ln\left(\frac{IL}{z}\right) - \mu)^2/2\sigma^2\right]}{\sqrt{2\pi}\sigma I} \tag{12.16}$$

μ and σ^2 in this equation are the first and second moments of the drainage flux pdf.

Alternatively, the soil can be sampled at different depths after a given amount of drainage to obtain a resident concentration profile. For example, Vanderborght et al. (1996) estimated the parameters of a convective lognormal transfer function using TDR probes to give a measure of the resident concentration at various depths in a time series. The time-integral normalized concentration at any depth z was given by

$$C^{rt^*}(z,t) = \frac{C^r(z,t)}{\int_0^\infty C^r(z,t)dt} \tag{12.17}$$

Given that resident concentrations of the solute have been measured at a particular depth at several times, the values for μ_L and σ_L for the CLT model calibrated at depth $z = L$ can be estimated by fitting the equation

$$C^{rt^*}(z,t) = \frac{1}{\sqrt{2\pi}\sigma_L t}\exp\left[-\frac{\left(\ln\left(\frac{tL}{z}\right) - \mu_L - \sigma_L^2\right)^2}{2\sigma_L^2}\right] \tag{12.18}$$

to the time series of time-integral normalized concentrations. This approach can be used in situations where there are time course measurements of resident concentrations obtained by techniques such as by TDR, soil sampling, or suction cup samplers.

A Square Pulse of Applied Solute

The integral form of Equation 12.13 can be used to predict the flux concentration of solute in the drainage at the calibration depth L for a step-up change in input concentration from 0 to C_o at the soil surface (Ellsworth et al., 1996). This equation applies for values of I from 0 to ΔI, the extent of the square pulse in terms of drainage I; that is, for $I \le \Delta I$

$$C^f(z,I) = \frac{C_o}{2}\left\{1 + \operatorname{erf}\left[\frac{\ln\left(\frac{IL}{z}\right) - \mu_L}{\sqrt{2}\sigma_L}\right]\right\} \tag{12.19}$$

At the end of the pulse, the flux concentration at depth L as drainage continues for $I > \Delta I$ is given by

$$C^f(z,I) = \frac{C_o}{2}\left\{\operatorname{erf}\left[\frac{\ln\left(\frac{IL}{z}\right) - \mu_L}{\sqrt{2}\sigma_L}\right] - \operatorname{erf}\left[\frac{\ln\frac{(I-\Delta I)L}{z} - \mu_L}{\sqrt{2}\sigma_L}\right]\right\} \tag{12.20}$$

μ_L and σ_L are the mean and standard deviation of the flux lognormal pdf. Hereafter, the subscript L will be understood for this pdf. Similarly, for a solute applied to the surface in net applied water ΔI, of concentration C_o, when the initial solute concentration in the soil is zero, the depth profile of the resident concentration $C^r(z,I)$ for $I \le \Delta I$ is given by

$$C^r(z,I) = \frac{C_o}{2L}\exp\left(\mu + \frac{\sigma^2}{2}\right)\left\{1 + \operatorname{erf}\left[\frac{\ln\left(\frac{IL}{z}\right) - \mu - \sigma^2}{\sqrt{2}\sigma}\right]\right\} \tag{12.21}$$

and for $I > \Delta I$

$$C^r(z,I) = \frac{C_o}{2L}\exp\left(\mu + \frac{\sigma^2}{2}\right)\left\{\operatorname{erf}\left[\frac{\ln\left(\frac{IL}{z}\right) - \mu - \sigma^2}{\sqrt{2}\sigma}\right] - \operatorname{erf}\left[\frac{\ln\left(\frac{(I-\Delta I)L}{z}\right) - \mu - \sigma^2}{\sqrt{2}\sigma}\right]\right\} \tag{12.22}$$

A Distributed Pulse Input

A distributed pulse has rarely been explicitly used in solute transport studies, although it is likely to occur when solutes applied to the soil surface are dissolved over a period of time, rather than immediately. Magesan et al. (1994a) used a lognormal function to model the dissolution of solid NaBr, applied as a tracer for Cl^- leaching, under natural rainfall in the field. The solution to a transfer function model using this distributed pulse input gave a reasonable simulation of Cl^- leaching over two winter seasons. However, there are difficulties in parameterizing such an input function independently of the leaching measurements.

A Step-Change Input

A step-up change in solute input occurs when the concentration of an applied solute increases from 0 to C_o at $I = 0$ and continues at this concentration indefinitely. Assuming a lognormal drainage flux pdf calibrated at depth L, the appropriate equation for calculating the concentration C^f at any depth z is given by Equation 12.19. This condition has been used extensively to study the movement of solutes and particulates such as bacteria through large undisturbed soil cores (White, 1985c). However, a step-up change in concentration of an externally applied solute can also be associated with a step-down change in concentration for an indigenous solute, when that solute is not present in the leaching water. This is a common situation for soil-generated solutes, such as nitrate and sulfate. The simplest case is when an indigenous solute is uniformly distributed through a defined soil volume at a resident concentration C_o^r in the transport volume. This condition is analogous to that of steady flow through the soil volume of influent solution containing solute at the same concentration as in the transport volume. Consider a step change in concentration of the influent solution from C_o to 0 at $I = 0$. Assuming a lognormal drainage flux pdf calibrated at depth L, the solute concentration C^f in the drainage at any depth z is given by

$$C^f(z, I) = \frac{C_O^r}{2}\left\{1 - \operatorname{erf}\left[\frac{\ln\left(\frac{IL}{z}\right) - \mu}{\sqrt{2}\sigma}\right]\right\} \tag{12.23}$$

White (1989) and White and Magesan (1991) used this approach to model nitrate leaching from soil cores, as did Heng et al. (1994) and Heng and White (1996) to model the leaching of soil sulfate in the field. However, resident solutes may not be uniformly distributed to the depth of interest (e.g., the bottom of the root zone), or indeed to the calibration depth of the pdf. A possible situation is that of an indigenous solute uniformly distributed from the soil surface to a depth Z below the surface. The initial conditions for leaching are

$$\begin{aligned} C_O^r &= C_O & \quad 0 < z \le Z \\ C_O^r &= 0 & \quad Z < z \le L \end{aligned}$$

where C_o^r is the initial resident concentration of solute in the transport volume to depth Z at $I = 0$ and L is the calibration depth. Consider a step change in concentration of the influent solution from C_o to 0 at $I = 0$. The layer of solute can be treated as a square pulse input, with the amount of solute per unit area that is available for leaching equal to M_o between the surface and depth Z; that is

$$M_O = C_O^r Z\theta_{st} \tag{12.24}$$

The leading edge (step-up) of the square pulse occurs at Z, and the trailing edge (step-down) occurs at the surface, so that depth-scaling of the pdf is required for the leading edge. The flux concentration C^f at depth z following the leaching of the layer of resident solute can therefore be described by

$$C^f(z,I) = \frac{C_O^r}{2}\left\{\operatorname{erf}\left[\frac{\ln\left(\frac{IL}{(L-Z)}\right)-\mu}{\sqrt{2}\sigma}\right] - \operatorname{erf}\left[\frac{\ln(I)-\mu}{\sqrt{2}\sigma}\right]\right\} \tag{12.25}$$

In the case of transient flow and depth-dependent water loss due to removal by plant roots, cumulative drainage with depth needs to be estimated using an appropriate model for crop water uptake. Jury and Scotter (1994) give other examples of how the stochastic-convective TFM can be solved for specific initial value and boundary value problems.

Exponential pdf

When the transport volume behaves as if it were well mixed, an exponential function may be appropriate to describe the change in drainage flux concentration of a indigenous solute with I. Such a system may occur in soils with closely spaced, shallow subsurface drains and subject to frequent water inputs. Utermann et al. (1990) used an exponential pdf to simulate pesticide transport through the saturated zone of a pipe-drained soil with a shallow water table. Scotter et al. (1991) successfully used an exponential pdf of the form

$$f^f(z,I) = a^{-1}\exp(-I/a) \tag{12.26}$$

to simulate losses of resident Cl^- to a mole-pipe drainage system in a silt loam soil under pasture during two successive winter periods. The value of a can be a fitted parameter, or calculated from the transport volume as (Magesan et al., 1994a)

$$a = L\theta_{ST} \tag{12.27}$$

where L is the calibration depth (drain depth or the bottom of the root zone). The equation to predict C^f, analogous to Equation 12.23, was

$$C^f(z,I) = C_O^r\left(1-\int_0^I f^f(z,I)dI\right) \tag{12.28}$$

$$= C_O^r \exp\left(-\frac{I}{a}\right) \tag{12.29}$$

Magesan et al. (1994a) also simulated the leaching of indigenous Cl^- and surface-applied Br^- from the same system used by Scotter et al. (1991), and found an exponential pdf modeled the Cl^- data as well as a lognormal pdf. The Br^- drainage concentrations could be simulated by an exponential pdf if the applied bromide was assumed to gradually enter the transport volume as a distributed pulse. Note that Equation 12.29 is identical to the equation developed by White (1987) for estimating NO_3^- concentrations during individual drainage events in a mole and pipe-drained soil. It is also consistent with the equation suggested by Raats (1978), for predicting the solute concentration in drainage following a step-increase in input concentration to C_o, i.e.,

$$\frac{C^f}{C_O} = 1 - \exp(-r) \tag{12.30}$$

where r is the number of pore volumes of drain discharge. Raats (1978) assumed that, provided the half drain spacing was > 5 times the depth to an impermeable layer, convective solute transport by saturated steady flow to the sub-surface drains would approximate to a well-mixed system.

Since an exponential pdf requires only one parameter, θ_{st}, to be estimated, and this parameter can be obtained from water content data, it has great potential for use in management models. For example, the nitrate leaching model NLEAP (Shaffer et al., 1991) uses an exponential pdf for predicting N leached for periods of time or cumulative drainage. The governing equation is

$$M = M_O\left(1 - \exp\left(-\frac{kI}{p}\right)\right) \tag{12.31}$$

where M is the amount of N leached, M_o is the amount of N available for leaching, k is the leaching coefficient, I is drainage, and p is soil porosity. The leaching coefficient k is used to adjust the porosity to account for immobile water, and the term p/k is thus equivalent to the transport volume $L\theta_{st}$. Therefore, the cumulative probability that nitrate will be leached $P(z,I)$, implicit in the NLEAP model, may be written as

$$P(z, I) = \frac{M}{M_O} = 1 - \exp\left(-\frac{I}{L\theta_{ST}}\right) \tag{12.32}$$

which on differentiation with respect to I gives

$$f^f(z, I) = \frac{1}{L\theta_{ST}} \exp\left(-\frac{I}{L\theta_{ST}}\right) \tag{12.33}$$

Note the comparison between this equation and Equation 12.26. An almost identical approach is used in the EPIC leaching model (Williams and Kissel, 1991), with the porosity being adjusted by subtracting the volumetric water content at permanent wilting point instead of using a leaching coefficient. Equation 12.32 is also identical to the equation used by White (1987) and White and Magesan (1991) to predict the amount of resident NO_3^- leached from soil (see below).

Bimodal pdf

Soils in which strong preferential flow occurs can sometimes exhibit a bimodal pdf (Beven and Young, 1988). The pdf then becomes the weighted sum of the respective pdfs for the fast preferential flow and slow matrix flow components, e.g.,

$$f^f(z, I) = \varepsilon f^f(z, I) + (1 - \varepsilon) f^f(z, I) \tag{12.34}$$

where the weighting factor ε is determined by the relative amounts of solute moving via the two major pathways. White et al. (1992) used this technique to decompose the BTC for a pulse input of Br^- to a large intact soil core, fitting two lognormal pdfs to the "fast" and "slow" components, and obtaining a better overall fit to the drainage flux concentrations than when a single lognormal function was fitted to all the raw data. Utermann et al. (1990) used a bimodal pdf to model pesticide migration through the unsaturated zone of a drained soil, with the early arrival times in drains and residual fraction remaining in the soil being very sensitive to the preferential flow component of the bimodal pdf.

The Burns pdf

Burns (1975) developed a simple functional model for the leaching of a surface-applied solute, as described by

$$F = \left\{ \frac{I}{(I + \theta \Delta z)} \right\}^{z/\Delta z} \tag{12.35}$$

where F is the fraction of the applied solute leached below depth z in a uniform soil of water content θ by drainage I, and Δz is a notional depth increment. Noting that F was equivalent to the cumulative probability $P(z, I)$ that a solute molecule applied to the surface will appear at depth z after the drainage I, Scotter et al. (1993) wrote Equation 12.35 in the form of a transfer function

$$P(z, I) = \exp\left(-\frac{z\theta}{I}\right) \tag{12.36}$$

This function has stochastic-convective properties and by differentiation can be used to obtain flux and resident pdfs for pulse inputs of surface-applied solutes, or for a solute that is initially distributed through the soil volume. The resident concentration form of the Burns equation for a solute present at $z = 0$ and $I = 0$ is

$$C^r(z, I) = \frac{M\theta}{I} \exp\left(\frac{-z\theta}{I}\right) \tag{12.37}$$

where $C^r(z,I)$ is defined as mass of solute per unit volume of soil, and M is the mass of solute per unit area at the surface at $I = 0$. The flux concentration form for the same initial condition is

$$C^f(z,I) = \frac{Mz\theta}{I^2}\exp\left(\frac{-z\theta}{I}\right) \tag{12.38}$$

Scotter et al. (1993) pointed out that the operationally defined transport volume θ_{st} could be substituted for θ in these equations.

For the initial condition of a solute spread throughout the soil to depth z with a uniform initial resident concentration C_o^r, the resident and flux concentration forms of the Burns equation are, respectively,

$$C^r(z,I) = C_o^r\left[1 - \exp\left(\frac{-z\theta}{I}\right)\right] \tag{12.39}$$

and

$$C^f(z,I) = \frac{C_o^r}{\theta}\left[1 - \left(\frac{z\theta}{I} + 1\right)\exp\left(\frac{-z\theta}{I}\right)\right] \tag{12.40}$$

Scotter et al. (1993) also derived equations for the boundary value problem—that of a solute pulse applied to the soil surface in solution at $I = 0$. Heng and White (1996) used Equations 12.38 and 12.40, modified to account for adsorption, to model the leaching of sulfate in a drained field soil for three successive years. The sulfate was derived from soil net mineralization and from a surface application of sulfate fertilizer. The net leaching outcome was obtained by superposition of the two linear transport processes.

The Effect of Soil Layers

The property of depth-scaling of the stochastic-convective model (Equation 12.10) implies perfect correlation of solute travel times in the vertical direction. This assumption works well for permeable homogeneous soils. However, many field soils show marked differences in properties between horizons which lead to heterogeneity with depth, and the possibility that travel times may not be perfectly correlated between horizons or soil layers. Jury and Roth (1990) suggested that uncorrelated or independent travel times might apply for solute transport to a pipe drain, where flow through the upper zone of unsaturated soil has different characteristics to that through the saturated zone around the pipe. However, flow through an underdrained field which has a network of pipes is likely to be two-dimensional, rather than one-dimensional, and the correlation of solute travel times unknown (Scotter et al., 1991). In this case, an exponential pdf, which is nonstochastic convective, can describe the leaching of resident solutes (White, 1987).

Dyson and White (1989a) constructed repacked columns comprising layers of soil aggregates and fine sand, to examine the effect of layering on the transport of surface-applied and indigenous solutes. Because of the marked differences in the physical properties of the two layers, they assumed that solute travel times were uncorrelated. Drainage flux concentrations could be accurately simulated by a TFM using a lognormal pdf, but resident concentration profiles were less well simulated below the layer interface. In these experiments, the CDE gave almost identical simulations to the TFM, which is to be expected since an assumption of uncorrelated travel times is implicit in the former model. Jury and Roth (1990) suggest that, in general, independence of travel times can be assumed when pronounced lateral flow occurs at interfaces, whereas perfect correlation might occur when the layers are permeable and flow rates are high.

Solute Reactions During Transport

The discussion so far has assumed that the solutes being transported were nonreactive and conservative. Many chemicals in soil are not conserved and they react with the soil matrix in various ways. The processes receiving most attention in solute transport studies are adsorption/desorption involving soil surfaces, and microbially mediated reactions such as decomposition and mineralization/immobilization.

Adsorption

Adsorption slows down the movement of a reactive solute relative to a nonreactive solute. In the stochastic-convective TFM, adsorption can be visualized as the lengthening of solute travel times, or an increase in the travel pathlengths. In all cases, the total resident concentration C^r of the solute in the soil is assumed to be partitioned according to

$$C^r = \rho_b C_a^r + \theta C_l^r \tag{12.41}$$

where ρ_b is the soil bulk density, C_a^r the adsorbed solute concentration (mass per unit mass of dry soil), and C_l^r the solute concentration in the liquid phase. For the simplest case of linear, reversible adsorption at equilibrium, we have

$$C_a^r = K_d C_l^r \tag{12.42}$$

where K_d is the distribution coefficient. The pdf of an adsorbed solute can then be derived from that of a nonreactive, conservative solute using the equation (Jury and Roth, 1990)

$$f_{ad}(z, I) = R^{-1} f(z, I/R) \tag{12.43}$$

where R is the retardation coefficient defined by

$$R = 1 + \rho_b K_d / \theta \tag{12.44}$$

Equation 12.43 can be written in terms of flux pdfs for the adsorbed and nonadsorbed solutes on the assumption that $C^f = C_l^r$. Heng et al. (1994) and Heng and White (1996) adopted this approach to simulate sulfate leaching in a drained field soil, comparing the results obtained for three parametric forms of the pdf; namely, lognormal

$$f_{ad}^f(I) = \frac{1}{\sqrt{2\pi}\sigma I} \exp\left[-\frac{(\ln I/R - \mu)^2}{2\sigma^2} \right] \tag{12.45}$$

exponential

$$f_{ad}^{f}(I) = (aR)^{-1} \exp(-I/aR) \tag{12.46}$$

and Burns

$$f_{ad}^{f}(I) = \frac{Rz\theta}{I^2} \exp\left(-\frac{Rz\theta}{I}\right) \tag{12.47}$$

The values of μ and σ in Equation 12.45 were obtained from the flux pdf for Cl^-. R was obtained from Equation 12.44 using a field adsorption/desorption isotherm for SO_4^{2-}, which was more likely to reflect the dynamic nature of the adsorption/desorption process than a laboratory-determined "batch" isotherm (see below). The alternative to a dynamic isotherm measurement, which is usually difficult to do independently of the transport process being studied, is to treat adsorption as rate-limited (Jury and Roth, 1990; Jury and Flühler, 1992), for which the governing equation is

$$\rho_b \frac{\partial C_a^r}{\partial t} = \alpha\left(K_d C_l^r - C_a^r\right) \tag{12.48}$$

where α is an empirically determined mass transfer coefficient.

Measurement of Adsorption Isotherms Under Laboratory and Field Conditions

Adsorption isotherms are often measured using the "batch" method under laboratory conditions. Small quantities of air-dried soil are shaken in a dilute salt solution containing graded concentrations of the solute, and the change in the concentration is measured after the reaction has approached equilibrium. The isotherm is normally plotted as the amount adsorbed versus the concentration remaining in solution. Hinz et al. (1994) showed that the type of isotherm can have a profound effect on the predicted BTC for a reactive solute. Many leaching studies have shown that a batch isotherm may not simulate field conditions accurately, because the soil to solution ratio is not the same in the batch method as in the field, and the breakdown of soil aggregates during shaking exposes a larger surface area for adsorption (Schweich and Sardin, 1981; Bond and Phillips, 1990).

Kookana et al. (1992) described methods for measuring adsorption isotherms from miscible displacement experiments. They obtained time-dependent sorption for two pesticides from the peak BTC after injecting a short pulse of the chemicals into the columns. Heng et al. (1994) constructed an adsorption isotherm for sulfate from the analysis of suction cup solutions and soil samples, simultaneously collected in the field on a number of occasions during winter when leaching occurred. Recently, Clothier et al. (1996) measured the transport volume of a soil and the adsorption isotherm of a reactive solute *in situ* using a disk permeameter. By measuring the BTCs of a conservative nonreactive tracer and a conservative reactive tracer simultaneously, the retardation factor can be calculated by comparing the average travel time or pathlength of the two solutes. Other parameters of the BTC, such as peak maxima or median pathlengths, can be used to calculate the retardation factor (Widmer et al., 1995; Widmer and Spalding, 1995).

Decomposition of Labile Solutes

Decomposition or decay of labile solutes is a major reaction occurring under field conditions. Decomposition is assumed to be a first-order process; i.e., the rate of disappearance of the solute is proportional to the mass of solute M present at any time; i.e.,

$$\frac{\partial M}{\partial t} = -k_c M \tag{12.49}$$

where k_c is a rate coefficient. Integrating Equation 12.49 between time $t = 0$ and t gives

$$M_t = M_o \exp(-k_c t) \tag{12.50}$$

Consider a narrow pulse of a labile solute injected into the transport volume at $I = 0$. Mass will not be conserved because of decomposition of the solute as it passes through the soil. Thus, the pdf of a conservative solute having the same physical transport characteristics as the labile solute is scaled according to the ratio M_t/M_0, giving the equation

$$f^f(z, I, k_c) = \exp(-k_c t) f^f(z, I; k_c = 0) \tag{12.51}$$

where f^f $(z, I; \mathbf{k}_c = 0)$ is the travel time pdf of the conservative solute.

Equations 12.43 and 12.51 can be combined (Jury and Roth, 1990) to give the travel time pdf for a solute which undergoes simultaneous adsorption and decay during transport; i.e.,

$$f_a^f(z, I, k_c) = \frac{\exp(-k_c t)}{R} f^f\left(z, \frac{I}{R}\right) \tag{12.52}$$

Utermann et al. (1990) and Suter et al. (1996) used this approach to model the leaching of pesticides through soil.

Incorporating Solute Sources and Sinks

Many solutes are subject to input to the transport volume other than through controlled applications, and output through pathways other than leaching. Inputs (sources) include rainfall, fertilizer applications, dung and urine, and mineralization of organic matter. Outputs (sinks) include plant uptake, microbial immobilization, gaseous losses, and biochemical or radioactive decay. An example of the incorporation of a decay sink in the TFM using a known functional relationship was given in the previous section. Where the input or output process is linear with respect to I (zero order), it can be incorporated in the TFM using the principle of superposition as demonstrated by Heng et al. (1994) and Heng and White (1996); i.e., in the case of soil sulfate

$$C_b = C_m + C_p + C_i + C_s \tag{12.53}$$

where C_b is a net source-sink term, and C_m = net mineralization, C_p = plant uptake, C_i = rainfall, and C_s = oxidation of elemental S fertilizer. The combined effect of the sources and sinks is treated as an input to the initial resident concentration in the transport volume, so that the contribution to drainage is given by (*cf.* Equation 12.28)

$$C^f(z,I) = C_b\left(1 - \int_0^I f^f(z,I)dI\right) \tag{12.54}$$

Where the relationship between the source (or sink) quantity and I changes substantially, as can happen in the case of net mineralization for example, through a marked change in temperature and/or moisture, the term C_b must be adjusted at $I \geq I_0$, where I_0 is the cumulative drainage at which the change takes effect. A comparison between measured and simulated SO_4^{2-} concentrations in the drainage from two fertilized and underdrained pasture soils using this approach is shown in Figure 12.3.

As with models such as LEACHM, which employs a finite difference solution for the CDE, numerical solutions of the TFM are possible. A numerical solution has advantages over the analytical approach when (1) the functions for the input of an applied solute, or the source-sink term for a labile solute, are complex (the functions need not be constrained to be linear with time or drainage), (2) the parametric form of the pdf cannot easily be determined, and (3) the adsorption isotherm for a reactive solute is not a simple function. Heng et al. (1994) used a finite difference form of Equations 12.6 and 12.23 combined, to model sulfate leaching when this solute had been surface-applied (and was assumed not to dissolve immediately), and was also being generated by net mineralization in the soil. The equation used was

$$C^f(m) = \sum_{n=1}^{m} C_{ent}(m-n+1)f(n)\Delta I + C_0^r\left[1 - \sum_{n=1}^{m} f(n)\Delta I\right] \tag{12.55}$$

where C_{ent} is a variable input concentration to the transport volume (including all source-sink terms) and m and n are integers. m is defined by

$$m = I / \Delta I \tag{12.56}$$

Sensitivity Analysis of Model Parameters

Parsimony in the number of parameters and input variables is most desirable in models intended for management purposes. However, the sensitivity of the output of a parsimonious model to changes in its parameter values should be determined. Heng and White (1996) examined the effect of changing R and θ on the simulation of sulfate leaching, using a TFM based on either the Burns pdf or a lognormal pdf. Their results for a lognormal pdf, shown in Figure 12.4, indicated that changing R would have a significant effect on the leaching of sulfate, especially when a lognormal pdf was used. However, the effect was not as pronounced when a range of values for θ_{st} were substituted for θ.

Comparison of Measured and Predicted Amounts of Solute Leached

Relatively few studies have compared amounts of solute leached as predicted from a TFM with measured amounts leached under field conditions, on a timescale rel-

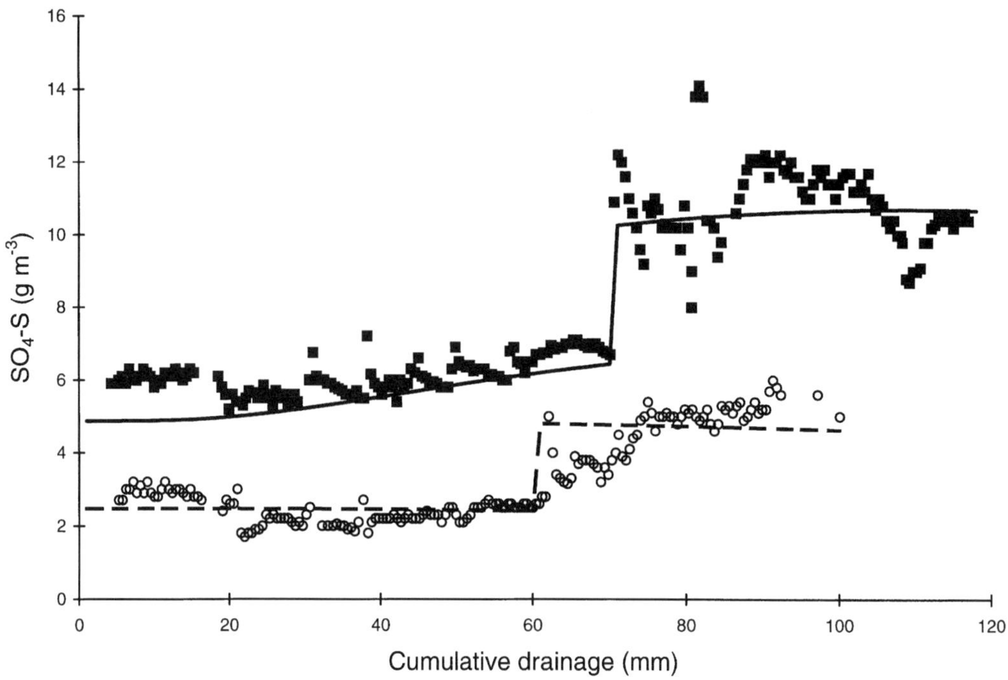

Figure 12.3. Measured and simulated drainage sulfate concentrations $C_{ex}(I)$ for winter and spring 1989 using a TFM with a lognormal pdf. Superphosphate (SO_4-S) fertilized paddock-measured (■) and predicted (solid line); elemental S (S^o) fertilized paddock-measured (o) and predicted (dashed line) (after Heng and White, 1996). The step change in slope of the simulated line is due to a step change in the source term for net mineralization at the end of winter.

evant to management. White (1987) found good agreement between measured and predicted nitrate leached from a drained clay soil under pasture for individual rainfall events over three winter seasons, using an equation identical to Equation 12.32. Similarly, Heng et al. (1994) and Heng and White (1996) observed good agreement between the measured and predicted amounts of sulfate leached from a drained pasture soil, using a TFM calibrated for Cl^- leaching, over three winter seasons. Edis et al. (1996, unpublished) compared measured and predicted amounts of indigenous NO_3^- leached from fallow and cropped soils over a year. Table 12.1 gives a summary of these results.

Extension of the TFM to Large Areas

Large-scale simulation models of processes in the unsaturated zone are important for predicting the effects of soil and land management changes at catchment and regional scales. While the application of the TFM has mostly been confined to relatively small-scale field studies, its parsimonious parameter requirements, easily obtained input variables, and versatility in incorporating various chemical and biological processes in a simple manner, offer much potential for it to be extended to larger scales. To satisfy the needs of models for large-scale, complex field systems, new methods of measurement and new or improved approaches for estimating parameters at various scales are needed.

To date, the key parameters of the TFM have been obtained by calibration from tracer leaching experiments at individual sites. While this may allow subsequent pre-

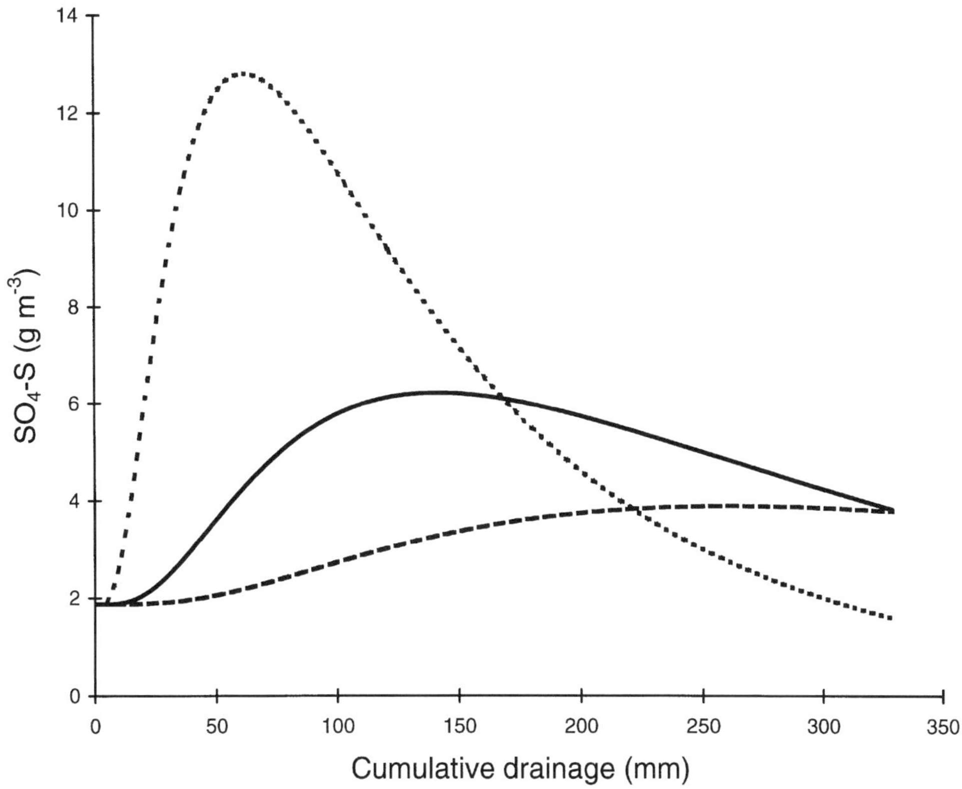

Figure 12.4. Sensitivity analysis for the TFM prediction of SO_4-S leaching for 1988 using a lognormal pdf (after Heng and White, 1996). (i) $R = 2.4$, $\mu = 4.8$ and $\sigma = 0.8$ (—), (ii) $R = 4.8$, $\mu = 4.8$ and $\sigma = 0.8$ (– –), and (iii) $R = 1$, $\mu = 4.8$, $\sigma = 0.8$ (- - - -).

Table 12.1. Comparison of Measured and Predicted Amounts of Solute Leached in Field Soils and Lysimeters Using a TFM.

Reference	Solute	Years	pdf	Measured (kg ha^{-1})	Predicted (kg ha^{-1})
Heng et al. (1991)	Cl	1988	Exponential	123	73
		1989	Exponential	132	81
Heng and White (1996)	SO_4–S	1988	Lognormal	17.0	17.9
		1989	Lognormal	9.4	8.2
		1990	Lognormal	12.3	13.1
	S^o	1988	Lognormal	3.4	6.1
		1989	Lognormal	3.5	3.2
		1990	Lognormal	9.5	9.3
Edis et al. (unpublished)[a]	NO_3-N	1994–5	Lognormal	182 (fallow)	199
		1994–5	Lognormal	24 (cropped)	17

[a] Lysimeter plots, 6 x 3 x 1.2 m deep, bromide calibration of the TFM, with a field estimate of net nitrification.

dictions of tracer transport at these sites, and the behavior of other solutes to be modeled at these sites, the approach is labor-intensive and not readily amenable to scaling up (extrapolation to other sites). It is therefore desirable that these parameters be independently measured, or estimated from surrogate variables using pedotransfer functions (Bouma et al., 1996). The *in situ* disk permeameter technique of Clothier et al. (1992, 1995) offers promise for estimating the transport volume, and simultaneously obtaining a dynamic adsorption isotherm for a reactive solute. However, intensive and systematic sampling to obtain spatially distributed input data for large-scale simulation models is difficult because the measurements are expensive and time-consuming. The use of pedotransfer functions may allow the transport parameters to be estimated without additional soil measurements. Simple, routinely measured soil properties such as texture, bulk density, particle size distribution and organic matter content can be used to predict soil water properties, such as the θ—water potential relationship (Campbell, 1985; Vereecken et al., 1992). The transport volume for the soil might then be derived from the characteristics of this relationship. However, this approach is still at a very developmental stage, and there are examples where the agreement between measured soil hydraulic properties and those predicted from pedotransfer functions has been poor (Espino et al., 1995).

SUMMARY

The use of the TFM, as for all models, is limited by the availability of input data and how easily the key parameters can be measured. There are also the underlying assumptions of stationarity and ergodicity which affect the robustness of the model for field use. The TFM shows great promise for use as a management tool because the input variables are easily measured and, depending on the form of the pdf and the nature of the target solute, there may be few parameters to estimate. The cumulative drainage form of the TFM requires a measure of cumulative drainage which can be calculated as precipitation in excess of evaporation (net applied water), from widely available meteorological data. The challenge in applying this approach in heterogeneous natural landscapes will be to partition the net applied water flux into surface runoff, subsurface lateral flow, and deep drainage below the root zone. Hydrological process models based on digital terrain models, such as TOPOG (O'Loughlin, 1986), may be useful for this purpose.

Solute input concentrations can be estimated from known application rates, of pesticides and fertilizers, for example, although there are problems in assessing the spatial variability of such applications. On a large scale, such inputs may be calculated approximately at the start of a leaching season from a nutrient balance for the previous period of crop or pasture growth. Such an estimate can be improved by strategic soil sampling; that is, sampling at an appropriate time to measure the pool of solute available for leaching in the soil. The most difficult aspects are those of measuring inputs from biotic sources, such as animal manure and organic wastes, and the dynamic source-sink terms for labile solutes, which are largely governed by biological transformations in the soil (De Willigen and Neeteson, 1985; Heng et al. 1994). Process models of the dynamics of N transformations, for example, have been developed for cropping systems (e.g., SUNDIAL) and coupled to a nitrate leaching model (Smith et al., 1996). Such models are less advanced for more complex grazed pasture systems (Scholefield et al., 1991), although Heng and White (1996) showed that a TFM with simplified functions for source-sink terms could simulate sulfate leaching successfully for a grazed pasture.

With increasing emphasis being placed on precision fertilizer management, we may expect research on solute transport modeling in the future to focus on (1) independent methods for estimating key model parameters from relevant soil properties that are routinely measured (through surrogate variables and pedotransfer functions); (2) quantifying the spatial expression of soil properties that influence water movement into and through the soil; and (3) improving the dynamic models of nutrient transformations for the main land use systems, and making these models operable at a range of scales in the field.

REFERENCES

Adams, E.E. and L.W. Gelhar. Field study of dispersion in a heterogeneous aquifer: 2: Spatial moments analysis. *Water Resour. Res.,* 28:3293–3307, 1992.

Addiscott, T.M. and D. Cox. Winter leaching of nitrate from autumn-applied calcium nitrate, ammonium sulphate, urea and sulphur coated urea in bare soil. *J. Agric. Sci. Cambridge,* 87:381–389, 1976.

Addiscott, T.M. A simple computer model for leaching in structured soils. *J. Soil Sci.,* 28:554–563, 1977.

Addiscott, T.M. Measuring and monitoring pollutant transport through soils and beyond. A study in uncertainty. In *Statistics for the Environment 2: Water Related Issues,* Barnett, V. and K.F. Turkman (Eds.), John Wiley & Sons, pp. 249–272, 1994.

Addiscott, T.M. and R.J. Wagenet. Concepts of solute leaching in soils: A review of modelling approaches. *J. Soil Sci.,* 36:411–424, 1985.

Addiscott, T.M. Simulation modelling and soil behaviour. *Geoderma,* 60:15–40, 1993.

Addiscott, T.M. Measuring and modelling nitrogen leaching: parallel problems. *Plant and Soil,* 181:1–6, 1996.

Andreini, M.S. and T.S. Steenhuis. Preferential paths of flow under conventional and conservation tillage. *Geoderma,* 46:85–102, 1990.

Angulo-Jaramillo, R., J.P. Gaudet, J.L. Thony, and M. Vauclin. Measurement of hydraulic properties and mobile water content of a field soil. *Soil Sci. Soc. Am. J.,* 60:710–715, 1996.

Baker, J.L., K.L. Campbell, H.P. Johnson, and J.J. Hanway. Nitrate, phosphorus and sulphate in subsurface drainage water. *J. Environ. Qual.,* 4:406–412, 1975.

Beven, K. and P. Germann. Macropores and water flow in soils. *Water Resour. Res.,* 18:1311–1325, 1982.

Beven, K.J. and P.C. Young. An aggregated mixing zone model of solute transport through porous media. *J. Contam. Hydrol.,* 3:129–143, 1988.

Biggar, J.W. and D.R. Nielsen. Spatial variability of the leaching characteristic of a field soil. *Water Resour. Res.,* 12:78–84, 1976.

Bolan, N.S., D.R. Scotter, J.K. Syers, and R.W. Tillman. The effect of adsorption on sulphate leaching. *Soil Sci. Soc. Am. J.,* 50:1419–1424, 1986.

Bond, W.J. and I.R. Phillips. Ion transport during unsteady water flow in an unsaturated clay soil. *Soil Sci. Soc. Am. J.,* 54:636–645, 1990.

Booltink, H.W.G. Field monitoring of nitrate leaching and water flow in a structured clay soil. *Agric. Ecosys. Environ.,* 52:251–261, 1995.

Bouma, J. and J.H.M. Wösten. Flow patterns during extended saturated flow in two undisturbed swelling clay soils with different macrostructures. *Soil Sci. Soc. Am. J.,* 43:16–22, 1979.

Bouma, J., L.W. Dekker, and C.J. Muilwijk. A field method for measuring short-circuiting in clay soils. *J. Hydrol.,* 52:347–354, 1981.

Bouma, J., H.W.G. Booltink, and P.A. Finke. Use of soil survey data for modeling solute transport in the vadose zone. *J. Environ. Qual.,* 25:519–526, 1996.

Buchter, B., L. Leuenberger, P.J. Wierenga, and F. Richard. Preparation of large core samples from stony soils. *Soil Sci. Soc. Am., J.* 48:1460–1462. 1984.

Burns, I.G. An equation to predict the leaching of surface-applied nitrate. *J. Agric. Sci., Cambridge,* 85:443–454, 1975.

Butters, G.L. and W.A. Jury. Field scale transport of bromide in an unsaturated soil. 2. Dispersion modelling. *Water Resour. Res.,* 25:1583–1589, 1989.

Cameron, K.C., D.F. Harrison, N.P. Smith, and C.D.A. McLay. A method to prevent edge flow in undisturbed cores and lysimeters. *Austr. J. Soil Res.,* 28:879–886, 1990.

Campbell, G.S. *Soil Physics with Basic; Transport Models for Soil-Plant Systems.* Elsevier, Amsterdam, 1985, p. 150.

Clothier, B.E., M.B. Kirkham, and J.E. McLean. In situ measurement of the effective transport volume for solute moving through soil. *Soil Sci. Soc. Am. J.,* 56:733–736, 1992.

Clothier, B.E., L.K. Heng, G.N. Magesan, and I. Vogeler. The measured mobile-water content of an unsaturated soil as a function of hydraulic regime. *Austr. J. Soil Res.,* 33:397–414, 1995.

Clothier, B.E., G.N. Magesan, L. Heng, and I. Vogeler. In situ measurement of the solute adsorption isotherm using a disc permeameter. *Water Resour. Res.,* 32:771–778, 1996.

Coats, J.H. and B.D. Smith. Dead-end pore volume and dispersion in porous media. *Soc. Pet. Eng. J.,* 33:73–84, 1964.

Cooke, G.W. and R.J.B. Williams. Losses of nitrogen and phosphorus from agricultural land. *Water Treat. Examin.,* 19:253–274, 1970.

Corwin, D.L., B.L. Waggoner, and J.D. Rhoades. A functional model of solute transport that accounts for bypass. *J. Environ. Qual.,* 20:647–658, 1991.

Costa, J.L., R.E. Knighton, and L. Prunty. Model comparison of unsaturated steady-state solute transport in a field plot. *Soil Sci. Soc. Am. J.,* 58:1277–1287, 1994.

Dagan, G. and E. Bresler. Solute dispersion in unsaturated heterogeneous soil at a field scale. I. Theory. *Soil Sci. Soc. Am. J.,* 43:461–467, 1979.

Darrah, P.R., R.E. White, and P.H. Nye. A theoretical consideration of the implications of cell clustering for the prediction of nitrification in soils. *Plant Soil,* 99:387–400, 1987.

Dekker, L.W. and J. Bouma. Nitrogen leaching during sprinkler irrigation of a Dutch clay soil. *Agric. Water Manage.,* 9:37–45, 1984.

de Willigen, P and Neeteson, J.J. Comparison of six simulation models for the nitrogen cycle in the soil. *Fertilizer Research,* 8:157–171, 1985.

Dyson, J.S. and R.E. White. A comparison of the convection-dispersion equation and transfer function model for predicting chloride leaching through an undisturbed, structured clay soil. *J. Soil Sci.,* 38:157–172, 1987.

Dyson, J.S. and R.E. White. A simple predictive approach to solute transport in layered soils. *J. Soil Sci.,* 40:525–542, 1989a.

Dyson, J.S. and R.E. White. The effect of irrigation rate on solute transport in soil during steady water flow. *J. Hydrol.,* 107:11–29, 1989b.

Edwards, W.M., M.J. Shipitalo, L.B. Owens, and W.A. Dick. Factors affecting preferential flow of water and atrazine through earthworm burrows under continuous no-till corn. *J. Environ. Qual.,* 22:453–457, 1993.

Ellsworth, T.R., P.J. Shouse, T.H. Skaggs, J.A. Jobes, and J. Fargerlund. Solute transport in unsaturated soil: Experimental design, parameter estimation, and model discrimination. *Soil Sci. Soc. Am. J.,* 60:397–407, 1996.

Espino, A., D. Mallants, M. Vanclooster, and J. Feyen. Cautionary notes on the use of pedotransfer functions for estimating soil hydraulic properties. *Agric. Water Manage.,* 29:235–253, 1995.

Flury, M., H. Flühler, W.A. Jury, and J. Leuenberger. Susceptibility of soils to preferential flow of water: A field study. *Water Resour. Res.,* 30:1945–1954, 1994.

Haigh, R.A. and R.E. White. Nitrate leaching from a small, underdrained, grassland clay catchment. *Soil Use Manage.,* 2:65–70, 1986.

Heng, L.K., R.E. White, N.S. Bolan, and D.R. Scotter. Leaching losses of major nutrients from a mole-pipe drained soil under pasture. *New Zealand J. Agric. Res.,* 34:325–334, 1991.

Heng, L.K., R.E. White, D.R. Scotter, and N.S. Bolan. A transfer function approach to modelling the leaching of solutes to subsurface drains. II. Reactive solutes. *Austr. J. Soil Res.,* 32:85–94, 1994.

Heng, L.K. and R.E. White. A simple analytical transfer function approach to modelling the leaching of reactive solutes through field soil. *Eur. J. Soil Sci.,* 47:33–42, 1996.

Hinz, L.K., L.A. Gaston, and H.M. Selim. Effect of adsorption isotherm type on predictions of solute mobility in soil. *Water Resour. Res.,* 30:3013–3021, 1994.

Hornberger, G.M., K.J. Beven, and P.F. Germann. Inferences about solute transport in macroporous forest soils from time series models. *Geoderma,* 46:249–262, 1990.

Jardine, P.M., G.K. Jacobs, and G.V. Wilson. Unsaturated transport processes in undisturbed heterogeneous porous media: I. Inorganic contaminants. *Soil Sci. Soc. Am. J.,* 57:945–953, 1993.

Jarvis, N.J., L. Bergstrom, and P.E. Dik. Modelling water and solute transport in macroporous soil. II. Chloride breakthrough under non-steady flow. *J. Soil Sci.,* 42:71–81, 1991a.

Jarvis, N.J., P.E. Jansson, P.E. Dik, and I. Messing. Modelling water and solute transport in macroporous soil. I. Model description and sensitivity analysis. *J. Soil Sci.,* 42:59–70, 1991b.

Jaynes, D.B., R.C. Rice, and R.S. Bowman. Independent calibration of a mechanistic-stochastic model for field-scale solute transport under flood irrigation. *Soil Sci. Soc. Am. J.,* 52:1541–1546, 1988.

Jaynes, D.B., S.D. Logsdon, and R. Horton. Field method for measuring mobile/immobile water content and solute transfer rate coefficient. *Soil Sci. Soc. Am. J.,* 59:352–356, 1995.

Jury, W.A., L.H. Stolzy, and P. Shouse. A field test of the transfer function model for predicting solute transport. *Water Resour. Res.,* 18:369–375, 1982.

Jury, W.A. Simulation of solute transport using a transfer function model. *Water Resour. Res.,* 18:363–368, 1982.

Jury, W.A. and G. Sposito. Field calibration and validation of solute transport models for the unsaturated zone. *Soil Sci. Soc. Am. J.,* 49:1331–1341, 1985.

Jury, W.A., G. Sposito, and R.E. White. A transfer function model of solute transport through soil 1. Fundamental concepts. *Water Resour. Res.,* 22:243–247, 1986.

Jury, W.A. and K. Roth. *Transfer Functions and Solute Movement Through Soil: Theory and Applications*. Birkhauser Verlag, Basel, 1990, p. 226.

Jury, W.A. and Flühler, H. Transport of chemicals through soil: mechanisms, models and field applications, *Advances in Agronomy*. Academic Press, 1992, pp. 141–201.

Jury, W.A. and D.R. Scotter. A unified approach to stochastic-convective transport problems. *Soil Sci. Soc. Am. J.,* 58:1327–1336, 1994.

Kneale, W.R. and R.E. White. The movement of water through cores of a dry (cracked) clay-loam grassland topsoil. *J. Hydrol.,* 67:361–365, 1984.

Kolenbrander, G.J. Nitrate content and nitrogen loss in drainwater. *Netherlands J. Agric. Sci.,* 17:246–255, 1969.

Kookana, R.S., R.G. Gerritse, and L.A.G. Aylmore. A method for studying nonequilibrium sorption in transport of pesticides in soil. *Soil Sci.,* 154:344–349, 1992.

Kreft, A. and A. Zuber. On the physical meaning of the dispersion equation and its solutions for different initial and boundary conditions. *Chem. Eng. Sci.,* 33:1471–1480, 1978.

Kung, K.J.S. Preferential flow in a sandy vadose zone: I. Field observation. *Geoderma,* 46:51–58, 1990.

Magesan, G.N., D.R. Scotter, and R.E. White. A transfer function approach to modelling the leaching of solutes to subsurface drains. I. Non-reactive solutes. *Austr. J. Soil Res.,* 32:69–83, 1994a.

Magesan, G.N., R.E. White, D.R. Scotter, and N.S. Bolan. Estimating drainage losses from subsurface drained soils. *Soil Use Manage.,* 10:87–93, 1994b.

Magesan, G.N., R.E. White, and D.R. Scotter. The influence of flow rate on solute concentrations in soil drainage. *J. Hydrol.,* 172:23–30, 1995.

Mallawatantri, A.P. and D.J. Mulla. Uncertainties in leaching risk assessments due to field averaged transfer function parameters. *Soil Sci. Soc. Am. J.,* 60:722–726, 1996.

McBratney, A.B., C.J. Moran, J.B. Stewart, S.R. Cattle, and A.J. Koppi. Modifications to a method of rapid assessment of soil macropore structure by image analysis. *Geoderma,* 53:255–274, 1992.

Miller, R.J., J.W. Biggar, and D.R. Nielsen. Chloride displacement in Panoche clay loam in relation to water movement and distribution. *Water Resour. Res.,* 1:63–73, 1965.

Miralles-Wilhelm, F. and L.W. Gelhar. Stochastic analysis of transport and decay of a solute in heterogeneous aquifers. *Water Resour. Res.,* 32:3451–3459, 1996.

Murphey, J.B., E.H. Grissinger, and W.C. Little. Fiber-glass encasement of large, undisturbed, weakly cohesive soil samples. *Soil Sci.,* 131:130–134, 1981.

Narasimhan, T.N. Models and modeling of hydrogeologic processes. *Soil Sci. Soc. Am. J.,* 59:300–306, 1995.

Nkedi-Kizza, P., J.W. Biggar, M.T. van Genuchten, P.J. Wierenga, H.M. Selim, J.M. Davidson, and D.R. Nielsen. Modeling tritium and chloride-36 transport through and aggregated oxisol. *Water Resour. Res.,* 19:691–700, 1983.

O'Loughlin, E.M. Prediction of surface saturation zones in natural catchments by topographical analysis. *Water Resour. Res.,* 22:794–804, 1986.

Pearson, R.J., W.P. Inskeep, J.M. Wraith, S.D. Comfort, and H.M. Gaber. Observed and simulated solute transport under varying water regimes: I. Bromide and pentafluorobenzoic acid. *J. Environ. Qual.,* 25:646–653, 1996.

Porro, I., P.J. Wierenga, and R.G. Hills. Solute transport through large uniform and layered soil columns. *Water Resour. Res.,* 29:1321–1330, 1993.

Quisenberry, V.L. and R.E. Phillips. Percolation of surface-applied water in the field. *Soil Sci. Soc. Am. J.,* 40:484–489, 1976.

Quisenberry, V.L., R.E. Phillips, and J.M. Zeleznik. Spatial distribution of water and chloride macropore flow in a well-structured soil. *Soil Sci. Soc. Am. J.,* 58:1294–1300, 1994.

Raats, P.A.C. Convective transport of solutes by steady flows II. Specific flow problems. *Agric. Water Manage.,* 1:219–232, 1978.

Richard, T.L. and T.S. Steenhuis. Tile drain sampling of preferential flow on a field scale. *J. Contam. Hydrol.,* 3:307–325, 1988.

Richter, G. and W.A. Jury. A microlysimeter field study of solute transport through a structured sandy loam soil. *Soil Sci. Soc. Am. J.,* 50:863–868, 1986.

Ritsema, C.J. and L.W. Dekker. Distribution flow, a general process in the top layer of water repellent soils. *Water Resour. Res.,* 31:187–1200, 1995.

Russo, D., W.A. Jury, and G.L. Butters. Numerical analysis of solute transport during transient irrigation. I. The effect of hysteresis and profile heterogeneity. *Water Resour. Res.,* 25:2109–2118, 1989.

Russo, D. and G. Dagan. On solute transport in a heterogeneous porous formation under saturated and unsaturated water flows. *Water Resour. Res.,* 27:285–292, 1991.

Scholefield, D., D.R. Lockyer, D.C. Whitehead, and K.C. Tyson. A model to predict transformations and losses of nitrogen in UK pastures grazed by beef cattle. *Plant Soil,* 132:165–177, 1991.

Schulin, R., M.T. van Genuchten, H. Flühler, and P. Ferlin. An experimental study of solute transport in a stony field soil. *Water Resour. Res.,* 23:1785–1794, 1987.

Schweich, D. and M. Sardin. Adsorption, partition, ion exchange and chemical reaction in batch reactors or in columns—A review. *J. Hydrol.,* 50:1–33, 1981.

Scotter, D.R., L.K. Heng, D.J. Horne, and R.E. White. A simplified analysis of soil water flow to a mole drain. *J. Soil Sci.,* 41:189–198, 1990.

Scotter, D.R., L.K. Heng, and R.E. White. Two models for the leaching of a non-reactive solute to a mole drain. *J. Soil Sci.,* 42:565–576, 1991.

Scotter, D.R., R.E. White, and J.S. Dyson. The Burns leaching equation. *J. Soil Sci.,* 44:25–33, 1993.

Seyfried, M.S. and P.S.C. Rao. Solute transport in undisturbed columns of an aggregated tropical soil: Preferential flow effects. *Soil Sci. Soc. Am. J.,* 51:1434–1444, 1987.

Shaffer, M.J., A.H. Halvorson, and F.J. Pierce. Nitrate leaching and economic analysis package (NLEAP): Model description and application. In *Managing Nitrogen for Groundwater Quality and Farm Profitability.* Follett, R.F., D.R. Keeney, and R.M. Cruse (Eds.), Soil Science Society of America, Inc., Madison, WI, 1991, pp. 285–322.

Smettem, K.R.J. Soil-water residence time and solute uptake. *J. Hydrol.,* 67:235–248, 1984.

Smith, C.M., P.E.H. Gregg, and R.W. Tillman. Drainage losses of sulphur from a yellow-grey earth soil. *New Zealand J. Agric. Res.,* 26:363–371, 1983.

Smith, J.U., N.J. Bradbury, and T.M. Addiscott. SUNDIAL: A PC-based system for simulating nitrogen dynamics in arable land. *Agron. J.,* 88:38–43, 1996.

Smith, M.S., G.W. Thomas, R.E. White, and D. Ritonga. Transport of *Escherichia coli* through intact and disturbed columns of soil. *J. Environ. Qual.,* 14:87–91, 1985.

Sposito, G., R.E. White, P.R. Darrah, and W.A. Jury. A transfer function model of solute transport through soil. 3. The convection-dispersion equation. *Water Resour. Res.,* 22:255–262, 1986.

Suter, H.C., L.K. Heng, R.E. White, and L.A. Douglas. Modelling the fate of Phosmet (an organophosphate pesticide) applied to undisturbed cores. In *Soil Science Raising the Profile,* Uren, N.C. (Ed.). Proceedings of the Australian and New Zealand National Soils Conference, Vol. 2, 1996, pp. 271–272.

Taylor, G.I. The dispersion of soluble matter flowing through a capillary tube. *Proc. London Math Soc.,* 2:196–212, 1953.

Thomas, G.W. and B.J. Barfield. The unreliability of tile effluent for monitoring subsurface nitrate-nitrogen losses from soils. *J. Environ. Qual.,* 3:183–185, 1974.

Tillman, R.W., D.R. Scotter, B.E. Clothier, and R.E. White. Solute movement during intermittent water flow in a field soil and some implications for irrigation and fertilizer application. *Agric. Water Manage.,* 20:119–133, 1991.

Toride, N., F.J. Leij, and M.T. van Genuchten. Flux-averaged concentrations for transport in soils having nonuniform initial solute distributions. *Soil Sci. Soc. Am. J.,* 57:1406–1409, 1993.

Toride, N. and F.J. Leij. Convective-dispersive stream tube model for field-scale solute transport: II. Examples and calibration. *Soil Sci. Soc. Am. J.,* 60:352–361, 1996.

Utermann, J., E.J. Kladivko, and W.A. Jury. Evaluating pesticide migration in tile-drained soils with a transfer function model. *J. Environ. Qual.,* 19:707–714, 1990.

Vanclooster, M., D. Mallants, J. Vanderborght, J. Diels, J. Van Orshoven, and J. Feyen. Monitoring solute transport in a multi-layered sandy lysimeter using time domain reflectometry. *Soil Sci. Soc. Am. J.,* 59:337–344, 1995.

Vanderborght, J., M. Vanclooster, D. Mallants, J. Diels, and J. Feyen. Determining convective lognormal solute transport parameters from resident concentration data. *Soil Sci. Soc. Am. J.,* 60:1306–1317, 1996.

van Genuchten, M.T. and P.J. Wierenga. Mass transfer studies in sorbing porous media. I. Analytical solutions. *Soil Sci. Soc. Am. J.,* 40:473–480, 1976.

Vereecken, H., J. Diels, J. van Orshoven, J. Feyen, and J. Bouma. Functional evaluation of pedotransfer functions for the estimation of soil hydraulic functions. *Soil Sci. Soc. Am. J.,* 56:1371–1378, 1992.

Vereecken, H., J. Diels, and P. Viaene. The effect of soil heterogeneity and hysteresis on solute transport: A numerical experiment. *Ecol. Modelling,* 77:273–288, 1995.

Warrick, A.W., J.W. Biggar, and D.R. Nielsen. Simultaneous solute and water transfer for an unsaturated soil. *Water Resour. Res.,* 7:1216–1225, 1971.

van Wesenbeeck, I.J. and R.G. Kachanoski. Effect of variable horizon thickness on solute transport. *Soil Sci. Soc. Am. J.,* 58:1307–1316, 1994.

White, R.E. The transport of chloride and non-diffusible solutes through soil. *Irrigation Sci.,* 6(1):3–10, 1985a.

White, R.E. A model for nitrate leaching in undisturbed structured clay soil during unsteady flow. *J. Hydrol.,* 79:37–51, 1985b.

White, R.E. The influence of macropores on the transport of dissolved and suspended matter through soil. *Adv. Soil Sci.,* 3:95–120, 1985c.

White, R.E. A transfer function model for the prediction of nitrate leaching under field conditions. *J. Hydrol.,* 92(3-4):207–222, 1987.

White, R.E. Leaching. In *Advances in Nitrogen Cycling in Agricultural Ecosystems,* Wilson, J.R. (Ed.), C.A.B. Wallingford, 1988, pp. 193–211.

White, R.E. Prediction of nitrate leaching from a structured clay soil using transfer functions derived from externally applied or indigenous solute fluxes. *J. Hydrol., Netherlands,* 107:31–42, 1989.

White, R.E., G.W. Thomas, and M.S. Smith. Modelling water flow through undisturbed soil cores using a transfer function model derived from ^{3}HOH and Cl transport. *J. Soil Sci.,* 35:159–168, 1984.

White, R.E., J.S. Dyson, Z. Gerstl, and B. Yaron. Leaching of herbicides through undisturbed cores of a structured clay soil. *Soil Sci. Soc. Am. J.,* 50:277–83, 1986a.

White, R.E., J.S. Dyson, R.A. Haigh, W.A. Jury, and G. Sposito. A transfer function model of solute transport through soil. 2. Illustrative applications. *Water Resour. Res.,* 22:248–254, 1986b.

White, R.E. and G.N. Magesan. A stochastic-empirical approach to modelling nitrate leaching. *Soil Use Manage.,* 7:85–94, 1991.

White, R.E., D.R. Scotter, and G.N. Magesan. Dynamics of nitrate leaching: measurement and modelling. In *Nutrient Management for Sustained Productivity,* Bajiua, M.S., N.S. Pasricha, P.S. Sidhu, M.R. Chaudhary, and V. Beri (Eds.), Punjab Agricultural University, Ludhiana, 1992, pp. 272–286.

White, R.E., S.R. Wellings, and J.P. Bell. Seasonal variations in nitrate leaching in structured clay soils under mixed land use. *Agric. Water Manage.,* 7:391–410, 1983.

Widmer, S.K. and R.F. Spalding. A natural gradient transport study of selected herbicides. *J. Environ. Qual.,* 24:445–453, 1995.

Widmer, S.K., R.F. Spalding, and J. Skopp. Nonlinear regression of breakthrough curves to obtain retardation factors in a natural gradient field study. *J. Environ. Qual.,* 24:439–444, 1995.

Wild, A. and K.C. Cameron. Soil nitrogen and nitrate leaching. In *Critical Reports on Applied Chemistry, Vol. 2.,* Tinker, P.B. (Ed.), Society of Chemical Industry, Blackwell Scientific Publications, Oxford, 1980, pp. 35–71.

Wilford, J.R., C.F. Pain, and J.C. Dohrenwend. Enhancement and integration of airborne gamma-ray spectrometric and Landsat imagery for regolith mapping—Cape York Peninsula. *Exploration Geophys.,* 23:441–446, 1992.

Williams, J.R. and D.E. Kissel. Water percolation: An indicator of nitrogen-leaching potential. In *Managing Nitrogen for Groundwater Quality and Farm Profitability,* Follett, R.F., D.R. Keeney, and R.M. Cruse (Eds.), Soil Science Society of America, Inc., Madison, WI, 1991, pp. 5–8.

Zhang, R., J. Yang, and Z. Ye. Solute transport through the vadose zone: a field study and stochastic analyses. *Soil Sci.,* 161:270–277, 1996.

LIST OF SYMBOLS

a	effective transport volume per unit surface area	[L^{-1}]
C^f	flux concentration	[$M\ L^{-3}$]
C^r	resident concentration	[$M\ L^{-3}$]
C^{rt*}	time-integral normalized concentration	
C_a^r	solute concentration in the adsorbed phase	[$M\ M^{-1}$]
C_l^r	solute concentration in the liquid phase	[$M\ L^{-3}$]
C_o^r	initial resident concentration	[$M\ L^{-3}$]
C_{ent}	time (or drainage) variable concentration input to the transport volume	[$M\ L^{-3}$]
C_b	net source-sink term concentration	[$M\ L^{-3}$]
C_m	net mineralization concentration	[$M\ L^{-3}$]
C_p	concentration in plant materials	[$M\ L^{-3}$]
C_i	concentration in rainfall	[$M\ L^{-3}$]
C_s	concentration of SO_4^{2-} from elemental sulfur	[$M\ L^{-3}$]
D	dispersion coefficient	[$L^2\ T^{-1}$]
$E\{I,L\}$	first moment of the drainage flux pdf	
f^f	flux pdf in terms of I	[L^{-1}]
f^r	resident pdf in terms of I/L	[T^{-1}]
F	fraction of applied solute leached	
I	cumulative drainage	[L]
k	leaching coefficient	[L^{-1}]
k_c	decay rate coefficient	[T^{-1}]
K_d	distribution coefficient	[$L^3\ M^{-1}$]
L	the calibration depth or length	[L]
M	mass of solute per unit area	[$M\ L^{-2}$]
p	soil porosity	[$L^3\ L^{-3}$]
P	pressure head	[L]
$P(z,I)$	cumulative probability density function	
q	soil water flux density	[$L\ T^{-1}$]
Q	rate of water flow	[$L^3\ T^{-1}$]
r	radius	[L]
R	retardation coefficient	
$\bar{t}$	median travel time	[T]
V_{st}	transport volume	[$L^3\ L^{-3}$]
z	depth of soil	[L]
α	mass transfer coefficient	[T^{-1}]
ε	weighting factor	
η	viscosity	[$M\ L^{-1}\ T^{-1}$]
μ	mean of the lognormal pdf	
v	pore-water velocity	[$L\ T^{-1}$]
$\bar{v}$	median pore-water velocity	[$L\ T^{-1}$]
θ	volumetric water content	[$L^3\ L^{-3}$]
θ_m	mobile water	[$L^3\ L^{-3}$]
θ_{st}	transport volume	[$L^3\ L^{-3}$]
ρ_b	bulk density	[$M\ L^{-3}$]
σ^2	variance of the lognormal pdf	

CHAPTER THIRTEEN

Field-Scale Solute Transport in the Vadose Zone: Experimental Observations and Interpretation

M. Flury, W.A. Jury, and E.J. Kladivko

INTRODUCTION

Chemical transport in porous media has long been conceptualized as a superposition of two processes: convection with moving water through the pore spaces of the soil or sediment matrix; and random mixing about the center of motion. The problem of predicting chemical transport under field conditions would be trivial if the water velocity pathways could be characterized experimentally with a high degree of spatial and temporal resolution. With this knowledge, chemical convection would represent the dominant mechanism of solute transport, and the remaining motion would be negligible at all but the smallest scales of interest. Unfortunately, highly detailed experimental information about velocity flow-paths in natural soil is not now available, nor in all likelihood will it ever be. Our devices for characterizing water flow in unsaturated soil are either too localized to detect all of the pathways, or integrate over too large a volume to allow underlying structure to be deduced.

In the absence of detailed information about the water flow-paths, the problem of modeling chemical transport through unsaturated soil is no longer trivial. When only a large-scale, volume-averaged water flux can be measured, a considerable portion of the chemical motion must be described as dispersive mixing about the mean convection. This strategy has proved to be useful in uniform media, such as repacked soil columns, because the dispersion in that case converges rapidly to a symmetric, homogeneous mixing process that can be described by a constant parameter (the dispersivity). Under field conditions, the situation has not proved to be that simple, either because there is a wide range of velocity pathways that remain relatively isolated from each other, or because a small portion of the flow pathways transport water at much higher than average flow rates. Dispersion then becomes a lumped parameter representing a host of processes, possibly involving diffusive mass transfer, unstable flow, rate-limited reaction with the solute matrix, small-scale convection, or any other type of motion that tends to cause solute spreading relative to the center of mass motion described by the convection process. Dispersion and convection are thus closely linked, the former growing in complexity and importance as the latter is rendered more simple by the limitations of monitoring. Under such circumstances, representation of dispersion requires more than just a single coefficient, and its process characterization expands into a network of interactive structures that have more demanding parameter requirements.

A second and related problem in solute transport is to extend observations made at one scale (say, by a localized monitoring device) to a higher scale, or equivalently, to deduce an underlying lower-scale transport or reaction process from a large-scale measurement that integrates over variations in the process of interest. A prime example of the latter is effluent monitoring of chemical migration through soil by a tile-drainage system. If the drain is intercepting all of the effluent, then it will record all of the solute migration, but in a form that is very difficult to resolve spatially because the effluent at any one time is a convergence of stream tubes that originate from all parts of the field surface above (Jury, 1975a,b).

Despite the complexity of the field regime, solute transport is still modeled most frequently as a combination of mean convection and random dispersive mixing. Moreover, the dispersion is generally either lumped into a single parameter, or is assumed to occur as a consequence of convection in isolated stream tubes of different flow velocity. The utility of these approximate approaches has received only limited examination until recently, and at the present time there are only a small number of field studies that characterize solute transport processes. The primary purpose of this review is to examine field observations of solute transport in the context of these limitations, to determine the presence or absence of consistent trends, and to explore the utility of lumped parameters, such as the dispersivity, for describing large-scale transport.

SOLUTE DISPERSION

Solute dispersion and its scale-dependence is an active research area in both saturated and unsaturated zone hydrology. However, the former setting has experienced much greater success at model formulation and validation than the latter. There are two fundamental differences between unsaturated flow in soils and saturated flow in aquifers. First, in soils water usually flows in the vertical direction, which is perpendicular to the natural layering of the medium. Thus, unsaturated flow pathways are likely to encounter frequent textural boundaries and changes in mean flow properties. In contrast, water flow in aquifers is generally parallel to the layering of the medium, and thus macroscopically homogeneous with mean properties that do not change along the direction of flow. Second, the hydraulic conductivity in soils is a nonlinear function of the water content, whereas in aquifers hydraulic properties are completely defined by the saturated hydraulic conductivity, its spatial variability, and its correlation structure. Because of these differences, the far more numerous experimental and theoretical results from solute dispersion in aquifers may not necessarily be applicable to unsaturated flow in soils.

Field experiments where solute dispersion in soils was studied have been summarized by Jury and Flühler (1992) and Beven et al. (1993). The focus of these field experiments was mostly on characterizing the spatial variability of the dispersion parameter (Biggar and Nielsen, 1976; van de Pol et al., 1977; Jaynes et al., 1988, van Ommen et al., 1989), and on evaluating the nature of the dispersion process (Jury et al., 1982; Brissaud et al., 1983; Butters and Jury, 1989; Ellsworth and Jury, 1991; Roth et al., 1991; Ellsworth et al., 1996). A common finding from the first category of studies was that measurements of the dispersion coefficient at different spatial locations of the same field or mapping unit are more lognormally than normally distributed, a feature characteristic of most hydraulic and solute transport parameters in the field (Jury et al., 1987).

Dispersion studies in unsaturated field soils have invariably been designed to probe the mechanisms of mixing. One of the foundational efforts of field experimentation in this area has been to examine the nature of longitudinal macrodispersion relative to flow of area-averaged solute concentration. Experiments of this type try to maintain a spatially uniform downward motion at the upper boundary, initiated by uniform flow to an extensive surface area. These studies have for the most part been designed to validate which of two approximate asymptotic dispersion-scale theories is dominant: the infinite-time convection-dispersion process, where the apparent dispersion coefficient is constant, or the zero-time stochastic-convective process, where the apparent dispersion coefficient increases linearly with travel distance or time (Jury and Roth, 1990). The first process is usually described by the convection dispersion equation (CDE), and the second process by the convective-lognormal transfer function model of Jury (1982).

Early research efforts to use the approximate models were not encouraging. Brissaud et al. (1983) measured deuterium (D_2O) movement in sandy loam soil with a gamma neutron probe, concluding that the dispersion process could be well described by the CDE. However, when they repeated the study on the same plot under similar experimental conditions several weeks later, they observed behavior that was incompatible with that predicted by the CDE.

In a bromide pulse experiment on 0.64 ha of Tujunga loamy sand in Southern California, Butters and Jury (1989) found a linear increase in apparent macrodispersion coefficient with distance over the top 3 m, followed by a region between 3 and 4.5 m where the dispersion decreased. In the first 3 m, the authors concluded that solute traveled in isolated stream tubes and did not have time to mix during the time of residence. They speculated that the decline in dispersion between 3 and 4.5 m could be the result of a narrow fine-textured silt layer at 3 m that could initiate enhanced lateral solute spreading. In a later experiment on the same Tujunga field site, Ellsworth and Jury (1991) added massive cubic plumes of solute approximately 1.5 m on a side to a field irrigated uniformly by sprinkler, and found that transport of the area-averaged solute concentration in each plume could be represented by downward convective drift and a constant macrodispersion coefficient.

The apparent discrepancy between the two studies may possibly be a consequence of the different sampling and analysis methodologies employed. Butters and Jury (1989) measured soil solution at 16 sites at each of 7 depths with suction cups, whereas Ellsworth and Jury (1991) sampled the massive solute plumes periodically with soil cores. Thus, the plume experiment integrated over a large depth interval with each observation. Both fields were irrigated by sprinkler after the solute was in the soil, but Ellsworth and Jury (1991) added their plumes to the soil by a low-energy dense network of drip emitters, which might have minimized initial formation of localized flow channels. In contrast, Butters and Jury (1989) added the solute pulse through the sprinkler system.

Ellsworth and Jury (1991) also used a fluid-coordinate transformation to account for vertical changes in mean soil water content, whereas Butters and Jury (1989) did not. However, a reanalysis of the earlier data using the fluid coordinate transformation did not change the conclusions made in the 1989 study (Fleming and Butters, 1995).

A later study focused specifically on the comparison between solution sampler and soil core monitoring of the same field experiment. Ellsworth et al. (1996) measured transport of Cl, Br, and NO_3 in a Pachappa fine sandy loam with solution samplers and soil excavation, adding the tracers by low-energy trickle emitters at different times to allow a single soil sampling to reveal several stages of migration. In contrast to the conflicting results found by the earlier field studies on the Tujunga field site, Ellsworth et al. (1996) observed that data from both sampling methods yielded consistent results, and indicated that the macrodispersion process increased linearly with residence time. They concluded that a stochastic-convective dispersion process was operative on their field over the depths they monitored, which they felt was a consequence of the absence of any changes in texture with depth.

Roth et al. (1991) measured Cl movement in a strongly layered, loamy soil in Switzerland with a two-dimensional network of suction cups. They found that the applied tracer split into a fast-moving fraction that migrated through structural voids, and a slow-moving fraction which traveled through the soil matrix. The rapidly moving portion was substantial, accounting for about 58% of the mass transfer, and was unpredictable from any of the measurements of the field properties that were made. In contrast, area-averaged movement of the slow-moving fraction could be well described as convective-dispersive transport with a constant macrodispersion coefficient, which they attributed to the pronounced horizontal layering in this soil causing enhanced lateral mixing. In another study in Switzerland, Schulin et al. (1987) found that neither constant nor linearly increasing macrodispersion models could explain the solute movement they observed in a stony soil, where Cl and Br were sampled by excavation along a trench.

Figure 13.1 shows two-dimensional Cl concentration profiles in a sandy and a loamy soil from the field study of Flury et al. (1995), and illustrates how variable solute transport can be in identical experiments at different locations. The authors applied a narrow pulse of Cl to the soil surface and leached the solute by sprinkler irrigation. After 30, 60, and 90 mm cumulative infiltration, the soil was excavated along a trench and Cl concentration distribution was determined. Figure 13.1 clearly depicts different dispersion and transport behavior in the two soils.

The results from the literature thus do not provide a consistent picture of the nature of the macrodispersion process. There is some evidence that macrodispersion might be represented as stochastic-convective (linearly evolving dispersion) in homogeneous sandy soils, and convective-dispersive (constant dispersion) in strongly layered soils. But as seen in the studies above, that pattern is by no means universal. Clearly, the dispersion process seems to be site-specific, but repeated studies on the same field have also yielded evidence of dependence on subtler factors having to do with the experimental conditions. It remains an open question whether there is a general trend observable in the dispersion behavior in soils and whether the dispersivity approaches an asymptotic value after a certain distance. In the following section we will focus on the scale dependency of the dispersivity in soils, by which we mean the value of the apparent macrodispersion parameter as a function of its distance from the point of origin.

Longitudinal Dispersivity

A considerable amount of information is available about longitudinal dispersivities in aquifers. The experimental data indicate that the longitudinal dispersivity increases with observation scale (Gelhar et al. (1992). In their comprehensive literature review, Gelhar et al. (1992) reported dispersivities ranging from 10^{-2} to 10^{4} m for length scales of 10^{-1} to 10^{5} m. Considering the reliability of the data, the longest length scale found was 250 m, and the trend of increasing dispersivity was less clear but still remained (Gelhar et al., 1992). Because the longitudinal dispersivities typically vary over 2 to 3 orders of magnitude at a given length scale, the relation between dispersivity and observation scale may not be represented with a single regression line, but rather with a family of curves (Gelhar et al., 1992).

The experimental trend reported by Gelhar et al. (1992) is consistent with theoretical results which postulate that the longitudinal dispersivity initially increases linearly with displacement distance and approaches a constant (convective-dispersive) asymptotic value (Gelhar et al., 1979; Dagan, 1984). Based on this theory, the longitudinal

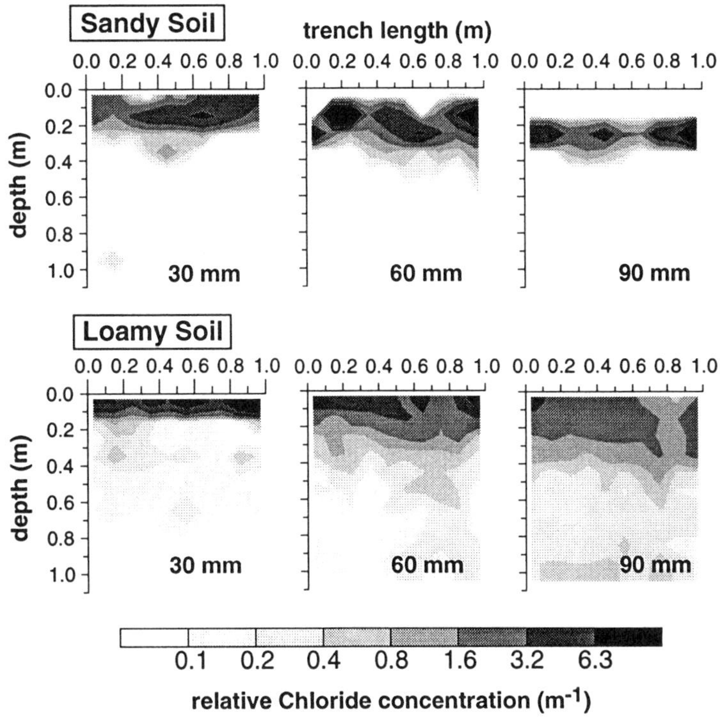

Figure 13.1. Two-dimensional Cl distribution in a sandy and loamy soil after 30, 60, and 90 mm cumulative infiltration (adapted from Flury et al., 1995).

dispersivity in aquifers can be calculated from the variance and the correlation scale of the logarithm of the hydraulic conductivity.

Available dispersivity data are much less plentiful for solute transport in soils than in aquifers. Field experiments where apparent dispersivity values have been reported are summarized in Table 13.1. A typical experiment in Table 13.1 consists of placing a series of suction cups at different depths in the soil and measuring the solute breakthrough as a function of time. The tracers, usually applied as a narrow pulse input to the soil surface, consist of Br, Cl, NO_3, BO_3, tritium, and organic compounds that have been found to move similarly to a conservative tracer. The tracers were generally leached through the soil by irrigation, and only in a few cases by natural rainfall. Different irrigation methodologies were used. Sprinkling and trickle irrigation resulted in unsaturated flow through soil, whereas ponding created saturated conditions.

The method of dispersivity calculation is indicated in the last column of Table 13.1. In most cases, the experimental data were analyzed with the conventional one-dimensional CDE:

$$\frac{\partial C}{\partial t} = D\frac{\partial^2 C}{\partial z^2} - v\frac{\partial C}{\partial z} \tag{13.1}$$

where D [L^2T^{-1}] is the dispersion coefficient and v [LT^{-1}] is the convective velocity. The CDE assumes a steady-state water flux, which, however, is seldom attained in field-scale solute transport experiments. A common remedy for this inconsistency has been to replace the ordinary time coordinate by cumulative infiltration or drainage I (Jury et al., 1982; Butters and Jury, 1989; Ellsworth and Jury, 1991; Roth et al., 1991):

Table 13.1. Dispersivity Measurements in Field Soils.

			Water Application		Sampling					
Reference	Soil	Tracer	Method	Rate ($cm\ d^{-1}$)	Method	Depth (m)	Study Area (m^2)	Dispersivity (cm)	Scale	Method of Calculation[d]
Biggar and Nielsen (1976)	clay loam	Cl, NO_3	continuous ponding	na	suction cups	0.3–1.8	6.5x6.5 (20 plots)	8.3	local	CDE
Bowman and Rice (1986)	clay loam	Br, Organics[b]	intermittent ponding	1.4	suction cups	0.3–3.0	37 (4 plots)	9.44–141	local	CDE
Brissaud et al. (1983)	sandy loam	D_2O	sprinkling	3.36–6.48	neutron probe	1.7	4.9	1.1, 2	local	CDE-I
Butters et al. (1989)	loamy sand	Br	bidaily sprinkling	1.1	suction cups	0.3–4.5	80x80 (16 locations)	3.2–15.8 5.6–29.1	local field	MOM
Butters et al. (1989)	loamy sand	Br	bidaily sprinkling	1.1	soil cores	0–26	6 cores	92	field	MOM
Ellsworth et al. (1991)	loamy sand	Cl, NO_3, BO_3	sprinkling	1.1	soil cores	0–6	1.5x1.5 (14 plots)	9.1	local	CDE-IF
Ellsworth et al. (1996)	sandy loam	Br, Cl, NO_3	sprinkling	1.3	suction cups	0.25, 0.65	2x2	1–2	field	CDE-IF
Ellsworth et al. (1996)	sandy loam	Br, Cl, NO_3	sprinkling	1.3	soil cores	0–2	2x2	1.1–2.7	field	CDE-IF
Fleming and Butters (1995)	clay loam	Br	sprinkling	96	suction cups	0.15–2	25	5.2–23	field	CDE-I
Fleming and Butters (1995)	clay loam	Br	sprinkling	96	soil cores	0–3.7	25	4.1–62.8	field	CDE-I
Flury (1993b)	sand	Br, Cl	sprinkling	12	soil cores	0–1.1	1.4x1.4	1	field	CDE
Hamlen and Kachanowski (1992)	sand	Cl	drip irrigation	84, 132	suction cups	0.2, 0.4	9.6x1.2	0.7–1.6 0.8–2	local field	MOM MOM
Jaynes (1991)	clay loam	Br	continuous ponding	30,33,41,67	suction cups	0.3–3.0	1.83x1.83	16–38	local	CDE
Jaynes and Rice (1993)	clay loam	Br, Organic[c]	intermittent ponding	4.75–5	suction cups	0.3–3.0	37	19.3–56.4	local	CDE
Jaynes and Rice (1993)	clay loam	Br, Organic[c]	drip irrigation	4.75–5	suction cups	0.3–3.0	37	4–12.8	local	CDE
Rice et al. (1986)	sandy loam	Cl	intermittent ponding	84, 117	soil cores	0–2.7	24.4x18.3	13.8–20.6	local	CDE
Roth et al. (1991)	loam	Cl	rainfall, sprinkling	max 9.6–19.2	suction cups	0.3–2.8	12x2	29	field	CDE-I
Schulin et al. (1987)	loamy sand	Br, Cl	rainfall	na	soil cores	0–3.1	94	2.1–8.2 2.8–12	local field	CDE
van de Pol et al. (1977)	clay, silty clay	Cl, Tritium	trickle irrigation	2	suction cups	0.15–1.5	8x8	9.4	local	CDE

van Ommen et al. (1989)	loamy sand (corn)	Br	rainfall	na	soil cores	0–0.7	20x260	3.7[e]	field	CDE-I
van Ommen et al. (1989)	loamy sand (grass)	Br	rainfall	na	soil cores	0–0.7	20x260	14.5[e]	field	CDE-I
van Wesenbeeck and Kachanowski (1991)	sand	Cl	trickle irrigation	84, 117	suction cups	0.4	6.4, 9.6 m-transect	1.7	local	MOM
								2.7	field	MOM

[a] Not specified or not available.
[b] *m*-Trifluoromethylbenzoic acid, *o*-trifluoromethylbenzoic acid, 2,6-difluorobenzoic acid, pentafluorobenzoic acid.
[c] *o*-Trifluoromethylbenzoic acid, 2,6-difluorobenzoic acid.
[d] CDE: Convection-dispersion equation, CDE-I: CDE with cumulative infiltration, CDE-IF: CDE with cumulative infiltration and fluid coordinate, MOM: Method of moments.
[e] Only first sampling used.

$$I = \int_0^t J_w(t')dt' \tag{13.2}$$

where J_w [LT^{-1}] is the water flux. The question of whether the steady-state water flux assumption is justified for moderately transient flux regimes has been discussed intensively in the literature. However, no consistent results have been reported, often depending on local experimental conditions (Wierenga, 1977; Beese and Wierenga, 1980; Jury et al., 1982; Russo et al., 1989; Jury et al., 1990).

In addition to the cumulative drainage transformation, Ellsworth and Jury (1991) used a fluid coordinate system, where the variable water storage capacity of the soil is considered. The space coordinate z is transformed to a new depth coordinate z^* by:

$$z^* = \int_0^z \theta(s)ds \tag{13.3}$$

where θ [L^3L^{-3}] is the volumetric water content in the soil. Based on the data analysis with the CDE or modified versions of it, the dispersivity can then be calculated as:

$$\alpha = D/v \tag{13.4}$$

The dispersivity can also be calculated from the moments of data collected from pulse input experiments. For sampling with suction cups, the dispersivity is given as a function of the time moments (Simmons, 1982; Jury and Sposito, 1985):

$$\alpha = \frac{z}{2}\frac{T_2 - T_1^2}{T_1^2} \tag{13.5}$$

where T_1, T_2 are the first two temporal moments of the breakthrough curve. For sampling with soil cores, the dispersivity is given as (Jury and Roth, 1990):

$$\alpha = \frac{1}{5}\left(Z_1 - \sqrt{6Z_1^2 - 5Z_2}\right) \tag{13.6}$$

where Z_1, Z_2 are the first two spatial moments of the concentration distribution.

When dispersivities were not explicitly given in the references, we calculated the values depending on the reported data with Eqs. 13.4, 13.5, or 13.6. We denote the latter two calculations as method of moments (MOM). The tenth column in Table 13.1 classifies the dispersivity values as field-scale or local, depending on whether area-averaged solute concentration was analyzed to produce a macrodispersivity, or whether separate plots were analyzed for dispersivity and averaged to produce a mean local value.

All available dispersivity values given in Table 13.1 are plotted as a function of observation distance in Figure 13.2. For suction cup measurements, the observation distance is the distance between the solute source, usually the soil surface, and the location of the suction cup. For soil core measurements, the observation distance is taken as the first moment of the resident solute concentration. The dispersivity values range from 0.7 cm to 141 cm and the length scales vary from 0.1 m to 14.8 m. There is no clear relation between dispersivity and observation distance apparent in Figure 13.2.

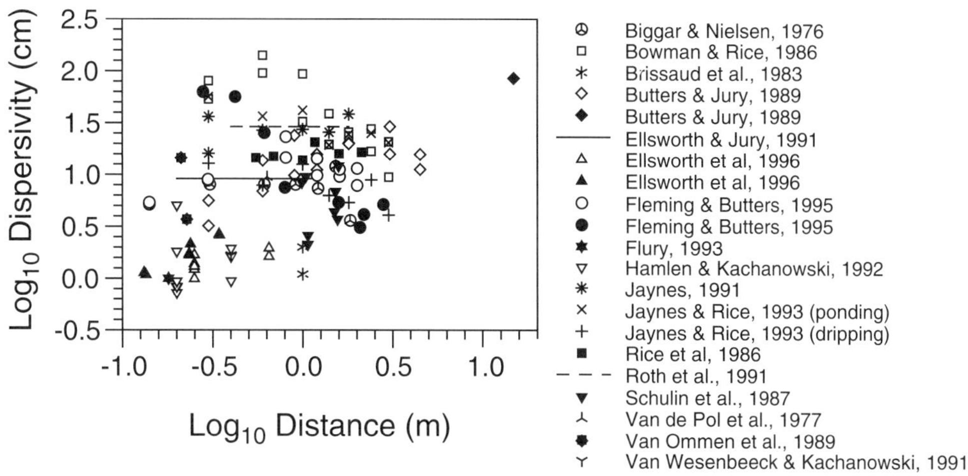

Figure 13.2. Longitudinal dispersivities measured in soils as a function of observation distance. Experimental conditions of the studies are listed in Table 13.1. Open symbols refer to concentration measurements versus time (breakthrough curves), solid symbols represent concentration-depth measurements (soil cores). Lines indicate range of reported data.

For obvious reasons, the observation scales in soils do not cover as great a distance range as in aquifers. Considering the variability of the data, the dispersivities from unsaturated transport in soils agree reasonably well with those from saturated transport in aquifers. The dispersivity-length plot (Figure 13.2) fits in the lower left corner of Figure 1 of Gelhar et al. (1992, p. 1966).

Transverse Dispersion

The importance of lateral mixing processes as a key to understanding field-scale solute transport has been recently emphasized in a review paper by Flühler et al. (1996). Lateral mixing governs the timescale for solutes to migrate into different stream tubes, and the relationship between this timescale and the travel time determines the nature of longitudinal macrodispersion (Jury and Roth, 1990). Thus, reliable estimates of transverse dispersion are essential to modeling solute transport in heterogeneous media where the velocity field varies spatially. Unfortunately, data on transverse dispersion of solutes in field soils are scarce, because measuring it requires a quantitative assessment of the extent of lateral movement during downward flow, a task attempted thus far only by Ellsworth et al. (1991). Qualitative experimental observations of substantial transverse mixing of solutes have been provided from the few studies where tracer plume migration has been monitored in two or three dimensions (Schulin et al., 1987; Ellsworth et al., 1991; Roth et al., 1991). Flury et al. (1995) found substantial transverse dispersion in a loamy soil, but almost no mixing in a sandy soil. In addition, several dye tracing studies have illustrated considerable lateral movement of water and solutes (Ghodrati and Jury, 1990; Kung, 1990; Flury et al., 1994). A summary of reports where lateral flow in the unsaturated zone was observed is given by Gelhar et al. (1985).

Type of Water Application

The method by which water is applied to the soil surface may strongly influence the transport regime, and thus also affect the parameter values estimated from experimental data. Bowman and Rice (1986) studied solute movement in a clay loam under ponded infiltration during a period of 10 months. Water was applied with a single 100-mm application per week or with two 50-mm applications per week. Measured mean tracer velocities were in the range of 64 to 74 mm d^{-1}, where the mean dispersivities ranged from 9 to 141 cm. Both parameters were independent of the irrigation regime. A similar observation was reported by Wild and Babiker (1976), who applied four different irrigation regimes to a loamy sand: 2.5 mm on 20 successive days, 5 mm on 10 successive days, 12.5 mm on three days, 25 mm on two days. The rate of water application was kept constant at 8 mm h^{-1}. The authors reported no significant differences for mean and modal depth of solute movement.

In contrast, Jaynes and Rice (1993) compared the effects of intermittent ponding and drip irrigation on solute transport for plots receiving a daily 50-mm water application. Mean velocities were slightly greater and more uniform under drip irrigation than under ponding, but the dispersion was on the average three times greater under ponding. The authors explained the higher dispersion as resulting from increasing variability of the hydraulic properties near saturation.

DYE TRACING STUDIES

Many experiments with soil cores and suction cups have offered indirect evidence that water flows in heterogeneous pathways. Due to the limited spatial resolution of cores and suction cups, the nature and cause of the observed heterogeneity usually cannot be determined with these measuring devices. As an alternative, dye tracing offers the advantage of visualizing the flow pathways of water in detail, revealing heterogeneous water flow patterns in a variety of soils (Saffigna et al., 1976; Bouma and Dekker, 1978; Ghodrati and Jury, 1990; Kung, 1990; Steenhuis et al., 1990; Flury et al., 1994).

Bouma and Dekker (1978) applied the cationic dye Methylene Blue (C.I. 52015) in solution to dry clayey soils having a blocky and prismatic structure, observing that water moved rapidly between the prismatic soil blocks. The predominantly planar voids were not stained completely, and discolorations were observed as small vertical bands along the wall. The stained area occupied about 2% of the available surface area of the prisms. Steenhuis et al. (1990), working with Brilliant Blue FCF (C.I. 42090) and Methylene Blue, found that earthworm burrows were dominant flow conduits in the upper part of a clay loam soil.

While movement of water through a small part of the soil volume is expected in structured soils, preferential flow of water has also been reported in structureless sandy soils. Saffigna et al. (1976) found nonuniform infiltration of Rhodamine WT in a Plainfield loamy sand planted with potatoes, even though the dye was applied uniformly with a sprinkler system. Foliage interception, stemflow, and hilling were identified as sources of the nonuniformity. The focus of this study was on the infiltration pattern near the soil surface, and the soil was only excavated to the 60-cm depth to examine the dye pattern. Kung (1990) conducted a much larger excavation of a potato field on Plainfield sand, adding Rhodamine WT dye to the furrow bottoms between the potato rows over a 4-m^2 surface area and later exposing the profile to the 6.6-m depth. Dye patterns showed very little lateral displacement in the uniform top layer, but significant lateral movement was observed in the deeper unsaturated zone along textural discontinuities. Water channeled into preferential pathways, and at 5.6 to 6.6 m depth water flowed through less than 1% of the soil matrix. The top 1.1 m of the soil was very uniform, but from 1.1 to 7.4 m depth the soil had distinct layers made of fluvio-glacial deposits. These layers were neither horizontal nor continuous, and consisted of coarse sand lenses interbedded in finer sand layers. The coarse lenses acted as barriers when dry, channeling the water around them into smaller and smaller pathways. Since the conductivity of the surrounding matrix was high, these pathways were still able to transport water without lateral spreading, a phenomenon Kung described as funnel flow.

Ghodrati and Jury (1990) compared the effects of different irrigation and surface treatment on infiltration patterns in a Tujunga loamy sand. The soil surface was either undisturbed or completely disturbed with a trencher, and a pulse of Acid Red 1 (C.I. 18050) solution was leached by either sprinkling or ponded irrigation. All plots showed irregular flow patterns, but the patterns in the undisturbed plots were less heterogeneous than in the disturbed soil because unstable flow initiated at the interface between the repacked soil layer and the undisturbed soil below it.

Fourteen different soils spanning all of the types used in crop production in Switzerland were tested with a common dye tracing experiment by Flury et al. (1994). The soils had a variety of different structures and textures that ranged from sandy to clayey. Forty-millimeters of Brilliant Blue FCF solution was applied to the surface with a sprinkler system. Most of the 14 soils showed irregular flow patterns to some degree, with the flow patterns in heavily structured soils strongly dependent on the water content of the soil.

The results from the dye tracing studies demonstrate that irregular flow patterns and preferential flow phenomena are abundant in field soils. The reported flow patterns also illustrate the difficulties associated with sampling solutes in field soils, since instruments with a localized range of influence may miss a flow event completely and those with a large range may not be able to deduce its small-scale structure.

WATER REPELLENCY

Water repellency of soil constituents may dramatically affect water and solute movement at the field-scale. Wallis and Horne (1992) concluded in their comprehensive review on soil water repellency that all soils may be water repellent to some degree, and that the importance of water repellency on water and solute transport in the field-scale has been underestimated. Water repellency and its spatial variability have been shown to cause nonuniform wetting and preferential flow in many field soils, which to the present time have eluded attempts to model except in a qualitative sense (Hendrickx et al., 1993; Ritsema et al., 1993; Dekker and Ritsema, 1994, 1995). Based on experimental observations, Ritsema et al. (1993) proposed a conceptual model for water flow in sandy water-repellent soils. According to this model, the initially uniform water infiltration is disrupted within the first few millimeters or centimeters of the water-repellent soil and moves laterally toward microdepressions and regions with lower water repellency, where fingers are formed. Water is transported along these preferential flow pathways until decreasing water repellency in the soil with increasing depth causes divergence of the flow lines. The lateral displacement of water and solutes within the first few centimeters of the soil may also be caused by crusted, sealed, or compacted soil surfaces, and probably occurs frequently in field soils (Ritsema and Dekker, 1995). This process, occurring at the onset of an infiltration or leaching event, may drastically affect all subsequent transport.

REACTIVE SOLUTES

Analysis of solute sorption during transport is outside the scope of this review, except in one respect. Strongly absorbed organic tracers have been used to offer indi-

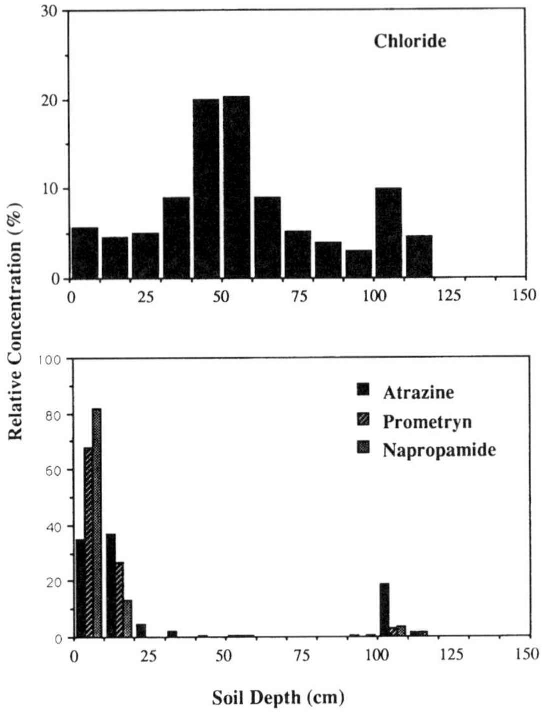

Figure 13.3. Chloride and herbicide distribution in a sandy loam soil after infiltration of 120 mm water. (Reprinted from Ghodrati and Jury, 1992, p. 119, with kind permission of Elsevier Science–NL, Sata Burgerhartstraat 25, 1055 KV Amsterdam, The Netherlands.)

rect evidence of preferential flow in soil, by observation of their migration well below depths expected if sorption to equilibrium occurred. An illustrative example of this phenomenon is presented by Ghodrati and Jury (1992). In this study, three herbicides with different sorption characteristics were applied to field plots and leached with sprinkling irrigation. All three herbicides were found simultaneously at a depth of about 1 m (Figure 13.3), indicating that little if any sorption was occurring in the fast flow channels responsible for the herbicide migration. Other studies have observed the same phenomenon (Jury et al., 1986; Clendening and Jury, 1990; Flury et al., 1995).

TILE DRAIN STUDIES

Tile drainage studies have shed additional light on the nature of solute movement under field conditions. Subsurface tile drainage is a common agricultural water management tool in areas with shallow groundwater or seasonally perched water tables. Although the drains remove water from the soil only when the zone near the tile is saturated, the flow in the tiles is a combination of unsaturated flow to the shallow water table and saturated flow to the tiles. These shallow, variably-saturated zones that are drained by tiles are very important for chemical transport in unsaturated soils.

Tile drains offer some unique advantages, as well as weaknesses, when studying preferential flow and other nonequilibrium transport processes. Tile drain outflow integrates water and chemical flow over a relatively large area in a field. In addition, when drains are installed in a systematic pattern, they drain a relatively well-defined area (from midplane to midplane) of the field, so quantitative calculations of total mass losses can be made. Tile drains are also prevalent in many regions and are readily available on research farms and production fields. One of the limitations of tile drain studies is the two-dimensional flow regime to the tile, where water and chemicals may flow primarily vertically through the unsaturated zone and then along curved streamlines to the tile. Consequently, chemical concentration in the tile consists of a combination of recent chemical applications near the tile and older chemical applications from further distances away. These different travel times may be estimated by modeling (Jury 1975a,b; Utermann et al., 1990), but the models do not generally account for preferential flow or nonequilibrium conditions in the saturated zones.

Tile drains have been used in the past to assess the impact of agricultural management practices on surface and groundwater quality (Logan et al., 1980; Baker and Johnson, 1981; Hallberg et al., 1986). Most of these studies were not designed to investigate preferential flow processes or other underlying mechanisms of chemical transport. However, a number of observations suggested that preferential flow might be occurring. For example, Bottcher et al. (1981) found small amounts of two chemicals with different sorption coefficients in drain outflow within four days of application. Several recent studies have focused specifically on preferential flow. Van Ommen et al. (1989) applied a nonsorbing tracer uniformly across a field of loamy sand in the Netherlands, and they observed some early breakthrough of the tracer. Everts et al. (1989) irrigated a silt loam soil with a solution of two nonsorbed tracers and two differently sorbed tracers, and they detected all four tracers in tile outflow within one hour of beginning the irrigation.

The general findings from tile drain studies include the following two main points: (1) sorbing chemicals (usually small amounts) arrive rapidly at the tile drain, often during the first rainfall and drainage event after chemical application, and (2) chemicals with different sorption coefficients all arrive at the same time, but usually at different concentrations. The arrival of sorbing chemicals at tile drain depth (usually 0.75 to 1.2 m depth) with very little net water flow, is not predictable by standard water flow and convection-dispersion descriptions and is usually interpreted to be preferential flow. And since chemicals with different sorption coefficients are usually assumed to be transported at different rates, nonequilibrium sorption is often invoked to explain the simultaneous, rapid arrival of multiple compounds. This hypothesis says that the transport of chemicals in the preferential flow zone is so fast that there is not time for equilibrium adsorption to occur, and thus some chemical is transported without any retardation (Grochulska and Kladivko, 1994).

Several recent studies will be used to illustrate these points. In the studies of Kladivko et al. (1991, 1994), four pesticides were applied to large field plots in spring at the time of corn planting. The time interval between pesticide application and the first large rainstorm and tile drainage event varied somewhat during the four years studied, with several small rainfalls occurring before effluent appeared. In all cases, pesticides were found in tile outflow during the first drainage event, and all chemicals arrived at the same time (Figure 13.4). Total water drainage in the first event was usually less than 1 cm, compared to a unit pore volume of about 26 cm for this system, which underscores how little water is needed to initiate preferential flow.

Czapar et al. (1994) applied four sorbing compounds plus a tracer to small tile plots and then irrigated one day later. Water flow and chemical concentration peaks occurred within 130 min after the start of irrigation in both years studied. All chemicals arrived at about the same time except for one with a much higher sorption coefficient and lower solubility, which was detected in only one sample. This suggests that the

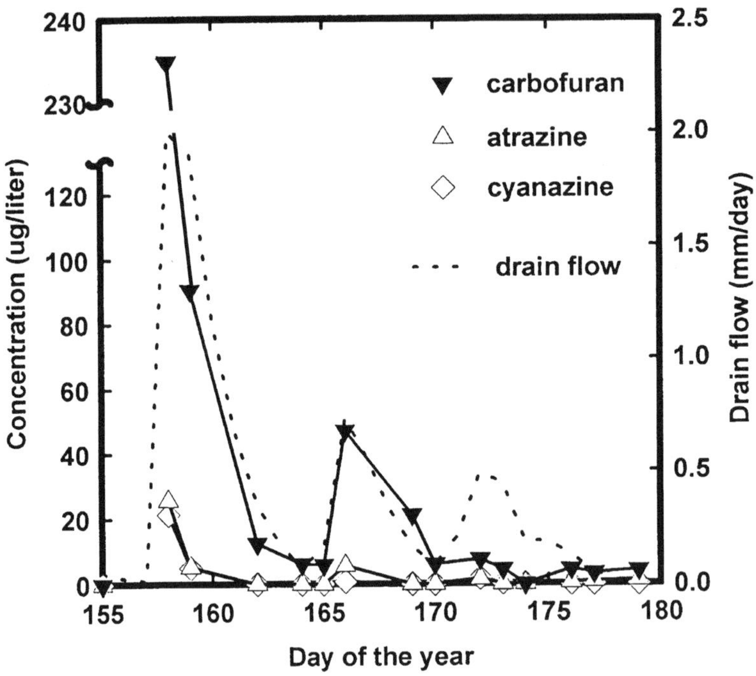

Figure 13.4. Tile drain outflow and pesticide concentrations from a 20 m spacing plot in the drainage study of Kladivko et al. (1994), from Day 155 to 180, 1990.

preferential flow and nonequilibrium sorption, as discussed earlier, may permit rapid movement of sorbing compounds only up to some limiting value of sorption affinity.

Milburn et al. (1995) studied atrazine leaching from a production corn field in New Brunswick, Canada. At one point in the study the producer had an accidental spill of atrazine on one of the plots, which was followed by a large rainfall several days later. A large concentration peak appeared in the tile outflow near the beginning of the resulting drainage event, again indicating preferential flow on a field-scale. Buttle and Harris (1991) measured metolachlor in surface runoff, streamflow, and tile outflow in a field site in Ontario, and found peak metolachlor concentrations in the first major drainage event after application (13 days).

Traub and Eberhard et al. (1994) applied chemicals in spring and again in autumn to tile-drained plots in Germany. The spring applications resulted in very little chemical leaching due to low water flow. Autumn application of two chemicals with different sorption coefficients showed the typical rapid arrival of both chemicals simultaneously. The authors concluded that the results supported the concepts of preferential flow and nonequilibrium sorption.

Not all tile drainage studies have provided evidence for preferential flow. As mentioned earlier, many drainage studies were not designed to study preferential flow and therefore did not sample water and chemical concentrations frequently enough to be able to detect rapid changes in concentration (Milburn and MacLeod, 1991). Buhler et al. (1993) concluded there was no basis for assuming preferential flow in their research conditions, both from their measured data and their knowledge of the soil. But since their primary sampling technique was to take periodic grab samples from tile lines, it is possible they missed the peak at the beginning of drainage events which would have indicated preferential flow. Other drainage studies focus primarily on nitrogen behavior, and preferential flow is probably less significant for nitrogen than for pesticides. Nitrogen has multiple sources and sinks in the soil throughout the year and it can distribute throughout the soil matrix water more quickly than pesticides, due to its lack of sorption.

One potential weakness of using tile drain fields to study preferential flow and nonequilibrium transport processes is the nature of the trench dug when installing the drains. Although the trench is backfilled with the original soil material in many cases, the disturbed soil will be looser than the surrounding soil for some time and therefore not representative of chemical transport through the undisturbed soil. With time the trench settles and should return to near normal flow conditions. Two studies have specifically tested their fields for preferential flow apart from the old trenchline. Richard and Steenhuis (1988) applied a tracer in a 3.5 m wide strip offset 2 m from the tile line, to avoid any remaining disturbed soil from the tile installation procedure. They also observed early breakthrough of the tracer. Kladivko et al. (1994) applied bromide and simazine to a 3 m wide strip offset 1.5 m from the tile drain. They detected the two compounds in drain outflow with less than 1 cm total net drainage, again supporting the interpretation that preferential flow is occurring in the field as a whole and not just in the part of the field disturbed during tile installation.

Some tile drain studies have also shown the phenomena of multiple chemical peaks arising from a single application of a sorbing compound. The multiple peaks are usually associated with multiple rainstorm events, with each new rainfall/drainage event resulting in a new peak of concentration (Kladivko et al., 1991, 1994; Traub-Eberhard et al., 1994). Concentrations were highest near the beginning of each event and decreased rapidly as the flow event continued. During subsequent events concentrations again started high and then dropped rapidly. In general there was some attenuation of peak concentrations with each successive event.

One proposed explanation for the observed multiple flushes of chemical is nonequilibrium sorption in the preferential flow paths. At the start of a flow event, pesticide in the existing soil solution is flushed rapidly through large pores and into the drain. Desorption is not rapid enough to maintain an equilibrium solution concentration in new rain water, so continued water flow through those pores contains much lower concentrations. When drainage ceases, desorption supplies chemicals to the water in the large pores, and another rainfall would cause a new flow event to contain a high initial concentration of pesticides. An example of a model that is based on this explanation is presented in Grochulska and Kladivko (1994).

Another type of multiple peak behavior was seen by Czapar et al. (1994). Although they had only one irrigation event, some of the tile lines showed multiple chemical peaks within the one flow event. They suggested that this may indicate some dead-end macropores with high solution concentrations that would sporadically empty and contribute chemicals to the drain.

INFLUENCE OF SAMPLING METHOD

Sampling Methods

Many of the unresolved problems in unsaturated zone hydrology are inherently related to the experimental methods available to investigate solute transport. There are currently four major methods for solute sampling used in the field: suction cups, tile drains, soil cores, and dye tracing. In addition, undisturbed field lysimeters offer realistic environments for solute transport observation, although they are difficult to construct and maintain. Each of these has certain advantages and disadvantages, as discussed in the following.

Suction Cups

Suction cups are probably the most popular measurement tools. They have the advantage that solutes can be measured over a prolonged period of time and that the transport velocities can be determined. The disadvantage is that the spatial distribution of a tracer can only be estimated roughly, depending on the density of suction cups used. With a few exceptions, suction cups are usually installed vertically, and disrupt the soil to an unknown degree. While suction cups have demonstrated utility in extracting solution from an intact matrix, they might not be suitable for monitoring fast chemical movement in soils containing macropores, because preferential flow is likely to be missed (England, 1974; Shaffer et al., 1979). The interpretation of suction cup measurements poses some fundamental theoretical problems. Suction cups yield neither resident nor flux concentrations, and their sphere of influence is generally unknown. In the sense of the measurement concept as introduced by Marle (1967) suction cups have an ill-defined weighting function. Implications of this uncertainty for the calculation of mass recovery in a tracer experiment have been discussed by Flury (1993a).

Tile Drains

Tile drains measure flux concentrations at the depth of the tile. One disadvantage of tiles is that samples are obtained only when soil near the tile is saturated. For many situations where tile drains are in use, however, the time periods of tile flow are of greatest importance for field-scale chemical transport, particularly the nonequilibrium behavior. Initial soil disruption during tile installation is more drastic than for suction cups, but tiles are long-term installations and with time the disturbance may heal. Tiles measure a much larger soil volume than most other techniques, and are excellent tools for studying preferential flow phenomena.

Soil Cores

Soil cores provide information about solute distribution at a given time. As sampling is destructive, no subsequent measurements can be taken at the same location. For repeated measurements over a period of time, holes created by the cores have to be refilled to prevent preferential flow. Because they intercept such a small fraction of the soil volume, soil cores are likely to miss preferential flow pathways, and the interpretation of solute distribution may be difficult when flow occurs in irregular patterns (Ghodrati and Jury, 1990). Measurement with soil cores, however, yields well-defined resident concentrations.

Dye Tracing

The best estimate of the spatial distribution of a tracer can be obtained with application of a dye tracer and subsequent excavation of the soil. This technique provides a detailed view of flow patterns at a given time, but nothing can be said about transport velocities and solute concentrations. To date, dye tracer studies have been analyzed only qualitatively, but attempts to quantify dye tracer concentrations in soil material are underway (Schincariol et al., 1993; Aeby et al., 1997).

Estimation of Transport Parameters

Because various solute sampling methods monitor different soil volumes in their vicinity, the type of sampling used in a given experiment may affect the interpretation of the results. Only recently have several sets of devices been used simultaneously to determine their performance. Measurements with solution samplers and soil cores have been compared by Fleming and Butters (1995) and Ellsworth et al. (1996).

Fleming and Butters (1995) added a narrow pulse of Br to the soil surface and leached it by precipitation and irrigation, monitoring its progress by three suction cups at six depths from 0.15 to 2 m and 18 soil core samples of 2.23 cm diameter to a maximum depth of 3.7 m. Solution samples were taken at 46 dates, and soil cores at 4 dates. The two methods provided consistent estimates of the mean convective velocity, but inconsistent solute dispersion values. The authors explained this inconsistency as arising from incomplete mass balance.

A contrary result was reported by Ellsworth et al. (1996), who applied Br, Cl, and NO_3 to a field plot and leached the solutes by sprinkler irrigation. These authors sampled soil solution with two sets of 12 suction cups at 0.25 and 0.65 m-depth and compared the results with those from soil samples obtained by destructive sampling the entire 2×2×2 m soil block. Mass recovery and mean convective velocity estimated from solution samplers were less than those estimated from soil excavation. However, the two sampling methods provided similar estimates for the dispersion.

These contradictory findings clearly demonstrate the difficulties associated with sampling at the field scale, and show that there is no ideal measurement instrument for solute sampling in natural media. The choice of the appropriate instrument has to be based on the specific research goal.

SUMMARY

The studies reviewed in this chapter do not paint a bright picture for the future of solute transport studies in unsaturated field soils. Despite a considerable effort at field experimentation, there are few cases where consistent patterns have been revealed. With respect to dispersion, it seems apparent that neither convective-dispersive nor stochastic-convective transport occur in a manner that can be predicted from a knowledge of soil properties alone. It is reasonable to conclude that soil layering enhances lateral mixing, and thus acts to promote convective-dispersive transport. But at the same time, layering has been implicated in creating unstable flow, or at least encouraging the development of preferential flow channels, neither of which are compatible with convective-dispersive transport.

A major obstacle for the progress in solute transport research is that the current measurement techniques often only provide indirect evidence of transport phenomena. The spatial and temporal resolution of measurements is either too small or too large to unequivocally identify the relevant transport mechanisms. Until improved sampling and measurement methodologies have been developed, the challenge is to assemble the puzzle of indirect evidence to a coherent picture.

REFERENCES

Aeby, P., J. Forrer, C. Steinmeier, and H. Flühler. Image analysis for determination of dye tracer concentrations in sand columns, *Soil Sci. Soc. Am. J.*, 61:33–35, 1997.

Baker, J.L. and H.P. Johnson. Nitrate-nitrogen in tile drainage as affected by fertilization, *J. Environ. Qual.*, 10:519–522, 1981.

Beese, F. and P.J. Wierenga. Solute transport through soil with adsorption and root water uptake computed with a transient and a constant-flux model, *Soil Sci.,* 129:245–252, 1980.

Beven, K.J., D.E. Henderson, and A.D. Reeves. Dispersion parameters for undisturbed partially saturated soil. *J. Hydrol. (Amsterdam),* 143:19–43, 1993.

Biggar, J.W. and D.R. Nielsen. Spatial variability of the leaching characteristics of a field soil. *Water Resour. Res.,* 12:78–84, 1976.

Bottcher, A.B., E.J. Monke, and L.F. Huggins. Nutrient and sediment loadings form a subsurface drainage system, *Trans. ASAE,* 24:1221–1226, 1981.

Bouma, J. and L.W. Dekker. A case study on infiltration into dry clay soil. I. Morphological observations, *Geoderma,* 20:27–40, 1978.

Bowman, R.S. and R.C. Rice. Transport of conservative tracers in the field under intermittent flood irrigation, *Water Resour. Res.,* 22:1531–1536, 1986.

Brissaud, F., A. Pappalardo, and P. Couchat. Gamma neutron method applied to field measurements of hydrodynamic dispersion. *J. Hydrol. (Amsterdam),* 63:331–343, 1983.

Buhler, D.D., G.W. Randall, W.C. Koskinen, and D.L. Wyse. Atrazine and alachlor losses from subsurface tile drainage of a clay loam soil. *J. Environ. Qual.,* 22:583–588, 1993.

Butters, G.L. and W.A. Jury. Field scale transport of bromide in an unsaturated soil. 2. Dispersion modeling. *Water Resour. Res.,* 25:1583–1589, 1989.

Butters, G.L., W.A. Jury, and F.F. Ernst. Field scale transport of bromide in an unsaturated soil. 1. Experimental methodology and results. *Water Resour. Res.,* 25:1575–1581, 1989.

Buttle, J.M. and B.J. Harris. Hydrological pathways of metolachlor export from an agricultural watershed. *Water Air Soil Pollut.,* 60:315–335, 1991.

Clendening, L.D. and W.A. Jury. A field mass balance study of pesticide volatilization, leaching, and persistence. In *Long Range Transport of Pesticides,* Kurtz, D.A. (Ed.), Lewis Publishers, Boca Raton, FL, 1990, pp. 47–60.

Czapar, G.F., R.S. Kanwar, and R.S. Fawcett. Herbicide and tracer movement to field drainage tiles under simulated rainfall conditions. *Soil Tillage Res.,* 30:19–32, 1994.

Dagan, G. Solute transport in heterogeneous porous formations. *J. Fluid Mech.,* 145:151–177, 1984.

Dekker, L.W. and C.J. Ritsema. How water moves in a water repellent sandy soil. 1. Potential and actual water repellency. *Water Resour. Res.,* 30:2507–2517, 1994.

Dekker, L.W. and C.J. Ritsema. Fingerlike wetting patterns in two water-repellent loam soils. *J. Environ. Qual.,* 24:324–333, 1995.

Ellsworth, T.R. and W.A. Jury. A three-dimensional field study of solute transport through unsaturated, layered, porous media. 2. Characterization of vertical dispersion. *Water Resour. Res.,* 27:967–981, 1991.

Ellsworth, T.R., P.J. Shouse, T.H. Skaggs, J.A. Jobes, and J. Fargerlund. Solute transport in unsaturated soil: Experimental design, parameter estimation, and model discrimination. *Soil Sci. Soc. Am. J.,* 260:397–407, 1996.

Ellsworth, T.R., W.A. Jury, F.F. Ernst, and P.J. Shouse. A three-dimensional field study of solute transport through unsaturated, layered, porous media. 1. Methodology, mass recovery, and mean transport. *Water Resour. Res.,* 27:951–965, 1991.

England, C.B. Comment on 'A technique using porous cups for water sampling at any depth in the unsaturated zone' by Warren W. Wood. *Water Resour. Res.,* 10:1049, 1974.

Everts, C.J., R.S. Kanwar, J.E.C. Alexander, and S.C. Alexander. Comparison of tracer mobilities under laboratory and field conditions. *J. Environ. Qual.,* 18:491–498, 1989.

Fleming, J.B. and G.L. Butters. Bromide transport detection in tilled and nontilled soil: Solute samplers vs. soil cores. *Soil Sci. Soc. Am. J.,* 59:1207–1216, 1995.

Flühler, H., W. Durner, and M. Flury. Lateral solute mixing processes—A key for understanding field-scale transport of water and solutes. *Geoderma,* 70:165–183, 1996.

Flury, M. Sampling efficiency for mass recovery calculations, in *Soil Monitoring,* Schulin, R., A. Desaules, R. Webster, and B. von Steiger (Eds.), Birkhäuser, Basel, 1993a, pp. 187–199.

Flury M. Transport of Bromide and Chloride in a Sandy and a Loamy Soil, Ph.D. Diss. No. 10185, ETH Zürich, 1993b.

Flury, M., H. Flühler, W.A. Jury, and J. Leuenberger. Susceptibility of soils to preferential flow of water: A field study. *Water Resour. Res.,* 30:1945–1954, 1994.

Flury, M., J. Leuenberger, B. Studer, and H. Flühler. Transport of anions and herbicides in a loamy and a sandy field soil. *Water Resour. Res.,* 31:823–835, 1995.

Gelhar, L.W., A.L. Gutjahr, and R.L. Naff. Stochastic analysis of macrodispersion in a stratified aquifer. *Water Resour. Res.,* 15:1387–1397, 1979.

Gelhar, L.W., A. Mantoglou, C. Welty, and K.R. Rehfeldt. *A Review of Field-Scale Physical Solute Transport Processes in Saturated and Unsaturated Porous Media,* EA-4190, Research Project 2485-5, Electric Power Research Institute, Palo Alto, CA, 1985.

Gelhar, L.W., C. Welty, and K.R. Rehfeldt. A critical review of data on field-scale dispersion in aquifers. *Water Resour. Res.,* 28:1955–1974, 1992.

Ghodrati, M. and W.A. Jury. A field study using dyes to characterize preferential flow of water. *Soil Sci. Soc. Am. J.,* 54:1558–1563, 1990.

Ghodrati, M. and W.A. Jury. A field study of the effects of soil structure and irrigation method on preferential flow of pesticides in unsaturated soil. *J. Contam. Hydrol.,* 11:101–125, 1992.

Grochulska, J. and E.J. Kladivko. A two-region model of preferential flow of chemicals using a transfer function approach. *J. Environ. Qual.,* 23:498–507, 1994.

Hallberg, G.R., J.L. Baker, and G.W. Randall. Utility of tileline effluent studies to evaluate the impact of agricultural practices on ground water. *Proc. of Conference on Agricultural Impacts on Ground Water, Aug. 11–13, Omaha, NE,* National Water Well Assoc., Dublin, OH, 1986, pp. 298–326.

Hamlen, C.J. and R.G. Kachanowski. Field solute transport across a soil horizon boundary. *Soil Sci. Soc. Am. J.,* 56:1716–1720, 1992.

Hendrickx, J.M.H., L.W. Dekker, and O.H. Boersma. Unstable wetting fronts in water repellent field soils. *J. Environ. Qual.,* 22:109–118, 1993.

Jaynes, D.B. Field study of bromacil transport under continuous-flood irrigation. *Soil Sci. Soc. Am. J.,* 55:658–664, 1991.

Jaynes, D.B. and R.C. Rice. Transport of solutes as affected by irrigation method. *Soil Sci. Soc. Am. J.,* 57:1348–1353, 1993.

Jaynes, D.B., R.S. Bowman, and R.C. Rice. Transport of a conservative tracer in the field under continuous flood irrigation. *Soil Sci. Soc. Am. J.,* 52:618–624, 1988.

Jury, W.A. Solute travel-time estimates for tile-drained fields: 1. Theory. *Soil Sci. Soc. Am. J.,* 39:1020–1024, 1975a.

Jury, W.A. Solute travel-time estimates for tile-drained fields: 2. Application to experimental studies. *Soil Sci. Soc. Am. J.,* 30:1024–1028, 1975b.

Jury, W.A. Simulation of solute transport using a transfer function model. *Water Resour. Res.,* 18:363–368, 1982.

Jury, W.A. and G. Sposito. Field calibration and validation of solute transport models for the unsaturated zone. *Soil Sci. Soc. Am. J.,* 49:1331–1341, 1985.

Jury, W.A. and H. Flühler. Transport of chemicals through soil: Mechanisms, models, and field applications. *Adv. Agron.,* 47:141–201, 1992.

Jury, W.A. and K. Roth. *Transfer functions and solute movement through soil. Theory and applications.* Birkhäuser, Basel, 1990.

Jury, W.A., D. Russo, G. Sposito, and H. Elabd. The spatial variability of water and solute transport properties in unsaturated soil. *Hilgardia,* 55:1–32, 1987.

Jury, W.A., H. Elabd, and M. Resketo. Field study of napropamide movement through unsaturated soil. *Water Resour. Res.,* 22:749–755, 1986.

Jury, W.A., J.S. Dyson, and G.L. Butters. Transfer function model of field-scale solute transport under transient water flow. *Soil Sci. Soc. Am. J.,* 54:327–332, 1990.

Jury, W.A., L.H. Stolzy, and P. Shouse. A field test of the transfer function model for predicting solute transport. *Water Resour. Res.,* 18:369–375, 1982.

Kladivko, E.J., G.E. van Scoyoc, E.J. Monke, K.M. Oates, and W. Pask. Pesticide and nutrient movement into subsurface tile drains on a salt loam soil in Indiana. *J. Environ. Qual.,* 20:264–270, 1991.

Kladivko, E.J., R.F. Turco, J. Grochulska, G.E. van Scoyoc, J.D. Eigel, and E.J. Monke. Pesticide and nitrate transport through a silt loam soil into subsurface tile drains, in *Proc. Int. Soil Tillage Research Org. (ISTRO) 13th Int. Conf. July 24–29,* Jensen, H.E. (Ed.), Aalborg, Denmark, 1994, pp. 221–225.

Kung, K.-J.S. Preferential flow in a sandy vadose zone: 1. Field observation, *Geoderma,* 46:51–58, 1990.

Logan, T.J., G.W. Randall, and D.R. Timmons. Nutrient content of tile drainage from cropland in the North Central Region. *No. Central Reg. Res. Publ.,* 268:16, 1980.

Marle, C. Écoulements monophasiques en milieu poreux. *Rev. Inst. Fr. Pet.,* 12:1471–1509, 1967.

Milburn, P. and J. MacLeod. Considerations for tile drainage water quality studies in temperate regions. *Appl. Eng. Agric.,* 7:209–215, 1991.

Milburn, P., D.A. Leger, H. O'Neil, K. MacQuarrie, and J.E. Richards. Point and nonpoint source leaching of atrazine from a corn field: Effects on tile drainage water quality. *Can. Agric. Eng.,* 37:269–277, 1995.

Rice, R.C., R.S. Bowman, and D.B. Jaynes. Percolation of water below an irrigated field. *Soil Sci. Soc. Am. J.,* 50:855–859, 1986.

Richard, T.L. and T.S. Steenhuis. Tile drain sampling of preferential flow on a field scale. *J. Contam. Hydrol.,* 3:307–325, 1988.

Ritsema, C.J. and L.W. Dekker. Distribution flow: A general process in the top layer of water repellent soils. *Water Resour. Res.,* 31:1187–1200, 1995.

Ritsema, C.J., L.W. Dekker, J.M.H. Hendrickx, and W. Hamminga. Preferential flow mechanisms in a water repellent sandy soil. *Water Resour. Res.,* 29:2183–2193, 1993.

Roth, K., W.A. Jury, H. Flühler, and W. Attinger. Transport of chloride through an unsaturated field soil. *Water Resour. Res.,* 27:2533–2541, 1991.

Russo, D., W.A. Jury, and G.L. Butters. Numerical analysis of solute transport during transient irrigation. 1. The effect of hysteresis and profile heterogeneity. *Water Resour. Res.,* 25:2109–2118, 1989.

Saffigna, P.G., C.B. Tanner, and D.R. Keeney. Non-uniform infiltration under potato canopies caused by interception, stemflow, and hilling. *Agron. J.,* 68:337–342, 1976.

Schincariol, R.A., E.E. Herderick, and F.W. Schwartz. On the application of image analysis to determine concentration distributions in laboratory experiments. *J. Contam. Hydrol.,* 12:197–215, 1993.

Schulin, R., M.T. van Genuchten, H. Flühler, and P. Ferlin. An experimental study of solute transport in a stony field soil. *Water Resour. Res.,* 23:1785–1794, 1987.

Shaffer, K.A., D.D. Fritton, and D.E. Baker. Drainage water sampling in a wet, dual-pore system. *J. Environ. Qual.,* 8:241–246, 1979.

Simmons, C.S. A stochastic-convective transport representation of dispersion in one-dimensional porous media systems. *Water Resour. Res.,* 18:1193–1214, 1982.

Steenhuis, T.S., W. Staubitz, M.S. Andreini, J. Surface, T.L. Richard, R. Paulsen, N.B. Pickering, J.R. Hagerman, and L.D. Geohring. Preferential movement of pesticide and tracers in agricultural soils. *J. Irrigation Drainage Eng.,* 116:50–66, 1990.

Traub-Eberhard, U., W. Kördel, and W. Klein. Pesticide movement into subsurface drains on a loamy silt soil. *Chemosphere,* 28:273–284, 1994.

Utermann, J., E.J. Kladivko, and W.A. Jury. Evaluating pesticide migration in tile-drained soils with a transfer function model. *J. Environ. Qual.,* 19:707–714, 1990.

Van de Pol, R.M., P.J. Wierenga, and D.R. Nielsen. Solute movement in a field soil. *Soil Sci. Soc. Am. J.,* 41:10–13, 1977.

Van Ommen, H.C., M.T. van Genuchten, W.H. vanderMolen, R. Dijksma, and J. Hulshof. Experimental and theoretical analysis of solute transport from a diffusive source of pollution. *J. Hydrol. (Amsterdam),* 105:225–251, 1989.

Van Wesenbeeck, I.J. and R.G. Kachanowski. Spatial scale dependence of in situ solute transport. *Soil Sci. Soc. Am. J.,* 55:3–7, 1991.

Wallis, M.G. and D.J. Horne. Soil water repellency. *Adv. Soil Sci.,* 20:91–146, 1992.

Wierenga, P.J. Solute distribution profiles computed with steady-state and transient water movement models. *Soil Sci. Soc. Am. J.,* 41:1050–1055, 1977.

Wild, A. and L.A. Babiker. The asymmetric leaching pattern of nitrate and chloride in a loamy sand under field conditions. *J. Soil Sci.,* 27:460–466, 1976.

CHAPTER FOURTEEN

Density-Coupled Water Flow and Contaminant Transport in Soils

R.S. Mansell, J.H. Dane, D. Shinde, and H.H. Liu

INTRODUCTION

Fluid density *(ρ)* and viscosity *(μ)* of aqueous solutions are commonly assumed to be constants in most model simulations of water flow and solute transport in porous media. However, these fluid properties may depend heavily upon concentration, *C*, of chemical component or components, and to a lesser extent upon temperature, *T*, (Gebhart et al., 1988). For many groundwater contamination problems, density and viscosity may vary significantly with solute concentration, in both space and time. The ratio (μ/ρ) of these two important properties, $\mu(C,T)$ and $\rho(C,T)$, are often reported as a single parameter, kinematic viscosity $\gamma(C,T)$. Density- and viscosity-dependent solute transport has been reported for variably-saturated soils and for saturated geologic formations. Circulation of fluids around salt domes (Bennett and Hanor, 1987), leaching of salts from soils during irrigation (Mulqueen and Kirkham, 1972), flow of landfill leachate into groundwater aquifers (Koch and Zhang, 1992), generation of salt water in lakes due to evaporation (Wood and Ostercamp, 1987), are but a few important problems involving the flow of heterogeneous fluids in heterogeneous porous media.

In the vadose zone, solution ρ and μ may be altered by diurnal/seasonal fluctuations in soil temperature and by variations in salt concentration due to the concentrating effect of selective water removal by evaporation and transpiration processes. In soils, extraction of water by plant roots tends to increase solution phase ρ and μ as a consequence of increased salt concentration; whereas infiltration of irrigation or rainwater tends to dilute the soil solution (Kutilek and Nielsen, 1994). In arid or semiarid regions with low annual rainfall, high evaporative demand, and shallow depth to groundwater, such as many irrigated areas of Australia and the western United States, salts tend to accumulate near the soil surface due to evapotranspiration. Periodic flushing with low-salt water is sometimes used to control the accumulation of salts. However, in humid regions with high annual rainfall, such accumulated salts are periodically flushed downward as infiltrated rainwater percolates into saturated groundwater zones of unconfined aquifers. Point sources of contamination, such as land disposal of municipal sewage sludge, and nonpoint sources, such as applications of fertilizer to agricultural fields, also may cause localized increases in salt concentrations within the vadose zone. Total dissolved solids concentrations for leachate from sanitary landfills typically range between 5,000 and 40,000 mg/L, resulting in a range of approximately 0.5 to 4.0% for relative fluid density differences $[\rho_l - \rho_o]/\rho_o$ or $\Delta\rho/\rho_o$ between the leachate density ρ_l and the ambient groundwater density ρ_o (Paschke and Hoopes, 1984).

Accidental spillage of light nonaqueous phase liquids (LNAPL), such as gasoline, and dense nonaqueous phase liquids (DNAPL), such as polychlorinated biphenyls (PCB), over land areas also may introduce density- and viscosity-driven fluid flow in the vadose zone. Injection of aqueous solutions of surfactants and cosurfactants into subsurface porous media to remove residual NAPLs may also result in density- and viscosity-dependent flow due to differences in fluid properties of the flushing solution, which often contains microemulsions and the ambient solution phase (Mansell et al., 1996).

Recent laboratory investigations of cosolvent flushing for removal of residual nonaqueous phase liquids (NAPL) from unconfined aquifers by Jawitz et al. (1997) show that the presence of a capillary fringe is an important factor in displacement efficiency. Steady horizontal flow was initially maintained in a two-dimensional flow container packed with silica sand prior to cosolvent displacement of resident water followed by water displacement of cosolvent. A lighter-than-water cosolvent (20% ethanol/80% water, $\rho = 0.97$ g cm^{-3} and $\mu = 1.61$ cp) was used in the experiments. During cosolvent displacement of water ($\rho_o = 0.997$ g cm^{-3} and $\mu_o = 0.895$ cp), the cosolvent tended to override the resident water and buoyant forces acted to preferentially move the cosolvent into the capillary fringe. During subsequent water flooding, the denser water tended to flow under the resident cosolvent so that the cosolvent became effectively entrapped within the capillary fringe, resulting in inefficient removal of cosolvent from the aquifer. Displacement of resident water in the sand with cosolvent tended to collapse the maximum thickness of the capillary fringe from 9 to 5 cm due to a large difference in air-liquid surface tension $\sigma = 40$ dynes cm^{-1} for the cosolvent relative to that for water ($\sigma_o = 72$ dynes cm^{-1}). Even though steady flow was initially enforced, temporal and spatial changes in capillary fringe thickness imparted a slight transient behavior to fluid flow within the porous medium during the displacements.

FLOW INSTABILITY

Instability during liquid flow through porous media can be attributable to fluid heterogeneity as well as to heterogeneity within the media. During flow through a porous medium, small flow disturbances are continuously generated due to heterogeneities of the medium (Moissis and Wheeler, 1990)

Instability Due to Media Heterogeneity

The stability of the interface between two miscible fluids in a porous medium is a topic of active research in many disciplines. Heterogeneity of porous media such as textural or structural layering can produce flow instability even for conditions of constant fluid density and viscosity, as in homogeneous fluids. Flows are considered as either being stabilized or destabilized by viscosity and density differences, with the interfacial velocity multiplying the viscosity difference (Glass et al., 1989). Fingers have been observed to develop under long-term ponded infiltration into an initially dry, layered soil system comprised of fine-textured material over coarse-textured material (Glass et al., 1989). The downward-moving wetting front in the top fine layer becomes unstable as it passes into the bottom coarse layer, causing the formation of viscous fingers in the coarse layer, due to textural- and structural-dependent flow instability.

Flow instability has been reported for infiltration of homogeneous liquids such as water into dry, seemingly homogeneous soils (Glass et al., 1989). This demonstrates that the concept of homogeneous media is basically hypothetical since microscale heterogeneity contributes to such flow instability. Unstable wetting fronts may cause preferential flow-paths or fingers in homogeneous soils that have neither macropores nor large variability of hydraulic properties. Such instabilities tend to occur under several conditions: (1) infiltration of ponded water with compression of air ahead of the wetting front, (2) redistribution of water in the profile, (3) water-repellent porous material (Hendrickx et al., 1993; Ritsema et al., 1993), (4) increasing water content with depth, and (5) continuous nonponding infiltration (Yao and Hendrickx, 1996). Since unstable wetting is a gravity-driven phenomenon, instabilities do not tend to occur during early stages of infiltration when capillary forces dominate over gravity. Coarse sands appear to be more sensitive to unstable wetting than finer sands. During water application with a sprinkler to small lysimeters, Yao and Hendrickx (1996) reported small diameter fingers (3–4 cm) in coarse sand and large diameter fingers (12 cm) in fine sand. For infiltration rates between 0.12 and 0.30 cm h^{-1}, finger diameters in the coarse sand remained more or less constant. However, wetting fronts were stable in the coarse sand for infiltration rates less than 0.12 cm h^{-1}. These results explain why unstable wetting is rarely observed in water-wettable field soils.

In homogeneous soils, when two superposed solutions (i.e., displacing and displaced) of unequal ρ are accelerated in one dimension across their interface, the moving front may be stable or unstable. Differences in ρ provide unbalanced forces and differences in μ account for unequal drag forces (Kutilek and Nielsen, 1994). During miscible displacement in porous medium, the stability of the moving interface between a displacing and a displaced solution depends upon the fluid densities, viscosities, intrinsic permeability of the medium (k), pore velocity of the displacement (υ), and direction of the displacement. For vertically downward flow, the moving interface between displacing and displaced solutions tends to be unstable if

$$k[\rho_2 - \rho_1]g < [\mu_2 - \mu_1]\upsilon \tag{14.1}$$

where subscripts *1* and *2* refer to the displaced and displacing solutions. When a dense, more viscous fluid displaces a less dense, less viscous fluid during vertically downward flow, instability may occur for particular velocities. Unbalanced forces in this case tend to accelerate the denser fluid into the underlying less dense fluid, but the viscous drag of the underlying fluid is unable to counterbalance this acceleration. According to Eq. 14.1, "fingers" of the more viscous fluid tend to invade those pore sequences occupied by the less dense fluid. Fingers of water moving ahead of the average wetting front into unsaturated soil may also develop during immiscible displacement of soil air by infiltrating rainwater (Kutilek and Nielsen, 1994). It is important to note here that the influence of soil structure and microscale heterogeneity within the pore space geometry are not explicitly considered in Eq. 14.1.

Unstable flow has been reported for groundwater contamination problems involving a dense plume becoming enclosed by and moving in an ambient body of less dense fluid (Schincariol and Schwartz, 1990; Oostrom et al., 1992a; Koch and Zhang, 1992). For contaminant plumes moving predominantly horizontally with groundwater flow, lobes of dense fluid move downward from the plume and the less dense ambient fluid moves upward into the plume to balance the movement of dense fluid. Current knowledge of the behavior of variable-density contaminant plumes is incomplete and is based primarily upon laboratory investigations involving one-dimensional columns and two-dimensional flow chambers (Oostrom et al., 1992b).

Published results of field studies for landfill leachate flow into aquifers are scarce (Kimmel and Braids, 1980).

Displacement of a fluid in a porous medium by a less viscous fluid is generally unstable (Moissis and Wheeler, 1990). Small perturbations in the flow tend to grow with time, producing large protrusions of the displacing fluid into the displaced fluid. Such protrusions are called fingers and this type of instability is referred to as viscous fingering.

Instabilities can occur during the migration of dense contaminant plumes in saturated porous media, significantly enhancing plume mixing (Schincariol et al., 1997). The stability of groundwater flow is reflected by how the flow system responds to small perturbations in pressure, solute concentration, and/or temperature. Perturbations in solute concentration can occur due to salt-laden leachate inflow from an unlined landfill. Disposal of industrial sources of hot water can cause temperature perturbations. Stable flow occurs when a perturbation decays in time or the position of the system tends to remain at its initial state. Unstable flow occurs when a perturbation grows or continues to displace the position of the system from its initial state (Schincariol et al., 1994). Heterogeneities in either the porous media or the fluid may contribute to unstable flow. With time, fluid mixing within unstable systems eventually produces stable density or viscosity gradients. Interfacial disturbances are continuously generated for flow through natural porous media due to heterogeneities in the medium. With miscible fluids, hydrodynamic dispersion tends to enhance flow stability, whereas with immiscible fluids, capillary effects tend to enhance flow stability (Schincariol et al., 1994). Situations with a more dense fluid overlying a less dense fluid tend to be potentially unstable and the greater the density difference, the more likely that instabilities will occur (List, 1965). The dimensionless parameters, the transverse λ_T, and longitudinal λ_L Raleigh numbers were given by List (1965) as

$$\lambda_T = \frac{gkl(\rho_I - \rho_O)}{\varepsilon\mu}\frac{1}{D_T} \tag{14.2}$$

$$\lambda_L = \frac{gkl(\rho_I - \rho_O)}{\varepsilon\mu}\frac{1}{D_L} \tag{14.3}$$

where D_T is the transverse dispersion coefficient, D_L is the longitudinal dispersion coefficient, ε is porosity, k is intrinsic permeability, and g is the acceleration of gravity. The ratio of Eq. 14.2 to 14.1 is also important since $\lambda_l/\lambda_T = D_T/D_L$. The symbol l is a characteristic length defined as

$$l = \left[\frac{\pi D_T X}{U}\right]^{1/2} \tag{14.4}$$

where X is the distance from the source where local stability is to be evaluated, and U is the Darcy flux. The dimensionless wave number of the disturbance ω,

$$\omega = \frac{2\pi l}{L} \tag{14.5}$$

where L is the wavelength of the unstable wave generated at the source, and was used by List (1965) to determine the stability of the interface between the plume and the ambient groundwater. His analysis demonstrated that instability growth was favored in systems with minimal dispersion and small wave numbers. Model parametric analysis by Schincariol et al. (1997) confirmed earlier stability theory by List (1965) that instability growth is promoted by porous media with large permeability values and systems characterized by large density contrasts. Instabilities were also shown to dissipate within zones of lower permeability. Stability was shown to be promoted by low to medium permeability, small density differences, and significant dispersion. Although anisotropy and heterogeneities on all scales tend to dampen instability growth, heterogeneity of the medium was concluded to provide the most likely source of initial perturbations to contaminant plumes. Schincariol et al. (1994) suggest that all stratified systems are inherently unstable. For small density contrasts, a flow system with no dispersive capabilities would probably be needed to demonstrate instability. In flow situations where a more dense, viscous fluid downwardly displaces a less dense, viscous fluid, density effects tend to be destabilizing whereas viscosity effects are stabilizing (Schincariol et al., 1997).

The stability of dense miscible plumes in homogeneous porous media has been shown to depend upon the magnitude of the relative density difference between the contaminant solution and the ambient groundwater (Schincariol and Schwartz, 1990; Oostrom et al., 1992a,b; Hayworth, 1993; and Liu and Dane, 1996a). Unstable plumes cause very large mixing zones that cannot be described by hydrodynamic dispersion alone. Schincariol and Schwartz (1990) demonstrated the development of flow instability with time (24, 36, 72, and 120 hours) after initiating an upstream injection of 2000 mg L^{-1} NaCl (1.0015 specific gravity at 20°C) during steady horizontal water flow in a two-dimensional flow container packed with a homogeneous porous media (Figure 14.1). The influence of source concentration in homogeneous media are shown in Figure 14.2 for 1,000 (1.0008 specific gravity), 2,000 (1.0015 specific gravity), and 10,000 (1.0071 specific gravity) mg L^{-1} NaCl after 54 hours. Heterogeneity in either or both the porous medium and the fluid may contribute to unstable flow. For two-dimensional flow, two dimensionless numbers, π_1 and R_a^*, were proposed by Güven et al. (1992) and Oostrom et al. (1992a) to predict the onset of gravitational instability during the movement of dense plumes in aquifers

$$R_a^* = \frac{H_{dp}}{\varepsilon D_T} K_{sat} \Delta\rho / \rho_o \tag{14.6}$$

$$\pi_1 = K_{sat} \frac{\Delta\rho / \rho_o}{q_x} \tag{14.7}$$

where K_{sat} is the saturated hydraulic conductivity in the vertical direction, $\Delta\rho = \rho_l - \rho_o$ is the density difference, q_x is the horizontal Darcy flux, R_a^* is a modified Rayleigh number, $H_{dp} = (Q_l/Q_x)h_s$ is the plume thickness, Q_L. is the leakage rate, Q_x is the total flow rate, and h_s is the thickness of the saturated zone. An analytical expression for the Rayleigh number (R_a) corresponding to one-dimensional vertical flow was given earlier by Wooding (1959) as

$$R_a = K_{sat} \frac{b^2}{D_L} \frac{\partial \rho}{\partial z} \tag{14.8}$$

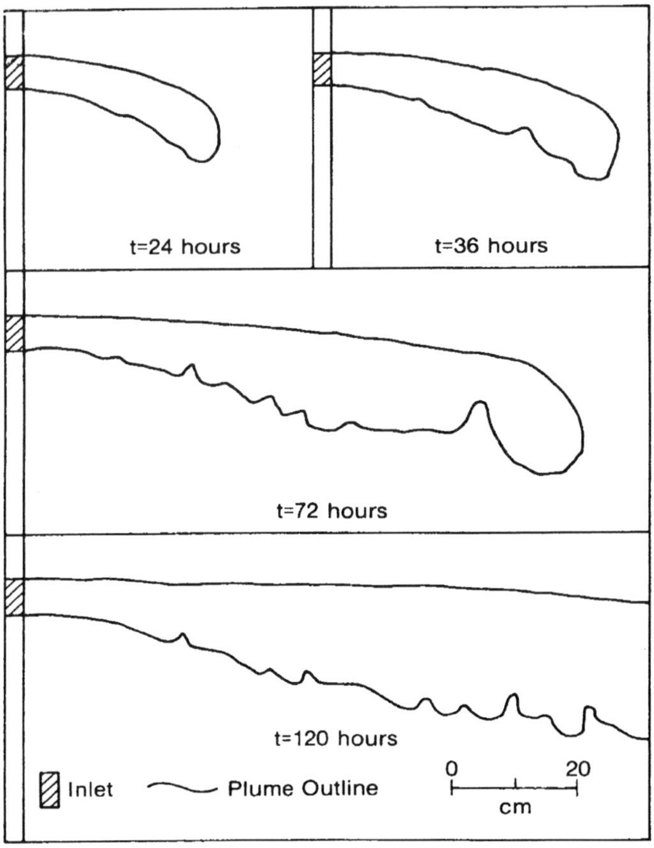

Figure 14.1. Instability development over time for a 2000-mg/L NaCl source in homogenous medium for four times (Schincariol and Schwartz, 1990).

where b is the radius of the cylindrical column and $\partial\rho/\partial z$ is a uniform upward vertical gradient for fluid density. The expression by Wooding (1959) applies to a long, vertical, homogeneous porous medium column saturated with pure water, closed at the bottom, and connected at the top to an open reservoir containing a dense aqueous solution ($\rho_l > \rho_o$). R_a was used to determine when free convection begins. Dense NaI or NaBr solutions were introduced by Oostrom et al. (1992a) from a line source located on top of a porous medium during experiments involving horizontal flow of groundwater in a two-dimensional flow container. The onset of plume instability occurred for $\pi_l = 0.3$, but a clearly defined transition point did not occur for R_a^*. The insensitivity of R_a^* to the onset of instability was attributed in part to the uncertainty in the measured value of D_T. Under most conditions, plumes were reported to be unstable for π_l values exceeding 0.35; however, for several specific cases stable plumes were reported for stability for $\pi_l > 0.35$ (Dane et al., 1994). Unstable plumes were characterized by the development of transient, three-dimensional, lobe-shaped perturbations at the bottom of the plume and also within the plume (Oostrom et al., 1992a). Such instabilities lead to enhanced mixing, dilution of contaminants, and the ultimate contamination of larger areas of aquifers. The transition from stable to unstable plume behavior also occurred for $\pi_l > 0.3$ in variable density flow experiments in homogeneous, layered, and lenticular porous media by Schincariol and Schwartz (1990). Contaminant solutions were injected into the porous medium from a source extending over a portion of

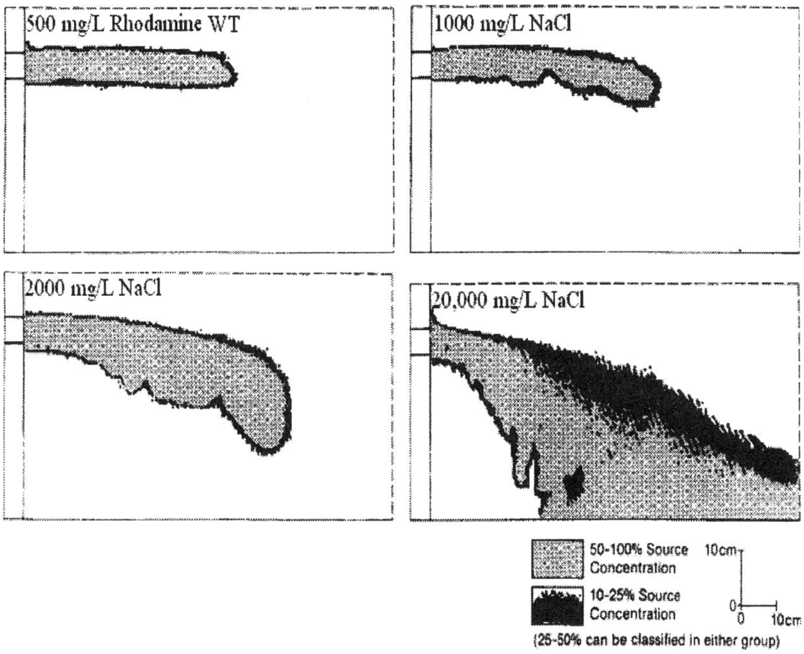

Figure 14.2. Binary images for various source concentrations in homogenous medium (Schineariol and Schwartz, 1990).

the upstream end of the flow container. Density differences as low as 0.0008 g cm^{-3} were observed to produce gravitational instabilities. Liu and Dane (1996a) utilized a criterion based upon the π_1 number and a modified Peclet number, P_e^*,

$$P_e^* = q_X \frac{\sqrt{k_Z}}{D_o} \qquad (14.9)$$

to predict gravitational instabilities in transverse mixing zones at the bottom of dense plumes in porous media where k_z is the intrinsic permeability in the vertical direction, and D_o is the molecular diffusion coefficient for the solute or salt. Experiments were performed in a two-dimensional flow container with imposed steady horizontal flow where dense NaI solutions were introduced from a line source on top of the porous media. The magnitude of P_e^* reflects the relative importance of mechanical dispersion and molecular diffusion. Although molecular diffusion generally plays a relatively minor role in overall mixing, Liu and Dane (1996a) indicate that molecular diffusion prevents the occurrence and development of small-scale instabilities. Heterogeneities in subsurface porous media exist at many scales and may serve as perturbations to generate experimentally observed gravitational instabilities. Heterogeneity of either or both subsurface porous media and fluids within porous media can result in flow instability. Even in homogeneous porous media, relatively small-scale heterogeneities may be important factors to cause instabilities (Dane et al., 1994). The role of transversal dispersivity appears to be more complicated than originally believed. The transverse dispersion coefficient, D_T, was proposed to play a dual role in the onset of gravitational instability: A large value for D_T should increase the potential for instability since it corresponds to a higher degree of heterogeneity, but a large value should also decrease the potential for instability by decreasing the relatively large-scale concentration gradient in the mixing zone. These results support previous indications that dispersion is a poorly understood phenomenon. Liu and Dane (1996) reported that the use of π_1 alone was sufficient for predicting leading edge instability in two-dimensional flow experiments. They postulated that the effect of small-scale perturbations on instability becomes insignificant at the leading edge because of relatively fast changes in flow and transport patterns near the edge. Effects of the dimensionality of flow domain on unstable behavior of dense miscible plumes has been numerically investigated by Liu and Dane (1997). They found that unstable behavior of dense plumes in three-dimensional (3D) domains are considerably different from that in the corresponding two-dimensional (2D) domains. They also stated that 3D dense plumes are less stable in cross-sections of flow domain, which are perpendicular to the main flow direction, than those parallel to the flow direction. Their work implies that 2D models may be less than adequate for investigating unstable flow and mixing in the subsurface, where porous media are actually three-dimensional in nature.

MATHEMATICAL DESCRIPTION OF DENSITY-DEPENDENT FLOW IN POROUS MEDIA

A local disturbance in fluid density in an extensive and quiescent ambient environment results in a gravitational buoyancy force described (Gebhart et al., 1988) as

$$B = g[\rho_o - \rho_l] = -g\Delta\rho \tag{14.10}$$

Therefore, if regions exist with fluids of different densities, upward and downward fluid motion will result due to positive and negative buoyancy forces, respectively. The gravitational force is, of course, vertical and always downward. Thus the buoyancy force B depends upon local differences in concentration $\Delta C = C_o - C$ and temperature $\Delta T = T_o - T$ where C_o and T_o are ambient concentration and temperature, respectively. Density of aqueous salt solutions tend to increase with C and decrease with T. Large values of B may result in conditions of instability. For example, consider a horizontal fluid layer located in a container between upper and lower bounding surfaces at different temperatures T_1 and T_2, respectively. If $T_2 > T_1$, the fluid layer is unstably stratified if ρ of the specific fluid decreases with T (Bear, 1972). Even a small disturbance may result in convective currents due to buoyancy; i.e., the tendency of the less dense fluid to rise. Local variations in salt concentration and/or temperature may provide spatial gradients in fluid density. Molecular diffusion and thermal conduction tend to counteract the rise of convective currents. If local inequalities in fluid density are maintained by external causes such as the leaching of chemical components from municipal landfills, convective currents appear in spite of the counteracting effect of diffusion and conduction (Bear, 1972). Total leachate concentrations of up to 50,000 mg L^{-1} have been reported for unregulated landfills (Koch and Zhang, 1992). Such concentrations were shown to provide density effects.

Assuming water to be incompressible, Bear (1972) provided a modified form of Darcy's equation to describe the influence of convective currents due to gravitational buoyancy on water flux in porous media

$$q_i = -\frac{k_{ij}\rho_o g}{\mu}\left[\frac{\partial H}{\partial x_j} + \left(\frac{\Delta\rho}{\rho_o}\right)\frac{\partial z}{\partial x_j}\right] \tag{14.11}$$

where hydraulic head $H(x,y,z,t) = P(x,y,z,t)/[\rho_o g] + z = h(x,y,z,t) + z$, $P = \rho_o gh$ is fluid pressure, h is fluid pressure head, and k_{ij} is the intrinsic permeability tensor for the porous media. Hydraulic head in Eq. 14.11 can be expanded to give the equation form

$$q_i = -\frac{k_{ij}\rho_o g}{\mu}\left[\frac{\partial h}{\partial x_j} + \left(\frac{\Delta\rho}{\rho_o} + 1\right)\frac{\partial z}{\partial x_j}\right] \tag{14.12}$$

Eq. 14.12 provides expressions for horizontal *(x)*, lateral *(y)*, and vertical *(z)* flux components

$$q_x = -\frac{k_x\rho_o g}{\mu}\left[\frac{\partial h}{\partial x}\right] \tag{14.13}$$

$$q_y = -\frac{k_y\rho_o g}{\mu}\left[\frac{\partial h}{\partial y}\right] \tag{14.14}$$

$$q_z = -\frac{k_z\rho_o g}{\mu}\left[\frac{\partial h}{\partial z} + 1 + \frac{\Delta\rho}{\rho_o}\right] \tag{14.15}$$

respectively, where $\rho(C,T)$ varies with space because of differences in C and/or T. The term $\Delta\rho/\rho_o$ in Eqs. 14.11, 14.12, and 14.15 account for gravitational buoyancy. If the ratio of the second and first terms of Eq. 14.11 greatly exceeds unity, i.e., $(\Delta\rho/\rho_o)/(\partial H/\partial z) >> 1$, then free convection driven by buoyancy dominates the flow regime. If $(\Delta\rho/\rho_o)/(\partial H/\partial z) << 1$, then forced convection, driven by hydraulic head gradient, dominates the flow regime (Bear, 1972). Examination of Eqs. 14.11–14.15 reveals that gravitational buoyancy is assumed to directly influence only the vertical flux component, whereas changes in density and/or viscosity directly influence hydraulic conductivity for all three major directions. Since hydraulic conductivity can be expressed as $K(h) = \rho gk(h)/\mu = gk(h)/\nu$ and if $K_o(h) = \rho_o gk(h)/\mu_o = gk(h)/\nu_o$ is chosen as a reference hydraulic conductivity corresponding to ρ_o and μ_o fluid properties, then the relative hydraulic conductivity is $K(h)/K_o(h) = [\rho/\rho_o]/[\mu/\mu_o] = \nu_o/\nu$. Thus, hydraulic conductivity tends to increase for situations when the relative density increase exceeds the relative viscosity increase and tends to decrease when the viscosity increase exceeds the density increase.

Eqs. 14.13 to 14.15 can be rewritten using the reference hydraulic conductivity $K_0(h) = k_{ij}\ \rho_o g/\mu_o$ corresponding to reference density ρ_o and viscosity μ_o parameters where $\omega(C,T) = \mu_o/\mu$ is a correction factor for viscosity,

$$q_x = -K_{xo}(h)\omega(C,T)\left[\frac{\partial h}{\partial x}\right] \tag{14.16}$$

$$q_y = -K_{yo}(h)\omega(C,T)\left[\frac{\partial h}{\partial y}\right] \tag{14.17}$$

$$q_z = -K_o(h)\omega(C,T)\left[\frac{\partial h}{\partial z} + 1 + \frac{\Delta\rho}{\rho_o}\right] \quad (14.18)$$

respectively. The conservation equation for density-driven fluid flow using volume balance in porous media has been expressed by Liu and Dane (1996) as

$$\frac{\partial\theta}{\partial t} = -\nabla\cdot(q) + Q_w \quad (14.19)$$

where ρ does not appear, θ is volumetric water content, and Q_w is a volumetric water sink/source term. Assuming incompressibility of both the fluid and the porous medium, this equation can be further expressed in terms of pressure head h

$$\psi(h)\frac{\partial h}{\partial t} = -\nabla\cdot(q) \quad (14.20)$$

where $\psi(h) = \partial\theta/\partial h$ is the specific water storage capacity for the porous media and the sink/source term has been omitted. Utilizing flux Eqs. 14.16–14.20 yields

$$\begin{aligned}\psi(h)\frac{\partial h}{\partial t} &= \frac{\partial}{\partial x}\left[K_{xo}(h)\,\omega(C,T)\frac{\partial h}{\partial x}\right] + \frac{\partial}{\partial y}\left[K_{yo}(h)\,\omega(C,T)\frac{\partial h}{\partial y}\right]\\ &+ \frac{\partial}{\partial z}\left[K_{zo}(h)\,\omega(C,T)\frac{\partial h}{\partial z}\right] + \frac{\partial}{\partial z}\left[K_{zo}(h)\,\omega(C,T)\right]\\ &+ \frac{\partial}{\partial z}\left[K_{zo}(h)\,\omega(C,T)\left(\frac{\Delta\rho}{\rho_o}\right)\right]\end{aligned} \quad (14.21)$$

where the last term on the right-hand side represents the influence of spatially variable density.

MATHEMATICAL DESCRIPTION OF SOLUTE TRANSPORT IN POROUS MEDIA DURING FLOW OF HETEROGENEOUS FLUIDS

Fluid flow and solute transport processes become coupled for heterogeneous fluids with variable density and viscosity (Gambolati et al., 1993). Upconing of groundwater toward pumping wells and seawater intrusion are common causes of water quality degradation in coastal aquifers (Galeati et al., 1992). Due to hydrodynamic dispersion, a mixing zone typically develops across the interface between freshwater and seawater. Many models disregard the mixing zone and replace it with a sharp interface so that seawater and freshwater are treated as distinct "immiscible" fluids (Bennett and Hanor, 1987). Such simplified models assume that fluid flow and solute transport equations are totally uncoupled with regard to fluid properties. When vertical flow and/or dispersion become important, the sharp interface, or uncoupled approach, may not be adequate (Galeati et al., 1992). Thus, density-dependent transport of salt in unconfined aquifers generally requires solving fluid flow and salt transport equations that are coupled through fluid density and viscosity, properties that are dependent upon the contaminant concentration C. In water-unsaturated porous media, the flow equation is nonlinear due to the dependence of (i) the hydraulic conductivity $K(\theta[h])$ on volumetric liquid content, θ, or pressure head, h, (ii) the dependence of density $\rho(C)$ upon C, and (iii) the dependence of viscosity $\mu(C)$ upon C. The flow equation is first solved numerically for h. The liquid Darcy flux, q, and θ are then calculated using results from the solution of the flow equation. The q and θ values are needed in the subsequent solution of the transport equation, which provide values for C and therefore for $\rho(C)$ and $\mu(C)$ needed for the solution of the flow equation. However, the coupled problem of heterogeneous fluid flow and salt transport in porous media is highly nonlinear compared to normal groundwater flow. Large fluctuations in salt concentrations provide changes in the aqueous phase density which strongly affect the flow. Iteration is commonly used to numerically solve nonlinear problems (Herbert et al., 1988). Galeati et al. (1992) used uncoupled, partially- and complete-coupling model modes to simulate the effect of "dewatering" (i.e., lowering the water table by pumping to allow construction of a canal) pumping on seawater intrusion within a vertical cross section through an unconfined aquifer. Partial coupling provided results very close to those obtained by complete coupling but at great savings in computer time. The decoupled approach, with a sharp interface between freshwater and seawater, provided substantially different results from those obtained with partial and full coupling.

A convective-dispersive equation for transport of salt or an inert solute during the flow of heterogeneous fluids in porous media was given by Bear (1972)

$$\frac{\partial C}{\partial t} = -\frac{1}{\theta}\nabla \cdot (D\theta \nabla C) - \upsilon \cdot \nabla C + Q_s \tag{14.22}$$

where $\upsilon = q/\theta$ is pore water velocity, D is the general hydrodynamic dispersion tensor, and Q_s is a solute sink/source term. The D tensor is defined as

$$D = D_o + \alpha_T \upsilon \delta_{ij} + (\alpha_L - \alpha_T)\frac{\upsilon_i \upsilon_j}{|\upsilon|} \tag{14.23}$$

where δ_{ij} is the Kronecker Delta function, α_T is transversal dispersivity, and α_L is longitudinal dispersivity (Bear, 1972). The fluid flow and solute transport equations are coupled with regard to both fluid density $\rho(C,T)$ and viscosity $\mu(C,T)$. For isothermal conditions Koch and Zhang (1992) used linear equations of state for aqueous salt solutions

$$\begin{aligned} \rho &= \rho_o + \xi C \\ \mu &= \mu_o + \chi C \end{aligned} \tag{14.24}$$

where ξ and χ are experimental constants. During experimental investigations of flow instability by introducing a concentrated NaBr solution into columns of sand saturated with distilled water, Liu and Dane (1996a) used values of ρ_o = 1000 kg m^{-3} for distilled water and ξ = 0.7 for calculating $\rho(C)$ as a linear function of the NaBr concentration in aqueous solution. The density of the concentrated salt solution influent was 1028 kg m^{-3} for C_{max} = 40 g L^{-1}. The transport of multispecies contaminant plumes in groundwater systems with variable fluid density and viscosity requires a more complicated theory to determine fluid properties of mixtures. Zhang and Schwartz (1995) proposed the following equations to describe density (ρ_{mix}) and viscosity (μ_{mix}) for aqueous mixtures involving common metallic ions

$$\rho_{mix} = \rho_w \left[1 - \sum_{k=1}^{n} \frac{c_k}{c_k^{ref}} \right] + \sum_{k=1}^{n} \frac{c_k}{c_k^{ref}} \rho_k^{ref} \tag{14.25}$$

$$\mu_{mix} = \sum_{k=1}^{n} \omega_k \ln \mu_k^{ref} + \left[1 - \sum_{k=1}^{n} \omega_k \right] \ln \mu_w^{ref} \tag{14.26}$$

where ρ_w is the density of pure water, n is the number of components, c_k is the concentration of component species k, c_k^{ref} is some upper reference concentration (normally that value is chosen to be much larger than the maximum c_k) of species k with a known density of ρ_k^{ref}, μ_k^{ref} is the viscosity of water with species k at reference concentration c_k^{ref}, ω_k is the weight fraction of species k, and μ_w^{ref} is the viscosity of pure water. For aqueous mixtures of NaCl these equations gave small relative errors of 0.01 and 0.2%, respectively, for density and viscosity estimates.

For situations where contaminant concentrations in the solution phase for porous media are relatively small, so that fluid properties are not greatly altered, mechanisms such as adsorption/desorption, cation exchange, precipitation, and microbial degradation are well known to potentially alter contaminant transport during water flow through the vadose zone. However, retention and transformation mechanisms were omitted from consideration here since contaminant concentrations in the solution phase were sufficiently large as to alter properties of the fluid.

MODELS AND APPLICATIONS

Forkel and Celia (1992) used a modified Picard iteration technique to approximate the nonlinear fluid flow equation and a simple Picard iteration technique to numerically solve the density- and viscosity-coupled flow and transport equation for transient conditions. An inner-iteration was thus used for the flow numerical solution and an outer-iteration for the coupled flow-transport system, all within the same time step.

A closed-form or exact solution for the simplified case of steady-state, density-dependent fluid flow and solute transport for vertical columns of homogeneous and nonhomogeneous soil was recently presented by Park (1996). Convection, molecular diffusion, and velocity-dependent dispersion transport mechanisms were included. Groundwater density was assumed to be a linear function of the solute concentration. Exact solutions for relatively simple cases can be used to verify numerical models designed to describe more complex density-dependent flow/transport cases.

Forkel and Celia (1992) utilized a one-dimensional model to investigate the influence of density- and viscosity-dependence upon coupled fluid flow and contaminant transport in unsaturated porous media. They used highly concentrated seawater, with a constant ρ and μ, as the leachate solution. Saltwater infiltration was simulated during (i) steady state flow (q = 7.8 cm h^{-1}), (ii) transient flow, and (iii) nearly hydrostatic conditions. For the steady-state flow case, the volumetric water content was 0.152 at all depths. For the transient flow case, a pressure head of –50 cm was imposed initially at all depths. At the beginning of the transient-flow simulation, the pressure head at the surface was changed to –1 cm. At the beginning of the simulations, saltwater concentrations of 3.25×10^{-7}, 1.5×10^{-3}, and 1.5×10^{-4} kg cm^{-3} were imposed at the top of the porous media for the steady flow, transient flow, and nearly hydrostatic cases, respectively. Zero concentrations were imposed initially for each case. For steady saltwater infiltration, concentration fronts simulated with variable- and constant-fluid properties moved with the same velocity. This was attributed to two opposing effects at work in the vicinity of the advancing concentration front—a lowering of the unsaturated hydraulic conductivity because of the increasing viscosity, and an increasing conductivity due to the increasing saturation of the porous medium. For transient saltwater infiltration, the primary driving force for saline water movement was the pressure gradient. Salt concentration fronts simulated with constant fluid properties moved faster than those with variable properties, but only slightly slower than those with variable density and constant viscosity. For the transient case, variable viscosity tended to retard salt transport more than for the nearly hydrostatic flow case. For the nearly hydrostatic case, density of the saltwater became the driving force so that pressure-driven flow was minimal. However, both density and viscosity were shown to have significant influence on the results for the initially hydrostatic case. The movement of the salt concentration front was faster for simulations utilizing variable density and viscosity compared to simulations with constant fluid properties.

Simulations by Koch and Zhang (1992) for density-driven flow in an unconfined aquifer that receives leachate infiltration from a landfill showed that relative fluid density differences larger than 0.5% affect the direction of plume migration even for short time periods. Density effects were most pronounced for situations involving small dispersivities. Homogeneous, isotropic porous media tends to have smaller dispersivity values compared to heterogeneous porous media. The larger the dispersivity value, the larger the plume spreading, and the smaller the overall influence of density variations on plume behavior. Density differences ranging from 0.3 to 0.5% had a negligible short-term effect on the transient behavior of the plume, but could not be ignored for long-term behavior (15 years). Koch and Zhang (1992) also simulated the migration of the low-density (860 kg m^{-3}) hydrocarbon benzene solvent from a leaking underground storage tank. Benzene is a major constituent of gasoline. It can be considered as a miscible phase due to a reasonably high solubility in water. Significant buoyant movement of benzene occurred only for concentrations greater than 20,000 mg/L. For sufficiently high concentrations, the benzene eventually rises to the surface of the water table where it can be pumped away through a well screened at that depth.

Zhang and Schwartz (1995) used a two-dimensional modeling approach to investigate the behavior of multispecies (common ions such as Na, Ca, Mg, Cl, SO_4, and a trace organic compound) contaminant plumes that can develop in shallow unconfined aquifers that receive dense leachate from unlined landfills such as in Babylon, New York (Kimmel and Braids, 1980). Retardation factors due to instantaneous sorption were assigned for individual components. For relatively large density contrasts between the landfill leachate and the ambient groundwater, pockets of dense water sank to or close to the bottom of the aquifer. Horizontal and vertical components of flow were determined by source concentrations, densities, and retardation factors of the contaminants. The authors state that when a variety of contaminants are present, individual plumes may be uniquely located, depending upon corresponding retardation factors. They also indicated that plumes near a leachate source may display an oscillatory behavior in variable density systems, even under conditions of constant loading.

Density-driven migration of hazardous liquid wastes in sloping aquifers has been investigated by Dorgarten and Tsang (1991) with two-dimensional analytical and numerical models for deep injection systems. Density-driven movement of liquid wastes in sloping aquifers were shown to be much stronger than plume migration due to regional groundwater flow. Even for a relatively high natural hydraulic gradient of

10^{-3} and an aquifer slope of only one degree, density-driven flow can be stronger than regional groundwater flow. During the initial stage of waste injection, spreading occurred in all directions due to density-induced stratification effects. Later, the waste mainly moved laterally along the slope of either the aquifer top or bottom, depending on the waste liquid density. Density-driven flow tends to accelerate waste transport if regional groundwater flow occurs in the same direction. For conditions of negligible regional hydraulic gradients, density-driven flow induces further groundwater motion after injection has stopped. A heavy waste plume moves downward in the injection zone, while the lighter water is pushed outward and upward. Potential upward migration was shown to be highest when the waste plume is lighter than the ambient groundwater. Such conditions are likely when fluid densities within the injection zone are large due to high salinities. Dorgarten and Tsang (1991) state that two-dimensional models tend to slightly overestimate density-induced movement since spreading of solute plumes in the third dimension is neglected. They indicate that a fully three-dimensional model is actually preferred. A similar conclusion was drawn by Liu and Dane (1997), who claimed that, for a given size of the contaminant source, the width of the porous medium had a significant effect on the occurrence and development of gravitational instabilities.

Accidental spills of DNAPLs such as the solvent trichloroethylene (TCE) or LNAPLs, such as gasoline, percolate as separate immiscible phases into the subsurface porous medium until they become immobilized by capillary forces, leaving zones of residual contaminant suspended in the pore structure. In water-unsaturated porous media, immobile NAPL is entrapped as films, pendular rings, wedges surrounding pendular rings, and filled pore throats. Upon reaching the groundwater table (phreatic zone), gasoline tends to spread laterally along the water table, but TCE usually continues to move downward through the saturated zone until a low permeability layer is encountered. Some of the TCE becomes immobilized within the water-saturated zone as entrapped droplets or disconnected pockets of liquid (Wilson et al., 1990). The residue provides a long-lasting source of contaminants subject to migration in both the aqueous liquid and vapor phases of the porous media. Because organic contaminants can be highly toxic at very low concentrations, even small amounts can pose a serious threat to aquifers. Vapors from residual organic liquids in the unsaturated vadose zone may migrate considerable distances through the aerated pore space and thus cause extensive contamination of the groundwater (Mendosa and Frind, 1990a). Advective-dispersive transport of dense organic vapors in the unsaturated zone has been described using flow and transport equations that are analogous to those used in density-dependent transport in water-saturated groundwater (Mendosa and Frind, 1990a,b). The two-dimensional model by Mendosa and Frind (1990a) includes diffusion, advection due to density gradients, and advection due to the vapor mass released by vaporization at the source. Inclusion of density-driven advection provided an improved match for TCE data from a three-dimensional laboratory experiment. Advection was shown to be a very effective transport mechanism for TCE in a highly permeable coarse sand. Model sensitivity analysis revealed that for compounds with high vapor pressures and molecular weights, mass transport by density-driven advection in highly permeable materials may greatly exceed that transported by diffusion alone. Relative to purely diffusive transport, the general effects of density-dependent advection include longer travel distances and larger areas of contamination, more rapid rates of vaporization at the liquid phase source, more vapor mass being transported into the unsaturated zone pore space, greater accumulation of contaminants adjacent to the water table, and a reduction in passive venting to the atmosphere. Increases in vapor pressure and molecular weight can result in large increases in advective transport (Mendosa and Frind, 1990b).

SUMMARY

Density- and viscosity-driven flows have been observed in both saturated groundwater aquifers and unsaturated porous materials in the vadose zone. Models for such flows require the solution of multidimensional flow of heterogeneous fluid and contaminant transport equations which are coupled by fluid density and viscosity. Differences in contaminant concentration or in temperature alter fluid density and viscosity, thus providing coupling between the flow and transport equations. Under certain situations, density-driven flow tends to be offset by differences in fluid viscosity. Density- and viscosity-driven flows can involve either stable or unstable interfaces between two miscible fluids. Unstable flow generates fingers or preferential flow-paths for contaminants. Mixing that occurs during unstable flow acts to eventually obliterate gradients in fluid density and viscosity. Hydrodynamic dispersion acts to oppose density-driven flow in subsurface porous media. Heterogeneities in natural porous media tend to enhance flow instabilities for heterogeneous fluids. In homogeneous saturated porous media, the value of the hydraulic conductivity remains relatively constant since the degree of fluid saturation remains near 100%. However, in variably-saturated, homogeneous porous media such as may occur in the vadose zone, the magnitude of hydraulic conductivity fluctuates with the degree of fluid saturation. Increases in fluid density and viscosity simultaneously act to increase and decrease the hydraulic conductivity, respectively. Spatial gradients of density and viscosity thus tend to provide complex flow patterns for contaminants in saturated and unsaturated porous media.

REFERENCES

Bear, J. *Dynamics of Fluids in Porous Media.* American Elsevier Publishing Company, Inc., New York, 1972, p. 653–662.

Bennett, S.S. and J.S. Hanor. Dynamics of Subsurface Salt Dissolution at the Welsh Dome, Louisiana Gulf Coast, in Dynamical Geology of Salt and Related Structures, Lerche, I. and J.J. O'Brien (Eds.). Academic Press, San Diego, 1987, pp. 653–677.

Dane, J.H., O. Güven, and B.C. Missildine. Flow Visualization Studies of Dense Aqueous Leachate Plumes in a Course and Fine Sand. Ala. Agric. Exp. Stn., Auburn University, Auburn, AL, Spec. Rep., 1994.

Dorgarten, H.-W. and C.-F. Tsang. Modeling the density-driven movement of liquid wastes in deep sloping aquifers. *Ground Water* 29:655–662, 1991.

Forkel, C. and M.A. Celia. Numerical simulation of unsaturated flow and contaminant transport with density and viscosity dependence, in *Computational Methods in Water Resources IX. Vol. 2: Mathematical Modeling in Water Resources,* Russell, T.F., R.E. Ewing, C.A. Brebbia, W.G. Gray, and G.F. Pinder (Eds.), Elsevier, New York, 1992, pp. 351–358.

Galeati, G., G. Gambolati, and S.P. Neuman. Coupled and partially coupled Eulerian-Lagrangian model of freshwater-seawater mixing. *Water Resour. Res.* 28:149–165, 1992.

Gambolati, G., C. Paniconi, and M. Putti. Numerical modeling of contaminant transport in groundwater, in *Migration and Fate of Pollutants in Soils and Subsoils,* Petruzelli, D. and F.G. Helfferich (Eds.), NATO ASI Series, G 32, Springer-Verlag, Berlin, 1993.

Gebhart, B., Y. Jaluria, R.L. Mahajan, and B. Sammakia. *Buoyancy-Induced Flows and Transport.* Harper and Row, New York, 1988.

Glass, R.J., T.S. Steenhuis, and J.-Y. Parlange. Mechanism for finger persistence in homogeneous, unsaturated, porous media: Theory and verification. *Soil Sci.* 148:60–70, 1989.

Güven, O., J.H. Dane, W.E. Hill, and J.G. Melville. Mixing and Plume Penetration Depth at the Groundwater Table. Elec. Power Res. Inst., Palo Alto, CA, Rep. EPRI TR-10056, 1992.

Hayworth, J.S. A Physical and Numerical Study of Three-Dimensional Behavior of Dense Aqueous Phase Contaminant Plumes in Porous Media. Ph.D. Dissertation, Department of Civil Engineering, Auburn University, Auburn, AL, 1993.

Hendrickx, J.M.H., L.W. Dekker, and O.H. Boersma. Unstable wetting fronts in water repellent field soils. *J. Environ. Qual.,* 22:109–118, 1993.

Herbert, A.W., C.P. Jackson, and D.A. Lever. Coupled groundwater flow and solute transport with fluid density strongly dependent upon concentration. *Water Resour. Res.,* 24:1781–1795, 1988.

Jawitz, J.W., M.D. Annable, and P.S.C. Rao. Miscible fluid displacement stability in unconfined porous media: Two-dimensional flow experiments and simulations. *J. Contaminant Hydrol.,* (in press), 1997.

Kimmel, G.E. and O.C. Braids. Leachate Plumes in Groundwater from Baylon and Islip Landfills, Long Island, New York. U.S. Geol. Serv. Prof. Pap. 1085, 1980.

Koch, M. and G. Zhang. Numerical simulation of the effects of variable density in a contaminant plume. *Ground Water,* 30:731–742, 1992.

Kutilek, M. and D.R. Nielsen. *Soil Hydrology.* Catena Verlag, Germany, 1994, 297–298.

List, E.J. The Stability and Mixing of a Density-Stratified Horizontal Flow in a Saturated Porous Medium. Rep. KH-R-11, Calif. Inst. of Technol., Pasadena, 1965.

Liu, H.H. and J.H. Dane. A criterion for gravitational instability in miscible dense plumes. *J. Contaminant Hydrol.,* 23:233–243, 1996a.

Liu, H.H. and J.H. Dane. Two approaches to modeling unstable flow and mixing of variable density fluids in porous media. *Trans. Porous Media,* 23:219–236, 1996b.

Liu, H.H. and J. H. Dane. A numerical study on gravitational instabilities of dense aqueous phase plumes in three-dimensional layered porous media. *J. Hydrology,* 194:126–142, 1997.

Mansell, R.S., R.D. Rhue, Y. Ouyang, and S.A. Bloom. Microemulsion-mediated removal of residual gasoline from soil columns. *J. Soil Contamination,* 5:309–327, 1996.

Mendosa, C.A. and E.O. Frind. Advective-dispersive transport of dense organic vapors in the unsaturated zone. 1. Model development. *Water Resour. Res.,* 26:379–387, 1990a.

Mendosa, C.A. and E.O. Frind. Advective-dispersive transport of dense organic vapors in the unsaturated zone 2. Sensitivity analysis. *Water Resour. Res.,* 26:388–398, 1990b.

Moissis, D.E. and M.F. Wheeler. Effect of the structure of the porous medium on unstable miscible displacement, in *Dynamics of Fluids in Hierarchical Porous Media,* Cushman, J.H. (Ed.), Academic Press, New York, 1990, pp. 243–271.

Mulqueen, J. and D. Kirkham. Leaching of a surface layer of sodium chloride into tile drains in a sand-tank model. *Soil Sci. Soc. Am. J.,* 36:3–9, 1972.

Oostrom, M., J.H. Dane, O. Güven, and J.S. Hayworth. Experimental investigation of dense solute plumes in an unconfined aquifer model. *Water Resour. Res.,* 28:2315–2326, 1992a.

Oostrom, M., J.S. Hayworth, J.H. Dane, and O. Güven. Behavior of dense aqueous phase leachate plumes in homogeneous porous media. *Water Resour. Res.,* 28:2123–2134, 1992b.

Park, N. Closed-form solutions for steady state density-dependent flow and transport in a vertical soil column. *Water Resour. Res.,* 32:1317–1322, 1996.

Paschke, N.W. and J.A. Hoopes. Buoyant contaminant plumes in groundwater. *Water Resour. Res.,* 9:1183–1192, 1984.

Ritsema, C.J., L.W. Dekker, J.M.H. Hendrickx, and W. Hamminga. Preferential flow mechanism in a water-repellent sandy soil. *Water Resour. Res.,* 29:2183–2193, 1993.

Schinariol, R.A., F.W. Schwartz, and C.A. Mendoza. On the generation of instability in variable density flow. *Water Resour. Res.,* 30:913–927, 1994.

Schincariol, R.A. and F.W. Schwartz. An experimental investigation of variable density flow and mixing in homogeneous and heterogeneous media. *Water Resour. Res.,* 26:2317–2329, 1990.

Schincariol, R.A., F.W. Schwartz, and C.A. Mendoza. Instabilities in variable density flows: Stability and sensitivity analyses for homogeneous and heterogeneous media. *Water Resour. Res.,* 33:31–41, 1997.

Wilson, J.L., S.H. Conrad, W.R. Mason, W. Peplinski, and E. Hagan. Laboratory Investigation of Residual Liquid Organics from Spills, Leaks, and the Disposal of Hazardous Wastes in Groundwater, EPA/600/6-90/004. U.S. Environmental Protection Agency, Robert S. Kerr Environmental Research Laboratory, Ada, OK, 1990.

Wood, W.W. and W.R. Ostercamp. Playa-lake basins on the southern high plains of Texas and New Mexico. II. A hydrologic model and mass-balance arguments for their development. *Geol. Soc. Am. Bull.,* 99:224–230, 1987.

Wooding, R.A. The stability of a viscous liquid in a vertical tube containing porous material. *Proc. R. Soc. London, Ser. A* 252:120–134, 1959.

Yao, T.-M. and J.M.H. Hendrickx. Stability of wetting fronts in dry homogeneous soils under low infiltration rates. *Soil Sci. Soc. Am. J.,* 60:20–28, 1996.

Zhang, H. and F.W. Schwartz. Multispecies contaminant plumes in variable density flow systems. *Water Resour. Res.,* 31:837–847, 1995.

CHAPTER FIFTEEN

Numerical Modeling of NAPL Dissolution Fingering in Porous Media

C.T. Miller, S.N. Gleyzer, and P.T. Imhoff

INTRODUCTION

Nonequilibrium phenomena have received considerable attention in recent years for a wide variety of subsurface systems and processes, and it is important to place these phenomena in an appropriate context. Deviations from true thermodynamic equilibrium are manifest as time-dependent changes in thermodynamic state variables, such as internal energy, entropy, temperature, pressure, volume, or chemical potential for a reference system that is considered to be at a scale much larger than the molecular scale. If nonequilibrium phenomena are important for a given system, an appropriate model must be formulated and solved to account for all such effects. The appropriate model and the degree of nonequilibrium in a process are a function of spatial scale. Failure to model nonequilibrium phenomena at an appropriate scale results in an inaccurate representation of the system.

Formulating a flow and transport model for a subsurface system requires that a relevant system of conservation laws be specified and closed using an appropriate set of constitutive relations. From a chemical perspective, a well-posed formulation requires constitutive relations that describe both the thermodynamic equilibrium of chemical potentials in the system and the rate at which an equilibrium state is approached. From a mathematical perspective, constitutive relations serve as projection operators, hiding the needless details of the small scale and expressing larger scale behavior in the simplest possible terms.

Constitutive relations are used to describe the essential behavior of a phenomenon of interest in an average sense at a scale of interest. Four spatial scales have been suggested for describing phenomena in the subsurface (Gray et al., 1993): the molecular scale, where materials are viewed as a discrete collection of particles or molecules; the microscale, which describes behavior within a single pore in a porous medium; the macroscale, the scale above the microscale where the porous medium properties of porosity, phase saturation, and permeability can be described as continuous functions of space; and the megascale, which is the scale of the system dimension (for example, the thickness of the geological unit of interest).

In subsurface hydrology, many processes are conveniently studied at the macroscale through laboratory experimentation. Unfortunately, constitutive relations developed at this scale are not always easily extendable to the larger scales that are typically of interest. This is due in large part to the heterogeneity of subsurface systems, which occurs at a variety of scales and affects most processes of concern. The focus of this chapter is the dissolution of nonaqueous phase liquids (NAPLs) into the aqueous phase—a process affected by both heterogeneity and flow mechanisms operative between the macro- and megascales.

Mass transfer from a NAPL to an aqueous phase is an important process used to describe transport phenomena in multiphase subsurface systems that contain multiple liquid phases. In general, two approaches have evolved to describe this process: local equilibrium and first-order mass transfer. While the former is convenient numerically, it is unrealistic for many systems.

Models of first-order mass transfer approaches at the macroscale have matured over the last several years, with several well-established advancements: (1) a nondimensional form of the process has been formulated (Miller et al., 1990); (2) controlled laboratory experiments have been performed for a variety of cm-scale systems to measure first-order mass transfer rate coefficients (Miller et al., 1990; Parker et al., 1991; Powers et al., 1992; Borden and Kao, 1992; Hatfield et al., 1993; Geller and Hunt, 1993; Imhoff et al., 1994; Powers et al., 1994); and (3) several correlation models have been advanced to relate a nondimensional form of the mass transfer rate coefficient, known as the Sherwood number, to other nondimensional system parameters (Miller et al., 1990; Imhoff et al., 1994; Powers et al., 1994). Recent work has extended these correlations for the case of nonisothermal systems (Imhoff et al., 1997).

These important advancements notwithstanding, a mature level of understanding for modeling NAPL-aqueous phase mass transfer does not exist for all systems of interest. Most notably, the constitutive relations developed for NAPL dissolution at the macroscale have not been extended to larger scales where the effects of heterogeneity are important. Under certain situations, including NAPL-aqueous phase mass transfer, explicit consideration of heterogeneity at scales less than the scale of interest must be included in the formulation and parameterization of appropriate constitutive relations.

One manifestation of the effect of heterogeneity and scale on NAPL-aqueous phase mass transfer is NAPL dissolution fingering, which we have considered in previous work (Imhoff and Miller, 1996; Imhoff et al., 1996). Dissolution fingering refers to the pattern of a NAPL dissolution front that results from even minor heterogeneities in the initial distribution of NAPL. These finger patterns may be predicted from theory, observed experimentally, and computed numerically. While these features do not occur in most of the experiments used to develop NAPL-aqueous phase mass transfer relations, they do occur at scales well below the typical discretization scale of field-scale numerical models. Existing NAPL-aqueous phase mass transfer relations do not explicitly account for dissolution fingering.

The objectives of this work are (1) to summarize fundamental considerations that lead to NAPL dissolution fingering; (2) to outline existing theoretical and experimental results that characterize NAPL dissolution fingering; (3) to formulate a mathematical model that can be used to investigate NAPL dissolution fingering; (4) to implement and apply the formulated model to verify and extend our understanding of NAPL dissolution fingering; and (5) to highlight the remaining questions that prevent a complete understanding of NAPL-aqueous phase mass transfer constitutive relations.

BACKGROUND

A detailed review of the extensive multiphase, mass transfer, and fingering literature is beyond the scope of this work; recent reviews of multiphase flow in porous media and numerical modeling of these systems are available (Adler and Brenner, 1988; Russell, 1995; Miller et al., 1997). Instead, we restrict our review to the three aspects of multiphase systems that are the most relevant to dissolution fingering: NAPL-aqueous phase mass transfer, fingering, and numerical simulation.

NAPL-Aqueous Phase Mass Transfer

Nonequilibrium NAPL-aqueous phase mass transfer is typically formulated using a first-order expression of the form

$$I_{na}^{\iota} = \rho_a a_{na} \hat{k}_{na}^{\iota} (\omega_a^{\iota *} - \omega_a^{\iota}) \tag{15.1}$$

$$I_{na}^{\iota} = k_{na}^{\iota} (C_a^{\iota *} - C_a^{\iota}) \tag{15.2}$$

where I_{na}^{ι} is the flux of species ι from the NAPL to the aqueous phase; ρ is a density; a_{na} is the specific interfacial area between the NAPL and aqueous phases; k_{na} is a mass transfer coefficient representing a mass transfer process that is written with an aqueous phase driving force referenced to a loss by the n phase and a gain by the a phase; ω is a mass fraction; the subscripts a and n are qualifiers for the aqueous and nonaqueous phases, respectively; superscript ι is a species qualifier; superscript * represents an equilibrium condition with the companion phase involved in the mass transfer, which is the NAPL phase for this case; $k_{na}^{\iota} = a_{na}\hat{k}_{na}^{\iota}$; and $C_a^{\iota} = s_a w_a^{\iota}$.

Research on nonequilibrium NAPL-aqueous phase mass transfer has focused on performing experiments to determine k_{na} and correlating k_{na} to experimentally measurable parameters affecting the system (Miller et al., 1990; Imhoff et al., 1994; Powers et al., 1994). Experiments have typically been performed at the centimeter scale, which is at or above the macroscale for some porous media, by bringing a system to residual NAPL saturation and observing the rate of NAPL dissolution as a function of system properties, including aqueous-phase velocity, media properties, and NAPL physicochemical characteristics. NAPL dissolution rates have usually been observed by measuring effluent concentrations of these small-scale, one-dimensional column experiments and solving an inverse problem to compute k_{na}. Alternatively, radiation attenuation methods have been used to measure changes in NAPL saturation nondestructively as a function of time, and these measurements used to compute k_{na}. Empirical correlations have been sought describing the NAPL dissolution experiments that are generally of the form

$$Sh = f(n, \theta_n, \mathrm{Re}, Sc, g_c) \tag{15.3}$$

where Sh is a Sherwood number, n is porosity, θ is a volumetric fraction, Re is a Reynolds number, Sc is a Schmidt number, and g_c represents other characteristics of the grain size, such as uniformity (Miller et al., 1990; Imhoff et al., 1994; Powers et al., 1994).

While the above experimental and analytic approaches have dominated the NAPL-aqueous phase mass transfer literature, they have several shortcomings: (1) relatively few media-NAPL combinations have been investigated; (2) little work has focused on complete dissolution; (3) only a portion of the nondimensional parameters thought to govern the mass transfer process have been investigated; and (4) the experimental database collected and analyzed to date is principally for systems that are orders of magnitude smaller and much more homogeneous than most natural systems (Miller et al., 1997).

One observation, first noted by Imhoff (1992) in controlled laboratory experiments with homogeneous materials, was that the length of the mass transfer zone increased as a function of time and total length of the residual NAPL zone that was eluted. Although such an observation can be parameterized and included in correlations for mass transfer coefficients (Imhoff et al., 1994), this approach provides neither a satisfying mechanistic description nor a basis to predict the effect of fingering on mass transfer for scales not investigated experimentally. Imhoff (1992) hypothesized that the NAPL dissolution process might be inherently unstable, leading to the development of a complex dissolution front. Later observations of this nonuniform dissolution process were made by Soerens et al. (1995) and Imhoff et al. (1995) while investigating surfactant- and alcohol-enhanced removal of NAPLs, respectively.

Fingering

Fingering occurs for fluid flow and species transport in multiphase porous media systems because of contrasts in a variety of system characteristics, including gravity forces, fluid viscosities, wettability, and dissolution. The common feature of fingering patterns is an instability in the system, which results in the development and persistence of complex patterns of fluid flow and species transport. Imhoff's observation (1992) regarding the characteristics of the NAPL-aqueous phase mass transfer zone was consistent with the development of a finger pattern, which we refer to as "dissolution fingering".

The development of a dissolution finger is depicted schematically in Figure 15.1. Consider the case in which a small spatial variation exists in the initial residual NAPL saturation, a condition that is consistent with detailed observations (Mayer and Miller, 1992, 1993). Further, consider that the porous medium is homogeneous with respect to macroscopic properties such as hydraulic conductivity and the saturation-conductivity relation—conditions that are characteristic of most laboratory mass transfer experiments but not most natural systems. A constant aqueous-phase pressure is imposed along the y coordinate at x=0 and x=L in Figure 15.1, driving aqueous phase flow in the positive x direction. Given the variations in NAPL saturation at initial time (t_1), the region with a lower than mean NAPL saturation, depicted as zone 1 in the figure, will have a higher than mean aqueous-phase velocity. Zone 1, with a higher velocity, will elute the residual NAPL at a rate that is more rapid than a region with a relatively lower velocity—zone 2 in the figure. As the residual NAPL is eluted, the relative permeability for the aqueous phase increases. Since the rate of NAPL elution is related to the aqueous phase velocity, and since the removal of NAPL in turn increases the relative permeability, zone 1 (with a lower than average initial residual NAPL saturation) is eluted more rapidly than zone 2. This pattern continues as elution proceeds, leading to the development of a dissolution finger.

Dissolution fingering is important because the mass transfer zone can attain a complex shape that may require long time and length scales to develop fully. The existence of such patterns demonstrates the problem of scale. Small-scale laboratory experiments may not encounter such patterns, and mass transfer relations determined from such experiments should not be expected to apply directly to discretization scales that are larger than the experiments on which they were based.

Dissolution fingering has been examined for the case of mineral dissolution (Chadam et al., 1986; Ortoleva et al., 1987b; Ortoleva et al., 1987a; Chadam and Ortoleva, 1990; Chadam et al., 1991), which is similar but not identical to the case of NAPL dissolu-

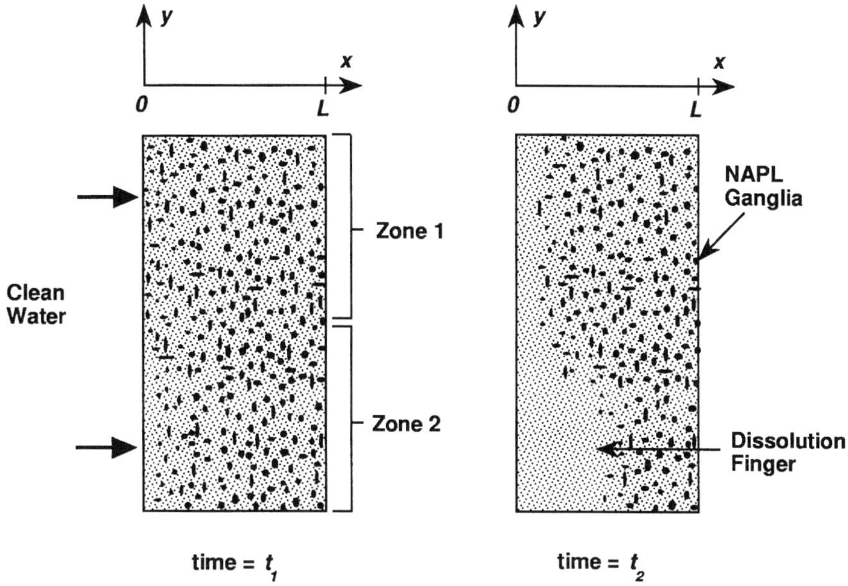

Figure 15.1. Hypothetical cross section through a porous medium, illustrating the formation of dissolution fingers. (From Imhoff, P.T. and C.T. Miller, *Water Resources Research,* 32(7)1919–1928, 1996, copyright by the American Geophysical Union.)

tion. The feedback mechanism for mineral dissolution is similar to the one described above for NAPL residual dissolution, but mineral dissolution is a diagenetic process that is often assumed to occur with low velocities; hence, molecular diffusion has been assumed to dominate over mechanical dispersion (Chadam et al., 1986). For residual NAPL dissolution, mechanical dispersion is important and typically anisotropic. Further, NAPL residual saturation morphology covers length scales that range from a single pore to multiple connected pores for large, complex NAPL ganglia (Mayer and Miller, 1992, 1993; Lowry and Miller, 1995). Conversely, mineral dissolution involves the dissolution of mineral grains that are typically on the order of only a few pore bodies (Chadam et al., 1986).

Numerical Simulation

An extensive body of literature exists on the simulation of unstable flows in porous media (Wei and Ortoleva, 1990; King et al., 1995; Simons, 1996). Most of these works address viscous fingering processes. To our knowledge, Chadam, Ortoleva and coworkers have published the only known numerical simulation work of two-dimensional dissolution fingering (Chadam et al., 1986; Ortoleva et al., 1987a,; Wei and Ortoleva, 1990; Chadam and Ortoleva, 1990). The best presentation of the numerical algorithm is given in Wei and Ortoleva (1990), where a fully implicit finite-difference method was used. In order to stabilize the numerical solution, an artificial dispersion term was introduced. Dissolution fingering was successfully simulated even though the numerical scheme was diffusive, underscoring the lack of stability in this process. Diffusive numerical schemes would have the same effect as physical diffusion or dispersion—dampening finger growth. The use of less diffusive and more accurate numerical methods would aid the study of such phenomena.

In simulating unstable flows in porous media, the numerical dispersion damping the instabilities should be considerably less than the physical dispersion (Moissis et al., 1988; Wei and Ortoleva, 1990; King et al., 1995). In order to reduce numerical dispersion, very fine grids and high-resolution numerical methods are often employed, which is why simulating unstable flows is computationally intensive. For example, according to King et al. (1995), each simulation of viscous fingering performed for an injection of two pore volumes on a 250×100 grid required about 20 hours of supercomputer time. For most NAPL dissolution problems related to subsurface groundwater remediation, the total number of injected aqueous phase pore volumes may be on the order of hundreds or even thousands. Thus, the potential amount of computational work necessary to study instabilities due to NAPL dissolution is expected to be much greater than for the simulation of viscous fingering.

LINEAR STABILITY ANALYSIS

Residual NAPL dissolution fingering was examined using a linear stability analysis (LSA) approach (Imhoff and Miller, 1996). To aid the present discussion, we summarize the approach taken in this previous work, outline its main conclusions, and list the open questions that remain.

The approach taken to perform a LSA was:

- conservation equations were formulated for an aqueous and a NAPL phase (here called flow equations) and for the mass fraction of the NAPL species in the aqueous phase (here called a species transport equation) for a two-dimensional domain that was infinite in areal extent;
- the conservation equations were closed by specifying a standard multiphase extension of Darcy's law, a saturation-conductivity relation, a multiphase form of a hydrodynamic dispersion tensor, and an interphase mass transfer relation;
- the conservation equations and constitutive relations were cast in the form of a moving boundary problem, where the boundary was the transition between a NAPL-free and a NAPL residual area, which was assumed to be narrow;
- the equations were put into a dimensionless form and examined in the realistic limit $\omega_a^{1^*} \to 0$;
- small-scale sinusoidal perturbations were introduced in the planar dissolution front and the growth rate of the perturbation, σ, was evaluated as a function of the wave number of the initial perturbation, m; and
- σ was investigated as a function of m for isotropic and anisotropic dispersion cases and for a range of aqueous-phase velocities and initial residual NAPL saturations to determine m that corresponded to maximum σ and the boundary at which σ changed signs—or the cut-off wavelength, indicating a transition point between stable and unstable systems.

The LSA of dissolution fingering yielded quantitative predictions showing that fingering is likely to occur in many systems of interest; the maximum dimensionless growth rate increases with velocity and anisotropy of dispersion; the fastest growing, or critical, wavelength is inversely related to the velocity and initial residual NAPL saturation; and the cutoff wavelength followed similar trends as noted for the critical wavelength. Based on this analysis, dissolution fingers are predicted to occur with a length scale normal to the mean direction of flow of less than a centimeter to meters depending upon the conditions.

The LSA results confirm the hypothesized phenomenon of dissolution fingering and support the preliminary experimental observations that a mass transfer zone increased in length as the dissolution front progressed through NAPL-contaminated media. However, this LSA approach assumes a thin mass transfer region and an infinite spatial domain. Both of these assumptions are violated. Further, this analysis provides no means for predicting the morphology of the fingers as they grow, and thus the details of the mass transfer zone; the effect of dimensionality of the system; or the influence of heterogeneity of one or more physicochemical parameters of the system.

PHYSICAL EXPERIMENTS

The results of the LSA were compared to detailed two-dimensional experiments and one three-dimensional experiment to further the understanding of dissolution fingering (Imhoff et al., 1996). In the two-dimensional experiments, media properties, aqueous phase velocities, initial residual NAPL properties and saturation, and gravity force contributions were varied and results analyzed using image analysis methods and, in one case, nondestructive X-ray analysis techniques. Figure 15.2 illustrates the fingering patterns in one of these experiments.

Results from the two-dimensional experiments showed very good agreement with the LSA. Fingers formed when the dissolution front was thin and when time exceeded values that were sufficiently long. The width of the fingers observed for cases in which fingering occurred also agreed with the LSA theory, as did the trends for each of the variables investigated. For the case in which detailed X-ray analysis was performed, fingers developed in regions as predicted from the distribution of the initial NAPL residual saturation.

The two-dimensional experimental results confirmed that dissolution fingering is a potentially important process that deserves further study and incorporation into mass transfer models. To address two issues not studied in these experiments, the effect of a third dimension and the length to which dissolution fingers may grow, a single experiment was conducted in a three-dimensional column. NAPL residual saturations were monitored using X-ray attenuation methods, and the experiment showed that the NAPL dissolution fingers grew to a length of 30 cm and were apparently limited by the length of the laboratory experimental apparatus, which was just over a meter long. Numerical simulations performed using a compositional simulator and typical mass transfer models showed relatively poor predictions of the observed elution patterns, which got progressively worse with time and length scale of the eluted region. These results strongly support the importance of dissolution fingering and the need to include it in constitutive relations and numerical models for NAPL elution, at least for some conditions.

NUMERICAL MODELING

Model Formulation

Conservation Equations

To complement existing theoretical and experimental work on NAPL dissolution fingering, a compositional multiphase flow and transport model was developed and applied. Our intent was to develop a numerical model that could be applied not only to a broad range of unconfirmed or unresolved aspects of dissolution fingering but also to a wide range of other issues involving flow and transport phenomena in multiphase systems with complex compositional characteristics. Other applications

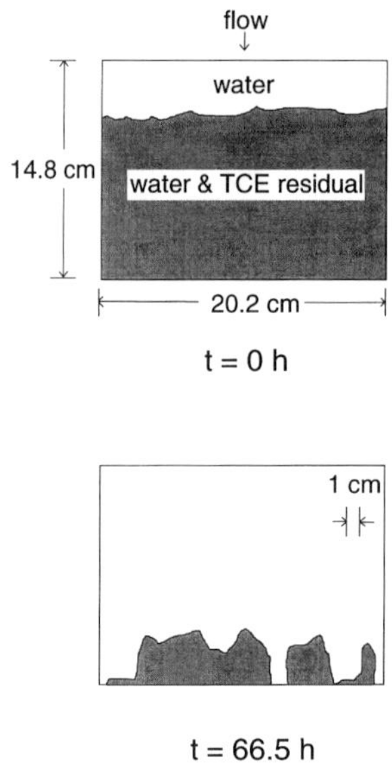

Figure 15.2. Sketches delineating regions containing TCE at residual saturation at three times during a dissolution experiment in a two-dimensional flow cell. (From Imhoff, P.T. et al., *Water Resources Research,* 32(7):1929–1942, 1996, copyright by the American Geophysical Union.)

of concern included cosolvent- and surfactant-enhanced remediation of NAPL-contaminated porous media. To accommodate these applications, consideration of bulk phase behavior, or flow, and phase composition was necessary. We refer to a formulation that accounts for the effects of phase compositions in a coupled sense as a compositional model.

Since our applications are inherently time-dependent, a transient form of the conservation equations was necessary. Further, many of the issues of interest required consideration of multiple spatial dimensions and heterogeneous media properties—we chose to formulate and develop our model in three spatial dimensions and fully account for heterogeneous properties. Thus, we consider a system comprised of n_p phases and n_s species in which capillarity, interphase mass transfer, and complex phase behavior may be important. A standard conservation equation for mass of a species ι in phase α is (Crichlow, 1977; Aziz and Settari, 1979; Abriola and Pinder, 1985; Abriola, 1988; Cohen and Mercer, 1993; Miller et al., 1997)

$$\frac{\partial}{\partial t}(\theta_\alpha \rho_\alpha \omega_\alpha^\iota) = -\nabla \cdot (\mathbf{j}_\alpha^\iota + \theta_\alpha \rho_\alpha \omega_\alpha^\iota \mathbf{v}_\alpha) + I_\alpha^\iota + R_\alpha^\iota + S_\alpha^\iota \tag{15.4}$$

where $\mathbf{j}$ is a flux vector comprised of diffusive and dispersive components; $\mathbf{v}$ is a macroscopic pore velocity vector; I represents the set of all possible interphase mass transfer pathways between binary groupings of phases; R represents chemical and biological reactions; and S is a source term that represents mass added to the system.

Summing Equation 15.4 over all species in a phase gives a species-summed equation for each phase that is referred to as a flow equation and is of the form

$$\frac{\partial(\theta_\alpha \rho_\alpha)}{\partial t} = -\nabla \cdot (\theta_\alpha \rho_\alpha \mathbf{v}_\alpha) + I_\alpha + S_\alpha \tag{15.5}$$

subject to the following identities

$$\sum_{\alpha} \theta_{\alpha} = 1, \sum_{\alpha} I_{\alpha}^{\iota} = 0, \sum_{\iota} \omega_{\alpha}^{\iota} = 1, \sum_{\iota} j_{\alpha}^{\iota} = 0, \sum_{\iota} R_{\alpha}^{\iota} = 0 \tag{15.6}$$

and definitions

$$\sum_{\iota} I_{\alpha}^{\iota} = I_{\alpha}, \sum_{\iota} S_{\alpha}^{\iota} = S_{\alpha} \tag{15.7}$$

We consider a three-phase case, specifically the case in which $\alpha = n, a, s$ (NAPL, aqueous, and solid phases). The extension of this formulation to three fluid phases is straightforward but is not treated further in this work. Thus, our set of conservation equations includes three equations of the form of Equation 15.5 and $n_p(n_s - 1)$ equations of the form of Equation 15.4 for the most general case considered. This system of conservation equations can be reduced—and usually needs to be, to yield a tractable problem—by making certain assumptions. Further, because many more unknowns exist than equations, closure of the system of conservation equations is needed, which requires specification of appropriate constitutive relations.

Fluid Flow Equations

We further specify our formulation for the case in which the solid phase is rigid, incompressible, and essentially inert, except for trace contaminants. These assumptions allow the solid-phase conservation equation to be eliminated. Further, we assume that all sources are accommodated through the specification of boundary conditions giving

$$\frac{\partial(\rho_{\alpha}\theta_{\alpha})}{\theta t} = -\nabla \cdot (\theta_{\alpha}\rho_{\alpha}\mathbf{v}_{\alpha}) + I_{\alpha}, \quad \text{for } \alpha = a, n \tag{15.8}$$

Instead of specifying a formal momentum balance equation, it is common to substitute a multiphase extension of Darcy's law into Equation 15.8, although the empirical nature and shortcomings of this approach are known (Hassanizadeh and Gray, 1993a, 1993b; Miller et al., 1997). We assume an isotropic media and a traditional multiphase form of Darcy's law of the form

$$\mathbf{q}_{\alpha} = \theta_{\alpha}\mathbf{v}_{\alpha} = -\frac{kk_{ra}}{\mu_{\alpha}}(\nabla p_{\alpha} + \rho_{\alpha} g \nabla z) \tag{15.9}$$

where $\mathbf{q}$ is a Darcy velocity vector, k is an intrinsic permeability, $k_{r\alpha}$ is a relative permeability for the α phase, μ is a viscosity, g is the acceleration due to gravity, and z is the vertical coordinate direction assumed to be aligned with gravity forces.

Combining Equations 15.8 and 15.9 yields

$$\frac{\partial(\theta_{\alpha}\rho_{\alpha})}{\partial t} = \nabla \cdot \left(\frac{\rho_{\alpha} kk_{r\alpha}}{\mu_{\alpha}}(\nabla p_{\alpha} + \rho_{\alpha} g \nabla z) \right) + I_{\alpha}, \quad \text{for } \alpha = a, n \tag{15.10}$$

which is a general flow formulation that requires specification of equations of state and constitutive relations for interphase mass transfer relations.

Equations of state are of the general functional form

$$\rho_\alpha = \rho_\alpha(p_\alpha, T_\alpha, \omega_\alpha^\iota), \quad \text{for } \iota = 1, \ldots, n_s \tag{15.11}$$

where T represents temperature. For the general formulation of our model, we assume isothermal conditions, linear compressibility of the fluid phases, and compositional effects, which are important for cosolvent and surfactant systems.

For computational efficiency of the fingering applications considered herein, we consider the limiting case of negligible compositional and compressibility effects on density, or constant phase densities, which yields

$$\frac{\partial \theta_\alpha}{\partial t} = \nabla \cdot \left(\frac{k k_{r\alpha}}{\mu_\alpha} (\nabla p_\alpha + p_\alpha g \nabla z) \right) + I_{\alpha\rho}, \quad \text{for } \alpha = a, n \tag{15.12}$$

where $I_{\alpha\rho} = I_\alpha / \rho_\alpha$.

For the case of the NAPL existing only as residual saturation, motion terms may be ignored giving

$$\frac{\partial \theta_n}{\partial t} = I_{n\rho} \tag{15.13}$$

or

$$\frac{\partial (n S_n)}{\partial t} = -\frac{k_{na}^n}{\rho_n} (C_a^{n*} - C_a^n) \tag{15.14}$$

where n is the porosity, and $S\alpha$ is the saturation of the α phase. This formulation is reasonable, since under the specified conditions interphase mass transfer to the aqueous phase is the only process that can change the NAPL volume fraction. It should be noted that this equation is a local equation, which does not depend upon spatial derivatives.

Equation 15.12 for the aqueous phase and Equation 15.13 can be summed, resulting in the elliptic equation

$$0 = \nabla \cdot \left(\frac{k k_{ra}}{\mu_a} (\nabla p_a + \rho_a g \nabla z) \right) + I_{a\rho} + I_{n\rho} \tag{15.15}$$

Since $I_{na} = I_a = -I_n$ then

$$0 = \nabla \cdot \left(\frac{k k_{ra}}{\mu_a} (\nabla p_a + \rho_a g \nabla z) \right) + \left(\frac{\rho_n - \rho_a}{\rho_a \rho_n} \right) I_{na} \tag{15.16}$$

While the general formulation and model requires specification of complete capillary pressure, saturation, and relative permeability relations, the fingering applications considered herein may be simplified, since the NAPL is in a state of residual saturation. For this case, capillary pressure effects are not important, and only the relation between fluid saturations and relative permeability of the aqueous phase is important. For this restricted application, we specify this relation as (Corey, 1994)

$$k_{ra} = \left(\frac{1 - S_n - S_{ia}}{1 - S_{ia}} \right)^{\gamma} \tag{15.17}$$

where S_{ia} is the irreducible saturation of the aqueous phase, and γ is an experimental parameter that depends upon pore size distribution characteristics.

Because $k_{ra} = f(x,t)$, Equation 15.16 changes with time, yet it is elliptic in form. This form may be viewed as a separation in timescales between the dissolution process of the residual NAPL, which occurs relatively slowly, and the changes in the Darcy velocities, which occurs relatively quickly—so as to be in a quasi-steady-state. The assumptions with this approach are emphasized, as is the fact that it is merely a special case of a more general formulation and model, but one that results in considerable computational savings for the problems considered herein.

Species Transport Equations

We now turn our consideration to the species transport equations. For the fingering applications, we assume that the NAPL is a single species that is sparingly soluble in the aqueous phase. We also assume that mass transfer between the solid phase and the fluid phases can be ignored. Under these conditions, only a single transport equation requires solution, which we state as

$$\frac{\partial}{\partial t}(\theta_a C_a^n) = \nabla \cdot (\mathrm{D}^n \cdot \nabla C_a^n - \mathrm{q}_a C_a^n) + I_a^n + R_a^n + S_a^n \tag{15.18}$$

where the superscript n represents the NAPL species, and the dispersion tensor is defined classically (Bear, 1979) as

$$D_{ij}^n = \delta_{ij} \alpha_t |\mathrm{q}| + (\alpha_l - \alpha_t) \frac{q_i q_j}{|\mathrm{q}|} + \theta_a D_m^n \tau_{ij}, \quad \text{for } i, j = 1, \ldots, 3 \tag{15.19}$$

where δ_{ij} is the Kronecker delta function; α_l and α_t are the longitudinal and transverse dispersivities; $|\mathbf{q}|$ is the magnitude of the Darcy velocity vector; D_m^n is the free liquid diffusivity of the solute in the aqueous phase; and τ_{ij} is a tortuosity tensor for the media.

Since a nonreactive NAPL without external sources in the aqueous phase is considered, the relevant species balance equation becomes

$$\frac{\partial}{\partial t}(\theta_a C_a) = \nabla \cdot (\mathrm{D} \cdot \nabla C_a - \mathrm{q}_a C_a) + I_{na} \tag{15.20}$$

in which I_{na} is specified by Equation 15.2, and k_{na} is given as

$$k_{na} = \beta_1 \theta_n^{\beta_2} Re^{\beta_3} \tag{15.21}$$

where β_i are experimentally determined parameters (Miller et al., 1990; Imhoff et al., 1994; Powers et al., 1994); and the superscript *n*, the species qualifier, has been dropped for this single-species transport case.

Numerical Solution

Algorithm

The solution algorithm used for the numerical model formulated in the last section consists of the following steps:

- solve a discrete form of Equation 15.16 for p_a for a current value of $S_a(\mathbf{x},t)$ and $k_{ra}[S_a(\mathbf{x},t)]$;
- use Equation 15.9 to compute $\mathbf{v}_a(\mathbf{x},t)$;
- use $\mathbf{v}_a(\mathbf{x},t)$ and $S_n(\mathbf{x},t)$ to compute $k_{na}(\mathbf{x},t)$ using Equation 15.21; and
- solve Equations 15.14 and 15.20 simultaneously to determine $S_n(\mathbf{x},t+\Delta t)$ and $C_a(\mathbf{x},t+\Delta t)$ using the available solution for $\mathbf{v}_a$.

This algorithm prevents the need to solve a nonlinear system of equations at the expense of lost accuracy by lagging $\mathbf{v}_a$ and k_{na} a time step. Since S_a, hence $\mathbf{v}_a$, will vary slowly with time, this approach results in substantial computational savings. The computational savings is important because of the significant computational burden associated with simulating unstable multiphase fluid flows. We do not advocate the use of this algorithm for a general compositional approach, since, under more general conditions, S_α may change relatively rapidly as a function of *t* in certain spatial locations.

Spatial and Temporal Approximations

An approximate numerical solution was implemented to solve finite-dimensional, discrete forms of Equations 15.16, 15.14, and 15.20, following the algorithm outlined previously. The discrete form of the equations was specified on an uniform, orthogonal spatial grid over a three-dimensional hexahedral domain using finite-volume methods for all equations, while variable size time steps were taken to meet stability constraints for the species transport equation. The general three-dimensional model developed was reduced to the two-dimensional case to provide a direct comparison with the theoretical and experimental work performed to date and to limit the computational burden of the simulations. For completeness, we detail the three-dimensional version of the model that was implemented.

The finite-volume method (FVM) was used to solve for p_a, S_n, and C_a in a cell-centered sense, where for example

$$p_{ijk}^{l} = \int_{x_{i-1/2}}^{x_{i+1/2}} \int_{y_{j-1/2}}^{y_{j+1/2}} \int_{z_{k-1/2}}^{z_{k+1/2}} p_a(x,y,z,t^l)\, dx\, dy\, dz \tag{15.21}$$

where $x_i = i\Delta x$, $y_j = j\Delta y$, $z_k = k\Delta z$, $x_{i\pm 1/2} = (i \pm 1/2)\Delta x$, $y_{j\pm 1/2} = (j \pm 1/2)\Delta y$, $z_{k\pm 1/2} = (k \pm 1/2)\Delta z$, $t^l = l\Delta t$; and Δx, Δy, Δz, and Δt are the spatial and temporal increments.

We follow a standard approach to discretize the elliptic equation to solve for p_a (Aziz and Settari, 1979)

$$
\begin{aligned}
&(\Delta x)^{-2}[\lambda_{i+1/2\,jk}(p^{l}_{i+1\,jk} - p^{l}_{ijk}) - \lambda_{i-1/2\,jk}(p^{l}_{ijk} - p^{l}_{i-1\,jk})] \\
&+ (\Delta y)^{-2}[\lambda_{ij+1/2\,k}(p^{l}_{ij+1k} - p^{l}_{ijk}) - \lambda_{ij-1/2\,k}(p^{l}_{ijk} - p^{l}_{ij-1k})] \\
&+ (\Delta z)^{-2}[\lambda_{ijk+1/2}(p^{l}_{ijk+1} - p^{l}_{ijk}) - \lambda_{ijk-1/2}(p^{l}_{ijk} - p^{l}_{ijk-1})] + \frac{\rho_a g}{\Delta z}(\lambda_{ijk+1/2} - \lambda_{ijk-1/2}) \\
&= \left(\frac{\rho_a - \rho_n}{\rho_a \rho_n}\right)(I_{na})^{l}_{ijk}
\end{aligned}
\tag{15.23}
$$

where $\lambda_{i\pm1/2jk}$, $\lambda_{ij\pm1/2k}$, and $\lambda_{ijk\pm1/2}$ are the aqueous phase mobilities at cell interfaces

$$
\begin{aligned}
\lambda_{i\pm1/2\,jk} &= \frac{k_{i\pm1/2\,jk}(k_{ra})^{l}_{i\pm1/2\,jk}}{(\mu_a)^{l}_{ijk}} \\
\lambda_{ij\pm1/2\,k} &= \frac{k_{ij\pm1/2\,k}(k_{ra})^{l}_{ij\pm1/2\,k}}{(\mu_a)^{l}_{ijk}} \\
\lambda_{ijk\pm1/2} &= \frac{k_{ijk\pm1/2}(k_{ra})^{l}_{ijk\pm1/2}}{(\mu_a)^{l}_{ijk}}
\end{aligned}
\tag{15.24}
$$

where the intrinsic permeabilities $k_{i\pm1/2jk}$, $k_{ij\pm1/2k}$, and $k_{ijk\pm1/2}$ are defined as harmonic means by

$$
\begin{aligned}
k_{i\pm1/2\,jk} &= \frac{2k_{ijk}k_{i\pm1\,jk}}{k_{ijk} + k_{i\pm1\,jk}} \\
k_{ij\pm1/2\,k} &= \frac{2k_{ijk}k_{ij\pm1k}}{k_{ijk} + k_{ij\pm1k}} \\
k_{ijk\pm1/2} &= \frac{2k_{ijk}k_{ijk\pm1}}{k_{ijk} + k_{ijk\pm1}}
\end{aligned}
\tag{15.25}
$$

and relative permeabilities are defined according to the upstream rule (Aziz and Settari, 1979)

$$
\begin{aligned}
(k_{ra})^{l}_{i\pm1/2\,jk} &= \begin{cases} (k_{ra})^{l}_{ijk}, & \text{if } p^{l}_{ijk} > p^{l}_{i\pm1\,jk} \\ (k_{ra})^{l}_{i\pm1\,jk}, & \text{if } p^{l}_{ijk} \le p^{l}_{i\pm1\,jk} \end{cases} \\
(k_{ra})^{l}_{ij\pm1/2\,k} &= \begin{cases} (k_{ra})^{l}_{ijk}, & \text{if } p^{l}_{ijk} > p^{l}_{ij\pm1k} \\ (k_{ra})^{l}_{ij\pm1k}, & \text{if } p^{l}_{ijk} \le p^{l}_{ij\pm1k} \end{cases} \\
(k_{ra})^{l}_{ijk\pm1/2} &= \begin{cases} (k_{ra})^{l}_{ijk}, & \text{if } p^{l}_{ijk} > p^{l}_{ijk\pm1} \\ (k_{ra})^{l}_{ijk\pm1}, & \text{if } p^{l}_{ijk} \le p^{l}_{ijk\pm1} \end{cases}
\end{aligned}
\tag{15.26}
$$

The FVM discretization of the NAPL phase mass balance is written in implicit form as

$$n_{ijk}\frac{(S_n)_{ijk}^{l+1}-(S_n)_{ijk}^{l}}{\Delta t}=-\frac{(k_{na})_{ijk}^{l}}{\rho_n}(C^*-C_{ijk}^{l+1}) \tag{15.27}$$

where $(k_{na})_{ijk}^l$ is the mass transfer rate coefficient, n_{ijk} is the porosity, and C_{ijk} is the aqueous phase concentration of the NAPL species for the (i, j, k) grid cell. The time-lagged nature of k_{na} is evident from this formulation, while the implicit formulation results in unconditional stability for this discrete equation with a truncation error of $O(\Delta t)$.

An explicit predictor-corrector FVM approach for the advective-dispersive-reactive transport equation is written as

$$n_{ijk}\frac{\left(1-(S_n)_{ijk}^{l+1}\right)C_{ijk}^{l+1}-\left(1-(S_n)_{ijk}^{l}\right)C_{ijk}^{l}}{\Delta t}-(k_{na})_{ijk}^{l}\left(C^*-C_{ijk}^{l+1}\right)=$$

$$-\frac{q_{x_{i+1/2\,jk}}C_{i+1/2\,jk}^{l+1/2}-q_{x_{i-1/2\,jk}}C_{i-1/2\,jk}^{l+1/2}}{\Delta x}$$

$$-\frac{q_{y_{ij+1/2k}}C_{ij+1/2k}^{l+1/2}-q_{y_{ij-1/2k}}C_{ij-1/2k}^{l+1/2}}{\Delta y}$$

$$-\frac{q_{z_{ijk+1/2}}C_{ijk+1/2}^{l+1/2}-q_{z_{ijk-1/2}}C_{ijk-1/2}^{l+1/2}}{\Delta z}$$

$$+\frac{D_{xx_{i+1/2\,jk}}\left(C_{i+1\,jk}^{l}-C_{ijk}^{l}\right)-D_{xx_{i-1/2\,jk}}\left(C_{ijk}^{l}-C_{i-1\,jk}^{l}\right)}{(\Delta x)^2}$$

$$+\frac{D_{yy_{ij+1/2k}}\left(C_{ij+1k}^{l}-C_{ijk}^{l}\right)-D_{yy_{ij-1/2k}}\left(C_{ijk}^{l}-C_{ij-1k}^{l}\right)}{(\Delta y)^2}$$

$$+\frac{D_{zz_{ijk+1/2}}\left(C_{ijk+1}^{l}-C_{ijk}^{l}\right)-D_{zz_{ijk-1/2}}\left(C_{ijk}^{l}-C_{ijk-1}^{l}\right)}{(\Delta z)^2} \tag{15.28}$$

$$+2\left[\frac{D_{xy_{i+1/2\,jk}}\left(C_{i+1/2\,j+1/2k}^{l}-C_{i+1/2\,j-1/2k}^{l}\right)-D_{xy_{i-1/2\,jk}}\left(C_{i-1/2\,j+1/2k}^{l}-C_{i-1/2\,j-1/2k}^{l}\right)}{\Delta x\Delta y}\right.$$

$$+\frac{D_{xz_{i+1/2\,jk}}\left(C^l_{i+1/2\,jk+1/2}-C^l_{i+1/2\,jk-1/2}\right)-D_{xz_{i-1/2\,jk}}\left(C^l_{i-1/2\,jk+1/2}-C^l_{i-1/2\,jk-1/2}\right)}{\Delta x\Delta z}$$

$$\left.+\frac{D_{yz_{ij+1/2\,k}}\left(C^l_{ij+1/2\,k+1/2}-C^l_{ij+1/2\,k-1/2}\right)-D_{yz_{ij-1/2\,k}}\left(C^l_{ij-1/2\,k+1/2}-C^l_{ij-1/2\,k-1/2}\right)}{\Delta y\Delta z}\right]$$

The interface velocity values are calculated using the pressure field obtained after solution of Equation 15.23 at the l-th time level using the following discrete form of Equation 15.9

$$q_{x_{i+1/2\,jk}}=-\lambda_{i+1/2\,jk}\frac{\left(p^l_{i+1\,jk}-p^l_{ijk}\right)}{\Delta x}$$

$$q_{x_{i-1/2\,jk}}=-\lambda_{i-1/2\,jk}\frac{\left(p^l_{ijk}-p^l_{i-1\,jk}\right)}{\Delta x}$$

$$q_{y_{ij+1/2\,k}}=-\lambda_{ij+1/2\,k}\frac{\left(p^l_{ij+1\,k}-p^l_{ijk}\right)}{\Delta y} \tag{15.29}$$

$$q_{y_{ij-1/2\,k}}=-\lambda_{ij-1/2\,k}\frac{\left(p^l_{ijk}-p^l_{ij-1\,k}\right)}{\Delta y}$$

$$q_{z_{ijk+1/2}}=-\lambda_{ijk+1/2}\left[\frac{\left(p^l_{ijk+1}-p^l_{ijk}\right)}{\Delta z}+\rho_a g\right]$$

$$q_{z_{ijk-1/2}}=-\lambda_{ijk-1/2}\left[\frac{\left(p^l_{ijk}-p^l_{ijk-1}\right)}{\Delta z}+\rho_a g\right]$$

The procedure for computing interface values $C^{l+1/2}_{i+1/2\,jk}$ in Equation 15.28 is a key factor in obtaining economical high-resolution simulation of contact discontinuities. We use an essentially nonoscillatory (ENO) type FVM scheme (Yee, 1987) to approximate the advective term, which we have developed. This scheme compares favorably to a variety of competing schemes, such as the Minmod (Van Leer, 1974; Huynh, 1989; Yokota and Huynh, 1990), Liu (Liu et al., 1994), and Roe (Roe, 1985) limiters.

We will report the detailed comparisons of these limiters in future work. The idea of the flux slope limiter is to select the differential approximation stencil adaptively, and using nodes in the local region that yield a smooth approximation of the derivative, preventing spurious oscillations in the vicinity of contact discontinuities. Although they prevent oscillations, ENO schemes retain higher-order approximation of the derivatives and have been effective for capturing shock features of a problem (Yee, 1987; Colella, 1990). The slope limiter used here provides a second-order accurate spatial discretization.

Our second-order ENO-type slope limiter is specified for interface values $C^l_{i+1/2jk}$ by

$$C^l_{i+1/2\,jk} = \begin{cases} \left(11C^l_{ijk} - 7C^l_{i-1\,jk} + 2C^l_{i-2\,jk}\right)/6, \text{ if } \delta_{i+1} > \max(\delta_i, \delta_{i-1}) \\ \left(2C^l_{i+1\,jk} + 5C^l_{ijk} - C^l_{i-1\,jk}\right)/6, \quad \text{otherwise} \end{cases} \tag{15.30}$$

where $\delta_i = C^l_{ijk} - C^l_{i-1jk}$, and $\delta_{i+1} = C^l_{i+1jk} - C^l_{ijk}$. In addition, the interface value $C^l_{i+1/2jk}$ is subject to the constraint

$$C^l_{ijk} \le C^l_{i+1/2\,jk} \le C^l_{i+1\,jk} \tag{15.31}$$

Similar formulas can be written for interface values $C^l_{ij+1/2k}$ and $C^l_{ijk+1/2}$. All interface concentrations are consistent and independent of the reference direction; therefore, negative half-step values follow directly from the above definitions.

The slope limiter given by Equations 15.30 and 15.31 performs well when the flow is oriented in the x direction. However, it does not account for the flux components in the two other directions. In the case of oblique flows, this shortcoming may lead to a significant corruption of the solution (Colella, 1990). In order to account for the multidimensional nature of flows in porous media, we incorporate the limiter described by Equations 15.30 and 15.31 in the following predictor-corrector equation

$$C^{l+1/2}_{i+1/2\,jk} = C^l_{i+1/2\,jk} - \frac{\Delta t}{2(\theta_a)^l_{i+1/2\,jk}} \left[\frac{\partial(q_x C_a)}{\partial x} + \frac{\partial(q_y C_a)}{\partial y} + \frac{\partial(q_z C_a)}{\partial z} \right]^l_{i+1/2\,jk} \tag{15.32}$$

where $(\theta_a)^l_{i+1/2\,jk} = \left((\theta_a)^l_{ijk} + (\theta_a)^l_{i+1\,jk}\right)/2$. The second term on the right-hand side of this formula is discretized in the same way as in Equation 15.28. A similar corner-transport scheme was used by Colella (1990) for two-dimensional pure advection problems. In this work, Colella presented a geometrical interpretation of his corner transport scheme and showed advantages for such an approach, including reduction of numerical dispersion and grid orientation error. Besides these advantages, the predictor-corrector implemented in this work increases the temporal order of approximation of the FVM for pure advection problems, which in turn generally leads to more accurate solutions. As follows from a Taylor series expansion, the use of the predictor-corrector approach described by Equations 15.28 and 15.32 provides a second-order correct temporal approximation for the pure advection problem, which also benefits advection-dominated problems.

In order to improve computational performance of the simulator described in this section, we used a selective adaptive-time-step method. This method is a modification of the conventional approach for calculating the time-step size at each time level based on the Courant condition (Aziz and Settari, 1979; Zheng, 1990)

$$\Delta t < C_r \min\left(\frac{\Delta x}{|q_x|}, \frac{\Delta y}{|q_y|}, \frac{\Delta z}{|q_z|}\right) \tag{15.33}$$

and a stability constraint for the explicitly approximated dispersion term

$$\Delta t < \frac{1}{2}\left(\frac{D_{xx}}{\Delta x^2} + \frac{D_{yy}}{\Delta y^2} + \frac{D_{zz}}{\Delta z^2}\right)^{-1} \tag{15.34}$$

where C_r is the Courant number that is less than one.

The approach most commonly used is based on the calculation of the maximum allowable time-step size by applying the above time-step constraints to every location in the domain and selecting the minimum value (Zheng, 1990). We modified this usual approach by choosing the time step based on the maximum time step satisfying the constraints given by Equations 15.33 and 15.34, but only within the area where $C^l_{ijk} > C_c$, with $C_c \leq 10^{-2} \max C^l_{ijk}$. Implementing the selective adaptive time-step method resulted in substantial computational savings.

Linear Algebra

The discrete finite-volume representation of Equation 15.16, given by Equation 15.23, results in the need to solve a simultaneous system of equations

$$[A]\{p_a\} = \{b\} \tag{15.35}$$

where $[A]$ is a known coefficient matrix, $\{b\}$ is a known vector, and $\{p_a\}$ is the solution vector sought. Because the governing partial differential equation is self-adjoint, this system of equations is symmetric. The standard approach for solving such equations is the preconditioned conjugate gradient (PCG) method. We used the line successive over-relaxation (LSOR) method, which is a simple solver that was popular before the maturation of the PCG method (Aziz and Settari, 1979). The rationale for this approach is that since S_n changes slowly, so does the solution sought, p_a. Because of this, LSOR converges rapidly if the previous solution for p_a is used as the initial estimate of the solution—typically within a couple of iterations. Given this convergence behavior, LSOR is a reasonable choice for this matrix-free implementation. The solution is termed matrix-free because the entire coefficient matrix $[A]$ is never constructed.

This solution approach is not optimal for general compositional models or other typical scientific computing applications. The special conditions of the fingering simulations performed in this work make it a reasonable choice. For general applications, PCG methods are far superior, especially when a good preconditioner is available. A multigrid preconditioner or solver are attractive choices for general applications as well.

SIMULATION RESULTS AND DISCUSSION

The numerical model was used to investigate dissolution fingering by first examining the effect of system parameters on finger development for an ideal geometry and

Table 15.1 Base-Case Simulation Variables.

Variable	Description	Value	Units
L	Length of the domain	5.00×10^{1}	cm
W	Width of the domain	2.50×10^{1}	cm
W_p	Perturbation width	2.00×10^{0}	cm
L_p	Perturbation length	5.00×10^{-1}	cm
S_{ni}	Initial NAPL saturation	2.00×10^{-1}	
S_{np}	NAPL saturation of perturbation	0.00×10^{0}	
n	Porosity	3.60×10^{-1}	
γ	Exponent of relative permeability relation	3.00×10^{0}	
S_{ia}	Irreducible water saturation	1.04×10^{-1}	
q_x	Darcy velocity of water injection	2.00×10^{2}	cm/day
α_l	Longitudinal dispersivity	7.20×10^{-2}	cm
α_t	Transverse dispersivity	1.44×10^{-2}	cm
D_m	Molecular diffusion	8.40×10^{-6}	cm^2/s
τ	Tortuosity (isotropy assumed)	6.60×10^{-1}	
ρ_n	NAPL density	1.46×10^{0}	g/cm^3
ρ_a	Water density	1.00×10^{0}	g/cm^3
μ_a	Water viscosity	1.00×10^{0}	cp
C^*	Aqueous solubility of NAPL	1.27×10^{-3}	g/cm^3
β_1	Mass transfer correlation parameter	1.86×10^{2}	s^{-1}
β_2	Mass transfer correlation exponent	8.70×10^{-1}	
β_3	Mass transfer correlation exponent	7.10×10^{-1}	

then simulating the development of dissolution fingers in a physical experiment. We begin with a discussion of the numerical experiments conducted for the sensitivity analysis, comparing these results to the LSA.

Sensitivity Analysis and Comparison with LSA

Grid Size

The base-state simulation variables used in the numerical simulation are given in Table 15.1. These values are representative of conditions encountered in the two-dimensional physical experiments discussed above (Imhoff and Miller, 1996). The domain modeled was 25-cm wide by 50-cm long, which was wide enough so that the left and right boundaries would have a minor effect on fingering in the center of the domain and sufficiently long that the long-term development of a dissolution finger could be observed. In these simulations, the top and bottom boundary conditions were a specified flux for the aqueous phase, while the lateral boundaries of the domain were no-flow. The permeability, porosity, and initial residual NAPL saturation field were homogeneous throughout the system, except for a perturbation in the residual NAPL saturation near the top boundary. Here, a region between 1 to 2-cm wide and 0.5-cm long was NAPL free at the start of the simulation. The development of this perturbation into a dissolution finger was observed in each computational run as the aqueous phase was flushed in the top boundary and the dissolved NAPL was eluted from the bottom boundary.

Figure 15.3 depicts the development of an initial perturbation that was 1-cm wide for two uniform rectangular grids: 50×100 (perturbation 1.0-cm long) and 100×200 (perturbation 0.5-cm long). The initial perturbation grew with time and eventually broke through the bottom boundary after approximately 160 pore volumes (PV) were

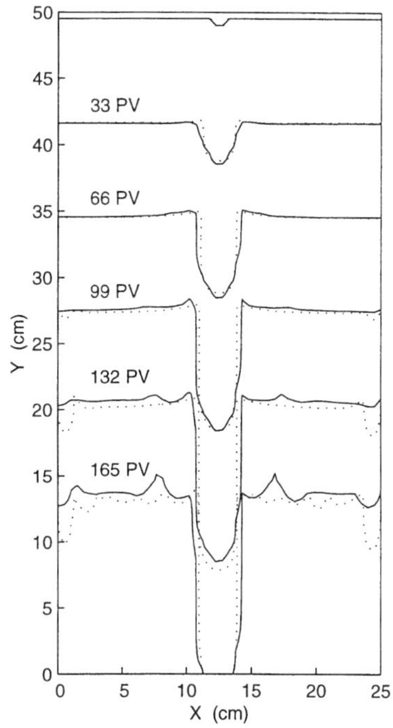

Figure 15.3. Numerical simulation of dissolution fingering for two discretizations: 100×200, dotted line; and 50×100, solid line. The initial perturbation in S_n was 1-cm wide and 0.5 cm long. Remaining parameter values are given in Table 15.1. Lines delineate contours of S_n =0.193.

flushed through the system. The initial perturbation width of 1 cm increased to approximately 4 cm at 33 PV, after which the finger did not widen. An asymptotic limit was not achieved for the finger length, though, suggesting that the dissolution finger could grow significantly longer. Differences between the coarse and fine grid discretization were minor, and the effect of the initial length of the perturbation was also not significant. For all subsequent simulations, 100×200 computational grids were employed, except for the relative permeability relation sensitivity analysis in which a 50×100 computational grid was found adequate.

The perturbations in the NAPL saturation field on the left and right boundaries of the system are due to the effect of the no-flow boundaries—the numerical approximations of these boundary conditions resulted in small perturbations in the NAPL saturation in these regions. Some asymmetry can also be observed at the boundaries due to the numerical discretization.

Width of Initial Perturbation

When the width of the perturbation in the initial NAPL saturation was reduced from 1 cm to 0.5 cm, the dissolution finger grew at a slightly slower rate, but the asymptotic limit of the finger width was ≈ 3 cm, the same as that observed when the initial perturbation was 1-cm wide. Increasing the width of the perturbation to 2 cm, though, led to significantly different results, shown in Figure 15.4. Bifurcation of the initial perturbation into two dissolution fingers occurred, illustrating the preferred finger width of ≈ 3 cm for this system. Similar bifurcation of dissolution fingers was observed in numerical experiments of mineral dissolution fingering (Wei and Ortoleva, 1990).

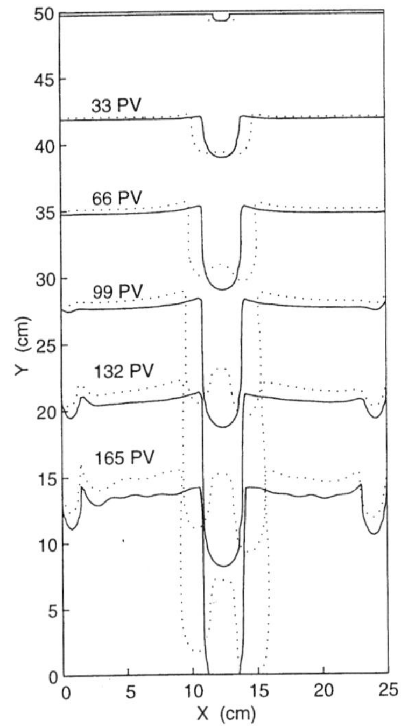

Figure 15.4. Numerical simulation of dissolution fingering for two widths in the initial perturbation in S_n: 2 cm, dashed line; and 1 cm, solid line. Remaining parameter values are given in Table 15.1. Lines delineate contours of S_n = 0.193.

It is apparent from these simulations that, for this system and for the range of perturbations examined, the preferred width of the dissolution finger is on the order of 3 cm. For this same system, the LSA predicts that the fastest-growing fingers—thus fingers that are most likely to be observed—have a wavelength of ≈ 2 cm (see Figure 6A of Imhoff and Miller, 1996), corresponding to a finger width of ≈ 1 cm. Thus, the numerical simulations indicate wider fingers than the LSA. It is important to note, though, that the LSA strictly applies only to the onset of fingering, when finger lengths are small. While finger widths in the two-dimensional cell ranged initially between 0.5 cm to 3.5 cm, they grew wider as they grew longer (Imhoff et al., 1996). The cell used in the two-dimensional experiments was not long enough to observe an asymptotic value for the finger width.

Transverse Dispersion

The LSA analysis of NAPL dissolution fingering indicated that transverse dispersion plays a significant role in fingering. Numerical simulations varying the transverse dispersivity are illustrated in Figure 15.5. The dissolution fingers grew wider and at a slower rate as the transverse dispersivity was increased from α_t = 0.0072 cm to α_t = 0.072 cm—the same trend that was predicted from LSA (see Figure 8 of Imhoff and Miller [1996]). To our knowledge, transverse dispersion between a region with trapped NAPL ganglia and clean media has not been studied systematically. Thus, the correct description of dispersion along the finger boundary is uncertain, although it plays an important role in finger development.

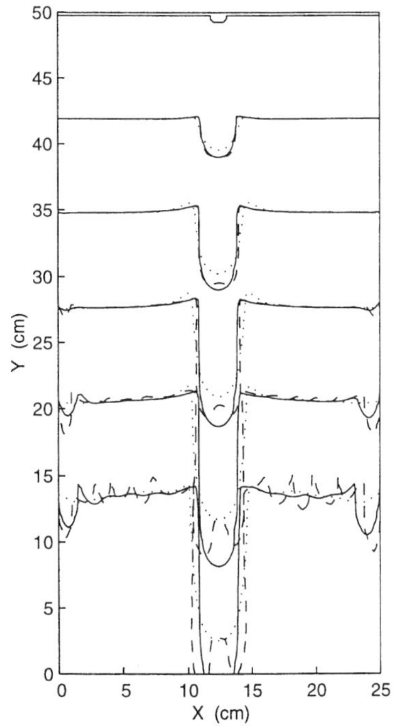

Figure 15.5. Numerical simulation of dissolution fingering for three transverse dispersivities: α_t = 0.072 cm, dotted line; α_t =0.0144 cm, solid line; and α_t = 0.0072 cm, dashed line. Remaining parameter values are given in Table 15.1. Lines delineate contours of S_n = 0.193.

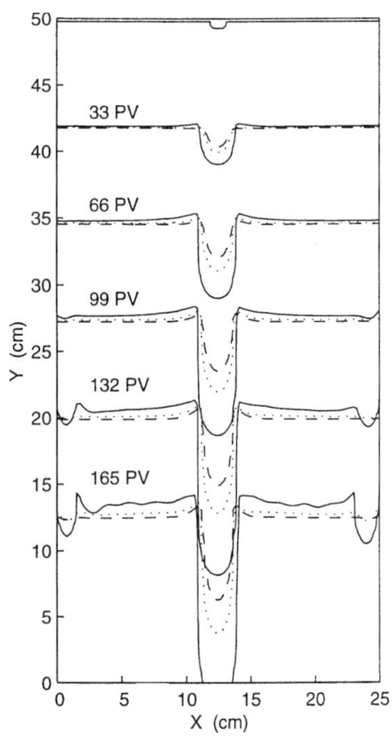

Figure 15.6. Numerical simulation of dissolution fingering for three values of the exponent γ for the relative permeability function: γ =1.5, dotted line; γ = 2, dashed line; and γ = 3, solid line. Remaining parameter values are given in Table 15.1. Lines delineate contours of S_n = 0.193.

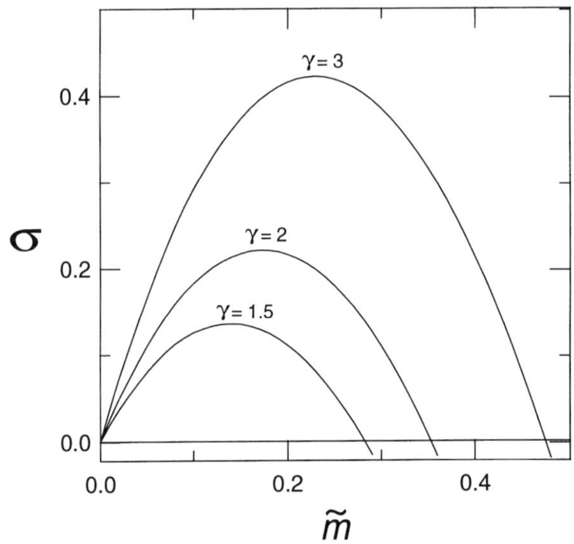

Figure 15.7. Dimensionless growth rate σ versus dimensionless wave number $\tilde{m}$ from the linear stability analysis for three values of the exponent γ for the relative permeability function. Remaining parameter values are given in Table 15.1.

Aqueous-Phase Relative Permeability

The aqueous-phase relative permeability is a function of pore structure as well as fluid and solid properties (Dullien, 1992). To investigate the importance of relative permeability on dissolution fingering, the exponent γ in the relative permeability function was varied between $\gamma = 1.5$ to $\gamma = 3$ in a series of numerical simulations. The results from these simulations are shown in Figure 15.6. As γ was decreased from 3 to 1.5, fingers grew more slowly and were slightly narrower. The same trend in growth rate is predicted from LSA and is illustrated in Figure 15.7. Here, the dimensionless growth rate of the fingers σ decreases as γ decreases from 3 to 1.5. Thus, dissolution fingering is quite sensitive to the aqueous phase relative permeability, which is to be expected, given the important role of aqueous-phase velocity in finger development.

Mass Transfer Rate

A fundamental assumption in the LSA was that the front over which NAPL-aqueous phase mass transfer occurred was thin, implying either a fast rate of NAPL dissolution or a slow aqueous-phase velocity (Imhoff and Miller, 1996). Dissolution fingers are perturbations on this front that grow with time. To assess the importance of NAPL-aqueous phase mass transfer on dissolution fingering, simulations were performed assuming local equilibrium between the nonaqueous and aqueous phases as well as rate-limited mass transfer, using the mass transfer parameters given for the base-state conditions in Table 15.1. The results are shown in Figure 15.8. The dissolution finger grew at a slightly faster rate when local equilibrium was assumed, but the finger was of similar width for both mass transfer models. Smaller-scale instabilities on the order of 1 to 2 cm in width developed on the dissolution front when local equilibrium was assumed. Such instabilities will be investigated using finer grids in future work.

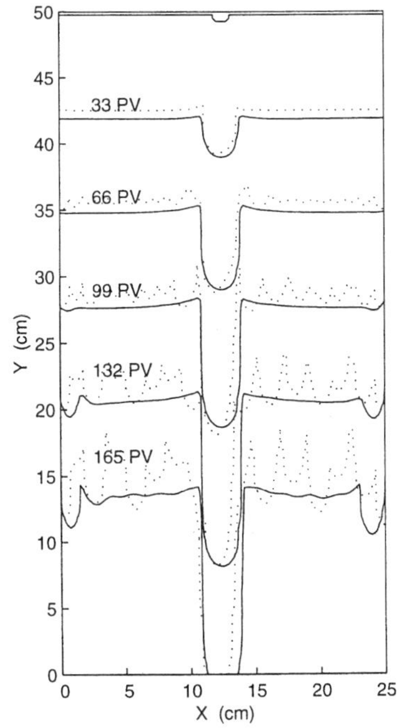

Figure 15.8. Numerical simulation of dissolution fingering assuming local equilibrium (dotted line) or first-order mass transfer (solid line) between the nonaqueous and aqueous phases. The parameters for the first-order mass transfer model are given in Table 15.1 along with the other parameters common to both simulations.

Comparison with a Physical Experiment

X-ray attenuation data were used to measure porosity and NAPL saturation on a 0.5×0.5-cm grid at three times during the two-dimensional experiment depicted in Figure 15.2. The conditions for this experiment are described in Imhoff et al. (1996) and are similar to the conditions listed in Table 15.1. Dissolution fingers formed in regions of the cell where the vertically-averaged initial residual NAPL saturation, as measured by X-ray attenuation, was lower than the cell-averaged value. Thus, fingers developed in regions where the initial aqueous phase velocity was slightly greater than the mean velocity, due to differences in the aqueous-phase relative permeability. The numerical model was used to simulate this experiment, using the parameters measured or estimated from the physical experiment (Imhoff et al., 1996) and a 0.25×0.25-cm discretization.

The dissolution front from the simulations is depicted in Figure 15.9, corresponding to the same times (44 hr) in the experiment shown in Figure 15.2. There is reasonable agreement between the experimental data and numerical simulation, although the numerical simulation did not capture the development of every finger. Two possible explanations for the disagreement between the physical experiment and the simulation results are (1) variations in porosity and NAPL saturation at scales smaller than the X-ray measurement scale may have occurred and been important, and these variations were not included in the numerical simulator; and (2) the models for relative permeability and/or transverse dispersion used in the numerical simulator may not be sufficiently accurate. Relative permeability and transverse dispersion were shown to have a significant effect on dissolution fingering in the sensitivity analysis. Despite some differences between the physical experiment and the simulation results, there is

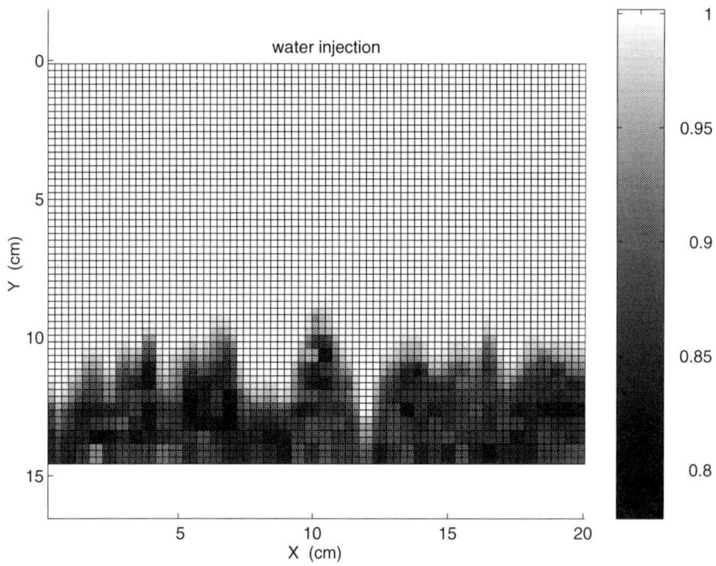

Figure 15.9. Numerical simulation of NAPL dissolution during a two-dimensional flushing experiment. Simulation parameters were taken from the experimental data reported by Imhoff et al. (1996).

enough agreement to give us confidence in the numerical simulator's usefulness for investigating dissolution fingering in porous media.

Unresolved Issues

The numerical simulations described in this chapter, and the LSA and physical experiments reported earlier (Imhoff and Miller, 1996; Imhoff et al., 1996) have shed considerable light on NAPL dissolution fingering. However, there are many aspects of dissolution fingering remaining that are worthy of investigation:

- the effect of dispersion on dissolution fingering;
- the effect of heterogeneity in porous media properties and residual NAPL saturation on finger formation;
- accounting for the third dimension on dissolution fingering;
- fingering and the validity of the LSA when the flushing solution is augmented with chemical agents (e.g., surfactants and alcohols); and
- the incorporation of the effects of dissolution fingering on NAPL-aqueous phase mass transfer in models where the system is discretized at significantly larger scales than the centimeter scale.

ACKNOWLEDGMENTS

This work was supported in part by: Army Research Office Grant DAAL03-92-G-0111; National Institute of Environmental Health Sciences Grant 5 P42 ES05948; U.S. Army Engineer Waterways Experiment Station Grant DACA39-95-K-0098; and the North Carolina Supercomputing Center.

REFERENCES

Abriola, L.M. Multiphase Flow and Transport Models for Organic Chemicals: A Review and Assessment. Tech. Rep. EPRI EA-5976, Electric Power Research Institute, Palo Alto, CA, 1988.

Abriola, L.M. and G.F. Pinder. A multiphase approach to the modeling of porous media contamination by organic compounds. 1. Equation development. *Water Resour. Res.,* 21(1):11–18, 1985.

Adler, P.M. and H. Brenner. Multiphase flow in porous media. *Annu. Rev. Fluid Mech.,* 20:35–59, 1988.

Armstrong, J.E., E.O. Frind, and R.D. Mcclellan. Nonequilibrium mass transfer between the vapor, aqueous, and solid phases in unsaturated soils during vapor extraction. *Water Resour. Res.,* 30(2):355–368, 1994.

Aziz, K. and A. Settari. *Petroleum Reservoir Simulation.* Applied Science Publ., Ltd., London, 1979.

Bear, J. *Hydraulics of Groundwater.* McGraw-Hill, New York, 1979.

Borden, R.C. and C.-M. Kao. Evaluation of groundwater extraction for remediation of petroleum-contaminated aquifers. *Water Environ. Res.,* 64(1):28–36, 1992.

Brusseau, M.L., R.E. Jessup, and P.S.C. Rao. Modeling the transport of solutes influenced by multiprocess nonequilibrium. *Water Resour. Res.,* 25(9):1971–1988, 1989.

Chadam, J., D. Hoff, E. Merino, P. Ortoleva, and A. Sen. Reactive infiltration instabilities. *J. Appl. Math.,* 36:207–221, 1986.

Chadam, J. and P. Ortoleva. Morphological instabilities in physico-chemical system. *Earth-Sci. Rev.,* 29:175–181, 1990.

Chadam, J., A. Peirce, and P. Ortoleva. Stability of reactive flows in porous media: Coupled porosity and viscosity changes. *SIAM J. Appl. Math.,* 51(3):684–692, 1991.

Cohen, R.M. and J.W. Mercer. DNAPL Site Evaluation. Tech. Rep. EPA/600/SR-93/022, U.S. EPA Robert S. Kerr Environmental Research Laboratory, Ada, OK, 1993.

Colella, P. Multidimensional upwind methods for hyperbolic conservation laws. *J. Computational Phys.,* 54:171–200, 1990.

Corey, A.T. *Mechanics of Immiscible Fluids in Porous Media.* Water Resources Publication, Highlands Ranch, CO, 1994.

Crichlow, H.B. *Modern Reservoir Engineering—A Simulation Approach.* Prentice Hall, Englewood Cliffs, NJ, 1977.

Dullien, F.A.L. *Porous Media: Fluid Transport and Pore Structure.* Academic Press, San Diego, CA, 1992.

Geller, J.T. and J.R. Hunt. Mass transfer from nonaqueous phase organic liquids in water-saturated porous media. *Water Resour. Res.,* 29(4):833–845, 1993.

Gray, W.G., A. Leijnse, R.L. Kolar, and C.A. Blain. *Mathematical Tools for Changing Scales in the Analysis of Physical Systems.* CRC Press, Inc., Boca Raton, FL, 1993.

Hassanizadeh, S.M. and W.G. Gray. Thermodynamic basis of capillary pressure in porous media. *Water Resour. Res.,* 29(10):3389–3405, 1993a.

Hassanizadeh, S.M. and W.G. Gray. Toward an improved description of the physics of two-phase flow. *Adv. Water Resour.,* 16(1):53–67, 1993b.

Hatfield, K., J. Ziegler, and D.R. Burris. Transport in porous media containing residual hydrocarbon. II: Experiments. *J. Environ. Eng. ASCE,* 119(3):559–575, 1993

Huynh, H.T. Second-order accurate nonoscillatory schemes for scalar conservation laws. In Proceedings of the Sixth International Conference on Numerical Methods in Laminar and Turbulent Flows, Pineridge Press, 1989, pp. 25–38.

Imhoff, P.T. Dissolution of a Nonaqueous Phase Liquid in Saturated Porous Media. Ph.D. thesis, Princeton University, 1992.

Imhoff, P.T., A Frizzell, and C.T. Miller. An evaluation of thermal effects on the dissolution of a nonaqueous phase liquid in porous media. *Environ. Sci. Technol.,* (in press) 1997.

Imhoff, P.T., S.N. Gleyzer, J.F. McBride, L.A. Vancho, I. Okuda, and C.T. Miller. Cosolvent-enhanced remediation of residual dense nonaqueous phase liquids: Experimental investigation. *Environ. Sci. Technol.,* 29(8):1966–1976, 1995.

Imhoff, P.T. P.R. Jaffe, and G.F. Pinder. An experimental study of complete dissolution of a nonaqueous phase liquid in saturated porous media. *Water Resour. Res.,* 30(2):307–320, 1994.

Imhoff, P.T. and C.T. Miller. Dissolution fingering during the solubilization of nonaqueous phase liquids in saturated porous media. 1. Model predictions. *Water Resour. Res.,* 32(7):1919–1928, 1996.

Imhoff, P.T., G.P. Thyrum, and C.T. Miller. Dissolution fingering during the solubilization of nonaqueous phase liquids in saturated porous media. 2. Experimental observations. *Water Resour. Res.,* 32(7):1929–1942, 1996.

King, M.J., M.J. Blunt, M. Mansfield, and M.A. Christie. Rapid evaluation of the impact of heterogeneity on miscible gas injection. In *New Developments in Improved Oil Recovery,* de Haan, H.J. (Ed.), The Geological Society, London, 1995, pp. 133–142.

Liu, J., M. Delshad, G.A. Pope, and K. Sepehrnoori. Application of higher-order flux-limited methods in compositional simulation. *Transp. Porous Media,* 16(1):1–29, 1994.

Lowry, M.I. and C.T. Miller. Pore-scale modeling of nonwetting-phase residual in porous media. *Water Resour. Res.,* 31(3):455–473, 1995.

Mayer, A.S. and C.T. Miller. The influence of porous medium characteristics and measurement scale on pore-scale distributions of residual nonaqueous-phase liquids. *J. Contaminant Hydrol.,* 11(3/4):189–213, 1992.

Mayer, A.S. and C.T. Miller. An experimental investigation of pore-scale distributions of nonaqueous phase liquids at residual saturation. *Trans. Porous Media,* 10(1):57–80, 1993.

Miller, C.T., G. Christakos, P.T. Imhoff, J.F. McBride, J.A. Pedit, and J.A. Trangenstein. Multiphase flow and transport modeling in heterogeneous porous media: Challenges and approaches. *Adv. Water Resour.,* (in press), 1997.

Miller, C.T., M.M. Poirier-McNeill, and A.S. Mayer. Dissolution of trapped nonaqueous phase liquids: Mass transfer characteristics. *Water Resour. Res.,* 26(11):2783–2796, 1990.

Moissis, D.E., C.A. Miller, and M.F. Wheeler. A parametric study of viscous fingering immiscible displacement by numerical simulation. In *Numerical Simulation in Oil Recovery,* Wheeler, M.F. (Ed.), Springer-Verlag, New York, 1988, pp. 227–247.

Ortoleva, P., J. Chadam, E. Merino, and A. Sen. Geochemical self-organization. II: The reactive-infiltration instability. *Am. J. Sci.,* 287:1008–1040, 1987a.

Ortoleva, P., E. Merino, C. Moore, and J. Chadam. Geochemical self-organization. I: Reaction-transport feedbacks and modeling approaches. *Am. J. Sci.,* 287:979–1007, 1987b.

Parker, J.C., A.K. Katyal, J.J. Kaluarachchi, R.J. Lenhard, T.J. Johnson, K. Jayaraman, K. Unlu, and J.L. Zhu. Modeling Multiphase Organic Chemical Transport in Soils and Ground Water. Tech. Rep. EPA/600/2-91/042, U.S. EPA Robert S. Kerr Environmental Research Laboratory, Ada, OK, 1991.

Powers, S.E., L.M. Abriola, and W.J. Weber, Jr. An experimental investigation of nonaqueous phase liquid dissolution in saturated subsurface systems: Steady state mass transfer rates. *Water Resour. Res.,* 28(10):2691–2705, 1992.

Powers, S.E., L.M. Abriola, and W.J. Weber, Jr. An experimental investigation of nonaqueous phase liquid dissolution in saturated subsurface systems: Steady state mass transfer rates. *Water Resour. Res.,* 30(2):321–332, 1994.

Powers, S.E., C.O. Loureiro, L.M. Abriola, and W.J. Weber, Jr. Theoretical study of the significance of nonequilibrium dissolution of nonaqueous phase liquids in subsurface systems. *Water Resour. Res.,* 27(4):463–477, 1991.

Roe, P.L. Some contribution of the modeling of discontinuous flows. *Lect. Appl. Math.,* 22:163–194, 1985.

Russell, T.F. Modeling of multiphase multicontaminant transport in the subsurface. *Rev. Geophys.,* 33(suppl. 2):1035–1047, 1995.

Simons, G.A. Extension of the "pore tree" model to describe transport in soil., *Ground Water,* 34(4):683–690, 1996.

Soerens, T.S., D.A. Sabatini, and J.H. Harwell. Column studies demonstrating the effects of heterogeneity on the kinetics of PCE dissolution into surfactant solutions. In *EOS Transactions, American Geophysical Union,* 76(46):F274, 1995.

Van Leer, B. Towards the ultimate conservative difference scheme, II. Monotonicity and conservation combined in a second order scheme. *J. Computational Phys.,* 14, 361–370, 1974.

Wei, C. and P. Ortoleva. Reaction front fingering in carbonate-cemented sandstones. *Earth-Sci. Rev.,* 29:183–198, 1990.

Yee, H.C. Construction of explicit and implicit symmetric TVD schemes and their applications. *J. Computational Phys.,* 68:151–179, 1987.

Yokota, J.W. and T.H. Huynh. A Nonoscillatory, Characteristically Convected, Finite Volume Scheme for Multidimensional Convection Problems. Tech. Rep., NASA Lewis Research Center, Cleveland, OH, 1990.

Zheng, C. MT3D. A Modular Three-Dimensional Transport Model for Simulation of Advection, Dispersion and Chemical Reactions of Contaminant in Groundwater Systems. Tech. Rep., Report to Environmental Protection Agency, Ada, OK, 1990.

CHAPTER SIXTEEN

Flow and Entrapment of Nonaqueous Phase Liquids in Heterogenous Soil Formations

T.H. Illangasekare

INTRODUCTION

Many organic pollutants that threaten the groundwater quality are in the form of nonaqueous phase liquids or NAPLs. Primary sources through which these contaminants are introduced to the subsurface are: seepage from unlined lagoons and surface impoundments, improper land filling, improper surface and subsurface disposal practices, leaks in pipes and underground storage tanks and other processing and transport equipment, and accidental spills. Many NAPLS are only sparingly soluble in water, and therefore migrate through the subsurface environment as a separate phase. Once entrapped in the soil pores, they act as long-term sources of soil and groundwater contamination.

Nonaqueous phase contaminants are identified according to their relative densities compared to water; lighter than water nonaqueous liquids (LNAPLs), and denser than water nonaqueous phase liquids (DNAPLs). Petroleum hydrocarbon products such as gasoline, fuel oils, and jet fuel are the most common LNAPLs that are of interest in groundwater contamination. When in contact with flowing water, the soluble components in these separate phase fluids (e.g., benzene, toluene, ethyl benzene, and xylene, or BTEXs) dissolve into the aqueous phase producing contaminant plumes. Common chlorinated solvents such as 1,1,1-trichloroethane (1,1,1-TCA), tetrachloroethylene (PCE), trichloroethylene (TCE), carbon tetrachloride, 1,2-dichloroethane (1,2-DCA), chloroform (TCM), etc., falls into the DNAPL category. In addition, halogenated benzenes, polychlorinated biphenyls, coal tar, creosote, and certain types of pesticides are also generally classified as DNAPLs. The solubilities of these chemicals are many orders of magnitude higher than the allowable maximum concentration limits (MCLs) specified in the drinking water standards that are currently enforced in the United States.

In heavily contaminated sites where large volumes of NAPLs have migrated, recovery has to be done before remediation. Primary recovery is generally accomplished by using pumping wells or drains. After primary recovery, the NAPL which stays behind in the soil becomes discontinuous and is immobilized by capillary forces, resulting in residual entrapment. Decontamination mainly involves excavating the contaminated soil and subsequent treatment, incineration, or land disposal. In sites where the contaminated zones are either under manufacturing plants or buildings where excavation becomes infeasible, or cost considerations exclude *ex situ* treatment or disposal of contaminated soils, *in situ* treatment provides an attractive alternative. These treatment schemes may involve pump-and-treat technologies, surfactant washing processes, air stripping, and bioremediation, among others.

The limitations of field application of laboratory validated remediation technologies can be attributed to a number of factors. These include complexities created by the natural soil heterogenities in the field; wastes that are complex chemical mixtures; complex physical, chemical, and biological interactions; nonavailability of efficient and cost-effective field characterization techniques; and limitations in the modeling tools needed in the design and evaluation of the field remediation techniques, among others.

In this chapter, we will present the basic physical processes that are important to the understanding and modeling of flow and entrapment of nonaqueous phase fluids in heterogenous porous media. We will primarily use experimental results to obtain an insight into the basic mechanisms. Incorporation of these mechanisms into numerical models becomes critical in using these models for the design and evaluation of various remediation schemes.

HETEROGENEITY IN MULTIPHASE FLOW

All underground porous formations in their natural state will exhibit variability of properties in space. These variations are referred to as heterogenities. As natural porous media are formed by mixtures of particles of varying sizes and shapes, the heterogenities will depend on the distribution of the particle sizes and their packing arrangement. Properties such as porosity, mean pore size, and permeability that depend on particle size and packing control the storage and flow of fluids. These properties in turn are used to quantify the variability in heterogeneous formations. The definition and determination of variability depend on the measurement scale of these properties. Natural heterogenities resulting from the spatial variation of soil types and properties affect the flow and entrapment behavior of nonaqueous phase fluids in soil formations. In this section, we will provide a brief summary of literature where the role of heterogeneity in flow and entrapment of NAPLs in porous media is discussed.

In natural soil formations, the movement of contaminants is complicated by soil heterogeneities such as fractures, macropores, and layering. Experimental studies by Schwille (1988), Kueper and Frind (1991), Campbell (1992), and Illangasekare et al. (1995a,b), have shown that soil layering can cause lateral spreading, preferential flow, and pooling of the organic liquid.

Schwille (1967) performed a number of qualitative experiments of oil infiltration into layered material. He observed that in stratified media, higher oil contents have been measured at the boundary between less permeable and more permeable layers. Huntley et al. (1994) investigated the distribution and mobility of a LNAPL in the subsurface at a field site. They found that the saturation of the LNAPL in most bore holes cannot be described as a smooth function. Variations from the expected smooth saturations correlated qualitatively with variations in soil texture. Significant heterogeneities occur on a relatively small scale at this site, which appear to impact the hydrocarbon saturation profile directly. Results similar to these field results have also been shown in laboratory experiments (Illangasekare et al., 1995a).

Schwille (1988) and Butts et al. (1993) report the initiation of unstable flow due to the presence of heterogeneities. Instabilities can cause the organic fluid to infiltrate the soil as one or more "fingers," rather than a uniform front (Held and Illangasekare, 1995a). Many researchers (e.g., Hill and Parlange [1972], Peters and Flock [1981], Glass et al. [1989]) have proposed schemes to predict the diameter and location of such fingers. However, mainly due to small-scale heterogeneities in natural porous media, the location and diameter of the infiltrating fingers are difficult to predict.

In field situations, heterogeneities may be the major factor in controlling the entrapment distribution of NAPL. NAPL entrapment is a complex phenomenon described in different terms by different scientific disciplines. In the most general sense, we will from now on refer to an "entrapped NAPL" as any NAPL that is no longer flowing. The NAPL may exist as discontinuous, stable pore-scale ganglia trapped in the porous media under capillary forces; also it may exist as an immobile continuous phase trapped by heterogeneities. We refer to this phenomenon as macroscale entrapment. NAPL immobilized by capillary forces is often referred to as a residual or irreducible saturation and has received the greatest attention in experimental mass transfer studies.

Two conceptual models, called bypassing and snap-off, have been proposed for NAPL entrapment at the microscopic scale (Marle, 1981). Bypassing is a phenomenon by which an organic fluid body becomes isolated due to tortuosity and velocity contrasts in different size pores. Snap-off, on the other hand, occurs when a volume of organic fluid is squeezed through pores with a small throat to body ratio, and inertial forces separate blobs from the main organic body. This mechanism is thought to cause small blobs occupying one or two pore bodies. These conceptual models provide an idealized picture of the microscopic scale; snap-off giving rise to isolated single blobs occupying one or two pore bodies, and bypassing causing complex ganglia elongated in the direction of flow.

Experimental studies on NAPL entrapment fall into two categories: those run on scales larger than the representative elementary volume, or REV, (macroscopic) and microscopic studies. The macroscopic studies have all used one-dimensional column methods. The microscopic studies have used etched glass micromodels or NAPL polymerization to quantify the microscopic blob distribution.

Hoag and Marley (1986), Szlag (1997), and Szlag and Illangasekare (1997) have experimentally investigated entrapment in sandy materials. Hoag and Marley (1986) studied the entrapment of gasoline in the vadose zone for sands, while Szlag and Illangasekare (1997) conducted entrapment studies in saturated sand columns with both DNAPLs and LNAPLs. Wilson et al. (1990) have conducted systematic studies in one-dimensional saturated columns on natural sandy materials and two-dimensional glass micromodels. In the saturated sand column experiments, under small groundwater gradients, they found that entrapped organic saturations were relatively insensitive to fluid properties but highly dependent on soil properties and heterogeneities. They also found that soil texture and grain size were very unreliable predictors of organic entrapment.

Continuing the work of Wilson et al. (1990), Conrad et al. (1992) concentrated on describing the microscopic blob structure using etched glass micromodels and polymerization of styrene within porous media. The glass micromodels are the experimental analogs of the computational network models. They noted a significant difference in blobs visualized in the etched glass micromodels and natural sands. They found that micromodel blobs occupy almost the entire pore body, leaving only a thin (immobile) film of water trapped between the NAPL and pore wall. Pore casts of NAPL blobs in a sand, however, showed a much higher water saturation in the pore body, which might be caused by surface roughness. In the Sevilleta sand used in their experiments, they noted that most of the blobs had complex three-dimensional shapes.

Chatzis et al. (1983) used similar techniques to look at the blobs trapped in sandstone cores and glass bead packs. Many more singlets were observed in the glass bead packs than in the sandstone cores. These singlets, which accounted for 58% of the blobs, only contained 15% of the residual NAPL. The majority of the NAPL saturation was held in blobs spanning five or more pore bodies.

Powers (1992) using similar methods as Chatzis et al. (1983) and Conrad et al. (1992) showed that well-graded sands entrapped more NAPL and held it in large ganglia, while uniform sands retained a smaller amount of NAPL and held it in small blobs. These results are consistent with the hypothetical models of bypassing and snap-off, respectively.

In the field scale, NAPL saturations are measured at the macroscale. The NAPL saturation is defined as the volume of NAPL divided by the volume of voids in the porous media. Many researchers (Pfannkuch, 1984; Schwille, 1988; Hunt et al., 1986a,b; Anderson et al, 1992) have identified three macroscale source geometries: (1) the floating lens (pancake) found in LNAPL spills. In a homogeneous sand, the leading edge may have NAPL saturations ranging between 15 and 75%, while the trailing edge may have saturations ranging from 5% to 25%; (2) cylinders or fingers (DNAPLs) at residual saturations of 5% to 25%; (3) suspended DNAPL pools on impermeable layers. The experimental work by Yates (1990) and Illangasekare et al. (1995a) and the conceptual studies by Hunt et al. (1986a,b) demonstrate that there are additional macroscopic source geometries; (4) zones of high LNAPL saturation trapped in coarse lenses beneath the water table; (5) thin pools of DNAPL trapped in coarse sand lenses; (6) DNAPL pools trapped on top of fine sand lenses, with significant low permeability to water.

Illangasekare et al. (1995a) have shown in laboratory flume studies with LNAPLs that pools, fingers, and zones at residual saturation are not the only, nor the most likely, source geometries in the groundwater zone. Macroscale heterogeneities, such as coarse sand lenses, form traps that may hold LNAPLs at saturations higher than the residual saturation. The nature of these heterogeneities is such that they are highly permeable in their native state. Once a high LNAPL saturation is established within them, they become less permeable to water and flow may actually diverge around them. NAPL may remain trapped due to the relatively high entry pressure (or capillary barrier effects) in the surrounding matrix.

Pore structure in microscopic studies will greatly influence entrapment of NAPLs and the available area for mass transport. The expense and difficulty of performing detailed pore-scale analysis and the lack of any statistical correlation between blob size distribution and NAPL dissolution precludes these methods from being widely used. Pfannkuch (1984) suggested that the capillary pressure-saturation curve may provide an estimate of the liquid-liquid exchange area in a multiphase porous media system. Powers (1992) attempted to identify correlations between the parameters describing the capillary suction saturation curve and the overall mass transfer coefficient. Ultimately she found that the mean grain size and soil uniformity coefficient were more robust predictors of dissolution of NAPL into the aqueous phase. Similarly, Szlag and Illangasekare (1997) in correlating NAPL entrapment with the parameters describing the capillary suction saturation curve found that the mean grain size, uniformity coefficient, and the Reynolds number of the water sweep were the best predictors of NAPL entrapment. Hoag and Marley (1986) were able to obtain an excellent correlation between NAPL entrapment and mean grain size for a narrow range of soil types and a single fluid, gasoline. In contrast, Wilson et al. (1990) investigating a broader range of media and fluids, saw little correlation with mean grain size.

NONWETTING FLUID DISPLACING WATER ACROSS AN INTERFACE

A common situation that is encountered in problems dealing with the behavior of multiphase fluids in heterogenous systems is the flow of a NAPL into a water-saturated soil across interfaces between two soil layers. In the case of nonwetting NAPL displacing the wetting water phase, the fluid pressure should exceed a critical entry pressure for the fluid to enter from a coarse soil to a finer soil. If the pressure is less than this entry pressure, the fluid will pond at the interface. To demonstrate and analyze this effect, Fairbanks (1993) conducted a series of experiments where a NAPL was pumped into water-saturated heterogenous soil columns and observed the change in water saturation due to NAPL displacement and the drop in pressure across the NAPL-water interface (capillary pressure). The testing apparatus consisted of a flexible wall permeameter and a flow pump. The system was also equipped with pressure cylinders for controlling fluid pressures, a differential pressure transducer for measuring capillary pressures, a ceramic high air-entry porous bottom plate and a data acquisition system. The permeameter has two hydraulically separated chambers. The soil sample is placed in the inner chamber between two caps surrounded by a latex membrane. The outer chamber is filled with water and is used to apply a confining pressure to the exterior of the soil specimen. This sample placement under confining pressure precluded leakage flow at the permeameter walls during fluid displacement. A precision pump that is able to control flow rates ranging from 2.69×10^{-5} mL/s to 0.135 mL/s was used to withdraw water from the bottom of the sample through the porous plate. A reservoir containing a lighter than water NAPL was connected to the top cap of the cylinder.

Figure 16.1 shows a plot of capillary pressure versus the water saturation for the case of a heterogenous system consisting of a layer of coarse #16 sand placed over a layer of a finer #70 sand (mean grain sizes d_{50} of 0.880 mm and 0.185 mm, respectively). As the water saturation decreased with the NAPL movement, the capillary pressure increased. As progressively smaller pores are desaturated, the capillary pressure curve followed the water retention curve of the #16 coarse sand. When the NAPL front reached the interface between the two soil layers, the capillary pressure increased abruptly, representing the higher pressure needed for the NAPL to enter the finer soil. This abrupt transition depends on the contrast in the properties of the two soils. Tests with other soil layer combinations showed that the interface between #8 sand (d_{50} = 1.5 mm) and #70 sand produced the sharpest transition, while the #30 (d_{50} = 0.490 mm) and #70 interface produced a more gradual transition. As the pore sizes of the two layers comes closer, this transition becomes discernible as was found in the #70 and #125 (d_{50} = 0.103 mm) interface.

The results demonstrate the importance of properly incorporating the entry effects between soil layers into modeling tools that simulate the flow behavior of NAPLS in heterogenous systems. As will be shown later, when the fluid pressures do not attain the critical entry value, the fluids can pond at interfaces, resulting in large entrapment saturations. The dependence of the entry pressure value on the properties of both soils indicates the need to characterize the interactive effects between the soils to predict entrapment in heterogenous systems.

RETENTION OF WETTING LIQUIDS IN LAYERED SOILS

In situations in which a water or a NAPL spills into a dry soil, the migrating fluid acts as wetting fluid with respect to the displaced nonwetting air phase. Under static conditions, the piezometric head is constant throughout all soils in a layered system. This requires that the capillary pressure head increase linearly with elevation despite the discontinuities of soil properties. At true static equilibrium, the distribution of a wetting phase in the unsaturated zone is given by the capillary pressure-saturation relation for every layer of a heterogeneous medium. For every soil in the layered system, a unique drainage capillary pressure curve exists. The wetting phase saturation

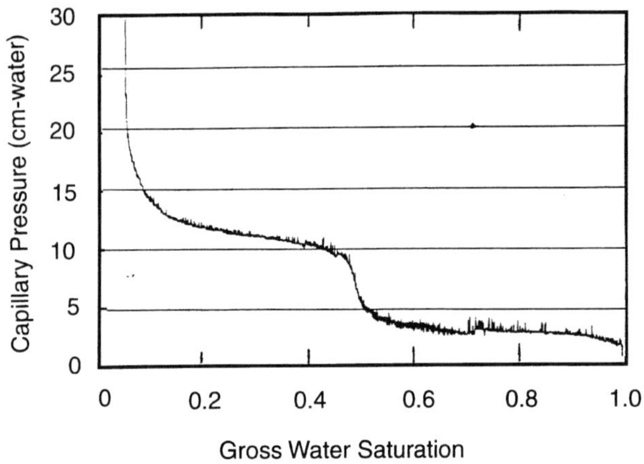

Figure 16.1. Representative curve of capillary pressure variation in a layered system.

versus height above a datum can be obtained by combining the capillary pressure curves for each soil.

In a layered porous media, the soil of least permeability has the dominant influence on flow (Collins, 1976). Although steep hydraulic gradients are often present, flow through a series of layers of unsaturated soil can be nearly zero, when coarse layers with large and empty pores with small unsaturated hydraulic conductivities are encountered. Unless the pressure head in the fine upper layer rises to a value near atmospheric, flow through the coarse layer will not occur, because the large pores of the coarse sand cannot be fully saturated at the capillary pressures of the upper region. This leads to the retention of liquids above a coarse layer and a nonstatic pressure distribution. A static pressure distribution may never be achieved, even though the flow may reach a negligible rate. A steady-state flow model was used by Walser et al. (1994) to describe the pressure and saturation distribution under very low flow rates, as may be occurring in layered media.

In the experiments conducted by Walser (1995), a plexiglass column was filled with a layer of #30 sand and on top, a layer of fine #125 sand. The sand was then brought to maximum saturation with a test NAPL (Soltrol 220) and then drained. The NAPL pressure was monitored continuously with six ring-tensiometers. A dual-gamma attenuation system (Illangasekare et al., 1994) was used to measure the NAPL saturations.

The results for the retention experiment shown here are based on measurements taken 14 days after drainage was started. Twenty-four hours after drainage was started, flow of Soltrol was too small to be measured. The saturation profile found experimentally and the static saturation profile calculated from the measured (capillary pressure-saturation profile) are shown in Figure 16.2. The experimental saturation profile shows an effective saturation of one at sample bottom and at the fine-coarse sand interface. In the lower layer of coarse sand, the saturation profile follows closely the saturation profile resulting from a static distribution. The static model distribution shows a saturation approaching residual for the fine sand throughout its thickness, and does not show the experimentally-found fully saturated zone near the interface. The measured capillary pressure does not approximate the linear static capillary pressure distribution. The experimental data show a capillary suction that remains constant with elevation in the coarse sand and increases in the fine sand (Figure 16.3). The apparent contradiction (that the saturation profile in the coarse sand is close to the static profile, while the pressure profile does not match the static profile) can be explained by

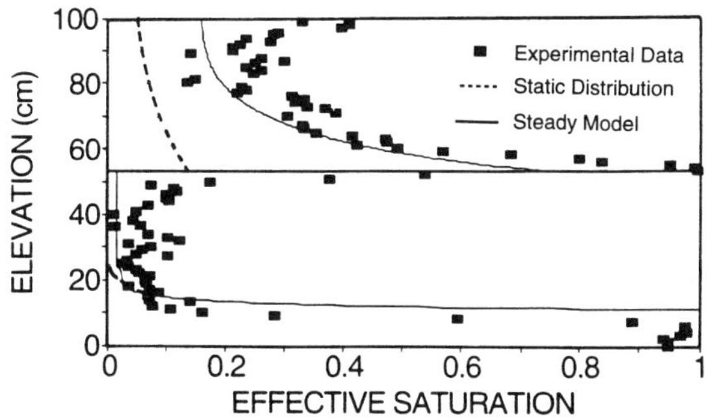

Figure 16.2. Experimentally measured saturation profile in a layered column in comparison with steady flow saturation profile.

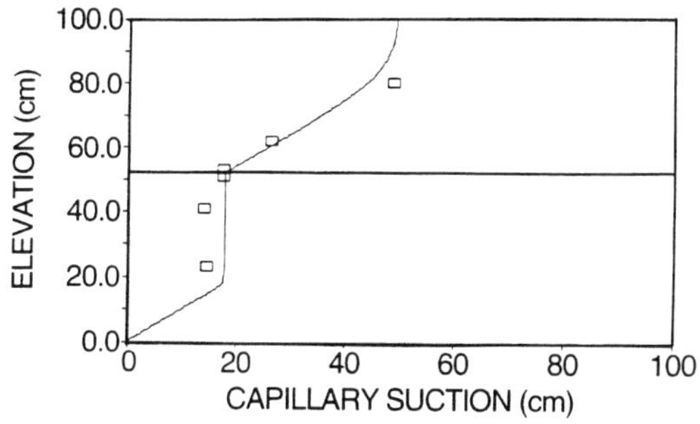

Figure 16.3. Experimentally measured capillary pressure distribution in comparison with steady flow pressure distribution.

noting that the pressures measured in the coarse sand are all in the region approaching residual saturation, where even large increases in capillary pressure will not change the saturation significantly. Thus, the saturation is not sensitive to differences in pressures, and the saturation profile matches the static case and also the steady-state case. The difference between static and steady-state case can here only be discerned in the pressure profiles.

Experimental saturation and pressure profiles were compared with profiles from the steady-state model (Figures 16.2 and 16.3). The steady-state saturation profile shows a discontinuity in saturation at the fine-coarse sand interface, which is similar to the experimentally observed jump in saturation. The model saturation profile also approximates the saturation profile in the coarse sand closely. The comparison of the pressure profiles shows that the steady-state model approximates the measured pressure data well for the assumed flow.

NONWETTING FLUIDS IN UNSATURATED LAYERED SOILS

The previous case where a wetting fluid was retained in layered soils helped to explain the capillary barrier effects through a pseudo-steady infiltration process. A more realistic situation that may be found in NAPL contaminated sites is the movement and retention of the nonwetting NAPL in the unsaturated zone. In this case, the nonwetting NAPL displaces a second nonwetting air phase while the soil grains are preferentially wetted by water. A set of experiments was conducted by Illangasekare et al. (1995b) in two-dimensional vertical tanks to study this process.

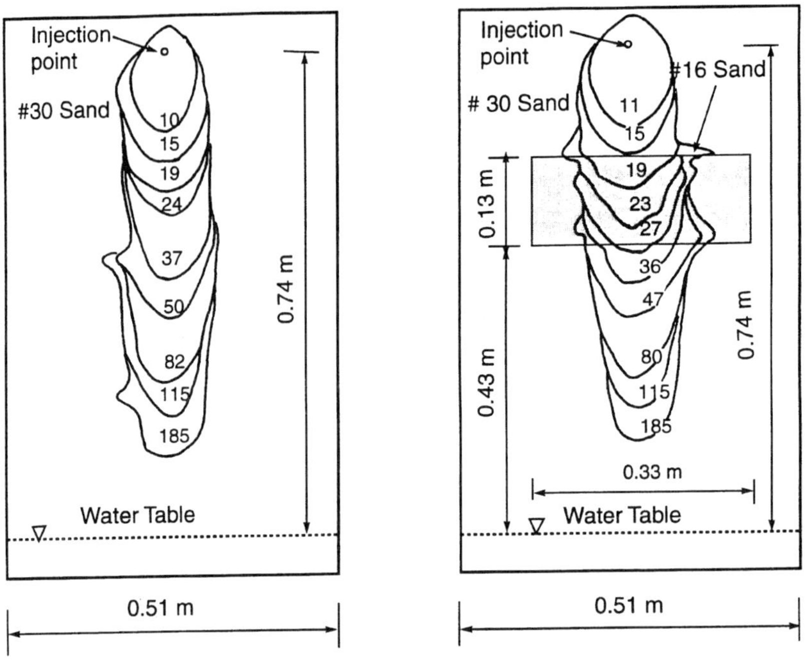

Figure 16.4. DBP migration through a lens in unsaturated soil (the numbers shown on the contour lines are the observation times in minutes after the spill).

Experiments were conducted in a 1.22 m × 1.83 m × 0.05 m two-dimensional soil flume equipped with a dual-gamma attenuation system to measure phase saturations. The flume was packed with three sand types to create various heterogenous configurations. Three test chemicals were used in the spill experiments. Two or more uniform sands were used in the experiments to create a heterogenous porous medium. Two denser-than-water organic fluids; namely, dibutyl phthalate (DBP) and 1,1,1-trichlorethane (TCA), were used as the test fluids. The results from three experiments are used here to demonstrate the flow and entrapment behavior in heterogenous unsaturated formations.

Coarse Lens in Unsaturated Soil

A coarse sand lens was placed in a fine sand formation. In Figure 16.4, the migration patterns of DBP through the homogeneous soil and the soil containing the coarse lens are compared. The contour lines in this figure represent the position of the advancing front observed visually through the walls of the flume as a function of time (in minutes). These observed fluid front profiles show that the presence of the coarse lens had virtually no influence on the transport of the organic. Some spreading of the DBP was observed at both the top and bottom interfaces of the lens.

Fine Lens in Unsaturated Soil

In this experiment a finer lens was placed in coarser formation. Unlike the previous experiment, the fine lens substantially altered the migration of the organic. The position of the advancing front of the DBP as a function of time is shown in Figure 16.5.

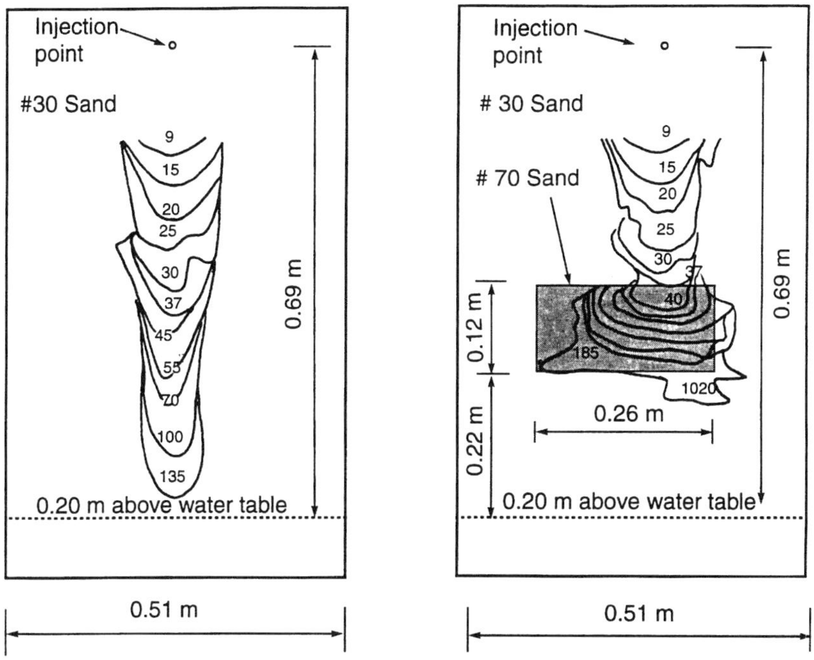

Figure 16.5. DBP migration through a fine lens in unsaturated soil (the numbers shown on the contour lines are the observation times in minutes after the spill).

At the instant the DBP contacted the fine lens, it rapidly entered the lens and began to spread laterally as well as vertically.

Layered Unsaturated Soils

A TCA spill was made on a packing configuration that consists of alternate layers of a fine and a coarse sand. The point of spill was located on the vertical center line at a depth of 12 cm below the soil surface.

Under static conditions, in the layered soil, the initial water saturation changed with the elevation as determined by the capillary pressure vs. saturation curve of each soil. The influence of the initial water saturation profile on the TCA migration pattern can be seen in Figure 16.6, where the position of the advancing front is plotted. At the beginning of the spill, the organic's initial movement was predominantly vertically downward. After only a couple of minutes, capillary forces in the #70 sand quickly spread the contaminant laterally as well as vertically. In 25 minutes, the TCA had reached the first coarse layer. At this time, the width of the plume was approximately equivalent to its vertical penetration, indicating that capillary forces controlled the TCA migration through the first fine layer.

The initial water saturation in the first coarse layer was lower than the water saturation in the fine soil above. As a result, for the two soils used, the relative permeability to the organic phase was greater in the coarse layer than in the fine layer above or below. Also, the intrinsic permeability of the #30 sand is an order of magnitude greater than the #70 sand. These two factors in combination make the effective permeability to the organic phase much higher in the coarse layer than in the fine layer. Consequently, movement through the coarse layers is controlled by gravitational forces with little to no spreading due to capillary forces.

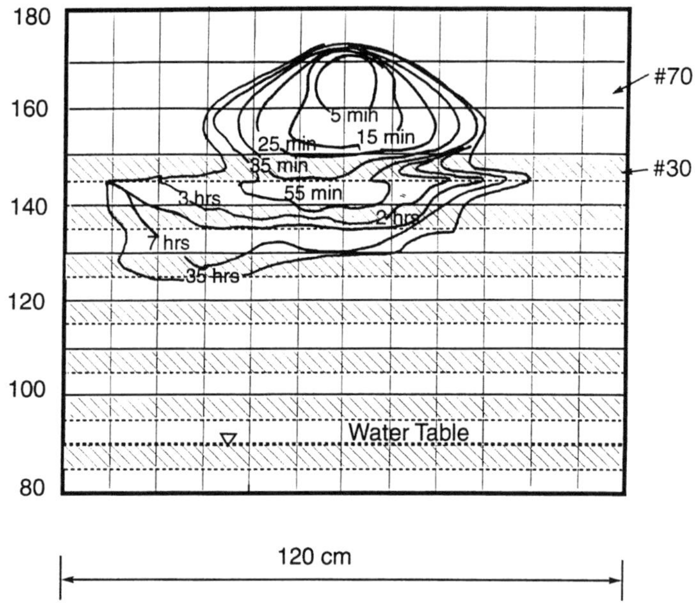

Figure 16.6. Migration pattern of a DNAPL in a layered unsaturated soil.

Upon reaching the next fine layer, the capillary forces of the fine soil pulled the TCA into the fine layer. The contour line of 55 minutes in Figure 16.6 illustrates this observation. Because of the much higher relatively effective permeability of the coarse soil compared to the fine soil, the fine layer does not have the capacity to transmit all the NAPL that infiltrated through the coarse layer. This resulted in the TCA ponding up at the base of the first coarse layer. The ponding of the organic led to the development of horizontal gradients that caused the TCA to spread horizontally along the base of the coarse layer. As can be seen in Figure 16.6, the subsequent migration was mostly vertical. It appears that the horizontal spreading was controlled by the effective permeability of the first coarse layer and the underlaying fine layer. The implication of this observation is that if one can determine the degree of spreading based on the physical properties of the first coarse layer and the underlying fine layer, subsequent transport quite possibly could be modeled as one-dimensional flow.

NONWETTING DENSE FLUIDS IN SATURATED LAYERED SOILS

When introduced into the subsurface, DNAPLs initially move downward through the unsaturated zone above the water table. If the spill is large enough and the depth to the water table is shallow, the contaminant will eventually reach the water table. Because DNAPLs are denser than water, they may continue to move downward through the saturated zone of the aquifer. However, to enter the saturated zone, the fluid must first build up adequate head to overcome the entry pressure of the water saturated soil at the water table. Until the contaminant attains the entry pressure, it will pool up at the top of a capillary fringe. Consequently, lateral spreading of the separate phase fluid is likely to occur at this location. If the lateral spreading is not too large, and enough of the spilled fluid is present at the capillary fringe to achieve the entry pressure, the contaminant will enter the saturated zone, possibly penetrating the entire

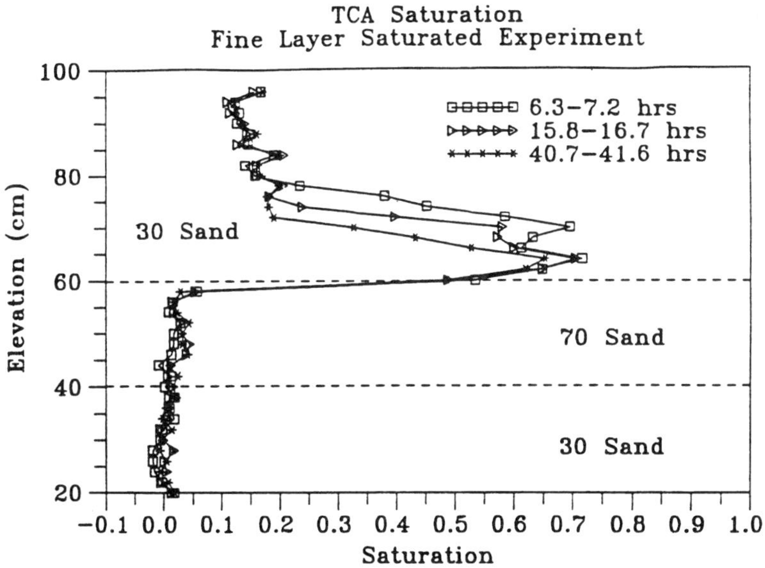

Figure 16.7. Fine-layer saturated experiment: 6.3 to 41.6 hours.

thickness of the aquifer. Results from two experiments reported by Illangasekare et al. (1995b) where a DNAPL was allowed to flow through a layered saturated zone are presented below.

Effects of a Fine Layer in the Saturated Zone

In the first experiment, a fine sand, #70 was used to create a sand layer in a #30 coarse formation. Saturation profiles along the vertical center line of the flume were recorded using a gamma system at various time periods. Sample saturation profiles at 4.9 to 51.2 hours after the spill are shown in Figure 16.7.

Due to the driving head at the point of injection, the TCA initially spread in all directions. Because the density difference between the water and the DNAPL is significant, the movement of the organic plume was mostly vertically downward through the #30 sand. Figure 16.8 shows, 15 hours after the spill, how the fine layer is acting as a barrier to the fluid and the lateral spreading at the interface of the two soils.

Spreading of the contaminant continued, but at a much slower rate than during the first three hours after the injection. The saturation data of Figure 16.7, clearly shows the TCA continued to drain from the coarse layer. Since the organic had practically ceased lateral spreading, but continued to migrate from the coarse layer, it must have been moving through the fine layer as single or multiple fingers. Indeed, Figure 16.8 shows a finger developing in the #30 sand layer beneath the #70 sand layer. The fingers are not visible in the #70 sand layer because they are believed to be much smaller than the thickness of the flume.

It is unknown when the instability developed, or the number, size, or frequency of the fingers that resulted. It appears the presence of the fine layer contributed to the initiation of the instability by spreading the contaminant along the top of the layer. Unable to achieve the entry pressure required to penetrate the fine layer, the organic began channeling through the largest pore spaces available.

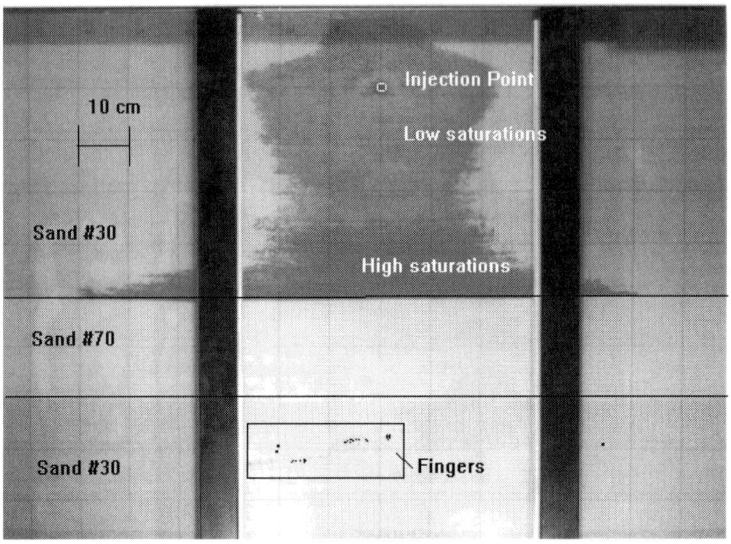

Figure 16.8. DNAPL distribution 15 hours after spill in saturated formation with fine layer.

Effects of a Coarse Layer in Saturated Zone

This experiment was set up identical to the previous experiment except that the fine #70 sand layer was replaced with a coarse #16 sand layer. The transport of the organic during the initial stages of the experiment was essentially identical to what occurred in the first experiment, only slower due to the reduced injection rate. Shortly after stopping the TCA injection, very little movement of the organic at the infiltrating front could be seen through the flume walls. Approximately one hour and 50 minutes into the spill, TCA fingers were seen developing in the coarse layer.

Clearly, the lack of stable displacement at the infiltrating front was because the organic became unstable and began to finger through both the #30 and #16 sand below. The instability is believed to have been triggered by the removal of the applied head when the injection was discontinued. After a short period, the applied force was dissipated as the TCA continued to migrate downward. It appears that gravitational forces alone were insufficient to maintain a stable displacement. Consequently, the TCA began to channel or finger through the largest available pore spaces.

After fingering through the #16 sand layer, the TCA was initially unable to achieve the entry pressure of the #30 sand and continue its downward migration. As a result, the organic began to pond on top of the #30 sand. Ponding led to the development of pressure gradients in the organic phase and subsequent stable flow in the form of spreading along the base of the coarse layer. This is shown in the photograph in Figure 16.9 that was taken two hours after the spill. Ponding also raised the pressure of the organic to the entry pressure of the #30 sand. This initiated unstable flow in the #30 sand and the gravel pack below.

Figure 16.10 presents the saturation profile after the TCA began to pond and spread at the base of the coarse layer. The organic is shown moving from the #30 sand in upper portions of the flume and ponding at the base of the #16 layer. Indications of fingering through the coarse layer and the #30 sand below the coarse layer can also be seen in these figures.

The fingers developed in this experiment were much smaller than the thickness of the flume or the resolution to be captured by the dual-gamma system. Consequently,

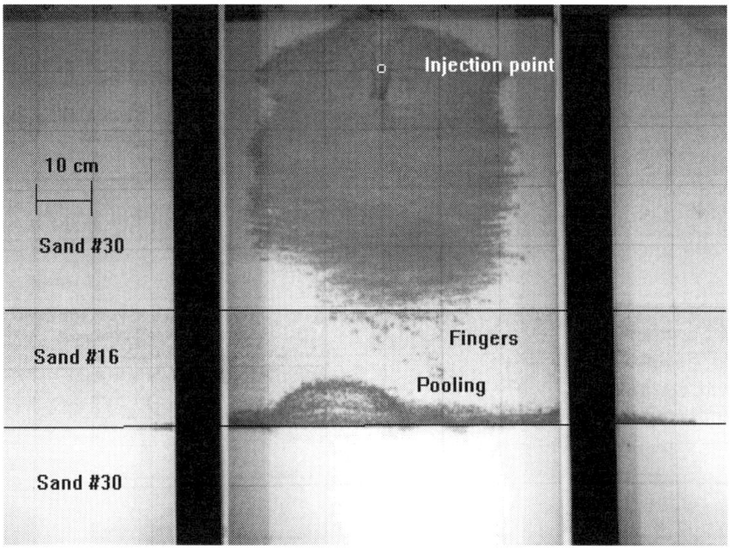

Figure 16.9. DNAPL distribution 2 hours after spill in saturated formation with coarse layer.

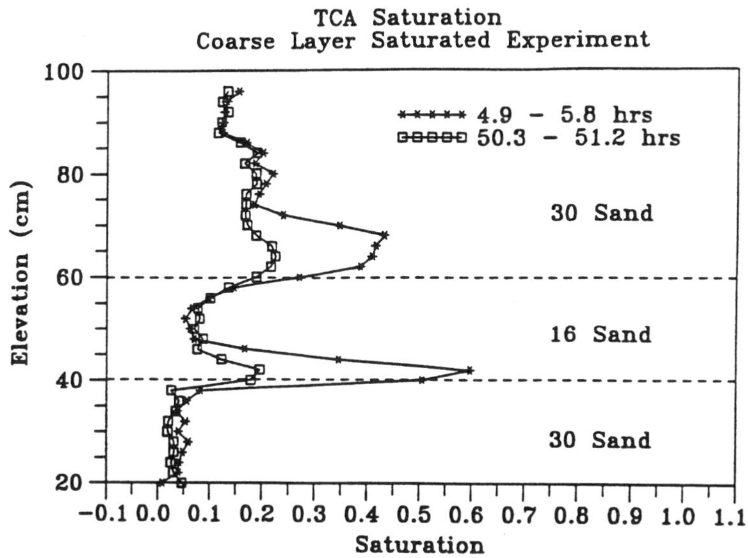

Figure 16.10. Saturation profiles of coarse layer TCA experiment 4.9–51.2 hours after spill.

the TCA saturation data in the unstable flow areas does not represent the saturation of the fingers themselves, but does represent the macroscopic saturation. TCA saturations above 5% occurred at several locations where visual inspection of the flume indicated the organic was not present.

UNSTABLE DISPLACEMENT OF DENSE FLUIDS IN SATURATED SOILS

Laboratory and theoretical studies on DNAPL infiltration imply an erratic flow even on homogeneous soils. Considerable skepticism about the validity of the classical macroscopically formulated flow and transport equations for unstable displacement and fingering exists in the literature (van Genuchten, 1991). With a fundamental understanding of the physical system and the processes that produce DNAPL fingering, realistic models can be developed for DNAPL spills and the appropriate decisions for groundwater remediation can be made. Held (1993) conducted a series of laboratory experiments with the objective of visualization of the displacement of a wetting fluid (water) by a nonwetting fluid (DNAPL) in sands. Relevant findings from this investigation are presented below.

The vertical test tank consisted of 16 individual segments. Three sands with mesh sizes of #8, #30, and #70 were used as test soils to pack the tank homogeneously. Trichloroethylene (TCE), 1,1,1-trichloroethane (TCA), and dibutyl pthalate (DBP) were selected as the test DNAPLs. In the above order they represent a decrease in density (1.46, 1.35, and 1.044 g cm^{-3}) and an increase in viscosity (0.57, 0.84, and 20.0 cP). The test fluids were dyed with 0.1 g L^{-1} of a water immiscible organic dye Nile Red that fluoresces under ultraviolet light After saturating the tank, the DNAPL was spilled below the water table using a point-injection device. The tank was disassembled no earlier than three days after the spill to obtain sections through a static DNAPL distribution. The DNAPL distribution was observed under ultraviolet light and the images were captured and digitized using an image processing system.

The finger initiation, plume shape, and preferential flow in fingers are controlled by fluid densities, viscosities, interfacial tension, and pore scale heterogenities in the soil matrix. The data obtained from these experiments were analyzed to understand the sensitivity of the fluid and soil parameters to displacement instability, plume shapes, and preferential flow. The data from ten experiments were analyzed and the details can be found in Held (1993) and Held and Illangasekare (1995a,b).

The displacement of the low viscosity but higher density TCE in the very coarse # 8 sand was predictably unstable. The critical length of finger separation (a critical wave length of minimum spacing for developing fingers) for this case was approximately 1 cm compared to 4 cm in the case of less dense TCA in medium # 30 sand. The increase in characteristic length scale is attributed to lower displacement velocities. The analysis of finger patterns showed that the macroscopic concept of a critical wavelength may not be applicable to DNAPL displacement in fine sands. In fine sands, capillarity was more pronounced, and finger clusters were dictated by the pore network, creating a complex, three-dimensional structure.

Fingering per se was initiated in all experiments as a result of the spill conditions, physical properties of the DNAPLs, and pore-scale perturbations at the displacement front. The porous media exhibit a spatially random pore network even for ideal homogeneity in the macroscopic sense. Finger patterns were characterized by shielding, spreading, and tip-splitting mechanism. As a result of shielding, multiple fingers develop into fewer fingers at lower elevations. The mechanism of spreading produced significant lateral movement of the infiltrating fluid. The term tip-splitting is used for the process of secondary finger development from a finger front.

Based on the experimental results it was possible to relate the plume shapes to the experimental parameters in general. Densities and also viscosities of TCE and TCA are alike, in contrast to DBP. Similar DNAPL distributions were developed for TCE and TCA in the very coarse sand, quite different from the DBP distribution. Hence, gravitational and viscous forces mainly govern the DNAPL movement in these cases. Viscous forces influence the displacement stability in the medium to a greater degree, leading to distinct finger patterns for every test fluid. Capillary forces seem to be the controlling mechanism for the complex fingering in the fine sand. Similar clusters and migration depth for TCE and TCA would be explained by the characteristic pore network and comparable surface tensions of these two DNAPLs.

In immiscible displacement in porous media, simultaneous flow of the two fluid phases takes place in a transition zone at the macroscopic displacement front (Blunt et al., 1992). The mobility of both phases is thereby reduced. For DNAPLs displacing water in porous media, capillary imbibition of the water restricts the expansion of the DNAPL distribution. An existing finger is likely to retain its shape, and the higher relative permeability of the organic phase inside the finger will preferentially conduct the DNAPL.

Preservation of fingers was seen in several experiments. For instance in the TCE in #30 sand, vertical channels persisted over a distance of half a meter, down the bottom of the tank. A temporal variation of finger shapes was yet apparent for the almost spherical plume in this spill experiment. Developed pathways were still used to conduct DNAPL, although the final distribution may not necessarily be projected from fingers that existed at early times.

In medium and coarse sands it was noted that sections close to the finger tips revealed a high DNAPL saturation. As fingers still developed after the source was cut off, the DNAPLs migrated or actually drained downward in response to the propagating front. Experiments conducted in glass tanks by Illangasekare et al. (1995b) that were also discussed in the previous section also demonstrated internal redistribution of the DNAPL by fingering. It was presumed that a static distribution was reached in the experiments. Consequently, the observed range of DNAPL saturations would need to be considered as a source for aquifer contamination by dissolution.

PREFERENTIAL FLOW OF LIGHT FLUIDS IN SATURATED SOILS

When LNAPLs after migrating through the unsaturated zone reach the water table, depending on the spill volume, the fluid can build up adequate pressure to overcome the entry pressure to displace the pore water in the saturated zone to penetrate the water table. If the fluid encounters a coarse formation below the water table, the fluid can enter the coarse soil at much lower entry pressure. Once the fluid enters the coarse formation it can migrate laterally, producing preferential flow channels below the water table. Armbruster (1990), Campbell (1993), and Illangasekare et al. (1995a) demonstrated this mechanism through a series of experiments conducted in a large soil tank. The results from one of these experiments are presented here.

In this experiment by Campbell (1993), a heterogenous packing configuration was created in a 9.8 m long two-dimensional tank by inserting a layer of coarse sand (#16) into a finer formation of # 30 sand. The water table was located above the coarse layer. The test LNAPL, Soltrol 220 was spilled to enter the coarse layer. The LNAPL proceeded to migrate preferentially along the coarse layer in the direction of groundwater flow. Visual observations showed that the upper interface between the two sands confined the NAPL flow within the coarse layer. The bouncy forces of the LNAPL staying below the water table were not sufficient to push the NAPL into the finer sand by overcoming the entry pressure. The transient saturation profiles recorded during the propagation of the NAPL through the coarser sand layer is shown in Figure 16.11. The LNAPL remained permanently entrapped below the water table within the coarse sand, producing much higher macroscale entrapment saturations.

FLOW AND ENTRAPMENT IN A RANDOM HETEROGENOUS FIELD

In all experiments that were presented so far, well-defined structured heterogenities such as layers and lenses were used to demonstrate the flow behavior under various conditions. However, the natural soil heterogenities in aquifers are complex and are

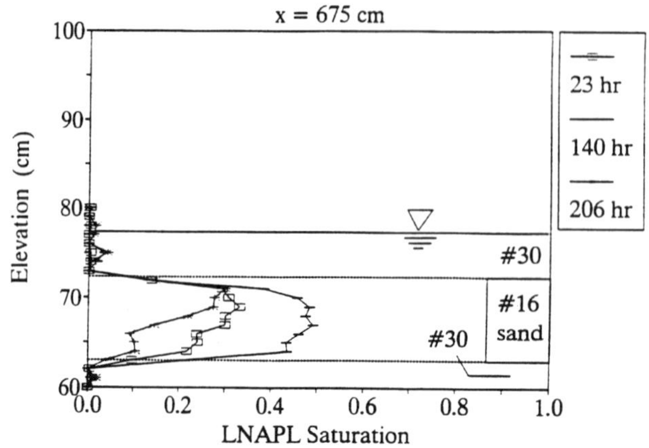

Figure 16.11. Saturation profile showing preferential flow of a LNAPL below water table.

difficult or impossible to characterize in a deterministic sense. In some applications investigators have relied on stochastic methods to characterize the spatial distribution of soil properties, and flow and transport parameters (Gelhar, 1993). In these methods the spatially changing parameters are treated as random variables and stochastic parameters of the distributions are used to characterize the variability.

A set of experiments were conducted by the author and coworkers with both LNAPLs and DNAPLs in large tanks packed to represent a randomly heterogenous configuration. Five test sands (#8, #16,#30,#70, and #110) were used. These sands represent a range of values of hydraulic conductivities (from 1.20 cm/s to 0.004 cm/s) and capillary pressure vs. saturation characteristics. A random field of saturated hydraulic conductivity was computer generated using a Fourier summation algorithm. The tanks were packed in layers that were 10 cm long and 2 cm thick to obtain a ln K variance of 2.86 and horizontal and vertical correlation lengths of 20 cm and 4 cm, respectively. The fluid distribution was observed visually and recorded using the gamma system. The results of an experiment in which TCA was spilled into saturated soil are presented.

The TCA was introduced below the water table at a rate of 50 mL/min for 54 min duration. The water table was located at an elevation 70 cm above the base of the tank. The DNAPL started to migrate and the migration pattern was controlled by spreading at interfaces, preferential flow through coarser layers, and fingering. The distribution of the DNAPL 90 min into the spill is shown in Figure 16.12. The results show that the random heterogeneity resulted in an almost random entrapment distribution of the chemical. The final NAPL distribution in this random distribution is a result of the interface and capillary barrier effects that are controlled by the macroscopic properties of the soil and the fingering that occur at the pore scale.

SUMMARY

The example laboratory spill simulations that were presented in this chapter demonstrate the importance of soil heterogeneities in the flow and entrapment behavior of nonaqueous phase organic chemicals in soils. The data collected during the spills were useful in understanding the basic mechanisms associated with the flow processes. The data can also be used in the validation of numerical models, as comprehensive field data are not easily available.

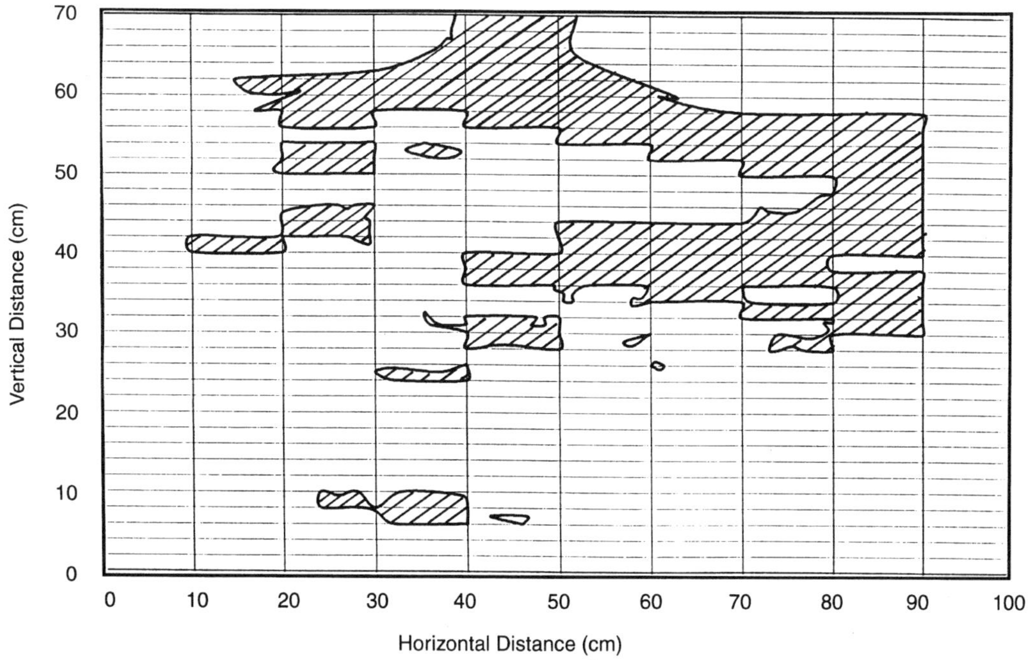

Figure 16.12. Entrapment of TCA in saturated zone of randomly heterogenous field.

A number of observations were made which have implications on the understanding of the fundamental processes, developing modeling strategies, and designing field characterization techniques.

1. The flow of fluid across interfaces between soil layers is controlled by critical entry pressure effects. These entry pressures not only control the rate of flow but also determine when and whether the contaminants will enter finer soil formations. Depending on whether the fluids enter a formation, the contaminant can pool and spread at the interfaces, thus producing entrapment saturations that are much higher than the residual values. The entry pressure values were found not to depend only on the finer soil but also on the coarser soil. This observation makes it necessary to develop techniques to obtain the characteristics of the transition zone at the interfaces. Special care should be taken to represent these interface behaviors in numerical models (Al-Sheriadeh and Illangasekare, 1993).
2. The initial water saturation has a major impact on the transport of nonaqueous phase organic fluids. In the unsaturated zone, the capillary forces of fine soil pulls the contaminant from the coarse soil and entraps it. However, in the saturated zone, the same fine soil established a capillary barrier to the organic phase.
3. Soil layering was found to significantly increase the spreading of immiscible contaminants in both the water-saturated and unsaturated regions. The organic infiltration rate and the contrast in permeability between the soils were found to affect the degree of spreading.
4. In layered systems, a static pressure distribution cannot explain the retention, and the retention in these types of systems can only be explained by treating it as a pseudo-steady infiltration process.
5. In the prediction of the true behavior of NAPLs in the heterogenous subsurface it is important to take into consideration the unstable flow that occurs at the pore scale and produce fingers.

REFERENCES

Al-Sheriadeh, M. and T.H. Illangasekare. Efficient immiscible multiphase flow model for heterogeneous porous media, *Proc. Groundwater Modeling Conference, IGWMC,* Golden Col., Colorado School of Mines, 1993, pp. 3-10 to 3-19.

Anderson, M.R., R.L. Johnson, and J.F. Pankow. Dissolution of dense chlorinated solvents into ground water: 1. Dissolution from a well-defined residual source. *Ground Water,* 30(2):250–256, 1992.

Armbruster, E.J., III. Study of Transport and Distribution of Lighter than Water Organic Contaminants in Groundwater, Thesis presented to the University of Colorado, Boulder, in partial fulfillment of the requirements for the degree of Master of Science, 1990, pp. 172

Blunt, M., M.J. King, and H. Scher. Simulation and theory of two-phase flow in porous media, *Phys. Rev.,* A, 46(12):7680–7699, 1992.

Butts, M.B, K.H. Jensen, D. Szlag, and T.H. Illangasekare. Fate of the miscible and immiscible compounds following a light oil spill, *Proc. Groundwater Modeling Conference, IGWMC,* Golden Col., Colorado School of Mines, June, 1993. pp. 3-1 to 3-9.

Campbell, J.H. Nonaqueous Phase Liquid Flow Through Porous Media: Experimental Study and Conceptual Sharp-Front Model Development, Thesis presented to the University of Colorado, Boulder, in partial fulfillment of the requirements for the degree of Master of Science, 1993, pp. 179.

Chatzis, I., N.R. Morrow, and H.T. Lim. Magnitude and detailed structure of residual oil saturation, *Soc. Pet. Eng. J.,* 24(2):555–562, 1983.

Collins, R.E. *Flow of Fluids Through Porous Materials.* Petroleum Publishing Company, Tulsa, OK, 1976.

Conrad, S.H., J.L. Wilson, W.R. Mason, and W.J. Peplinski. Visualization of organic fluid trapped in aquifers. *Water Resour. Res.,* 28(2):467–478, 1992.

Fairbanks, T. M. Light Non-Aqueous Phase Fluid Barriers in Initially Water Saturated Heterogenous Porous Media, Thesis presented to the University of Colorado, Boulder, in partial fulfillment of the requirements for the degree of Master of Science, 1993, p. 164.

Gelhar, L.W. *Stochastic Subsurface Hydrology.* Prentice Hall, Englewood Cliffs, NJ, 1993, p. 390.

Glass, R.J., Y.-J. Parlange, and T.S. Steenhuis. Wetting front instability 1. Theoretical discussion on dimensional analysis, *Water Resour. Res.,* 25(6):1195–1207, 1989.

Held, R.J. and T.H. Illangasekare. Fingering of dense nonaqueous phase liquids in porous media: 1. Experimental investigation, *Water Resour. Res.,* 31(5):1213–1222, 1995a.

Held, R.J. and T.H. Illangasekare. Fingering of dense nonaqueous phase liquids in porous media: 2. Analysis and classification, *Water Resour. Res.,* 31(5):1223–1231, 1995b.

Held, R.J. Investigation Of fingering of Dense Nonaqueous Phase Liquids in Saturated Porous Media, Thesis presented to the University of Colorado, Boulder, in partial fulfillment of the requirements for the degree of Master of Science, University of Colorado, Boulder, 1993, p. 172.

Hill, D.E. and Y.-J. Parlange. Wetting front instability in layered soil, *Soil Sci. Soc. of Amer. Proc.,* 36:697–702, 1972.

Hoag, G.E. and M. Marley. Gasoline residual saturation in unsaturated uniform aquifer materials. *J. Environ. Eng.,* 112:586–604, 1986.

Hunt, J.R., N. Sitar, and K.S. Udell. Nonaqueous phase liquid transport and cleanup: I. Analysis of mechanisms. *Water Resour. Res.,* 24(8):247–1258, 1986a.

Hunt, J.R., N. Sitar, and K.S. Udell. Nonaqueous phase liquid transport and cleanup: II. Experimental studies. *Water Resour. Res.,* 24(8):1259–1269, 1986b.

Huntley, D., R.N. Hawk, and H.P. Corley. Nonaqueous phase hydrocarbon in a fine-grained sandstone: 1. Comparison between measured and predicted saturations and mobility, *Ground Water,* 32:626–634, 1994.

Illangasekare, T.H., D. Znidarcic, G.S. Walser, and J.W. Weaver. An experimental evaluation of two sharp front models for vadose zone non-aqueous phase liquid transport, Rep. EPA/600/R/-94/197, Ada, OK, 1994.

Illangasekare, T.H., D.N. Yates, and E.J. Armbruster. Effect of heterogeneity on transport and entrapment of nonaqueous phase waste products in aquifers: An experimental study, *ASCE J. Env. Eng.*, 121(8):572–579, 1995a.

Illangasekare, T.H., J.L. Ramsey, K.S. Jensen, and M.B. Butts. Experimental study of movement and distribution of dense organic contaminants in heterogenous aquifers, *J. Contaminant Hydrol.*, 20:1–25, 1995b.

Kueper, B.H. and E.O. Frind. Two-phase flow in heterogeneous porous media 1. Model Development, *Water Resour. Res.*, 27(6):1049–1057, 1991.

Marle, C.M. *Multiphase Flow in Porous Media*. Gulf Publishing Co. Houston, TX, 1981, p. 257

Peters, E.J. and D.L. Flock. The onset of instability during two phase immiscible displacement in porous media, *Soc. Pet. Eng. J.*, 21:249–258, 1981.

Pfannkuch, H.O. Determination of the contaminant source strength from mass exchange processes at the petroleum groundwater interface in shallow aquifer systems, in *Proceedings of the NWWA Conference on Petroleum Hydrocarbons and Organic Chemicals in Groundwater*, 1984. National Water Well Association, Dublin, OH, 1984, pp. 111–129.

Powers, S.E. Dissolution of Nonaqueous Phase Liquids in Saturated Subsurface Systems, Thesis presented to the University of Michigan, Ann Arbor, in partial fulfillment for the degree of Doctor of Philosophy, 1992.

Schwille, F. Petroleum contamination of the subsoil—A hydrological problem, in *The Joint Problems of the Oil and Water Industries*, Hepple, P. (Ed.), The Institute of Petroleum, London, England, 1967.

Schwille, F. *Dense Chlorinated Solvents in Porous and Fractured Media*. Translated by J. F. Pankow. Lewis Publishers, Inc., Boca Raton, FL, 1988, p. 146.

Szlag, D. Dissolution of Nonaqueous Phase Liquids in Sandy Soils, Thesis presented to the University of Colorado, Boulder, in partial fulfillment for the degree of Doctor of Philosophy, 1997.

Szlag, D. and T.H. Illangasekare. Entrapment of nonaqueous phase liquids in sandy soils, submitted for publication in *Groundwater*, 1997.

van Genuchten, M.T. Progress and opportunities in hydrologic research, 1987–1990, in U.S. National Report International Union of Geod. and Geophysics, 1987–1990, *Rev. Geophys.*, 29:189–192, 1991.

Walser, G.S. Vadose Zone Infiltration, Mobilization and Retention of Non-Aqueous Phase Liquids, Thesis presented to the University of Colorado, Boulder, in partial fulfillment for the degree of Doctor of Philosophy, 1995, pp. 255.

Walser, G.S., T.H. Illangasekare, and A.T. Corey. Retention of liquid contaminants in layered soils, *Proc. of 9th Annual Conference on Hazardous Waste Remediation,* June 8–10, Montana State University, Bozeman, MT, 1994, pp. 35–48.

Wilson, J.L., S.H. Conrad, W.R. Mason, W. Peplinski, and E. Hagan. Laboratory Investigation of Residual Liquid Organics, EPA/600/6-90/004, 1990.

CHAPTER SEVENTEEN

Strategies for Describing Preferential Flow: The Continuum Approach and Cellular-Automaton Fluids

L. Di Pietro

INTRODUCTION

During the last 20 years, a considerable and increasing amount of research has been devoted to preferential flow. Yet macropore flow, bypass flow, rapid drainage, all referring to a fast infiltrating response to rain along certain soil pathways, is still a challenge for the scientific community. Under certain boundary and initial conditions, preferential flow governs the soil-water regime. The impact of rapid drainage on water use efficiency and deep percolation of pollutants is nowadays widely recognized. To some extent, bypass flow occurs in most soils where roots, earthworm channels, cracks, and fissures of varying sizes are nearly always present. Structured clay soils, plow layers of agronomic soils, and forest soils with low bulk densities are just a few examples. In these soils rapid drainage and irregular flow patterns are ubiquitous, and Darcy's regular, smooth flow is an idealized simplification. Preferential flow does not always occur, because input intensities, vertical continuity of meso- and macropores, and initial soil water may combine to determine its occurrence. However, the factors initiating preferential flow are not yet quantified.

All the soil pores may effectively transport water, from the smallest voids up to the largest fissures. The mean water velocity in each void is determined by the balance between the forces driving the flow and the resistance forces opposing it. These forces depend on the geometrical configuration of the solid boundaries confining water flow, void sizes being a determinant factor. As a consequence, the average velocity of the flow varies with regard to the geometry and the size of the voids. Water advances rapidly in wider pores because they exert less resistance. The geometric arrangement of voids determines the degree of heterogeneity of the water front.

At the microscopic pore scale, water follows heterogeneous patterns, but this heterogeneity is not observable at the macroscopic scale. When the pores are sufficiently large to distinguish them and when they are carrying water, the irregular advance of the water front is perceived. From the point of view of fluid dynamics, preferential flow obeys the same laws as micropore flow. From a macroscopic point of view, the main difference resides in the unequal advance of the water front in the larger pores, avoiding the smaller ones, and the complexity of micropore flow is smoothed out as instruments are not sensible enough to detect it. In the macroscopic perception, microscopic flow is just a slight perturbation of thermodynamic equilibrium conditions. When mesopores and macropores are present, instabilities and nonequilibrium dynamics cannot be smoothed out. Macroscopic instruments are able to detect them.

The quantification of preferential flow is problematic from a theoretical point of view. So far, a generally acceptable model is not available. It is like a mischievous ghost, sometimes appearing and sometimes not, and knowledge about it is still rather intuitive. Research efforts toward a general theory of flow in heterogeneous media are needed since the impact of preferential flow on the environment is substantial.

THE CONTINUUM APPROACH TO PREFERENTIAL FLOW

On Scales and Macroscopic Transport Equations

The behavior of a physical system, such as fluid transport, may be described at various scales. The system is completely defined, at each scale, by a set of conservation laws for the relevant physical variables. These equations are called *the constitutive equations* of the system at the corresponding scale. The system evolves in a space-time domain, and the physical variables are point to point functions of the space and time coordinates. The constitutive equations involve a number of coefficients, which are scalar, vector, or tensor quantities called *the transport coefficients*. These coefficients depend on the underlying structure of the space-time domain (e.g., symmetries, heterogeneities), and on the physical laws at scales below the observation scale (Wolfram, 1986). At any scale of description, the complexity of the system observed at more detailed scales is smoothed out, but microscopic information appears in the transport coefficients as well as in the form of the constitutive equations.

The *continuum approach to hydrodynamics* is derived from the large-scale behavior of fluid molecules. Each point of the fluid represents a volume which is large with respect to the molecular paths. To be consistent with the terms usually used in the theory of transport in porous media, this scale will be called *the microscopic scale*. It is assumed that the physical variables vary slowly from point to point of the continuous space-time domain. The formulation of the basic conservation principles for mass and linear momentum leads to a set of continuous differential equations, which are called the *microscopic point transport equations*. At this scale, the interaction of the fluid with other phases in the system is introduced as boundary conditions (e.g., no slip-conditions between solid-liquid boundaries, jump conditions at the moving boundaries between two fluid phases). This approach is impractical for describing transport in porous media. The complicated and mostly unknown geometry of the interface boundaries obstructs direct solutions to the constitutive equations. The scale of interest for many practical applications is many orders of magnitude greater than the pore scale.

Usually, a process of scaling-up from the microscopic scale is used in order to adequately deal with measurable and observable macroscopic quantities in porous media. This is *the macroscopic scale* at which the porous media is treated as a continuous space in which fluids flow. Scaling-up involves averaging. A point in the scaled-up space-time domain represents a volume of the smaller scale space-time domain. This volume must be large compared with the typical dimensions of the microspace. The set of scaled-up physical variables are obtained by averaging the corresponding small-scale variables over this volume. The latter is a local spatial average, as the time coordinate is not rescaled.

In order to define the macrospace as a continuum and to establish constitutive relations between the scaled-up variables, the averaging volume must have the following properties (Baveye and Sposito, 1984; Bear and Bachmat, 1990):

1. The invariance under translations and rotations of the space-time domain. This property implies the invariance of the metric tensor of the space-time domain.
2. The indifference property: the averaging volume is the same for all the physical variables of interest.

At the microscopic scale, a porous medium may be viewed as a distribution of solid in the continuous and empty space. The scaling-up process implies a partition of the space into contiguous volumes. The macrocontinuum space is constructed by mapping each volume into a point. The new space is not empty. Each point contains implicitly the information of the microscopic solid distribution within the averaging volume. This averaging volume is usually called the representative elementary volume (REV). When at a given scale, a REV complies with 1, the continuum approach of a porous medium is possible. If 2 is also fulfilled, macroscopic constitutive relations may be obtained.

The Macroscopic Transport Equations for Unsaturated Flow in Homogeneous Porous Media

Under the assumption of continuity, the *macroscopic transport equations* have been derived from the point equations by using different techniques: simple volume averaging (Bachmat and Bear, 1986; Bear and Bachmat, 1986; Whitaker, 1986), convolutions with weighting functions (Baveye and Sposito, 1984), or more general filters (Cushman, 1984). The macroscopic equations have also been obtained by the multiple-scales or perturbation method (Saez et al., 1989).

In the unsaturated zone of soils, we must consider the transport of two fluid phases (i.e., water and air) which are simultaneously flowing. Each fluid partially occupies the void space, and they have a common microscopic interface, representing a moving boundary across which momentum may be exchanged. Under the assumption that the continuum hypothesis may be applied to the porous medium, a system of coupled macroscopic transport equations, one for each fluid, must be considered.

If it is assumed that the diffusive and dispersive microscopic fluxes are negligible compared to the advective ones, and that the momentum transfer at the interface water-air is negligible, the macroscopic momentum balance equation for the water phase, written in terms of the volumetric water flux $\mathbf{u} = \theta\mathbf{v}$, ($\theta$ being the volumetric water fraction and $\mathbf{v}$ the macroscopic water velocity), is:

$$\underbrace{\theta\left[\frac{\partial(\mathrm{u}/\theta)}{\partial t}+\frac{\mathrm{u}}{\varepsilon}\cdot\nabla(\mathrm{u}/\theta)\right]}_{[1]}=\underbrace{-\frac{\theta}{\rho}\mathrm{A}\cdot[\nabla p+\rho\nabla(gz)]}_{[2]}+\underbrace{\upsilon\nabla^2\mathrm{u}}_{[3]}-\underbrace{\upsilon\mathrm{B}\cdot\mathrm{u}}_{[4]} \qquad (17.1)$$

where ρ is the macroscopic fluid density,
p is the macroscopic fluid pressure,
g is the acceleration due to gravity,
z is the spatial coordinate parallel to the gravity force,
υ is the kinematic viscosity coefficient of the fluid,
A is a second order tensor related to the microscopic tortuosity of the void space,
B is a second order tensor which depends on the microscopic solid-void distribution.

Equation 17.1 is the macroscopic analog of the Navier-Stokes point equations (Landau and Lifchitz, 1971), except for the damping term [4]. The term [1] represents the total inertial force per unit volume of porous medium, part [2] is the total force exerted on the fluid by pressure and gravity, and term [3] is the viscous resistance exerted within layers of the fluid due to internal friction. This term is of diffusive type, and shows that the velocity field will diffuse with a coefficient of diffusion υ. Term [4] represents the viscous resistance to flow due to the presence of the microscopic solid boundaries. From a macroscopic point of view, the solid distribution exerts a force on the fluid in the porous continuum proportional to its velocity, and all the information on the microscopic distribution is contained in the coefficient **B**.

The mass balance equation, written for the water phase, leads to the macroscopic continuity equation:

$$\frac{\partial \theta}{\partial t} + \nabla \cdot \mathbf{u} = 0 \tag{17.2}$$

Equations 17.1 and 17.2, with the boundary conditions for each particular problem, define the flow system.

The Applicability of the Macroscopic Approach to Soils

In Soil Science, we are interested in describing transport phenomena in spatial domains having a typical length of at least a few centimeters. If all soils could be treated as continua at this scale, one would only need to solve the macroscopic transport equations with the appropriate boundary conditions. Unfortunately, this approach is limited to some restricted soils, or portions of soils, because the basic assumption, existence of a REV with the properties 1 and 2, usually fails.

To emphasize this point, the implications of property 1 are discussed. This property is closely related to the homogeneity of the space. Let us imagine a space where the density of points is not uniform. The volume in such a space is not invariant under linear transformations. In a region where the density of spatial points is higher, the volume will be contracted, because the distance between two points is smaller. A number of particles or molecules traveling in this space will be separated from each other in the dilated regions of the space and will be concentrated in the denser regions. We now increase the scale of observation, to see the molecules as a continuum fluid. In the regions where the microspace density is high, there will probably be a large enough number of molecules to be seen as a point of the continuous fluid. At other points, mapped from regions of lower densities, the number of particles may not be high enough as they may be very far from each other. At these points, the macroscopic observer will see holes in the fluid, and at this scale the continuum approach fails. In theory, a much larger scale may exist at which the intrinsic heterogeneity of the microspace will not be seen, and the averaging volume adequately increases.

In the case of porous media, we map a solid-void distribution into points of the macrospace. The ideal case to apply the continuum hypothesis would be a uniform solid-void distribution. It would suffice to find the minimum volume having a sufficient number of voids and solid particles to form a unity of porous medium. At any larger scale, the porous medium would be seen as a continuum. If the microscopic distribution is not uniform, we must at least assure that at the scale of observation we shall not see holes due to the low density of the solid. Bear and Braester (1972) stated that the invariance property implies that any microscopic variable, and, in particular, the porosity, must be stationary under first-order changes in the averaging volume. This is equivalent to requiring a slow and smooth variation of the averaged variables in space. In a local volume averaging technique, time averaging is not involved. However, to apply continuum approaches, it is implicitly supposed that two successive configurations (at times t and $t + dt$) evolve slowly.

Cushman (1990) described porous media as containing multiple nested or continuous length and time scales. He introduced the concept of structural and functional hierarchy. In a soil containing nested or continuous length scales, we shall not be able to find a unique averaging volume, and to apply the macroscopic transport equations to the whole domain. Let us classify natural soils at the macroscopic scale into three main groups:

1. soils with one length scale
2. soils with two separated length scales
3. soils with continuous or successively nested length scales

We shall hereafter analyze the applicability of the macroscopic transport equations for each group of soils.

Macroscopic Approach to Soils with One Length Scale

The average size of voids is many orders of magnitude smaller than the typical length of the macroscopic scale of observation. The soils are macroscopically homogeneous, the continuum hypothesis applies, and the macroscopic equations can be used to solve transport problems, (e.g., sandy soils). The dispersion of the porosity is small, and its variation may be represented by a continuous function of the spatial coordinates. No functional differentiation of the various void sizes is observed at the macroscopic scale, and the underlying information on the microscopic structure is contained in the transport coefficients. For a macroscopic observer, the flow is smooth and continuous, and the hydrodynamic processes approach the thermodynamic equilibrium. These soils exhibit no preferential flow.

In this case, Darcy's empirical law may be derived from Equation 17.1, under the restriction that the resistance to flow due to the presence of the solid is dominant with respect to inertial effects and to momentum diffusion due to viscosity within the fluid. Then, the terms [1] and [3] in Equation 17.1 may be dropped which leads to the well-known relation:

$$\mathrm{u} = -\mathrm{K} \cdot \nabla H \tag{17.3}$$

where

$$H = \frac{p}{\rho g} + z \tag{17.4}$$

is total water potential, and

$$\mathrm{K} = \frac{\mathrm{A} \cdot \mathrm{B}^{-1} \theta \rho g}{\upsilon} \tag{17.5}$$

is the hydraulic conductivity tensor. This tensor captures all the microscopic effects due to the presence of fluid-solid boundaries. The combination of Equation 17.3 with Equation 17.2 yields Richards' transport equation (Richards, 1931).

Macroscopic Approach to Soil with Two Separated Length Scales

In this type of soils cracks or fractures are distributed within a homogeneous matrix. The macropores have a mean aperture which is several orders of magnitude greater than the mean length characterizing the pore size of the matrix . There are hardly any soils having two well-separated scales; however, this idealization may apply in certain cases like tropical vertisols. The structural hierarchy implies a functional hierarchy. The hydrodynamic response of the two distinct length scales is quite different. Within the cracks, preferential flow may occur. The timescale of this process is very different from that of the flow in the matrix. While the infiltration and redistribution of water within the matrix is a slow process in the order of days to months, the rapid drainage response to rain may last from minutes to hours (Germann, 1986).

In this case, the continuum hypothesis does not apply to the entire medium, because macroscopic discontinuities are present. Many authors (Barenblatt et al., 1960; Beven and Germann, 1981; Jarvis and Leeds-Harrison, 1987; Peters and Klaveter, 1988; Zimmerman et al., 1993) have treated transport in these double porosity media by conceptualizing them as superposition of two interacting continua, one representing the cracks and the other the microporous matrix. Chen (1989) reviewed these models. Usually, two macroscopic continuity equations, one for each continuum, are used to describe the system, coupled by a source-sink term accounting for the volume exchange of fluid between the two continua. These equations are:

$$\frac{\partial \theta_{1,2}}{\partial t} = -\nabla \cdot \mathbf{u}_{1,2} \pm S \tag{17.6}$$

where the index 1 and 2 indicate the matrix and the fractures, respectively, and is the source-sink term. To solve this coupled system, we need to relate and with measurable macroscopic physical variables in each continuum. Within the matrix, it is supposed that Equation 17.3 is valid, and thus for the matrix flux it holds

$$\mathbf{u}_1 = \mathbf{K}_1 \cdot \nabla H_1 \tag{17.7}$$

Let us analyze Equation 17.1 for flow within fractures. The damping term [4] may be neglected if the resistance force due to solid walls is very small. Even if there is not a convincing experimental evidence, the inertial term is also assumed to be small with respect to the diffusive viscous term. This assumption is equivalent to admitting quasi-steady flow conditions and that small Reynolds number of the flow within the fractures. Under these assumptions, Equation 17.1, written for the fractures, results in

$$\nabla^2 \mathbf{u}_2 = \mathbf{R} \cdot \nabla H_2 \tag{17.8}$$

where $\mathbf{R} = \theta g \mathbf{A}/\upsilon$ is a second-order tensor, depending on the microscopic configuration of the fluid and of the tortuosity of the void space of the fractures within the REV. If we assume that the total potential only depends on the vertical coordinate (z-axis) and that the fractures are vertical tubes of simple geometry, Equation 17.7 reduces to

$$\frac{\partial^2 u_{2z}}{\partial y} + \frac{\partial^2 u_{2z}}{\partial z} = R_{zz} \frac{\partial H_2}{\partial z} \tag{17.9}$$

which may be solved with the boundary conditions $u_{2z} = 0$ on the walls of the tubes.

If the tubes are cylindrical or planar, the solution of Equation 17.9 leads to Poiseuille's law for the volume flux in fractures. But even in this simple case, the flux depends on the unknown coefficient R_{zz}.

Many models assume that the macropore system consists of interconnected tubes of simple geometry, possibly with varying apertures. Poiseuille's law is assumed to be valid for the fractured system. Empirical or numerically calculated coefficients are introduced to account for the variability of the fracture apertures (Narasimhan, 1982; Wang and Narasimham, 1985; Moreno et al., 1990).

The fluid exchange between the two continua is generally assumed to be proportional to the potential difference between the two media. The potential is assumed to be well-defined everywhere in each continuum, and that quasi-steady conditions hold for the inter-continuum flow. Two fluxes and two potentials are assigned to each point of the space in this approach, leaving the double continuum approach ill-defined from the physical point of view. However, it may produce solutions of practical interest like the case of fractured rock reservoirs in the saturated zone, where the assumptions about the fluxes in the fractures are not too far from reality.

Many of these assumptions are not generally valid for the unsaturated zone. It is difficult to accept, for instance, the quasi-steady conditions for flow in the wider cracks. The cracks may carry water very deep in a short time after heavy rains. In that case, the inertial term of Equation 17.1 cannot be neglected. Furthermore, it is difficult to define instantaneous value of the potential in the fracture under these rapid transient conditions. Alternatively, if it is considered that the flux within the fractures is a single-valued function of the fluid volumetric fraction, Equation 17.6 may be transformed into a kinematic wave equation (Lighthill et al., 1955; Majda, 1984), which may be solved by the method of characteristics (Abbot, 1966). By assuming that the relation

$$u_2 = b\theta_2{}^a \tag{17.10}$$

holds for the fractures, Germann (1985) and Germann and Beven (1985) obtained traveling wave solutions for the flux within the macropores for a square pulse input. It has been shown that the kinematic approach applies reasonably well to drainage hydrographs at high degrees of saturation, and that the exponent a may serve as a measure to assess the degree of preferential flow (Germann and Di Pietro, 1996). Yet generalization of this approach needs further investigations to assess the physical significance of the coefficient b, and the relation of both parameters with the structural characteristics of the porous medium.

Macroscopic Approach to Soils with Continuous or Successively Nested Length Scales

In the general case, soil porosity shows a nearly continuous distribution of void sizes, ranging from micrometers to centimeters. The frequency distributions of void size classes are not homogeneous; the larger sizes usually represent only a small portion of the total porosity. Packing voids, pore necks, vughs, irregular channels of different sizes are usually present in soils. The hydraulic behavior varies in principle across void sizes. There is experimental evidence demonstrating that the boundary and initial conditions determine the degree of participation of the different pore classes in total water transport (Di Pietro and Lafolie, 1991). Any classification of the pore space based on two-dimensional size only is impractical from the point of view of hydraulic behavior. Classifications based on the static hydraulic behavior of pore classes (e.g., with respect to the hydraulic conductivity) may not be realistic due to the dependence of flow on boundary conditions.

A macroscopic observer will recognize holes of varying sizes, geometries, and vertical connectivities. Some voids, even if not seen, may be sufficiently large for not having enough solid within the hypothetical volume of a REV. The microscopic space in these regions may be dilated to such an extent that the continuum hypothesis has to be discarded. At the macroscopic scale, heterogeneities and instabilities during the flow, i.e., different degrees of preferential flow may be observed. Scaling-up from the macroscopic to a megascopic scale is not practical. A megascopic REV sufficiently large to integrate the effects of macroheterogeneity is required. A suitable REV would probably extend over several meters. Even if it would exist, it will be difficult to determine meaningful megascopic transport coefficients.

CELLULAR AUTOMATON FLUIDS

Many complex problems in hydrodynamics are approached with numerical methods. First, the physical phenomenon is usually described with a mathematical model, a set of differential equations, and with the initial and boundary conditions. The field of continuous variables is discretized to be handled by computers, and the solutions to the equations are derived from numerical algorithms. Many extremely complex physical systems have a high number of degrees of freedom or arbitrarily varying boundary conditions. These solutions cannot be obtained in these cases, even with powerful computers and with sophisticated numerical methods. Flow in heterogeneous media is an example of a complex physical system, in which numerical approximations are intractable unless we impose highly restrictive conditions.

Let us now imagine that we have a virtual robot behaving like a viscous fluid, programmed to interact with solid boundaries and to dynamically react to external forces like a real fluid. With such an automaton, we could conceive virtual experiments, compute flow variables, and by an inductive method try to generalize the governing flow laws, just as we would do with the results of real experiments. These kinds of virtual fluids exist. They are usually called cellular-automaton fluids or lattice-gases. Both terms will be used indifferently.

Introduction to Cellular-Automata

A physical automaton is a robot which responds to signals according to a set of rules. A virtual automaton is a kind of robot programmed on a computer. Cellular-automata are mathematical systems consisting of many identical components or cells, each one being a simple virtual robot. The entire system is capable of simulating complex behavior (Wolfram, 1984). Each cell is defined by:

- its position vector r_i on a discrete space
- a finite number of states S_j, with $1 < j < N$,
- a set of transition rules, mapping the state of the cell at time t into the state of the cell at time $t+1$.

The transition rules are local functions of the state of the cell at time t and of the states of the neighboring cells. The imposed transition functions are either deterministic or probabilistic.

The evolution of the cellular automaton involves sequential processing of the initial information associated with the states of the cells. Wolfram (1984) reviewed the general behavior and the mathematical properties of this type of model.

Cellular automata are entirely encoded with bit operations, because they are complete discrete dynamic systems. The architecture of cellular automata follows the one of a simple parallel computer and thus they are well-suited for simulations in parallel machines.

Since their conception by Von Neuman (1966), cellular-automata have served as models for studying a wide variety of physical, biological, and chemical systems. Their application in hydrodynamics began in 1986 with a cellular automaton which behaves at a large-scale like a viscous fluid in the incompressible limit, the FHP lattice-gas (Frisch et al., 1986). Since then, numerous lattice-gases have been developed and applied to nonlinear hydrodynamics, such as free-boundaries in flow systems (Clavin et al., 1988), nonmiscible flow of mixtures (Rothman and Keller, 1988), phase transitions (Appert and Zaleski, 1990; 1993), aggregation and crystal growth in suspensions (Bremond and Jeulin, 1994), brownian motion (Ladd et al., 1988), and coupled reaction-diffusion equations (Dab et al., 1991).

Structure and Properties of Cellular-Automaton Fluids

Common structure and properties of cellular-automaton fluids follow. For further details see the article by Frisch et al. (1987). In a cellular-automaton fluid, the cellular space is a discrete and regular lattice $\mathcal{L} \in \Re^D$. Each cell consists of a site and its k-links to neighboring sites. The cells may be occupied by utmost M identical particles, residing either within the site, or within the links. A particle occupying the α-link is associated with a velocity c_d, and a particle residing at a site is associated with zero velocity. The combinations of occupancies define the set of N possible states associated with each cell. The entire automaton configuration is determined by the actual configuration of each cell at each time step.

The system evolves in discrete time steps, obeying imposed transition rules. Each transition involves the propagation of particles to neighboring sites in the direction of their velocities, and the collision with other particles. Collision is a local exchange of the linear moments of the particles. The collision rules conserve the total mass and linear momentum, like real elastic collisions. Extended models include nonlocal interactions. In those, particles separated from one another by a preset distance may exchange momentum, simulating long-range attractive or repulsive molecular interactions.

The evolution of automaton is analyzed statistically. The large-scale dynamic equations describing the behavior of these automata are obtained as a statistical average in space and time of the corresponding microscopic variables under the assumption of statistical equilibrium. This is analogous to the process of scaling-up from the molecular to the microscopic continuum description in real fluids. The macroscopic dynamic equations of these automata correspond to the point-to-point Navier-Stokes equations in the incompressible limit, when the lattice has suitable symmetries to ensure the isotropy of the tensor quantities (Wolfram, 1986).

The geniality of these systems is that from an extremely simplified version of molecular dynamics the same macroscopic behavior of real fluids is obtained. The full details of real molecular dynamics are not needed to create a microscopic model with macroscopic hydrodynamic behavior. The large-scale equations derive only from the microscopic conservation laws and from the symmetries of the underlying space. The collision rules will determine the magnitude of the transport coefficients; in this case, the magnitude of the viscosity coefficient, but they will not change the form of the macrodynamic equations.

The existing 2D models of a single fluid are all variations of the original FHP lattice-gas (Frisch et al., 1986). The cellular space is built as an hexagonal lattice. At most, six moving particles may reside in a cell at any time. The existing versions differ in the number of particles at rest, and in the collision rules. As a consequence, they produce different viscosity coefficients. As no 3D lattice with the required symmetries exists, the cellular space of 3D automata is a 4D regular polytope, the face centered hypercube (FCHC) where one dimension is projected onto a two-layer surface (Sommers and Rem, 1992). In this case, up to 24 moving particles may occupy the cells.

Generalizations of the lattice-gas models into two or more species have also been introduced as an efficient way to model the dynamics of interfaces coupled with fluid motion. The method consists in distinguishing two or more species among the population of lattice-gas particles by labeling or "coloring" them. Each family of particles may have its own dynamic evolution controlled by specific transition rules. The interaction rules between particles of different types determine the behavior of the mixture. For instance, attractive, repulsive or reactive interactions may be introduced. Labeling of particles results in an extension of the hydrodynamic behavior of the automaton with respect to advection, diffusion, and reactions at the moving interfaces. An interacting liquid-gas lattice model for liquid-gas phase transition has also been constructed (Appert and Zaleski, 1990; 1993). The interface shows surface tension. It may be used to study the simultaneous flows of the two phases, at equilibrium.

Application of Cellular-Automaton Fluids to Transport in Porous Media

To apply cellular-automaton to transport in porous media it is necessary to introduce the solid boundaries which interact with the virtual fluid and the external body forces, such as gravity. The solid is represented in the lattice by labeling selected sites. Lattice particles are not allowed to hop into these sites. Usually, no-slip conditions are imposed on particles arriving at a solid site; they are bounced back in the opposite direction with a preset probability. This probability may be tuned from 0 to 1, to simulate different wetting properties. The geometry of the solid may be arbitrarily complex as the solid distribution and the boundary interactions are easily implemented. External forces of varying intensities are also easily introduced by randomly deviating some lattice particles parallel to the "force" direction. The intensity of the force is proportional to the probability that a particle deviates.

The scale of practical interest in soil transport problems is the macroscopic scale, yet the transport coefficients depend on the microscopic configuration of boundaries. To determine the coefficients, we have to either measure or estimate them. For instance, in homogeneous media, we must determine the value of the hydraulic conductivity. In heterogeneous media with two separate scales, we have already seen that we need to know additional parameters involved in the macroscopic equations dealing with the fracture domain. But how to measure, for instance, the coefficient **R** of Equation 17.8, or **A** and **B** of Equation 17.1? The instruments are not available, and we do not know enough about their relation with the microstructure to estimate them. Lattice-gas automata may supply the required information, because they allow us to study hydrodynamic phenomena at the pore scale, but they also recover the laws at a macroscopic scale. To illustrate the potential of lattice-gas possibilities in this domain, we summarize some examples.

Flow at the Pore Scale

Cellular automata have been used to study flow at the pore scale for single and multiphase fluid flow, and for steady-state and transient regimes. Flow velocity fields of simple one-dimensional flow between parallel solid walls under a pressure gradient were analyzed by Hayot (1987), Balasubramanian et al. (1987), and Rothman (1988), with the FHP lattice-gas, and later by Di Pietro et al. (1994), with the interacting liquid-gas automaton (Appert and Zaleski, 1990; 1993). They observed the initial transient regime during which the initial flat velocity profile evolves into the parabolic dynamic equilibrium profile of steady Poiseuille's flow at the hydrodynamic scale.

The interacting liquid-gas model was used to study evaporation and infiltration of liquid films in pores of different diameters. In the evaporation studies, capillary properties at the liquid-gas/solid interfaces, such as surface tension, equilibrium pressures, meniscus formation, and evaporation rates were investigated under static and dynamic conditions (Pot, 1994). The evaporation fluxes in partially saturated tridimensional structures with irregular pore size distributions have also been analyzed and estimated (Appert et al., 1994). The liquid flow regimes in pores, including surface wave patterns, were analyzed for different initial conditions in the infiltration studies as a function of the diameter of the pore, and the balance between the capillary and the gravity forces (Di Pietro, 1993; Di Pietro et al., 1994). Figure 17.1 illustrates different stages of the advancement of a liquid in cylindrical tubes of varying radii and impermeable solid walls as simulated by the 3D interacting liquid-gas automata.

Stockman et al. (1990) investigated with the nonmiscible lattice-gas automata (Rothman and Keller, 1988) the viscous segregation of two nonmiscible fluids of different viscosities flowing in pore channels for varying initial conditions. Simultaneous flow and solid aggregation in suspensions were analyzed (Bremond and Jeulin, 1994) with a two-species extension of the FHP automata.

Recovering Macroscopic Laws

Darcy's macroscopic law was obtained for the calculated mean fluxes from numerical automata experiments (Balasubramanian et al., 1987; Rothman, 1988; Di Pietro, 1996) by simulating a saturated pressure-driven flow through various arrangements of porous solids. By analyzing the velocity field in the porous medium, Rothman (1988) showed that the majority of the flow followed just a few winding paths, many of the regions being almost stagnant. Di Pietro (1996) was able to calculate the saturated hydraulic conductivity of various porous media from the slope of the linear flux-potential gradient curves. While in the previously cited works, the porous media were simulated by explicitly designing the geometry of the solid aggregates, Balasubramanian et al. (1987) represented the porous medium with a spatial distribution of punctual scatters within the flow region. They found a relation between the coefficient **B** of the damping term [4] of Equation 17.1 with the density of scatters in the flow region. They obtained Darcy's law in the limit where viscous effects within the fluid were negligible. This is an example of how to interpret the macroscopic transport coefficients with these virtual dynamic experiments.

Numerical experiments of infiltration under gravity in unsaturated homogeneous aggregated media (Di Pietro, 1993; Di Pietro et al., 1994) are further applications of lattice-gases with those to calculate moisture profiles and infiltration rates. Cumulative infiltration followed closely Philip's law of infiltration (Philip, 1957). Figure 17.2 shows two infiltration stages in a random porous medium, simulated by the 2D interacting lattice-gas automata.

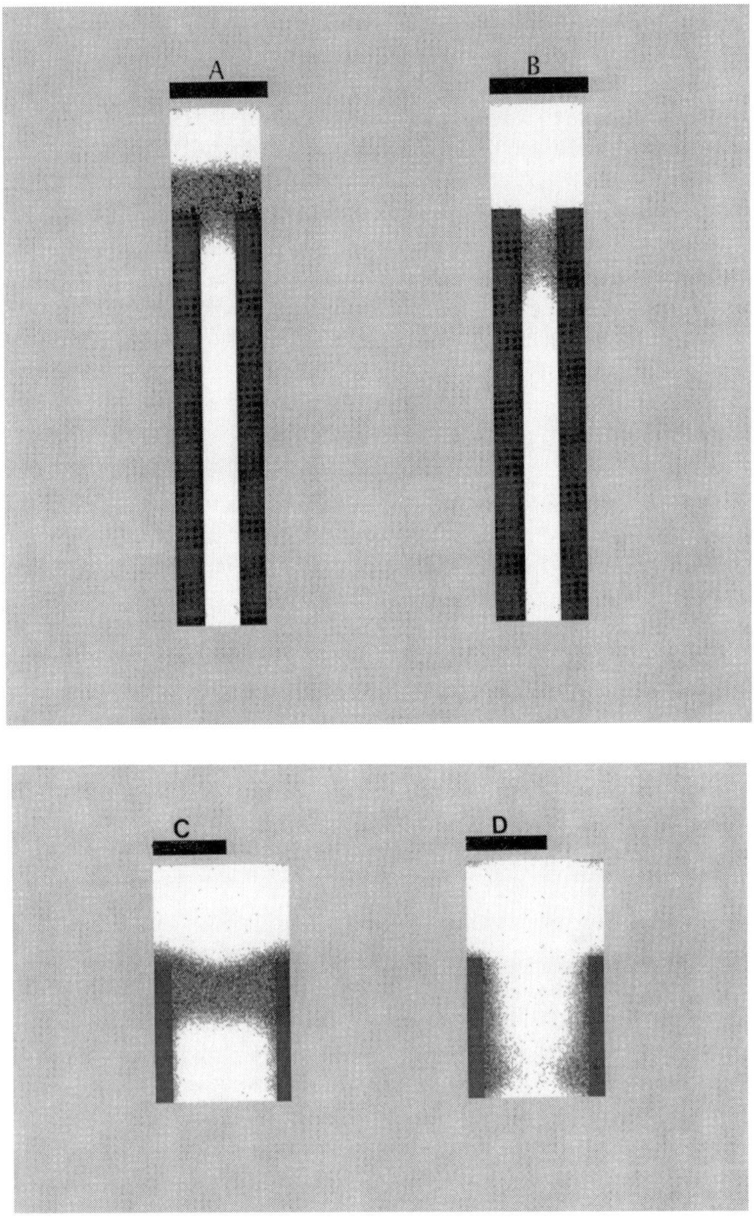

Figure 17.1. Two stages of the advancement of a liquid in a cylindrical pore under gravity, simulated by the 3D interacting liquid-gas automaton. The figures correspond to a central vertical cut. Solid walls are represented in black, liquid in dark grey, and gas in white with sparse grey dots. Cases A and B correspond to 600 and 5,000 time steps, respectively, and to a pore radius of 10 lattice sites. The formation of the meniscus due to the effect of capillarity is shown. Cases C and D correspond to 750 and 2,500 time steps and to a pore radius of 40 lattice sites. The liquid tends to advance in films along the walls. (Numerical experiments from Di Pietro and Germann, unpublished).

Using Automaton Fluids for Analyzing Preferential Flow

Troubles arise when trying to measure flow variables and structural properties in laboratory and field experiments. As discussed by Cushman (1986), the instrument

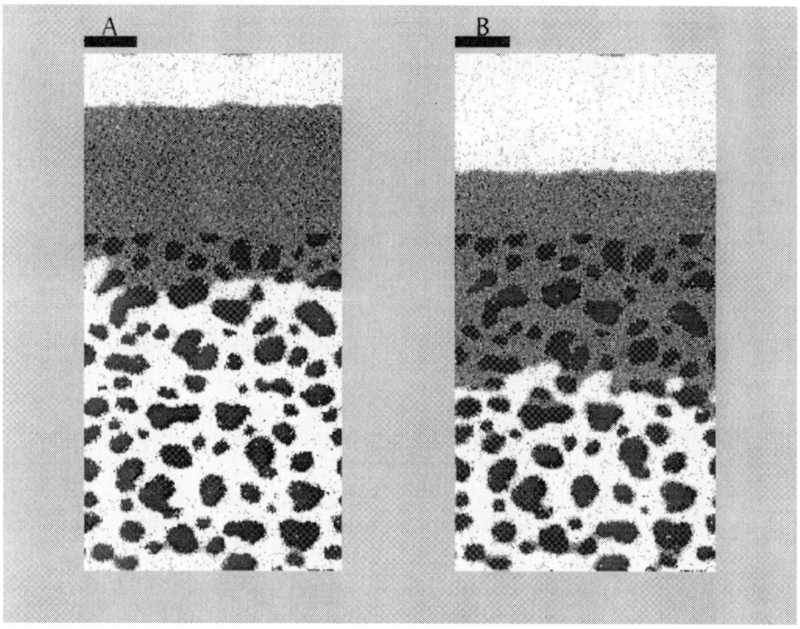

Figure 17.2. Two infiltration stages under gravity in a random porous medium, simulated by the 2D interacting liquid-gas automaton. Cases A and B correspond to 2000 and 15,000 time steps, respectively. (Numerical experiments from Di Pietro, 1993, unpublished).

scale must be compatible with the scale of the process. Usually, the instruments are not adapted to detect flow instabilities and preferential flow pathways within complex porous structures because they do not have the required temporal and spatial resolutions. For instance, local water content measured with a neutron probe results from the integration over a volume of many cubic centimeters, and during 10 to 120 seconds. If there is a macroheterogeneity within this volume, water may bypass it before the neutron probe could detect it. Other measuring techniques are static or indirect, such as dye tracer techniques or drainage hydrograph analysis. They may provide useful but incomplete information about the transport process. When analyzing the traces left by dye in soil pathways, the flux cannot be quantified, and water rapidly passing through other pathways without leaving stains cannot be identified. From the drainage hydrographs, the amount of rapid drainage can globally be quantified, but the mechanisms of the flow cannot be assessed.

Cellular automata techniques may provide complementary information about the dynamics of transport in multiple-scale media, as we can simultaneously compute flow variables and structural properties at different scales. The computation of flow variables may be viewed as an almost ideal monitoring tool with nearly unlimited temporal and spatial resolution.

Automata techniques have recently been applied to the analysis of preferential flow (Di Pietro, 1993; Di Pietro et al., 1994; Di Pietro, 1996). The interacting lattice-gas model (Appert and Zaleski, 1990; 1993) was adapted for simulating liquid and a gas flow in two- and three-dimensional heterogeneous porous media. The effect of single parallel cracks of various apertures within a microporous matrix was investigated for contrasting boundary and initial conditions. Some of the parameters regulating the flow regime and the amount of rapid drainage (the amount of liquid supplied, the permeability of the microporous matrix, and the crack aperture), were determined

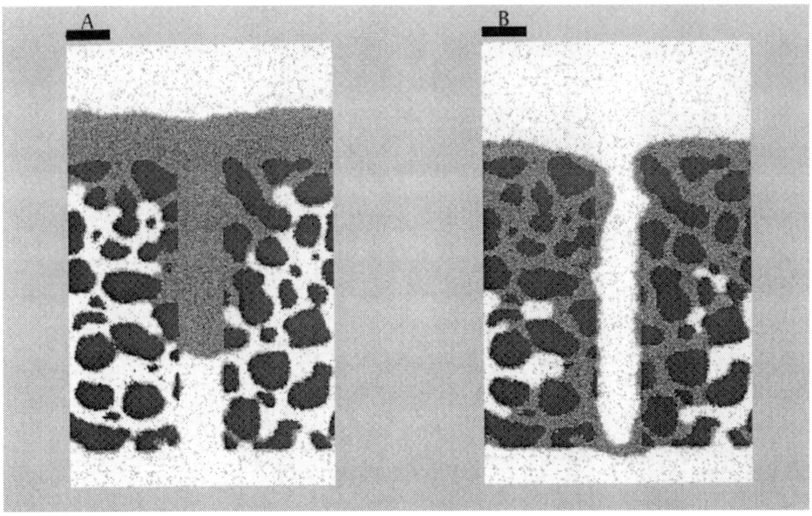

Figure 17.3. Infiltration in a porous medium with a planar central crack. Solid is represented in black, liquid in dark grey, and gas in white with sparse grey dots. Cases A and B correspond to 5000 and 49,000 time steps, respectively. (Numerical experiments from Di Pietro, 1993, unpublished).

for several flow conditions. Figure 17.3 shows some patterns of rapid infiltration through macropores simulated by the automaton model. The bimodal behavior of liquid flow was observed and fluxes were estimated. From virtual experiments of this type, we hope to find general relationships between the structural parameters of the porous medium and the degree of preferential flow.

THE FUTURE

A practical limitation of lattice-gas models is that they must be run at moderate mean lattice-gas velocities relative to the speed of sound in the lattice, in order to avoid spurious high-order nonlinear terms in the large-scale equations. Theoretical and experimental difficulties for describing preferential flow were summarized. Cellular automaton fluids were introduced as promising new tools to approach the subject. Automaton fluids will not provide the multiscale flow theory, but they may help to construct it. They provide information otherwise not obtainable. Interpretation of morphological features and data, combined with the results of virtual dynamic experiments will help formulate a descriptive theory of preferential flow.

Preferential flow is not an exception. Soils are generally multiscale porous media in that preferential flow occurs to some degree. Scientists and engineers, still applying models only based on Richards' equation to solve transport field problems, frequently observe deviations from the expected water balances or calculated fluxes. The deviations are usually observed after high intensity rains. It is also known that when calculating the hydraulic conductivity—moisture relation from laboratory or field experiments, the presence of macroheterogeneities affects parameter estimation and is responsible for spatial variability. Current conception of water and solute transport in soils should evolve to admit that the quasi-equilibrium continuum description is an ideal extreme, and that including stochastic variability in the transport coefficients does not suffice. Dual-porosity models are a first attempt at generalization, but still two-scale soils represent only a small portion of natural soils. A general theory of soil-water transport, including physical nonequilibrium patterns has not yet been formulated and research effort must be focused in this direction.

ACKNOWLEDGMENTS

The author would like to thank Peter Germann and Vijay Singh for their helpful remarks.

REFERENCES

Abbot, M.B. *The Method of Characteristics*. American Elsevier, New York, 1966.

Appert, C. and S. Zaleski. Lattice gas with a liquid-gas transition. *Phys. Rev. Lett.*, 64:1–4, 1990.

Appert, C. and S. Zaleski. Dynamical liquid-gas phase transition. *J. Phys. II France*, 3:309–337, 1993.

Appert, C., V. Pot, and S. Zaleski. Evaporation in a three dimensional liquid-gas model, in *Proceedings of Computational Methods in Water Resources*, Peters, A., G. Wittum, B. Herrling, and V. Meissner (Eds.), Kluwer Academic Publishers, Heidelberg, 1994, pp. 918–924.

Bachmat, Y. and J. Bear. Macroscopic modelling of transport phenomena in porous media 1: The continuum approach. *Trans. Porous Media*, 1:213–240, 1986.

Balasubramanian, K., F. Hayot, and W.F. Saam. Darcy's law from lattice-gas hydrodynamics. *Phys. Rev. A*, 36:2248–2253, 1987.

Barenblatt, G.E., I.P. Zheltov, and I.N. Kochina. Basic concepts in the theory of homogeneous liquids in fissured rocks. *J. Appl. Math. Mech.*, 24:1286–1303, 1960.

Baveye, P. and G. Sposito. The operational significance of the continuum hypothesis in the theory of water movement through soils and aquifers. *Water Resour. Res.*, 20:521–530, 1984.

Bear, J. and C. Braester. On the Flow of Two Immiscible Fluids in Fractured Porous Media, in *Proceeding of the First Symposium on Fundamental of Transport in Porous Media*. Elsevier, New York, 1972, pp. 177–202.

Bear, J. and Y. Bachmat. Macroscopic modelling of transport phenomena in porous media 2: Applications to mass, momentum and energy transport. *Trans. Porous Media*, 1:241–269, 1986.

Bear, J. and Y. Bachmat. *Introduction to Modeling of Transport Phenomena*. Kluwer Academic Publishers, Dordrecht, 1990.

Beven, K. and P.F. Germann. Water flow in soil macropores II: A combined flow model. *J. Soil Sci.*, 32:15–29, 1981.

Beven, K. and P. Germann. Macropores and water flow in soils. *Water Resour. Res.*, 18:1311–1325, 1982.

Bremond, R. and D. Jeulin. Morphogenesis Simulations with Lattice Gas, in *Mathematical Morphology and Its Applications to Image Processing*, Serra, J. and P. Soille (Eds.), Kluwer Academic Publisher, Dordrecht, 1994.

Chen, Z.X. Transient flow of slightly compressible fluids trough double porosity, double permeability systems. A state-of-the-art review. *Trans. Porous Media*, 4:147–184, 1989.

Clavin, P., P. Lallemand, Y. Pomeau, and G. Searby. Simulations of free boundaries in flow systems by lattice-gas models. *J. Fluid Mech.*, 188:437–464, 1988.

Cushman, J.H. On unifying the concepts of scale instrumentation and stochastics in the development of multiple transport theory. *Water Resour. Res.*, 20:1668–1678, 1984.

Cushman, J.H. On measurement, scale, and scaling. *Water Resources Research*, 22:129–134, 1986.

Cushman, J.H. An Introduction to Hierarchical Porous Media, in *Dynamics of Fluids in Hierarchical Porous Media*, Cushman, J.H. (Ed.), Academic Press, London, 1990.

Dab, D., J.P. Boon, and Y.X. Li. Lattice-gas automata for coupled reaction-diffusion equations. *Phys. Rev. Lett.*, 66:2535–2538, 1991.

Di Pietro, L. "Transfert d'Eau en Milieu Structuré: Analyse Bibliographique, Expérimentation et Essai de Caractérisation des Ecoulements." Mémoire de Diplôme d'Etudes Approfondies en Hydrologie, Université de Sciences et Techniques du Languedoc, Montpellier II, France, 1990.

Di Pietro, L. "Transferts d'Eau dans des Milieux à Porosité Bimodale: Modélisation par la Méthode des Gaz sur Réseau." Thèse de Doctorat en Mécanique, Génie Mécanique et Génie Civil, Université de Sciences et Techniques du Languedoc, Montpellier II, France, 1993.

Di Pietro, L. Application of a lattice-gas numerical algorithm to modeling water transport in fractured porous media. *Trans. Porous Media,* 22:307–325, 1996.

Di Pietro, L. and F. Lafolie. Water flow characterization and test of a kinematic-wave model for macropore flow in a highly contrasted and irregular double-porosity medium. *J. Soil Sci.,* 42:551–563, 1991.

Di Pietro, L., A. Melayah, and S. Zaleski. Modeling water infiltration in unsaturated porous media by interacting lattice-gas cellular automata. *Water Resour. Res.,* 30:2785–2792, 1994.

Frisch, U., B. Hasslacher, and Y. Pomeau. Lattice-gas automata for the Navier-Stokes equation. *Phys. Rev. Lett.,* 56:1505–1508,1986.

Frisch, U., D. d'Humières, B. Hasslacher, P. Lallemand, Y. Pomeau, and J. Rivet. Lattice gas hydrodynamics in two and three dimensions. *Complex Systems,* 1:649–707, 1987.

Germann, P. Kinematic wave approximation to infiltration and drainage into and from soil macropores. *Trans. ASAE,* 28:745–749, 1985.

Germann P.F. Rapid drainage response to precipitation. *Hydrol. Process,* 1:3–13, 1986.

Germann, P.F. and K. Beven. Kinematic wave approximation to infiltration into soils with sorbing macropores. *Water Resour. Res.,* 21:990–996, 1985.

Germann, P.F. and L. Di Pietro. When is porous media flow preferential? A hydrodynamic perspective. *Geoderma,* 74:1–21, 1996.

Hayot, F. Unsteady, one-dimensional flow in lattice gas automata. *Phys. Rev. A,* 35:1774–1777, 1987.

Jarvis, N.J. and P.B. Leeds-Harrison. Modelling water movement in drained clay soil. 1: Description of the model, sample output and sensitivity analysis. *J. Soil Sci.,* 38:487–418, 1987.

Ladd, A.J.C., M.E. Calvin, and D. Frenkel. Application of lattice-gas cellular automata to the Brownian motion of solids in suspension. *Phys. Rev. Lett.,* 60:975–978, 1988.

Landau, L. and E. Lifchitz. *Mécanique des Fluides.* Editions MIR, Moscow, 1971.

Lighthill, M., F. Whitman, and G. Whitman. On linematic waves 1: Flood movement in long rivers. *Proc. Royal Soc. London, Serie A,* 229:281–316, 1955.

Majda, A. *Compressible Fluid Flow and Systems of Conservation Laws in Several Space Variables.* Applied Mathematical Sciences 53, Springer Verlag, New York, 1984.

Moreno, L., C.F. Tsang, Y. Tsang, and I. Neretnieks. Some anomalous features of flow and solute transport arising from fractures aperture variability. *Water Resour. Res.,* 26:2377–2391, 1990.

Narasimhan, T.N. Multidimensional numerical simulation of fluid flow in fractured porous media. *Water Resour. Res.,* 18:1235–1247, 1982.

Peters, R.R. and E.A. Klaveter. A continuum model for water movement in an unsaturated fractured rock mass. *Water Resour. Res.,* 24:416–430, 1988.

Philip, J. The theory of infiltration 1: The infiltration equation and its solution. *Soil Sci.,* 83:345–357, 1957.

Pot, V. "Etude Microscopique du Transport et du Changement de Phase en Milieux Poreux, par la Méthode des Gaz sur Réseau." Thèse de Doctorat en Sciences de l'Eau et Mécanique, Université Pierre et Marie Curie, Paris VI, France, 1994.

Richards, L. Capillary conduction of liquids through porous medium. *Physics.*, 1:318–333, 1931.

Rothman, D.H. Cellular-automaton fluids: A model for flow in porous media. *Geophysics*, 53:509–518, 1988.

Rothman, D.H. and J.M. Keller. Immiscible cellular-automaton fluids. *J. Stat. Phys.*, 52:1119–1127, 1988.

Saez, A.E., C.J. Otero, and I. Rusinek. The effective homogeneous behavior of heterogeneous porous media. *Trans. Porous Media*, 4:213–238, 1989.

Sommers, J.A. and P.C. Rem. Obtaining Numerical Results from the 3D FCHC-Lattice Gas, in *Workshop on Numerical Methods for the Simulation of Multi-Phase and Complex Flow*, Verheggen, T.M.M. (Ed.), Springer, Berlin, 1992.

Stockman, H.W., C.T. Stockman, and C.R. Carrigan. Modelling viscous segregation in immiscible fluids using lattice-gas automata. *Nature*, 348:523–525, 1990.

Von Neuman, J. *Theory of Self-Reproducing automata*. University of Illinois Press, Urbana, 1966.

Wang J.S.Y. and T.N. Narasimham. Hydrologic mechanisms governing fluid flow in a partially saturated, fractured, porous medium. *Water Resour. Res.*, 21:1861–1872, 1985.

Whitaker, S. Flow in porous media I: A theoretical derivation of Darcy's law. *Trans. Porous Media*, 1:3–25, 1986.

White, R.E. The influence of macropores on the transport of dissolved and suspended matter through soil. *Adv. Soil Sci.*, 3:95–120, 1985.

Wolfram, S. Cellular automata as models of complexity. *Nature*, 311:419–424, 1984.

Wolfram, S. Cellular automaton fluids 1: Basic theory. *J. Stat. Phys.*, 45:471–526, 1986.

Zimmerman, R.W., G. Chen, T. Hagdu, and G.S. Bodvarsson. A numerical dual-porosity model with semianalytical treatment of fracture-matrix flow. *Water Resour. Res.*, 29:2127–2137, 1993.

CHAPTER EIGHTEEN

Using GIS and Geostatistics to Model Nonequilibrium Flow at a Farm Scale

A.S. Rogowski

INTRODUCTION

Under natural conditions, soil water movement at a field or catchment scale is usually a physical nonequilibrium process. It takes place through a network of pores, which may or may not be connected. Differences in soil properties that affect flow, as well as spatially varying patterns of connectivity, and geometry of pores, all contribute to the nonequilibrium nature of water movement within the landscape. Spatial distributions of precipitation, plant water demand, root density, and antecedent conditions change with time, depending on season, location, and climatic conditions. An opinion is also emerging that plant growth and soil variability are related, and often constitute an outward manifestation of the physical nonequilibrium process. A perceived diversity in natural ecosystems frequently reflects local soil variability (Cruse and Dinnes, 1995). Therefore, to understand implications and limitations of diversity there is an underlying need to understand ramifications of soil variability at a scale of an ecosystem (Rudzitis, 1996). Our ability to represent such a diverse and time varying process is rather limited and simplifications are required. For example, under saturated conditions water movement is usually conceptualized in terms of Darcy's law as outflow flux q in time t, from a soil column with a cross-sectional area α and length l, subjected to a head difference Δh,

$$\frac{q}{t} = \left(K_{sat} \frac{\alpha}{\ell} \right)(\Delta h) \tag{18.1}$$

A constant K_{sat}, known as saturated hydraulic conductivity, is assumed to represent characteristic permeability of the material in the column. In reality it is a fudge factor expressing our inability to describe the geometry and connectivity of the soil pore network in a deterministic manner (Journel, 1996). As such, K_{sat} is neither constant nor characteristic, and varies appreciably both in space and time depending on system complicating factors. Some of these factors are detailed in Youngs (1995) and include local effects of soil air, thermal gradients, hysteresis, swelling and shrinking, and differences in soil aggregation and anisotropy. Layers with a high colloid content, and locations where there are marked differences in structure and texture may also contribute to nonequilibrium behavior. Philip (1980) points out that any local change in texture (to either coarser or finer) will reduce the water flow rate (Baver et al., 1972).

Soil Variability

Although K_{sat} is a standard deterministic model input parameter and can be readily measured at a point, it often varies spatially in unpredictable ways reflecting nonequilibrium conditions that control local outflow response over an area. To better define this type of variability we can model the outflow deterministically at a number of points where K_{sat} is well known, and then describe the distribution of spatial recharge flux stochastically. Such distribution could then be evaluated in terms of soil attributes associated with internal geometry and connectivity of the soil pore network operating at a variety of scales and in different time frames. The objective would be to replace spatial K_{sat} distribution with a number of equiprobable scenarios of recharge flux (Journel, 1996). Stochastic images are considered *equiprobable* if, for a given simulation algorithm with its specific computer code and statistics, each image is uniquely indexed by a seed number that starts the algorithm. The seed numbers are drawn from a uniform [0,1] probability distribution, hence each image has the same chance to be drawn. (Journel, 1995; Borgault et al., 1995). Such approach still reflects our ignorance of a true geometry and connectivity of flow pathways and the specific nonequilibrium processes operating at a local-to-regional scale within a soil system in the field. However, when modeling the spatial effects of water flow and pollutant transport, it allows us to incorporate known soil physical and morphological properties as conditioning data.

Under unsaturated conditions, water flow in soil is now typically described with the Richards equation (Richards, 1931) in term water content *(θ)*, matric potential *(ψ)*, hydraulic conductivity *(K)*, and depth *(z)*:

$$\frac{\partial \theta}{\partial t} = \frac{\partial}{\partial z}\left[K(\psi)\left(\frac{\partial \psi}{\partial z} - 1\right)\right] \tag{18.2}$$

From a modeling perspective, there is a need for a functional representation of soil moisture retention curve and hydraulic conductivity from dryness to saturation. Numerous papers have been written detailing different ways to represent and model these soil hydraulic functions. Much of this work is summarized in the proceedings of a recent workshop on indirect methods for estimating the hydraulic properties of unsaturated soils (van Genuchten and Leij, 1992). While valid at a local scale, or under laboratory conditions, $\theta(\psi)$, $K(\theta)$, and $K(\psi)$ approximations cannot be easily implemented under the nonequilibrium situations operating at various scales in nonhomogeneous field materials (Corey, 1992). Nor are we likely to predict how these algorithms will respond to temporal changes in soil pore geometry and connectivity at a field or catchment level.

John Philip's studies over the past several decades have shed invaluable light on many different aspects of soil water flow (1992 John Philip Symposium, SSSAJ 59:285–319). Although these studies provide a much needed theoretical framework, they apply primarily to homogeneous, or very simple heterogeneous systems (Philip, 1980). Practical, large-scale applications of soil physics to hydrology are hampered by nonequilibrium field conditions which can vary greatly in space and time. For example, as Youngs (1995) points out, our current perception of infiltration process relies primarily on Darcy's law and the solutions of Richards' equation of flow in unsaturated soil. When assumptions of Richards' equation do not hold, other factors may come into play, which modify the classical treatment and require a different approach.

Current Methodology

Direct methods of measuring hydraulic conductivity, retention curve, and other hydraulic properties are well-established and documented in the literature (Klute, 1986; Topp et al., 1992). They are also relatively easy to implement in the field for well-defined boundary and initial conditions. Similarly, a large number of functional representations for estimating moisture retention curve $\theta(\psi)$ and conductivity $K(\theta)$ are now also available, and a great many models have been documented, tested, and validated (Mualem, 1992). Numerous attempts on record describe soil pore space geometry and hydraulic properties in terms of other readily measured properties such as texture, aggregate size distribution, or fractal dimension (Arya and Dierolf, 1992; Tyler and Wheatcraft, 1992). However true, field applications are few, and most $\theta(\psi)$ and $K(\theta)$ models apply to uniform systems that operate at a point in a lab or plot scale environment (Greminger et al., 1985). We do not know how to express distributions of hydraulic conductivity or a retention curve under the spatially varying nonequilibrium conditions for a field or catchment.

Much attention has recently centered on correlating soil hydraulic properties such as conductivity $K(\theta)$ and water retention data $\theta(\psi)$ with structure, texture, and other profile attributes including organic matter and bulk density (Saxton et al., 1983; Rawls et al., 1992). This approach appears attractive from the standpoint of modeling physical nonequilibrium and variability in hydraulic properties at a large scale. Needed values are easily measured in the field, or are readily available from a comprehensive soil survey database. Every time a soil profile is described in the field, a new set of exhaustive data on local nonequilibrium conditions is acquired. Unfortunately, such data are seldom georeferenced, and information on spatial and temporal variability is nearly always lacking. Moreover, any location mapped as a given mapping unit does not necessarily correspond to the specified soil (Soil Survey Staff, 1993). Soil survey descriptions are for typical or modal profiles which may be far removed from a particular site, and include small areas (<2 ha) of other soils (Rogowski and Wolf, 1994).

Soil Morphology

While a concerted effort has been underway for some time now to evaluate and model hydraulic properties of soils, a parallel effort to document and classify factors of soil formation has been unfolding for about as many years (Jenny, 1980; Hoosbeek and Bryant, 1992). It is rather ironic that hydraulic modelers, while seeking a comprehensive description of water flow in a nonequilibrium environment of field soils, usually overlook the exhaustive documentation that pedogenesis provides in terms of soil profile description. Conversely, soil morphologists seeking to explain processes active in soil formation, are apt to pay little attention to the physics and energetics associated with water flow and biochemical equilibria within a profile. Implications of species diversity in natural ecosystems which may mirror local soil variability are seldom taken into account, because few natural ecosystems survive when subjected to systematic monoculture and imposed artificial boundaries.

The extent of soil profile maturity, position and definition of impeding layers, internal soil structure and color, as well as distribution with depth and forms of elements such as phosphorus (Smeck, 1973, 1985), and to a lesser extent nitrogen, sulfur, and organic carbon (Walker and Adams, 1959), serve as indicators of water movement and equilibrium status of a soil pedon. All information encoded within a soil profile is by nature exhaustive, descriptive of pore geometry and connectivity, and readily accessible to document spatial variability of soil properties through an appropriately georeferenced sampling scheme. The need for, and significance of georeferencing is that it fixes sample locations relative to one another in space and time, so the spatial analysis can be performed and same locations can be revisited at a later date.

There are, of course, exceptions. For example, Bouma has made extensive use of pedogenesis information to model hydraulic properties of soils (Wosten and Bouma, 1992). Webster applied techniques of spatial statistics to soil survey data collection (Webster and Oliver, 1990). Bierkens and Burrough (1993) used a conditional simulation approach to describe distribution of map impurities and categorical data. Rogowski adapted geostatistical techniques to model uncertainty associated with soil hydraulic attributes (Rogowski 1996a,b,c; Rogowski and Wolf, 1994; Rogowski and Hoover, 1996; Wolf and Rogowski, 1991). Runge (1973) and Smeck et al. (1983) modeled soil formation based on entropy and available energy of percolating water. Moore et al. (1993) predicted catenary soil development based on water pathways in landform environment.

Future Challenges

Despite these advances there is still a need to include additional tenets of spatial analysis in soil studies, particularly under the nonequilibrium field conditions. We must recognize that a soils' continuum is only a thin veneer in an overall spatial extent of a landscape (Daniels and Hammer, 1992). However, the geometry and connectivity of pore space, as well as the state of physical and thermodynamic equilibrium within this thin veneer largely controls the flow in a soil system at any scale (Hall and Olson, 1991). Consequently, physical nonequilibrium processes operating at a scale of centimeters with depth, can affect distribution of flow both locally and over a large area. Because of the inherent complexity of the system, practical problems encountered under field conditions can easily get bogged down and lost in excessive detail. Thus, a clear perception is required as to which issues are more important at a given scale.

A justification often advanced for studying soil at a single location by taking measurements at a point, is that it somehow enables us to generalize the behavior over an area, and predict a spatial response. This is true as long as homogeneous conditions prevail. However, under the nonequilibrium heterogeneous field conditions, the methodology for scaling-up, or extrapolating point measurements to larger areas is largely lacking. As demand for clean water continues to rise, accurate large-scale modeling and accounting for losses due to evapotranspiration and streamflow, in addition to strictly enforced protection of groundwater supplies and quality, will become increasingly important. Thus a future challenge will be to model a farm field, or catchment area response in a way that would reflect local nonequilibrium conditions, and at the same time consider potential impacts at other scales. This chapter will discuss how this may be accomplished, and illustrate it with a practical example pertaining to soil water movement.

MODELING OF UNCERTAINTY

Soil Perspective

It has been customary among the research community, when modeling water flow within a deterministic framework, to group, average, manipulate, and scale field- or lab-observed soil data in an effort to bring out similarities common to an area, layer, or volume (Beckett and Webster, 1971; Nielsen et al., 1983; and Neuman, 1990). Paralleling this effort but directed more toward the needs of a user, there has developed another source of available information in the form of published soil surveys and interpretations (Soil Survey Staff, 1983). The soil survey approach generally reflects a systematic variation related to pedogenesis and is more descriptive in nature, but includes no information concerning random variation and does not distinguish between continuous and categorical properties of a soil map unit (Arnold and Wilding, 1991). Soil properties within a map unit are seldom uniform, and may exhibit considerable variability (Rogowski and Wolf, 1994). Such variability is often incorporated into soil descriptions as inclusions, associations, or complexes which are seldom marked on a map, or quantified in an attribute table, unless larger than about 2 ha $(140\ m)^2$ in area. Problems arise when users attempt to incorporate soil survey data into more detailed models, and applications such as precision farming require a prediction of physical equilibrium status and spatial attribute behavior at a finer scale.

The advent of GIS (Geographic Information Systems) has allowed rapid manipulation and display of large quantities of related spatial data. Distributions of thermal gradients, locations with predominantly swelling or shrinking soil, and spatial zones that vary in aggregation and anisotropy, can now be represented as separate digital overlays (Burrough, 1986). Extraction of layers with high colloid content, and delineating locations with marked differences in structure, texture, or any other soil property, can be easily accomplished in GIS. Such overlays are subsequently discretized into appropriate grids which represent both the resolution and support scale of the process under study. Resolution defines the measurement scale, by specifying the smallest cell or grid size at which an image is to be viewed, processed, or analyzed. Similarly, size, shape, and orientation of a plot, or block of soil describe *a support;* i.e., an averaging area or volume in geostatistics (Olea, 1991). Neither resolution nor support lay any claim to being representative elementary areas or volumes (REV) commonly used in hydrology literature, they simply define a grid size at which the analysis is carried out. Thus a grid size in GIS corresponds to a support domain in geostatistics. As a result, modeling of location-specific variability and uncertainty and stochastic simulation of nonequilibrium conditions has become a reality.

Geostatistical Approach

In many areas of soil science and hydrology a single overriding concern is the optimal utilization and quantification of information residing in the soil databases. To deal with this concern a number of different methods have been suggested (Mausbach and Wilding, 1991). In more recent applications, elements of geostatistics, primarily kriging, cokriging, and structural semivariogram analysis have also been used, (Oliver and Webster, 1990; Wolf and Rogowski, 1991; Rogowski and Wolf, 1994, Borgault et al., 1995). Geostatistics is an applications-based methodology for modeling of spatial uncertainty. It was founded on a premise that values of many attributes in earth sciences, although variable in space, may be correlated with one another. The extent of correlation will depend on the separation distance and on sample location. Geostatistical methods currently applied to modeling of spatial uncertainty do not claim to be unique. They are merely tools to be used until better ones become available (Journel, 1986).

Spatial dependence is usually modeled with a semivariogram, which describes the average magnitude of differences among sampled values, based on the separation distance between sampling locations. Interpolation by kriging predicts values at unsampled locations based on surrounding data and a semivariogram model. Kriging in general features good local accuracy, but has a tendency to smooth the natural variability, potentially masking patterns of continuity. Although still used for spatial interpolation, currently its more important role is in building of models of spatial dependence. At present, the major thrust in geostatistics is conditional modeling of spatial uncertainty, which should precede a selection of an optimal predictor (Journel, 1996). In soil science when applying geostatistical methods to a practical field problem, such as, for example, distribution of cumulative recharge under the nonequilibrium field conditions, three questions usually arise:

- Is the process stationary and ergodic?
- Are the parameters of a distributed process likely to be correlated in space?
- Can we detect potential presence of spatial connectivity?

The answer to these questions is *yes*. Stationarity and ergodicity are properties of a probabilistic model not of a process (Myers, 1989). A random function model, must be assumed to be both stationary and ergodic to make any inferences from limited (usually a single realization) available data (Journel, 1993). Most processes associated with water flow are correlated in space, but little effort is made to georeference field measurements. Advances in geostatistics currently emphasize uncertainty, risk, and connectivity assessment (Journel, 1986, 1989; Gomez-Hernandez and Srivastava, 1990; Journel, 1993, 1994, 1996). Most recent progress has centered around reservoir (petroleum) modeling utilizing simulated annealing optimization and stochastic conditional simulation techniques to describe connected structural features potentially contributing to flow (Deutsch and Cockerham, 1994; Srivastava, 1993, 1994a; Deutsch and Hewett, 1996). The purpose of conditional simulation is to construct, based on limited conditioning data, multiple equiprobable spatial realizations, or outcomes, of an attribute. The conditioning data can be either measured, or estimated from related properties. This has led to the development of a new approach to spatial variability, capable of probabilistic spatial imaging of connectivity and simultaneous handling of diverse information sources (Journel, 1994, 1996). The approach can be used to quantify the uncertainty associated with data inputs to distributed models, while incorporating available information about local variability (Rogowski and Hoover, 1996; Rogowski, 1997).

Potentially the above geostatistical approach could be useful when modeling physical nonequilibrium conditions. In practical terms, a flow model such as Darcy's law (Eq. 18.1), or Richard's equation (Eq. 18.2), when processed by a computer program, predicts soil water movement on a cell by cell basis, for cells where input parameters are known and boundary conditions can be specified. In a GIS context a grid cell may represent part of a layer in soil profile, a portion of a farm field, or a digital elevation model (DEM) subdivision in a catchment. Conditional simulation can then be used to approximate a physical nonequilibrium by generating multiple equiprobable scenarios (realizations) of the process in cells other than where hydraulic parameters are known. In general, the computations are carried out on a regular grid. The realizations produce a range of outcomes known as a space of uncertainty (Deutsch, 1994). The space of uncertainty defines a domain of outcomes most likely to contain a true response of the system. Exhaustive data sets that contain hydraulic properties for all pixels of a GIS grid are generally not available, but the overlays of such properties can be constructed from a limited local knowledge using a nonparametric conditional simulation approach. The approach produces multiple realizations of a spatially distributed random variable (RV) that will honor measured values at sampled locations and preserve the overall spatial structure of the distribution, while reflecting the spatial continuity, or connectivity, of the attribute. Simulations constructed so as to honor sampled values are said to be conditioned by the measured (conditioning) data. If sampling of conditioning data is representative, the approach could approximate a true response to physical nonequilibrium conditions.

Soil parameter distributions can often be modeled based on limited data from many sources; e.g., direct measurements, soil survey interpretations, profile descriptions, texture, and landform position. There is a need for a combined utilization of diverse forms of soil information. This is usually done by considering so-called *hard* and *soft* data values. For example, field-measured values of hydraulic conductivity would be considered hard while ranges of permeability obtained from a soil survey profile description may be thought of as soft. Ability to use the two sources of information together is a desired effect. Because actual measurements and observations are sparse, they are usually augmented with soft data. It is unlikely that there is a unique statistical model that fits all situations. This is where stochastic modeling can be used to advantage to generate a number of equiprobable scenarios which are consistent with both the hard and soft field data (Journel, 1996). For example, prediction of spatial uncertainty approximating physical nonequilibrium conditions can be accomplished by simulating multiple realizations conditional to data values, and comparing their statistics. A good simulation technique will generate credible realizations of input parameters which utilize all available information.

Nonequilibrium flow processes in a soil continuum are likely to be very sensitive to how well connected the locations are with extremely low, or high permeabilities within a profile (Deutsch and Hewett, 1996). Present trends in nonparametric geostatistics attempt to quantify the model of uncertainty associated with connectivity directly from observed data. The new techniques appear to closely parallel current needs in soil science and utilize both kriging and stochastic simulation (Rogowski, 1996b). There appears to be a need for implementation of stochastic modeling of water flow and pollutant transport at a scale of a farm or catchment, when field soils are likely to be in a state of nonequilibrium due to local variability. To account for this variability a stochastic simulation of spatially distributed cumulative recharge flux may consist of some processes that would allow drawing alternate spatial distributions of flux from both the hard and soft data values with a given probability of occurrence. In addition to generating multiple equiprobable realizations of recharge, a simulation approach may reproduce spatial patterns of connectivity associated with the occurrence of extreme flux values (Journel, 1989, 1994).

Sequential Indicator Simulation

A conditional simulation procedure currently in widespread use is the sequential indicator simulation (SIS), (Journel, 1989, 1994, 1996; Deutsch and Journel, 1992; and Rogowski, 1996b). It is recommended to characterize strong connectivity of either the high or low values (Journel and Alabert, 1989). The procedure uses indicator variables and is distribution-free (not limited to a specific distribution). This enables it to process different types of information together, without constraints of a specific model, and based strictly on field observations. It addresses joint uncertainty of the probability that at a number of sampling locations u_j, $j=1, 2,...,N$, the property of interest will be greater than a certain threshold value q_t. Prevailing nonequilibrium conditions can then be nonspecifically described as being either more or less than a threshold value.

To model different portions of an irregular distribution, the sequential indicator simulation partitions observed population into several classes, each separated by a threshold. For each observation it then defines an indicator transformation with respect to that threshold. For a random variable Q at a location x and threshold value q_t, corresponding indicator transformation $i\ (q_t, x)$ is,

$$\begin{aligned} i(q_t, x) &= 0, \text{ if } Q(x) > q_t \\ &= 1, \text{ if } Q(s) \leq q_t \end{aligned} \tag{18.3}$$

Given a set of values distributed over an area the procedure converts them into an indicator distribution for each threshold. It then defines an algorithm for adding an indicator value at an unsampled location consistent with a spatial structure of observed population for a particular threshold. The simulated value becomes a part of the distribution for the subsequent step. Because actual values of the random variable Q are not known at all points, a model of Q has to be built. To do so, first a model of spatial dependence (semivariogram) is constructed from available conditioning data, then spatial distribution of Q values is obtained by kriging. Maximum uncertainty about a value $q(u)$ at an unsampled location u is given by a prior cumulative distribution function (cdf) of $Q(u)$,

$$F\{(u,q)|(n)\} = P\{Q(u) \leq q\} \tag{18.4}$$

where P denotes probability. The expression $F\{(u;q)\backslash(n)\}$, is also referred to as a conditional cdf of $Q(u)$, since it is conditioned by measured values of $q(u)$ and a covariance model,

$$P\{Q(u) \leq q | \text{extra information } (n)\} \tag{18.5}$$

$$F(u; q|(n)) = \sum_{a=1}^{n} \lambda_a(u) \bullet I(u; q_t), \quad (0,1) \tag{18.6}$$

In Eq. 18.6 λ_a refers to variable weights from kriging, and $I(u;q_t)$ to an indicator transformation of data about a threshold q_t,

$$\begin{aligned} I(u; q_t) &= 0, \text{ if } Q(u) > q_t \\ &= 1, \text{ if } Q(u) \leq q_t \end{aligned} \tag{18.7}$$

Equations 18.4 through 18.7 have also been formulated in terms of spatial dependence between $Q(u)$ and $Q(u+h)$ as a two-point connectivity function derived from a bivariate distribution of random variable Q as probability of nonexceedence of a threshold q_t (Journel, 1989),

$$F_I(q_t, q_t';h) = P\{Q(u) \leq q_t, Q(u+h) \leq q_t'\} \tag{18.8}$$

The bivariate cdf $F_I(q_t, q_t';h)$ can also be expressed as an indicator covariance for the separation vector h,

$$F_I(q_t, q_t';h) = E\{I(u;q_t) \bullet I(u+h;q_t')\} \tag{18.9}$$

where $E\{I(u;q_t)\}$ is the expected value. Similarly, a complement indicator transform $J(q_t;u)=1-I(q_t;u)$ defines a joint probability of exceedence,

$$F_J(q_t, q_t'; h) = P\{Q(u) > q_t, Q(u+h) > q_t'\} \qquad (18.10)$$

$$F_J(q_t, q_t'; h) = E\{J(u; q_t) \bullet J(u+h; q_t')\} \qquad (18.11)$$

where the indicator covariance (Equation 18.11) is also a measure of spatial connectivity; i.e., a probability that values $Q(u)$ and $Q(u+h)$ will jointly exceed a threshold q_t. When $F_J(q_t, q_t'; h)$ is high, the probability that locations u and $u+h$ are connected is likely to be large.

To model soil water recharge under field conditions we might need to account for and explain spatial connectivity of extreme (low or high) values. In general, spatial connectivity models such as SIS define probabilities for two (or more) attribute values located at sites u and u' to jointly exceed a threshold q_t (Equations 18.8 to 18.11); i.e., probability that high values at the two locations define a continuum of a given property. Their indicator correlogram $\rho(u;p)$ can be viewed as an index of two point connectivity. A correlogram $\rho(u;p)$, is a plot of correlation coefficient between two random variables as a function of separation distance (Olea, 1991). The higher the $\rho(u;p)$, the higher the probability that two values $q(u)$ and $q(u+h)$ jointly exceed (or not exceed) a threshold value q_t. Multiple point connectivity $\Phi(m)$ in a given direction is then expressed as the expected value of a product of m indicator variables (Journel and Alabert, 1989),

$$\Phi(m) = E\left\{\prod_{j=1}^{m} I[u+(j-1)h; q_t]\right\} \qquad (18.12)$$

APPLICATIONS

Selection of Conditioning Data

To illustrate a practical application problem, we used SPAW, a deterministic Soil-Plant-Atmosphere-Water model, and a sequential indicator simulation (SIS) approach to predict a distribution of soil water recharge on a farm under the physical nonequilibrium conditions due to soil variability. The first step was to assemble all available information into GIS overlays describing field conditions. The GIS overlays are based on a 5 m grid and illustrate site topography (Figure 18.1), distribution of slopes (Figure 18.2a) and distribution of soil polygons for the eight primary mapping units present at the farm (Figure 18.2b). The assumption was that we had no information about the soil properties other than what was in a standard soil database. Additional SPAW input included antecedent water content adjusted to field capacity in the spring. Although climatic and plant inputs to the SPAW model developed by Saxton et al. (1992) are likely to vary spatially, at a farm scale they may be assumed constant, while variability in soil properties, is more localized and thus more likely to affect the recharge. We will therefore concentrate on nonequilibrium-inducing spatial differences in cumulative recharge brought about by variation in soil properties.

Texture, aggregation, and porosity are closely related to a physical nonequilibrium status of a soil. Site-specific soil texture information is nearly always available from published soil surveys, and can be checked in the field. At times, multiple descriptions of the same mapping unit provide a needed insight into its variability. In this particular case, 60 pedons, corresponding to the eight mapping units found at the

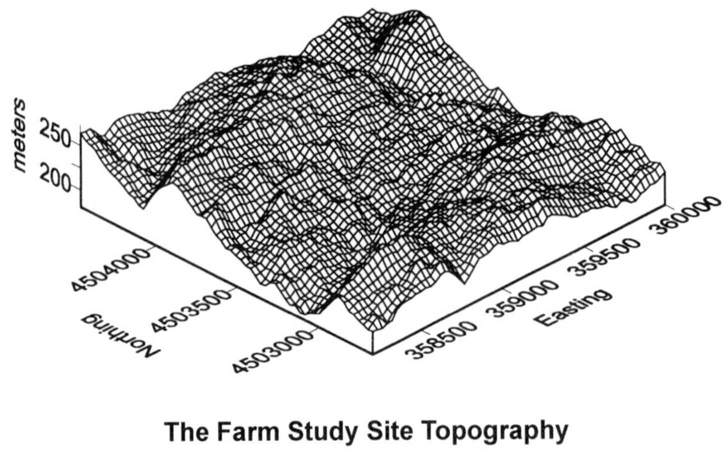

Figure 18.1. A three-dimensional 1:5 vertical exaggeration of topography on a 1,200×1,200 m farm study site, based on a standard 30 m U.S. Geological Survey digital elevation model (DEM).

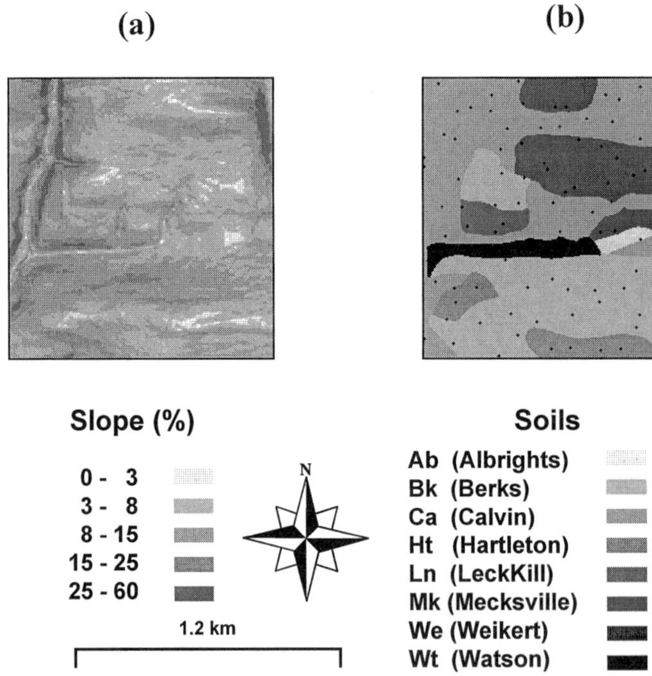

Figure 18.2. Geographic information system (GIS) overlay of slope (a), and polygons of soil mapping units (b), prepared from a soils map and a detailed 5 m digital elevation model (DEM) of the farm study site; assigned locations of the conditioning pedons are shown on the soils map.

farm, were extracted from the Pennsylvania State University Soils Database (Ciolkosz and Thurman, 1992). These were augmented with 41 pedons of related soils considered as inclusions, for a total of 101 pedons. The selected pedons were then located at random at the nodes of the 5 m grid within the polygon of the same mapping unit on the farm. Figure 18.3a shows the statewide distribution of the representative pedons

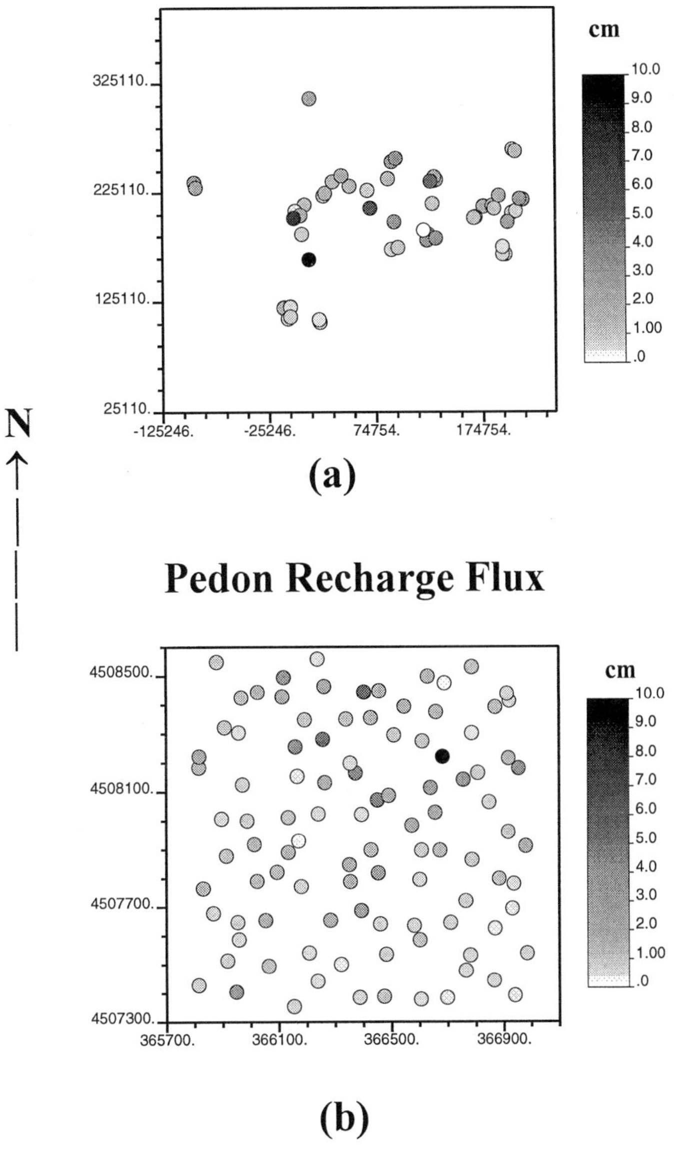

Figure 18.3. The statewide locations of related soil pedons (a), and corresponding locations of assigned conditioning sites (b) on the farm study site; values expressed as daily recharge flux q_v (cm) computed with a SPAW (Soil-Plant-Atmosphere-Water) model of Saxton et al. (1974, 1983, 1992).

and Figures 18.2b and 18.3b specify the assigned on-the-farm locations of pedons selected for analysis.

Textural information and depth to each layer were tabulated for each pedon, and saturated hydraulic conductivity K_{sat} was estimated from soil texture (Saxton et al., 1983). The K_v (vertical) and K_h (horizontal) conductivity components were then computed as harmonic and arithmetic means, respectively, weighted by a layer thickness. Daily precipitation and potential evapotranspiration (PET) data were downloaded from a nearby weather station, and plant cover was assumed to be corn.

Modeling Soil Water Recharge

In general, a deterministic flow model such as SPAW will predict recharge subject to the uniform initial and boundary conditions which are specified *a priori*. A tacit assumption expects soil properties within a study area to be uniform and behave in a similar manner. Unfortunately, under the field conditions this assumption does not hold (Rogowski, 1996d). Soil hydraulic properties are not the same, but differ from point to point reflecting physical nonequilibrium properties associated with normal field locations. Such variability may be due to local topography, changes in slope and aspect, land use, and differences in soil properties. It is possible to display and account for some of these differences by creating respective GIS overlays showing their distribution. However, spatial differences in soil properties defy conventional treatment, and need to be modeled stochastically because they are generally not known in sufficient detail to be mapped exhaustively. To remedy the situation we combine a deterministic flow model (SPAW) with a sequential indicator simulation (SIS) and predict a spatial distribution of soil water recharge within a GIS framework.

To model the variability and generate appropriate conditioning data, SPAW model may be run with, for example, a weekly time step on each of the 101 profiles resulting in multiple sets of conditioning values, one for each week of the run. The model apportions incoming precipitation among soil layers using a subroutine based on Darcy's law (Eq. 18.1) in a finite difference form (Saxton et al., 1992),

$$q_v = K(\theta)\frac{[\Delta h + Z]}{Z}(\Delta t) \qquad (18.13)$$

where $K(\theta)$ is the mean hydraulic conductivity of the two layers, Δh is the matric potential head difference between them, Z is the distance between layer midpoints, and Δt is the time increment. Both $K(\theta)$ and Δh are considered to be functions of their respective water content θ. The portion of precipitation that reaches the lower profile boundary becomes the vertical recharge flux component q_v. Weekly outputs consist of cumulative recharge flux q_v and soil water content estimate for each layer. Knowing q_v we can estimate a corresponding horizontal flow component q_h at each location from the ratio $n = K_h/K_v$, which describes profile anisotropy (Zaslavsky and Rogowski, 1969),

$$q_h = q_v \times n \times \tan\alpha \qquad (18.14)$$

where α represents % land slope (Figure 18.2) from the 5 m DEM, $\tan\alpha$ is the (slope gradient)$^{-1}$, and K_v and K_h correspond to vertical and horizontal hydraulic conductivity components of K_{sat}.

Simulation of Spatial Distribution

A single realization of cumulative recharge flux distribution in Figure 18.4 illustrates a typical SIS output. The realization comprises a simulated grid of 58310 values that honor recharge flux modeled at 101 pedon locations with a deterministic simulation model SPAW. The SPAW output acts as conditioning data in SIS. Figure 18.5 compares the respective cdfs and the corresponding statistics for the conditioning data and the simulated values. The distribution in Figure 18.4 is just one realization of a large number of simulations which are a standard output of SIS. The equiprobable realizations, assumed to approximate the nonequilibrium conditions in the field, are

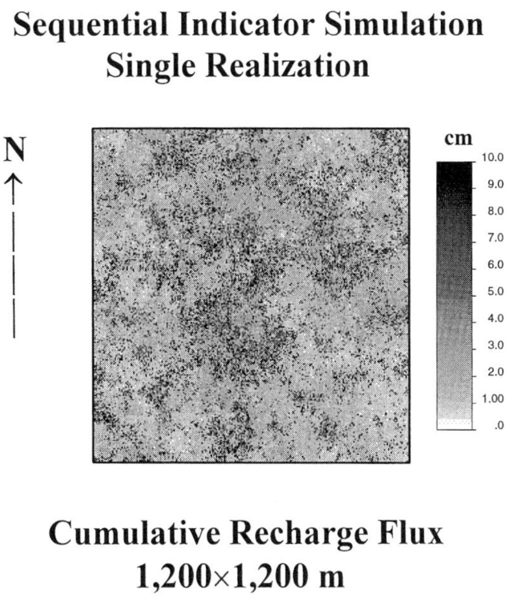

Figure 18.4. A single sequential indicator simulation (SIS) of a cumulative recharge flux q_v (cm) distribution on a 1,200×1,200 m farm study site.

based on the distribution structure of conditioning data derived from the threshold indicator semivariograms. Taking the individual realizations together we can compute the expected values and quantiles of the simulated flux at each location. More importantly, we can use them to delineate vulnerable zones by describing areas which have the probability of exceeding (or being less than) some critical value. Since all realizations honor the conditioning data, a physical nonequilibrium problem can be expressed in probabilistic terms.

Detailed examination of individual distributions will help establish a degree of spatial connectivity among the high and low values of recharge. For example, the simulation in Figure 18.4 in general resembles that of its expected values shown in Figure 18.6a. But a typical cdf for a single simulation in Figure 18.5a, is a more nearly correct representation of conditioning data than the histogram of expected values (not shown), because it better portrays the high values. The significance of SIS ability to characterize connectivity of the extreme data values becomes quickly apparent when we recognize the dominating influence that a few locations may have on the overall rate of flow and pollutant transport.

Results of multiple simulations (SIS) are presented in Figure 18.6 as GIS overlays of expected values for: (a) vertical (q_v), and (b) horizontal (q_h) components of cumulative flux. The expected value overlays, expressed as quartiles of the cumulative distribution for that day, represent point averages of 25 realizations, similar but not identical to kriged representations. The q_h-*values* are in general 4 to 5 times less than the corresponding q_v-*values*. When we compare the q_h distribution with the corresponding distribution of q_v-*values,* the overlay of slopes (Figure 18.2a) and area topography (Figure 18.1), the comparison suggests a dominating influence of the slope gradient ($tan\ \alpha$).

Expected values of simulations may also illustrate the potentially connected zones. For example, results in Figure 18.6a suggest that areas with high q_v may be the sites of preferential recharge. Comparing simulated distributions with corresponding terrain properties in Figures 18.1 and 18.2, the primary recharge area appears to be the cen-

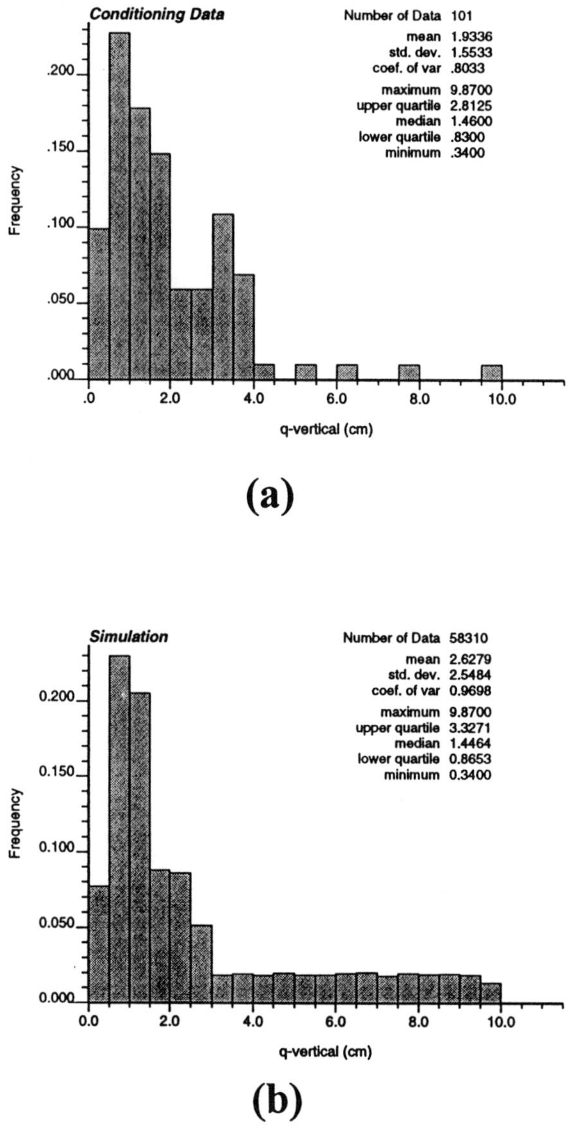

Figure 18.5. Comparison of the cumulative distribution function (cdf) for the conditioning data (a), with the corresponding (cdf) for a single sequential indicator simulation (SIS) of a daily recharge flux q_v (cm) on a 1,200×1,200 m farm study site (b).

tral sloping zone on the north side of the stream. Similarly, above average q_h values in Figure 18.6b indicate principal interflow, or runoff producing locations situated on steep areas adjacent to streams. While individually each pixel represents flow partitioning in a 5×5 m grid, collectively the expected values (Figure 18.6) reflect a spatial pattern of connectivity.

Practical Implications

The above application illustrates a possible approach to managing water movement and pollutant transport under physical nonequilibrium conditions in the field.

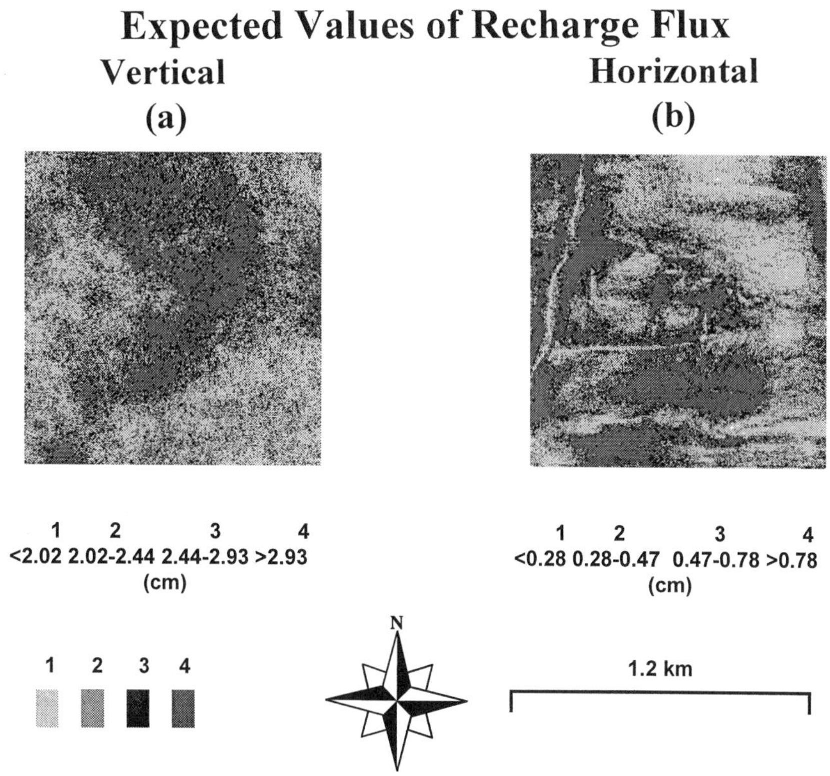

Figure 18.6. Distribution of the expected values of cumulative vertical q_v (a), and horizontal q_h (b) recharge flux (cm) on a 1,200×1,200 m farm study site.

The analysis assumes that the only available information is what can be obtained from a standard soil survey database and an on-site inspection. Generation of equiprobable scenarios based on realistic conditioning information provides a series of stochastic images potentially reflecting nonequilibrium field conditions. Differences among images can then be used to describe a degree of uncertainty associated with a given location, or value of an attribute. As a consequence, the analysis provides a means of scaling-up measured or estimated point values to apply to larger areas or volumes. Since the approach can accommodate different types of information, the predictions can be updated as new information becomes available. We suggest the approach be used in conjunction with a continuous deterministic model such as, for example, SPAW, capable of daily time steps. Under these circumstances the SIS procedure can provide a probabilistic documentation of changes occurring in time, subject to differences in point measurements which are customary with the environmental data.

Assessment of uncertainty under the field or catchment conditions will likely open the door to modeling of nonequilibrium processes at a variety of scales. However, because at present there is no single agreed-on model of uncertainty it may be necessary to provide some general guidelines. Among the three recently proposed by Journel (1994) are:

- recognition that all assessments of uncertainty are not reality, but only models of reality,

- requirement that a model reproduces all information which is critical to the establishment of the intrinsic properties of the process, and
- confidence in the validity of data used to condition model values.

It should, however, be remembered that simulation modeling and model parameters are but a convenient approximation, and that any data points used be considered in the context of real world values (Srivastava, 1994b). The practice of conditional simulation encourages calibration of a random function model with information from locations other than the local pool, which often may be too limited. Thus a good model of uncertainty is theoretically sound, relies heavily on local data, but leaves room for improvement with additional data from other sources. While calibration is certainly encouraged, it may not always be necessary. Familiarity with the process, experience, and good judgment can often lead to the correct choice of the uncertainty model and proper selection of its parameters (Srivastava, 1994b).

REFERENCES

Arnold, R.W. and L.P. Wilding. The need to quantify spatial variability. In *Spatial Variabilities of Soils and Landforms.* Mausbach, M.J. and L.P. Wilding (Eds.) SSSA Special Publication No. 28, Soil Science Society of America , Inc., Madison, WI, 1991.

Arya, L.M. and T.S. Dierolf. Predicting soil moisture characteristics from particle-size distributions: An improved method to calculate pore radii from particle radii. In *Proc. Int. Workshop on Indirect Methods for Estimating the Hydraulic Properties of Unsaturated Soils,* van Genuchten, M.Th., F.J. Leij, and L.J. Lund (Eds.), University of California, Riverside, CA, 1992.

Baver, L.D., W.H. Gardner, and W.R. Gardner. *Soil Physics,* 4th ed. John Wiley & Sons Inc., New York. 1972, p. 487.

Beckett, P.H.T. and R. Webster. Soil variability: A review. *Soils Fert.,* 34(1):1–14, 1971.

Bierkens, M.F.P and P.A. Burrough. Modeling of map impurities using sequential indicator simulation. pp. 637–648. In *Geostatistics Troia '92,* Kluver Academic Publishers, Boston, 1993, p. 1088.

Borgault, G., A.G. Journel, S.M. Lesch, J.D. Rhoades, and D.L. Corwin. Geostatistical Analysis of Soil Salinity Data Set. pp. 53–114. In *Applications of GIS to the Modeling of Non-Point Source Pollutants in the Vadose Zone.* Proceedings of the Bouyoucos Conference, Riverside, CA, p. 690, 1995.

Burrough, P.A. *Principles of Geographical Information Systems for Land Resources Assessment.* Clarendon Press, Oxford. p. 194, 1986.

Ciolkosz, E.J. and N.C. Thurman. *Soil Characterization Laboratory Database System.* Agronomy Serial No. 124. Agronomy Dept., Pennsylvania State University, University Park, 1992, p. 59.

Corey, A.T. Pore size distribution. In *Proc. Int. Workshop on Indirect Methods for Estimating the Hydraulic Properties of Unsaturated Soils,* van Genuchten, M.Th., F.J. Leij, and L.J. Lund (Eds.), University of California, Riverside, CA, 1992, pp. 37–44.

Cruse, R.M. and D.L. Dinnes. Spatial and temporal diversity in production fields. In *Exploring the Role of Diversity in Sustainable Agriculture,* Olson, R., C. Francis, and S. Kaffka (Eds.), Soil Science Society of America, Inc., Madison, WI, 1995.

Daniels, R.B. and R.D. Hammer. *Soil Geomorphology.* John Wiley & Sons, Inc., New York, NY. 1992.

Deutsch, C.V. Algorithmically defined random function models. In *Geostatistics for the Next Century.* R. Dimitrakopaulos (Ed.), Kluver Academic Publishers, Boston. 1994, pp. 422–435.

Deutsch, C.V. and P.W. Cockerman. Practical considerations in the application of simulated annealing to stochastic simulation. *Math. Geol.,* 26:67–82, 1994.

Deutsch, C.V. and T.A. Hewett. Challenges of reservoir forecasting. *Math. Geol.,* 28(7):829–842, 1996.

Deutsch, C.V. and A.G. Journel. *GSLIB, Geostatistical Software Library and User's Guide.* Oxford Univ. Press, New York, 1992, p. 340.

Gomez-Hernandez, J.J. and R.M. Srivastava. ISIM3D: An ANSI-C three dimensional multiple indicator conditional simulation program. *Comput. Geosci.,* 16:395–440, 1990.

Greminger, P.J., Y.K. Sud, and D.R. Nielsen. Spatial variability of field measured soil water characteristics. *Soil Sci. Soc. J.,* 49(5):1075–1081, 1985.

Hall, G.F. and C.G. Olson. Predicting variability of soils from landscape models. pp. 9–24. In *Spatial Variabilities of Soils and Landforms.* Mausbach, M.J. and L.P. Wilding (Eds.), SSSA Special Publication No. 28, Soil Science Society of America, Inc., Madison, WI, 1991.

Hoosbeek M.R. and R.B. Bryant. Towards the quantitative modeling of pedogenesis—A review. *Geoderma,* 55:183–210, 1992.

Jenny, H. *The Soil Resource—Origin and Behavior.* Springer-Verlag, New York, 1980, p. 377.

Journel, A.G. Geostatistics: Models and tools for the earth sciences. *Math. Geol.,* 18:119–140, 1986.

Journel A.G. Fundamentals of geostatistics in five lessons. *American Geophysical Union. Short Course in Geology,* 8:40, 1989.

Journel, A.G. Geostatistics: roadblocks and challenges. pp. 213–224. In *Geostatistics Troia '92,* Kluver Academic Publishers, Boston, 1993, p. 1088.

Journel, A.G. Modeling uncertainty: some conceptual thoughts. pp. 30–43. In *Geostatistics for the Next Century.* Dimitrakopaulos, R. (Ed.), Kluver Academic Publishers, Boston, 1994, p. 497.

Journel, A.G. A return to the equiprobability of stochastic realizations. In Report #8, *Stanford Center for Reservoir Forecasting,* Stanford University, Stanford, CA, 1995.

Journel, A.G. Modeling uncertainty and spatial dependence: Stochastic imaging. *Int. J. Geographic Information Systems,* 10(5):517–521, 1996.

Journel, A.G. and F. Alabert. Non-Gaussian data expansion in the earth sciences. *Terra Nova.* 1:123–134, 1989.

Klute, A. (Ed.). *Methods of Soil Analysis. Part 1. Physical and Mineralogical Methods,* 2nd edition. American Society of Agronomy, Inc., Madison, WI, 1986, p. 1188.

Mausbach, M.J. and L.P. Wilding. *Spatial Variabilities of Soils and Landforms.* SSSA Special Publication No. 28 , Soil Science Society of America, Inc., Madison, WI, 1991, p. 270.

Moore, I.D., P.E. Gessler, G.A. Nielsen, and G.A. Peterson. Soil attribute prediction using terrain analysis. *Soil Sci. Soc. Am. J.* 57:443–452, 1993.

Mualem, Y. Modeling the hydraulic conductivity of unsaturated porous media. In *Proc. Int. Workshop on Indirect Methods for Estimating the Hydraulic Properties of Unsaturated Soils,* van Genuchten, M.Th., F.J. Leij, and L.J. Lund (*Eds.*), University of California, Riverside, CA, 1992, pp. 15–36.

Myers, D.E. To be or not to be...stationary? That is the question. *Math. Geol.,* 21(3):347–362, 1989.

Neuman, S.P. Universal scaling of hydraulic conductivities and dispersivities in geologic media. *Water Resour. Res.,* 26(8):1749–1758, 1990.

Nielsen, D.R., P.M. Tilllotson, and S.R. Vieria. Analyzing field-measured soil-water properties. *Agric. Water Manage.,* 6:93–109, 1983.

Oliver, M.A. and R. Webster. Kriging: A method of interpolation for geographical information systems. *Int. J. Geogr. Inf. Syst.,* 4:313–332, 1990.

Olea, R.A. *Geostatistical Glossary and Multilingual Dictionary.* Oxford Univ. Press, New York, 1991.

Philip, J.R. Field heterogeneity: Some basic issues. *Water Resour. Res.,* 16(2):443–448, 1980.

Rawls, W.J., L.R. Ahuja, and D.L. Brakensieck. Estimating soil hydraulic properties from soils data. In *Proc. Int. Workshop on Indirect Methods for Estimating the Hydraulic Properties of Unsaturated Soils,* van Genuchten, M.Th., F.J. Leij, and L.J. Lund (Eds.), University of California, Riverside, CA, 1992, pp. 329–340.

Richards, L.A. Capillary conduction of liquids through porous media. *Physics,* 1:318–333, 1931.

Rogowski, A.S. Quantifying soil variability in GIS applications. II. Spatial distribution of soil properties. *Int. J. Geogr. Inf. Syst.,* 10(4):455–475, 1996a.

Rogowski, A.S. Quantifying the model of uncertainty and risk using sequential indicator simulation. pp. 143–164. In *Field Variability and Risk Assessment.* Nettleton, W.D., A.G. Hornsby, R.B. Brown, and T.L. Coleman (Eds.), SSSA Special Publication No. 47. Soil Science Society of America, Inc., Madison WI , 1996b, p. 164.

Rogowski, A.S. Incorporating soil variability into a spatially distributed model of percolate accounting. pp. 57–64. In *Spatial Accuracy Assessment in Natural Resources and Environmental Sciences: Second International Symposium,* Mowrer, H.T., R.L. Czaplewski, and R.H. Hamre (Eds.), May 21–23, 1996, Fort Collins, CO, 1996c.

Rogowski, A.S. GIS modeling of recharge on a watershed. *J. Environ. Qual.,* 25:463–474, 1996d.

Rogowski, A.S. Catchment infiltration II: Contributing areas. *Trans. GIS,* (in press), 1997.

Rogowski, A.S. and J.K. Wolf. Incorporating variability into soil map unit delineations. *Soil Sci. Soc. Am. J.,* 58:163–174, 1994.

Rogowski, A.S. and J.R. Hoover. Catchment infiltration I: Distribution of variables. *Trans. GIS,* 1(2):95–110, 1996.

Rudzitis, G. *Wilderness and the Changing American West.* John Wiley & Sons, Inc., New York, 1996, p. 220.

Runge, E.C.A. Soil development sequences and energy models. *Soil Sci.,* 115(3):183–193, 1973.

Saxton, K.E., H.P. Johnson, and R.H. Shaw. Modeling evapotranspiration and soil moisture. *Trans. ASAE,* 17(4):673–677, 1974.

Saxton, K.E., W.J. Rawls, J.S. Romberger, and R.I. Papendick. Estimating generalized soil water characteristics from texture. *Soil Sci. Soc. Am. J.,* 50:1031–1036, 1983.

Saxton, K.E., M.A. Porter, and T.A. McMahon. Climatic impacts on dryland winter wheat by daily soil water and crop stress simulations. *Agric. For. Meteorol.,* 58:177–192, 1992.

Smeck, N.E. Phosphorus: An indicator of pedogenetic weathering processes. *Soil Sci.,* 115(3):199–206, 1973.

Smeck, N.E. Phosphorus dynamics in soils and landscapes. *Geoderma,* 36:185–189, 1985.

Smeck, N.E., E.C.A. Runge, and E.E. Makintosh. Dynamics and genetic modeling of soil systems. In *Pedogenesis and Soil Taxonomy 1. Concepts and Interactions,* Wilding, L.P., N.E. Smeck, and G.F. Hall (Eds.), Elsevier, Amsterdam, 1983, pp. 23–49.

Soil Survey Staff. *Soil Survey Manual.* U.S. Gov. Print. Office, Washington, DC, 1993.

Srivastava, M.H. Reservoir characterization with probability field simulation. In *Annual Technical Conference Society of Petroleum Engineers,* SPE Paper No. 24753, 1993, pp. 927–938.

Srivastava, M.H. An annealing procedure for honoring change of support statistics in conditional simulation. pp. 277–290. In *Geostatistics for the Next Century,* Dimitrakopaulos, R. (Ed.), Kluver Academic Publishers, Boston, 1994a, p. 497.

Srivastava, M.H. Comment on "Modeling uncertainty: Some conceptual thoughts: By A.G. Journel. pp. 44–45. In *Geostatistics for the Next Century.* Dimitrakopaulos, R. (Ed.), Kluver Academic Publishers, Boston, 1994b, p. 497.

Topp, G.C., W.D. Reynolds, and R.E. Green (Eds.). *Advances in Measurement of Soil Physical Properties: Bringing Theory into Practice.* SSSA Special Publication No. 30. Soil Science Society of America, Inc., Madison, WI, 1992.

Tyler, S.W. and S.W. Wheatcraft. Fractal aspects of soil porosity. In *Proc. Int. Workshop on Indirect Methods for Estimating the Hydraulic Properties of Unsaturated Soils,* van Genuchten,

M.Th., F.J. Leij, and L.J. Lund (Eds.), University of California, Riverside, CA, 1992, pp. 53–63.

van Genuchten, M.Th. and F.J. Leij. On estimating the hydraulic properties of unsaturated soils. In *Proc. Int. Workshop on Indirect Methods for Estimating the Hydraulic Properties of Unsaturated Soils,* van Genuchten, M.Th, F.J. Leij, and L.J. Lund (Eds.), University of California, Riverside, CA, 1992, pp. 1–14.

Walker, T.W. and A.F.R. Adams. Studies on soil organic matter: 2. Influence of increased leaching at various stages of weathering on levels of carbon, nitrogen, sulfur, and organic and total phosphorus. *Soil Sci.,* 87(1):1–10, 1959.

Webster, R. and M.A. Oliver. *Statistical Methods for Soil and Land Resource Survey.* Oxford University Press, Oxford. 1990, p. 316.

Wolf J.K. and A.S. Rogowski. Spatial distribution of soil heat flux and growing degree days. *Soil Sci. Soc. Am. J.,* 55:647–657, 1991.

Wosten, J.H.M. and J. Bouma. Applicability of soil survey data to estimate hydraulic properties of unsaturated soils. In *Proc. Int. Workshop on Indirect Methods for Estimating the Hydraulic Properties of Unsaturated Soils,* van Genuchten, M.Th., F.J. Leij, and L.J. Lund (Eds.), University of California, Riverside, CA, 1992, pp. 463–472.

Youngs, E.G. Developments in the physics of infiltration. *Soil Sci. Soc. Am. J.,* 59:307–313, 1995.

Zaslavsky, D. and A.S. Rogowski. Hydrologic and morphologic implications of anisotropy and infiltration in soil profile development. *Soil Sci. Soc. Am. J.,* 33(4):594–599, 1969.

Index

A

B

C

D

G

H

M

N

Q

R

S

T

U